Satyendra Nath Bose
His Life and Times
Selected Works (With Commentary)

Satyendra Nath Bose
His Life and Times

Selected Works (With Commentary)

Edited by

Kameshwar C. Wali

Syracuse University, USA

World Scientific

NEW JERSEY · LONDON · SINGAPORE · BEIJING · SHANGHAI · HONG KONG · TAIPEI · CHENNAI

Published by

World Scientific Publishing Co. Pte. Ltd.
5 Toh Tuck Link, Singapore 596224
USA office: 27 Warren Street, Suite 401-402, Hackensack, NJ 07601
UK office: 57 Shelton Street, Covent Garden, London WC2H 9HE

British Library Cataloguing-in-Publication Data
A catalogue record for this book is available from the British Library.

SATYENDRA NATH BOSE — HIS LIFE AND TIMES
Selected Works (With Commentary)

ISBN-13 978-981-279-070-5
ISBN-10 981-279-070-5
ISBN-13 978-981-279-071-2 (pbk)
ISBN-10 981-279-071-3 (pbk)

Printed by FuIsland Offset Printing (S) Pte Ltd, Singapore

Foreword

The year 2005 was celebrated worldwide as the centenary of Einstein's miraculous year, 1905, as well as for his life and work. During a period of twelve months (1905), he published five papers that changed the face of physics — two of which laid the foundations of the Special Theory of Relativity; a paper on the quantum nature of radiation in explaining the Photo-Electric effect; papers on the theory of Brownian motion, and on molecular dimensions and the calculation of Avogadro's number. In 1924, almost two decades after the miraculous 1905, and almost a decade after his famous theory of gravitation in 1915, Einstein was involved in another important discovery of 20th century physics — the birth of the new quantum statistics to become known as the Bose–Einstein statistics. It all began with a letter and a short article. A relatively unknown young man, Satyendra Nath Bose, from Dacca University in East Bengal (now Bangladesh) wrote to Einstein, claiming he had derived Planck's law for black body radiation without recourse to classical electrodynamics. Bose further wrote that if Einstein thought it was important, he would be grateful if he (Einstein) could arrange its publication in *Zeitscrift fur Physik*, as he, Bose, did not know German well enough.

Einstein translated the paper himself and got it published with a note added saying, "*Bose's derivation of Planck's law appears to me an important step forward. The method used here also yields the quantum theory of ideal gas, as I shall show elsewhere*." Indeed in a matter of weeks, Einstein published his paper extending Bose's work to monoatomic gas and followed it up with another paper within months, predicting according to the new theory the possibility of a new state of matter, the Bose–Einstein condensate. Highlighting the importance of this development, Abraham Pais in his celebrated book, "*Subtle is the Lord*," on Einstein says, "*For Einstein this period was only an interlude. He was already engrossed in his search for a unified theory. Such is the scope of his oeuvre that his discoveries in those six months do not even rank among his five main contributions, yet they alone would have sufficed for Einstein to be remembered forever*."

Einstein's recognition of Bose's work had an immediate impact on Bose's life and career. He was granted study leave from his university for two years "*with a stipend, a separation allowance for the family, with sumptuous travel allowance with round trip fare*." Although Bose was primarily a theoretical physicist at the time, he spent his first year abroad gaining first-hand experience working in the crystallographic laboratories of Maurice de Broglie and the radioactivity research laboratories of Madam Curie before going to Berlin to meet Einstein in October of 1925. Berlin was humming with the excitement of the dawn of the new quantum mechanics with the visits and seminars of Heisenberg and Schrodinger. Einstein introduced Bose to several prominent physicists including Otto Hahn, Lise Meitner, Fritz Haber and Walter Gordon. He came to know Hermann Mark and worked closely in his laboratories of X-ray crystallography.

Bose's two-year stay in Europe marked, in many ways, a turning point in his career as a teacher and researcher. He realized the importance of experimental research in the advancement of science. On his return, he occupied himself mostly with setting up an X-ray crystallography laboratory at Dacca University. He designed and built his own equipment, requiring that his students do the same. On other fronts he devoted himself almost exclusively to teaching and guiding research. He also held various administrative positions as Department Head, Dean of the Faculty of Science and Provost. In 1945, he left Dacca to return to his alma mater, the University of Calcutta as the Khaira Professor of Physics. He retired from the University of Calcutta in 1956, spent a year as the Vice Chancellor at the Viswa-Bharati University and finally lived the remainder of his years in Calcutta until his death in 1974.

Bose, a brilliant and inquisitive scholar throughout his career, lived during a period of great change in India. He witnessed the partition of Bengal, his home in 1905. Instigated by Lord Curzon, one of the most imperialist viceroys of India, this action resulted in the rise of a strong nationalist movement against the British. Bose was one of a group of accomplished intellectuals motivated by national fervor to excel academically and creatively to prove their competitiveness with their counterparts in the western world. He was active politically throughout his life, serving on many committees to further scientific and industrial development in India after independence.

Bose was an inspiring figure to many outside of the physics community. His love for stimulating conversations, wide ranging interests in all human endeavors, his intellectual fervor, his scientific accomplishments beyond physics and not least, his association with Einstein established his legendary reputation in India.

I had the occasion to see S.N. Bose briefly in 1952 as a post-graduate student at the Banaras Hindu University working in the spectroscopy laboratory of R.K. Asundi. Bose's unannounced visit made us students gather quickly around the tea table. My only recollection is that we students sat quietly while two stalwarts (Asundi and Bose) conversed. I remember vividly a handsome, dark skinned Bengali with a head full of gray hair and dark penetrating eyes. We students were also invited on a traditional evening boat trip on the Ganges, yet we remained at a distance with no personal encounter.

Years later, in the 1970's, an anecdotal story came to me whilst writing *CHANDRA*, the biography of S. Chandrasekhar. Chandra recounted a story told by Sir C.V. Raman, his famous uncle during a boat trip on the Hugli River in Calcutta in 1928. Raman told Chandra how S.N. Bose, after attending Raman's lecture, was the first to recognize the importance of his discovery.

Bose presciently remarked to Raman, "You have made an important discovery. It will be known as Raman Effect and you will get a Nobel Prize."

As part of the Einstein centenary celebrations in 2005, I was privileged to present talks about the discovery of Bose–Einstein statistics at the American Physical Society meetings in Tampa, Florida and Lincoln, Nebraska, and at the invitation of the editors of *Physics Today*, I wrote a brief article, "The Man behind Bose Statistics," published in the October 2006 issue. That led to this assignment from Dr. K.K. Phua, of World Scientific Publishing Co., to edit a book on Bose, his life and his works. I have collected together all his physics and mathematical physics papers along with selected articles based on his lectures and addresses. Also included are some miscellaneous pieces translated from his Bengali writings. Some published articles by others provide a historical perspective on the discovery of Bose–Einstein statistics. The remaining articles reflect upon and shed light on Bose as a teacher, an intellectual and a national figure.

Acknowledgments

I am grateful to Sushanta Dattagupta and Arup Kumar Raychaudhuri, former and current Directors of the S.N. Bose National Center for Basic Sciences, Calcutta, India for granting permission to reproduce articles and pictures from *The Man and His Works Part I: Collected Scientific Papers* and *Part II: Life, Lectures and Addresses, Miscellaneous Pieces**, S.N. Bose, National Center for Basic Sciences, Calcutta (1994), to John Stachel and Supurna Sinha for many helpful discussions and providing me documents and pictures, and Falguni Sarkar for permission to use pictures and material from his article "Ode to grandpa: Tribute to S.N. Bose." My special thanks to Etienne Eisenmann for information concerning Jacqueline Eisenmann with pictures and letters, Sauer Tilman for the files from the *Einstein Archives of the Hebrew University, Jerusalem* and all the contributors. Many thanks also to Achala Wali for editing parts of this book and Heather Kirkpatrick for assembling the articles, pictures and preparing the manuscript. Finally, I like to thank the Senior Editor, Lakshmi Narayanan and her colleagues at World Scientific for the excellent job they have done in producing this volume.

*The numbers at the bottom left-hand corner of the pages of articles reproduced in this volume refer to the actual page numbers of articles in these two volumes.

Contents

Foreword v

Satyendra Nath Bose — His Life and Times
K. C. Wali xv

I. Selected Scientific Papers

A. Early Papers — Classical Physics

On the Influence of the Finite Volume of Molecules on the Equation of State [*Phil. Mag.* Ser. 6, **36**, pp. 199–203 (1918)]
M. N. Saha and S. N. Basu 4

On the Herpolhode [*Bull. Cal. Math. Soc.* **11**, pp. 21–22 (1919)]
S. Basu 7

On The Equation of State [*Phil. Mag.* Ser. 6, **39**, p. 456 (1920)]
M. N. Saha and S. N. Basu 9

B. Quantum Mechanics and Quantum Statistics

On the Deduction of Rydberg's Law from the Quantum Theory of Spectral Emission [*Phil. Mag.* **40**, pp. 619–627 (1920)]
S. N. Basu 12

Plancks Gesetz und Lichtquantenhypothese [*Z. Physik* **26**, pp. 168–171 (1924)]
Bose 21

Planck's Law and the Light-Quantum Hypothesis [English translation of *Z. Physik* **26**, pp. 168–171 (1924)]
Bose 26

Wärmegleichgewicht im Strahlungsfeld bei Anwesenheit von Materie [*Z. Physik* **27**, pp. 384–393 (1924)]
S. N. Bose 30

Thermal Equilibrium in the Radiation Field in the Presence of Matter [English translation of *Z. Physik* **27**, pp. 384–393 (1924)]
S. N. Bose 40

A Note on Dirac Equation and the Zeeman Effect [*Ind. J. Phys.* **17**, pp. 301–308 (1943)]
S. N. Bose and K. Basu 49

C. The Ionosphere — Mathematical Physics

Anomalous Dielectric Constant of Artificial Ionosphere [*Sci. & Cult.* **3**, pp. 335–337 (1937)]
S. N. Bose and S. R. Khastgir 61

On the Total Reflection of Electromagnetic Waves in the Ionosphere [*Ind. J. Phys.* **12**, pp. 121–144 (1938)]
S. N. Bose 63

On the Complete Moment-Coefficients of the D^2-Statistic [*Sankhya — Ind. J. Stat.* **2**, pp. 385–396 (1936)]
S. N. Bose 87

On the Moment-Coefficients of the D^2-Statistic and Certain Integral and Differential Equations Connected with the Multivariate Normal Population [*Sankhya — Ind. J. Stat.* **3**, pp. 105–124 (1937)]
S. N. Bose 100

Studies in Lorentz Group [*Bull. Cal. Math. Soc.* **31**, pp. 137–147 (1939)]
S. N. Bose 122

The Complete Solution of the Equation: $\nabla^2\phi - \frac{\partial^2\phi}{c^2\partial t^2} - k^2\phi = -4\pi p(xyzt)$ [*Proc. Nat. Inst. Sc. India* **7**, pp. 93–102 (1941)]
S. N. Bose and S. C. Kar 133

On an Integral Equation Associated with the Equation for Hydrogen Atom [*Bull. Cal. Math. Soc.* **37**, pp. 51–61 (1945)]
S. N. Bose 143

D. Unified Field Theory

The Identities of Divergence in the New Unitary Theory [English translation of *Comptes rendus de L'Academie des Sciences* **236**, pp. 1333–1335 (1953)]
Note of S. N. Bose, presented by L. de Broglie 158

A Unitary Field Theory with $\Gamma_\mu \neq 0$ [English translation of *Le Jour de Phys et le Radium* (Paris), **14**, pp. 641–644 (1953)]
S. N. Bose 161

Certain Consequences of the Existence of the Tensor g in the Affine Relativistic Field [English translation of *Le Jour de Phys et le Radium* (Paris), **14**, pp. 645–647 (1953)]
S. N. Bose 169

The Affine Connection in Einstein's New Unitary Field Theory [*Ann. Math. USA* **59**, pp. 171–176 (1954) (Princeton University, NJ)]
S. N. Bose 175

Solution of a Tensor Equation Occurring in the Unitary Field Theory [English translation of *Bull. Soc. Math.* **83**, pp. 81–88 (1955)]
S. N. Bose 182

II. Satyendra Nath Bose

A. Lectures and Addresses

Science in Education [*Man and His Work, Part II*, pp. 127–133]
Proc. The Bengal Education Week, **I**, pp. 56–64 (1936) 194

Visva-Bharati [*Man and His Work, Part II*, pp. 141–144]
Convocation address delivered at Visva-Bharati on 15 January 1957 201

Higher Education and Us [*Man and His Work, Part II*, pp. 145–150]
Convocation address delivered at Calcutta University in 1962 205

Man in the Scientific Age [*Man and His Work, Part II*, pp. 157–161]
Address at a Symposium in Tokyo, Japan 1962 211

The Mother Tongue [*Man and His Work, Part II*, pp. 162–167]
Address delivered in Bengali at the "Angrezi Hatao Conference" [Banish English Conference] at Hyderabad in October 1962 216

Education and Science [*Man and His Work, Part II*, pp. 168–174]
Convocation address delivered at Ranchi University on 11 May 1963 222

On His Seventieth Birthday [*Man and His Work, Part II*, pp. 175–180]
Address given in Bengali at the Mahajati Sadan Auditorium, Calcutta on 1 January 1964 229

The Krishnan Memorial Lecture, 1971 [*Man and His Work, Part II*, pp. 188–195]
Lecture delivered at the National Physical Laboratory on 12 March 1971 235

Education and the New Age [*Man and His Work, Part II*, pp. 196–199]
Address delivered in Bengali at the Rabindra Bharati University on 10 May 1971 243

At the Calcutta University Convocation 1973 [*Man and His Work, Part II*, pp. 200–203]
Convocation address delivered in Bengali at Calcutta University on 16 June 1973 247

Address at Seminar on Scientific Contributions of S. N. Bose [*Man and His Work, Part II*, pp. 204–206] *Speech delivered at the Calcutta Mathematical Society on 29 December 1973* 251

B. Miscellaneous Pieces

A Scientist's Apologia [*Man and His Work, Part II*, pp. 219–223] 256

The Progress of Science in Ancient India [*Man and His Work, Part II*, pp. 224–229] 261

The Early Phase of My Explorations in Science [*Man and His Work, Part II*, pp. 244–249] 267

Einstein (1935) [*Man and His Work, Part II*, pp. 279–284] 273

In the Company of Madame Curie [*Man and His Work, Part II*, pp. 290–291] 279

III. Historical Perspectives; Reflections

A Loss of Identity: The Birth of Quantum Statistics ['*Subtle is the Lord*' *The Science and Life of Albert Einstein* (Clarendon Press, Oxford, 1982)]
A. Pais 284

Bose Statistics: A Historical Perspective [*The Man and His Work, Part I*, pp. 35–70]
P. Ghose 296

Who Discovered the Bose–Einstein Statistics? [*Symmetries in Physics (1600–1690); Proc. 1st Int. Meeting*, Saint Feliu de Guixoles, Catalonia, Spain (September 20–26, 1983) pp. 222–250]
S. Bergia 332

Addendum
P. Ghose 361

Nobel Lecture: Bose–Einstein Condensation in a Dilute Gas, the First 70 Years and Some Recent Experiments* [*Rev. Mod. Phys.* **74**, 875 (2002)]
E. A. Cornell and C. E. Wieman 369

Experimental Proof of the Spin of the Photon
C. V. Raman and S. Bhagavantam 388

Evidence for the Spin of the Photon from Light Scattering
C. V. Raman and S. Bhagavantam 398

S. N. Bose: The Man [*Talk presented at a seminar on Albert Einstein, Dhaka, 30 May, 2005; Proc. Bangladesh Academy of Sciences*]
P. Ghose 400

Satyendranath Bose: Co-Founder of Quantum Statistics [*Am. J. Phys.* **40**, 1212 (1972)]
W. A. Blanpied 414

Einstein and Bose [*Einstein from 'B' to 'Z'* (Birkhäuser, 2002), pp. 519–538]
J. Stachel 422

A Conversation with Satyendranath Bose about Five Decades Ago — Some Recollections [*Current Science* **78**, pp. 636–638 (2000)]
S. Ramaseshan 442

Satyen Bose in Berlin [*The Physicist, Bulletin of the Bangladesh Physical Society* (1995), pp. 17–21]
P. Sinha 445

Satyendra Nath Bose: Historical Notes and Reflections
E. C. G. Sudarshan 457

Satyendra Nath Bose — Some Vignettes of a Many-sided Personality
D. N. Bose 462

Contributors 467

Satyendra Nath Bose — His Life and Times (1894–1974)

Never accept an idea as long as you are yourself not satisfied with its consistency and the logical structure on which the concepts are based. Study the masters. Those are the people who have made significant contributions to the subject. Lesser authorities cleverly bypass the difficult points.

Satyendra Nath Bose

Considering Bose's legendary association with Einstein and a remarkable discovery to his credit, one might expect him to be better known in the international sphere. However, little is known about him outside India. Quite often he is thought to be from Germany, name misspelt without the "umlaut" on "o". Even in India, he is often confused with Jagadish Chandra Bose, who was the first 19th century scientist from the country to have received worldwide recognition for his experiments on wireless transmission as well as his experiments with plants, demonstrating such vital characteristics in them as growth, nervous system and other distinguishing features of animal life.

The first quarter of the 20th century was indeed a golden period in the history of India, and the province of Bengal in particular. Considered more progressive than other provinces, it was the first to accept and adopt British ways of life and literature. Bengal was also the first to rebel against the colonial masters, especially, when in 1905 (coincidentally the Einstein miracle year), Lord Curzon partitioned Bengal into two, West and East Bengal to curb the rising nationalist movement. Bose, a brilliant student throughout his career, joined a group of accomplished intellectuals motivated strongly to excel in arts and sciences and other creative endeavors to demonstrate their national fervor and to prove that they were competitive with their counterparts in the western world.

Beginnings

Satyendra Nath (Bose) was born on January 1, 1894 in Kolcuta (formerly known as Calcutta, the capital of British India). He was the eldest child and the only son in a family of six daughters of Surendra Nath and Amodini Bose. Bose's family, with two generations of English education, belonged originally to the village Bara Jagulia in the district of Nadia, known for its traditional learning of Sanskrit scriptures and its scholars. According to family records, the family lived in this village for over three centuries. With the advent of British Raj and the emergence of Calcutta, Nadia lost its importance and the family moved to Calcutta. The house in which Satyendra Nath was born and grew up is supposed to have been built by his great-great-great-grandfather and rebuilt with extensive modifications by his father.

Satyendra Nath's grandfather, Ambikacharan, was probably the first "English-educated" in the family. He worked as an accountant in the British military commissariat and spent the most part of his life in northwestern parts of India. Surendra Nath, his eldest son, after completing his primary education in the village, studied in Meerut Government School. His education was interrupted due to the sudden death of his father. With the family responsibilities on his shoulders, he took an accountant's examination, qualifying himself as an accountant. He was working as an accounts clerk in the East India Railways when Satyendra was born. Although working in the British civil service, Surendra Nath was a part of the rising English-educated middle class that was deeply influenced by the mid-nineteenth century rationalism and individualism of European culture. At the same time, he was also equally becoming a part of the rising nationalism against the British colonization. With a broad range of interests including mathematics, science and philosophy, and inspired by the nationalist ideal that the future of India demanded the development of science and technology, he became the founder of an independent chemical and pharmaceutical company in 1903. Amodini, Satyendranath's mother was the daughter of Motilal Chowdhary, a renowned lawyer and a classmate of Bankimchandra Chatterjee, one of the early and famous novelists of Bengal. With only a nominal education in a primary school, she adapted well in the extended family of the Boses, and was apparently a woman of great dignity, full of warmth and affection, and extremely proud of her only son.

Satyendra grew up in a normal extended family of a loving mother, younger sisters, uncles and aunts and a strict father who commanded more respect than overt love as was customary in families of India. He traveled often with his father on his executive tours of East India Railways in Assam and Northern Bengal. The scenic beauty of Assam and the river Brahmputra were some of his rare memories of childhood. His schooling began at the age of five in schools that were situated close to where they lived, but finally he switched over to the Hindu School, a school of great tradition and distinction in the first decade of the 20th century. From all accounts, Satyendra was recognized as a gifted child by all his teachers. In spite of his weak eyesight since childhood, he was a voracious reader of Bengali, Sanskrit and English literature, well versed in the poetry of Lord Tennyson, Rabindranath Tagore and Kalidasa. His mathematics teacher in the Hindu School recognized him as a genius in mathematics and proclaimed that he would be one day known as another Laplace or Cauchy.

Presidency College, *Swadeshi* Movement

By the time he joined the Presidency College in Calcutta in 1909, Bose, despite his natural aptitudes for languages and humanities, had decided to pursue science. That was partly motivated by the prevailing nationalistic sentiment ignited by the partition of Bengal in 1905. By partitioning Bengal, Lord Curzon, one of the most imperialistic, reactionary viceroys of British India, had intended to curb Bengal's influence on the rising national struggle for independence. Instead this created a furor against the Raj, and initiated a strong *swadeshi* movement of self-sufficiency, symbolized by the boycott of foreign goods, a public bonfire of

British mill-textiles, reliance on home spun industries, and street gatherings singing patriotic songs. *Rākhi-bandhan*, a popular festival in North India in which a sister ties a colored thread around a brother's wrist as a mark of assurance of protection from all sorts of dangers, was converted into a solidarity pledge between all peoples, Hindus and Muslims exchanging *rākhis*.

Among the young, it generated a strong national awareness and dedication to the service of the country. Talented students like Satyendra Nath thought science, engineering and technology were the ways to lead the country to progress and prosperity. They had the ideals of people like Jagadish Chandra Bose (J.C. Bose) and the noted chemist, Acharya Prafullachandra Ray (P.C. Ray), who had gone abroad and returned home with international recognition and fame. With meager salaries and poorly equipped laboratories they were carrying on original research. They also had the example of Srinivasa Ramanujan orchestrated by the well-known Cambridge mathematician G.H. Hardy.

The year 1909 was remarkable in the history of the Presidency College and in the scientific history of Bengal. The college's entering class included a group of extremely talented students who would make a mark in their own original research in many areas and play a dominant role in the scientific and economic development of free India. The brightest among them were Bose and Meghnad Saha. Saha who was to become a close friend and collaborator in research joined the Presidency College two years later.

United in their nationalistic ideals, the group was a part of overt and clandestine activities in the *swadeshi* movement. Some of the students risked their careers. Some became members of the National Revolutionary Movement — one of several independence-minded groups in British colonial India — based in the state of Bengal. Upon the outbreak of the First World War, its members began to organize an armed rebellion in India. In that noted politically charged atmosphere, several people, notably Manabendra Nath Roy (1887–1954, popularly known as M.N. Roy) and Abani Mukherjee (1891–1937), left the country to secure arms from Germany and America.

Bose was in contact with Abani Mukharjee directly or indirectly and helped him in many ways. He, however, was under strict orders from his father not to get involved in any activities that might spell the end of his brilliant career. As the eldest son and only son, he was made aware of his responsibilities to the family. He obeyed his father, but remained a strong supporter of those who participated in the Swadeshi movement, helping in whatever way he could, acting as a courier, providing shelter to fugitives from police pursuit, or helping out with money. Throughout his college years he took an active part in running the night schools for the children of working class that came to be known as the Working Men's Institute. He strongly believed in the enlightenment and self-awareness through education of the masses more vital to independence than popular simple slogans like, "Down with the British, down with English education."

Marriage, Career Ups and Downs

Bose completed his B.Sc. degree in 1913 and his M.Sc. in 1915, securing the first place in both the examinations; Saha took the second place. Both Saha and Bose, mostly self-taught, had glimpses of exciting discoveries in modern atomic physics and relativity and had developed a strong passion for theoretical physics. Both, however, faced a bleak situation, with no opportunities for further study and research or any jobs to speak of. In the mean time, when he was only 20 years old and still a student doing M.Sc. Bose married Ushabati, who was the only daughter of a renowned and rich doctor, Jogindranath Ghosh. Bose was not keen on getting married so early, but he could not go against his mother's wishes. Marriage in India at an early age, when still in college, was very common then and remains common now. Bose, however, was exceptional in neither accepting a dowry nor seeking monetary help from his wife's family when faced with difficult financial situations. If he wanted to, he could very well have received support from his father-in-law for a trip abroad for higher studies, a prevalent practice. The couple's first child was born soon after Bose received his M.Sc. degree.

Unemployed both Bose and Saha spent a year, earning what they could from private tutoring. Bose tried to work towards a doctorate degree in mathematics under Ganesh Prasad who had just joined the University of Calcutta. Prasad (1876–1936), a D.Sc. from Allahabad University, used to teach at Queen's College in Banaras before coming to Calcutta. Known to set hard questions that went beyond the expectations of normal students, Prasad had an aura of scholarship and intimidation. Bose failed to establish a rapport with him. Recalling, Bose writes:

> The students flocked to him for training in search. They were the best science students of Calcutta University, but they had not fared so well in his exams. But the fault lay with the teachers at Presidency College — at least which is what Ganesh Prasad thought. The young students had to stomach adverse comments about their former teachers, too scared to answer back. After my M.Sc. I too presented myself before Ganesh Prasad who was also my examiner though I had not fared as badly as the others [on Prasad's exam questions]. Dr. Prasad was kind to me at first but I was notorious for plain speaking. I found it difficult to bear his tirade against my teachers. I had dared to counter his adverse criticisms. This infuriated him. He said — you may have done well in the examination but that does not mean you are cut for research. Disappointed I came away. I decided to work on my own.[1]

He also decided to apply for a teaching position when he came across an advertisement by the government of Bihar — a state neighboring Bengal. With letters of recommendation from Principal James and Debendranath Mullick [who had taught at Hooghly, Patna and Presidency Colleges and was President of the Physics and Mathematics section of the Indian Science Congress in 1919], Bose applied for the position but failed to get it because he was over qualified. He also applied for a position at the Alipore Meteorological Office but was turned down for the same reason.

Just as the situation looked bleak, a splendid opportunity presented itself to Bose and Saha. It came from Sir Asutosh Mookerjee, the Vice Chancellor of the University of Calcutta, who was a distinguished mathematician with strong interest in physics. However, having had no opportunities for continuing research and advanced studies, he had opted for law and had risen to eminence as the high court judge. With generous financial help from two other lawyers, Taraknath Palit and Rashbehari Ghosh, he established a University College of Science as part of the Calcutta University with emphasis on teaching advanced courses and basic research. It started functioning in 1916. As Bose recounts,

> "One of our classmates, Sailen Ghosh, popularly known as 'Mama' [lit. uncle], had done well in physics. He too was without a job. But he was very resourceful. He went up to Sir Asutosh with a suggestion: Why not start teaching subjects in the University that are not taught at Presidency? The logicians had in the meantime thought of various plans for the physics department in the University. But who was going to teach?
>
> One day we were called up by Sir Asutosh. Meghnad, Sailen and I went up the steep stairs to the library, to the special chamber where Sir Asutosh sat. We were naturally meek and submissive and overawed by his august presence. He had heard that the younger generation wanted more modern subjects to be introduced in the University curriculum. He asked, 'What subjects are you competent to teach, boys?
>
> 'Sir, we will try our best to teach whatever you want us to.' He smiled. We had only heard of the many new discoveries in physics, most of them made in Germany — new developments and new discoveries. Planck, Einstein, Bohr — we Bengalis had only heard of them. To know more about them one had to read books in German or research journals in other languages. During the war most of these journals did not come to India.
>
> At long last, as the first step to a new career, we were given a special allowance of Rs. 125 per month. Meghnad was assigned to study quantum theory and I had to learn Einstein's relativity theory. We came away committing ourselves to being prepared to teach within a year. But where were we to get the books from? There were some books in English on relativity; we got hold of them. But where could we get hold of the writings of Boltzmann, Kirchhoff, Planck..?"[2]

Help came again from an unexpected source. Bose knew Dr. P.J. Bruhl, who was teaching engineering physics at Sibpur in the Bengal Engineering College. M.Sc. students from the Presidency College used to do their required "practical work" in his laboratory. Bruhl was from Germany with a doctorate in botany. He had come to Calcutta seeking a warm and sunny climate because of his ill health due to tuberculosis. The scholarship he had to study flora in India was suspended when he got married. He decided to stay in India, switched to physics, and got a job to teach engineering physics.

He had an excellent collection of science books, advanced texts and journals on physics in German including Planck's *Theorie de Wärmstrahlung*, Laue's *Relativitätsprinzip.* Bose and Saha could not have asked for more. They borrowed whatever books they could lay their hands on — Planck, Boltzmann, Wilhelm Wien and others. Saha had taken pains to learn German;

Bose knew French and was taking lessons in German. Together they prepared themselves to teach advanced modern physics and forged a collaboration in research.[3]

In 1919, soon after the war ended and Einstein's theory received worldwide acclamation due to the confirmation of one of its predictions, namely, the deflection of light due to gravitational field, the young self-taught stalwarts took upon themselves the translation of the original papers on special and general relativity. Bose and Saha's book was the first such collection in English.[4]

Satyendra Nath, like other aspiring young scientists, was eager to go abroad to Europe, but he was not fortunate enough for one reason or the other to acquire such an opportunity. Being married was one of his obstacles. His friend and colleague Saha was more enterprising. He was awarded the Doctor of Science degree of Calcutta University and on that basis won a scholarship to go abroad. In the mean time, while teaching thermodynamics and astrophysics, Saha got interested in thermal ionization and developed the celebrated Saha's theory of high temperature ionization and its consequences in astrophysics. The scholarship afforded him the opportunity to work in the laboratories of A. Fowler in London and Herman Walther Nerst in Germany to verify his theory.

In 1921, Bose got an offer of a Readership at a higher salary in the newly established university in Dacca in East Bengal (presently called Dhaka University, that is now in Bangladesh). The Vice Chancellor of the University, P.J. Hartog, was a visionary committed to excellence. He hand-picked Bose, who soon found himself with the task of building a whole new department — including setting up laboratories — to teach advanced courses for B.Sc. honors and M.Sc. students. He taught thermodynamics and Maxwell's theory of electromagnetism. The library was just being replenished with books and journals. And just as his first group of students was graduating in 1923, he received notification that his appointment could not be extended beyond a year. A conflict between the government of India and the provincial Government of Bengal had resulted in the cut-off of funds for the university. Bose became involved in a long struggle with the university to keep his appointment. It was under such troubled circumstances that he wrote his famous paper and sent it for publication in late 1923 or the beginning of 1924 to *Philosophical Magazine.* Six months later, when the editors of the journal informed him that (regrettably) the referee's report on his paper was negative, he sent his rejected paper to Einstein with an accompanying letter (See Fig. 1), which begins[5]

> Respected Sir,
>
> I have ventured to send you the accompanying article for your perusal and opinion. I am anxious to know what you think of it. You will see that I have tried to deduce the the coefficient $8\pi\nu^2/c^3$ in Planck's law independent of the classical electrodynamics only assuming that the ultimate elementary regions in the phase-space has the content h^3. I do not know sufficient German to translate the paper. If you think the paper is worth publishing, I shall be grateful if you arrange for its publication in *Zeitscrift für Physik.*

Figure 1. Satyendra Nath Bose in Paris in 1925. It was from Dacca in 1924 that Bose sent Einstein this letter accompanying his manuscript on Planck's blackbody radiation law. (Photo courtesy of Falguni Sarkar; letter from ref. 8.)

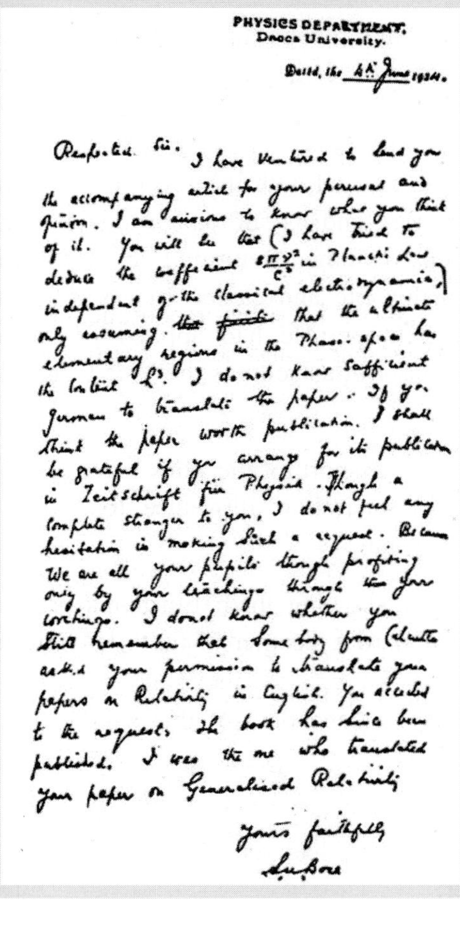

PHYSICS DEPARTMENT,
Dacca University.

Dated, the 4th June 1924.

Respected Sir, I have ventured to send you the accompanying article for your perusal and opinion. I am anxious to know what you think of it. You will see that I have tried to deduce the coefficient $\frac{8\pi\nu^2}{c^3}$ in Planck's Law independent of the classical electrodynamics, only assuming that ~~the finite~~ the ultimate elementary regions in the Phase-space has the content h^3. I do not know sufficient German to translate the paper. If you think the paper worth publication I shall be grateful if you arrange for its publication in Zeitschrift für Physik. Though a complete stranger to you, I do not feel any hesitation in making such a request. Because we are all your pupils though profiting only by your teachings through the writings. I do not know whether you still remember that somebody from Calcutta asked your permission to translate your papers on Relativity in English. You acceded to the request. The book has since been published. I was the one who translated your paper on Generalised Relativity.

Yours faithfully
S.N. Bose

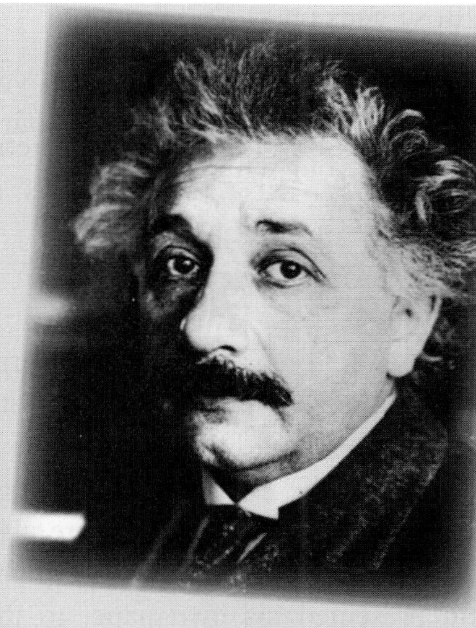

Figure 2. In July 1924, Albert Einstein sent S.N. Bose a postcard (left) that accepted Bose's paper on radiation statistics and acknowledged its importance. (Card from ref. 1; 1920s photo of Einstein courtesy of AIP Emilio Segré Visual Archives.)

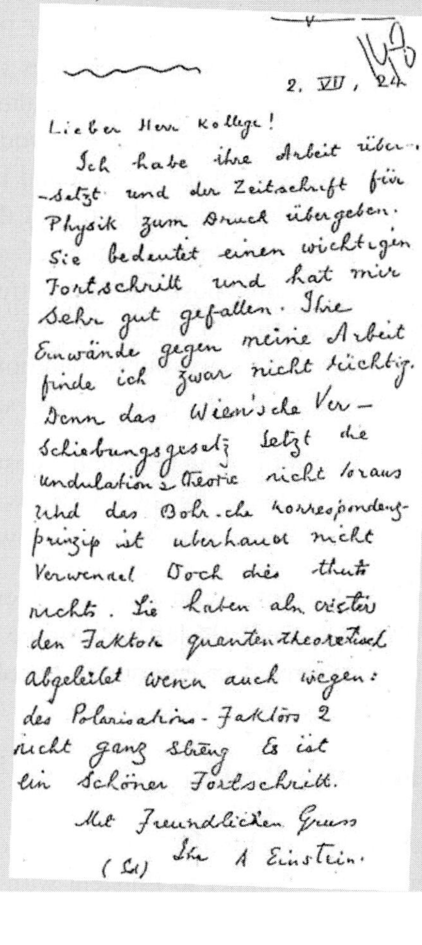

2. VII. 24

Lieber Herr Kollege!

Ich habe Ihre Arbeit übersetzt und der Zeitschrift für Physik zum Druck übergeben. Sie bedeutet einen wichtigen Fortschritt und hat mir sehr gut gefallen. Ihre Einwände gegen meine Arbeit finde ich zwar nicht richtig. Denn das Wien'sche Verschiebungsgesetz setzt die Undulationstheorie nicht voraus und das Bohr'sche Korrespondenzprinzip ist überhaupt nicht verwendet. Doch dies thut nichts. Sie haben als erster den Faktor quantentheoretisch abgeleitet wenn auch wegen des Polarisations-Faktors 2 nicht ganz streng. Es ist ein schöner Fortschritt.

Mit freundlichen Gruss
(Sd) Ihr A Einstein.

Bose goes on:

> Though a complete stranger to you, I do not feel any hesitation in making such a request. Because we are all your pupils though profiting only by your teachings through your writings.
>
> Yours Faithfully,
> S.N. Bose

Einstein's reply came in a postcard dated 2 July 1924 (Fig. 2) in which he wrote, "I have translated your paper and given it to *Zeitscrift für Physik* for publication. It signifies an important step forward and pleases me very much." In a note appended to his translation and published with the paper, Einstein says that Bose's derivation "appears to me an important step forward. The method used here also yields the quantum theory of an ideal gas, as I shall show elsewhere." Indeed, on 10 July 1924, within a week or so after receiving Bose's paper, Einstein presented his own paper to the Prussian academy. Titled "On the Quantum Theory of Monoatomic Gas," it was an extension of Bose's work. He followed up that paper with two more in 1925, the second of which is well known for its prediction of a possible new state of matter whose existence took 70 additional years to demonstrate — the Bose–Einstein condensate.

Bose's successful derivation of Planck's law and Einstein's recognition of its importance is one of the most exciting episodes of the 20th century physics. It has all the twists and turns and ironies of failed attempts on the part of some of the great physicists of the period and the success of a relatively unknown young man from Dacca University in East Bengal. It involved a great struggle on many profound philosophical issues associated with the fundamental nature of matter and radiation. I will not dwell here on this history and its subsequent developments since it is the subject matter of several papers in this volume. Returning to Bose's life, Einstein's 2 July postcard had a tremendous impact on his life and career. It was influential enough for him to obtain a two-year study leave in Europe. He had applied for such a fellowship in January, but he had not received any response. But as soon as the Vice Chancellor saw the postcard, the problem was solved. As Bose says

> That little thing [Einstein's postcard] gave me a sort of passport for the study leave — study leave for two years with a good stipend, a separation allowance for the family, sumptuous travel allowance with a round trip fare.[6]

He also got a visa from the German consulate just by showing them Einstein's postcard. No fee required! He left for Europe in early September aboard a steamer of the Lyod Triestino Line and arrived in Paris in 18 October 1924.

Paris and Berlin

Before he left for Paris, on 15 June, only eleven days after he had sent his first paper, Bose sent a second paper to Einstein with the following letter:

I send herewith another paper of mine for your kind perusal and opinion. I hope my first paper has reached your hand. The result to which I have arrived seems rather important (to me at any rate). You will see that I have dealt with the problem of thermal equilibrium between Radiation and Matter in a different way, and arrived at a different law for the probability of the elementary processes [presumably, different from the law proposed in Einstein and Ehrenfest 1923], which seems to have simplicity in its favor. I have ventured to send you the typewritten paper in English; it being beyond me to express myself in German (which will be intelligible to you). I shall be glad [to (crossed out in letter)] if its publication in *Zeit. für Physik* or any other German journal can be managed. I myself know not how to manage it. In any case, I shall be grateful if you express any opinion on the papers, and send it to me at the above address.

Yours Truly,
S.N. Bose
(Einstein Archive, Doc. 6-128)

In Paris, Bose stayed in a pension at 17 Rue du Sommerard in a building that was the home of India Association, a meeting place for students from India. The association helped visitors, students, and student activists involved in anti-British national movement with accommodations and other needs. Bose knew several of them from Calcutta and spent the first few days' sightseeing in Paris.

During his study leave Bose was expected to learn something new in the great laboratories of Europe to take back home. The Vice Chancellor Hartog had written to Earnest Rutherford and William Braggs, whom he knew personally, introducing Bose and requesting an opportunity for Bose to work in their laboratories. Rutherford had replied regretfully that he could not because of lack of space. Bragg was non-committal. Paris offered the possibility of working in the X-ray and crystallography laboratories of Maurice de Broglie or the institute where Madame Curie worked on radioactivity. So Bose stayed on in Paris before venturing to go to Berlin to meet Einstein.

Bose came to know Paul Langevin who had read his paper and so received him warmly. When he expressed interest in working in radioactivity at the Institute of Madame Curie, Langevin wrote a letter of introduction with which Bose went to see her. Madame Curie greeted him affectionately and said that there was no way she could disregard a recommendation from a person like Langevin. Bose would certainly get an opportunity to work in her laboratory, but not right then. Speaking in chaste English for about ten minutes, she asked Bose to get to know the French language first to be able to work in her laboratory. Bose had no opportunity to tell her that he knew French, he had been at it for ten years.[7] Bose came away resigned to wait, and chose to work in de Broglie's X-ray laboratory. Later though, at the beginning of the following year, he was granted facilities to work in her Institute, but Bose does not mention any further contact with Madame Curie.

As mentioned earlier, around the same time Bose sent his first paper to Einstein, he had also sent a second paper from Dacca to Einstein. In March 1924, M.N. Saha paid a friendly visit to Dacca and stayed with Bose. He also brought attention to new attempts at the derivation of Planck's law due to Wolfgang Pauli, and Einstein and Paul Ehrenfest. This set Bose thinking about the interaction between radiation and ordinary matter and led to the second paper he sent to Einstein. On 26 October, shortly after his arrival, he wrote to Einstein:

Dear Master,

My heartfelt gratitude for taking the trouble of translating the paper yourself and publishing it. I just saw it in print before I left India. I have sent you about the middle of June a second paper entitled "Thermal Equilibrium in the Radiation Field in the presence of Matter."

I am rather anxious to know your opinion about it, as I think it to be rather important. I don't know whether it will be possible also to have this paper published in Zeit. für. Physik. I have been granted study leave by my University for 2 years. I have arrived just a week ago in Paris. I don't know whether it will be possible for me to work with you in Germany. I will be glad however if you grant me the permission to work under you, for it will mean for me the realisation of a long-cherished hope.

I shall wait for your decision as well as your opinion of my second paper here in Paris. If the second paper has not reached you by any chance, please let me know, I shall send you the copy I have with me.

With respect

Yours Sincerely,
S.N. Bose (Prof.) (26 October 1924, Einstein Archive, DOC. 6-129)

Bose's second paper begins with a critical review of the previous derivations (those of Einstein, Pauli, and the generalization of Pauli by Einstein and Ehrenfest) that, according to him, contained special assumptions that were unnecessary. For Bose, the methods of statistical mechanics alone were sufficient to study the thermal equilibrium of radiation and matter independent of any specific mechanism of the elementary processes on which the energy exchange depended. Bose had obtained a general relation, which was valid for all special assumptions made by the previous authors about the elementary processes and their probabilities.

Einstein's reply came on 3 November,

Most esteemed colleague!

Friendly thanks for your letter of 26.10. I am happy to have the opportunity to meet you personally. Your papers have appeared already some time ago;

unfortunately the offprints were sent to me instead of you. You can have them any time. I am not in agreement with your elementary law of probability for the interaction between radiation and matter, and have given the reason in a note that appeared together with your article. To wit, your law is not compatible with the following two conditions:

1. *The coefficient of absorption is independent of the density of the radiation.*
2. *The behavior of an oscillator in a radiation field must result from the statistical laws as a limiting case.*

We can discuss this in more detail when you come here.
With friendly greetings

Yours
(Einstein Archive, DOC. 6-130)

Bose took some time to reply to Einstein. He wrote on 27th January from his Paris residence,

Respected Master,

I received your kind note of 3rd November in which you mentioned your objections against the elementary law of probability. I have been thinking about your objections all along and so did not answer immediately. It seems to me there is a way out of this difficulty, and I have written down my ideas in the form of a paper which I send under a separate cover. It seems that the hypothesis of negative Einstrahlung stands, which, as you have yourself expressed, reflects the classical behaviour of a resonator in a fluctuating field. But the additional hypothesis of a spontaneous change, independent of the state of the field, seems to me not necessary. I have tried to look at the radiation field from a new standpoint and have sought to separate propagation of Quantum of energy from the propagation of electro-magnetic influence. I seem to feel vaguely that some such separations are necessary if Quantum theory is to be brought in line with Generalised Relativity Theory.

The views about the radiation-field, which I have ventured to put forward, seem to be very much like what Bohr has recently expressed in May Phil. Mag. 1924. But it is only a guess, as I cannot say honestly to have exactly understood all he means to say, about virtual fields and virtual oscillators.

I am rather anxious to know your opinion about it. I have shown it to Professor Langevin here and he seems to think it is interesting and worth publishing.

I cannot exactly express how grateful I feel for your encouragement and the interest you have taken in my papers. Your first postcard came at a critical

moment, and it has more than any other made this sojourn to Europe possible for me. I am thinking of going to Berlin at the end of this winter, where I hope to have your inestimable help and guidance.

Yours Sincerely,
S.N. Bose
(Einstein Archive, DOC. 6-131)

Bose never published this third paper, and no trace of it exists in the Hebrew University's Albert Einstein Archives. It is regrettable that, either because of Einstein's critical remarks or because of the advent of new quantum mechanics, Bose's second paper has also received little attention.

Fortunately, Partha Ghose, has provided an in-depth analysis of both these papers.[8] Recalling a long conversation he had with Bose in December 1973 or January 1974, Ghose writes that Bose had his own approach to quantum theory based on his views about the interconnectedness of spontaneous and induced emissions, which formed the subject matter of his second and third papers. He did not agree with Einstein's proposal to treat them as independent physical processes. Stationary states in atoms have zero widths both in the old and the new quantum theories and therefore have infinite life-times, whereas in reality they do decay and jump down to lower energy states with a finite life-time.

What is the origin of these finite widths? In Bose's view, they are a consequence of the interaction between atom and environment, as he attempted to show in his second and third papers. But Einstein did not agree and dismissed his views with the remark that in a world consisting of a single excited hydrogen atom and nothing else, the atom would obviously emit a photon and come down to its ground state! Einstein did not explain what would excite an atom in such a world! Bose was very disappointed by this simply because the world does have much more than a single hydrogen atom, and inevitably every atom does have an environment. What further disappointed him was that later on, stationary states were seen to acquire finite widths in Dirac's Quantum Electrodynamics through interactions with the vacuum, which was equivalent to an environmental interaction, and this came to be accepted even though Quantum Electrodynamics led to unacceptable divergences. Presumably Bose's way of solving the problem did not produce divergences, but this is something that Ghose didn't have the opportunity to discuss further, because it was apparently quite late by then and it was time for his lunch. He was feeling quite exhausted and weak, and asked Ghose to come back another day. Ghose went back to Santiniketan where he was working, and before he could go back to him, he passed away in February 1974.

Ghose recalls two other comments Bose made that are significant. For him it was clear that the world was basically non-linear as was evident from general relativity. A linear theory like the quantum theory could never be the real story. He implies this in his letter to Einstein (*...if quantum theory is to be brought in line with Generalized Relativity*). Also, implied in his second paper, the existence of Bohr-like stationary states and the vanishing of cross-sections at certain energies (Ramsauer effect) were evidence for "no interaction" (absence of radiation from stationary states) where classical theory leads one to expect such interactions.[9]

After spending a year in Paris, Bose went to Berlin in October 1927 and on his arrival wrote to Einstein,

> *Dear Sir,*
>
> *I have just arrived in Berlin, and am very anxious to see you. I shall be glad if you kindly write when it will be convenient for you.*
>
> *Yours Sincerely,*
> S.N. Bose
> (8 October, 1925, Einstein Archive, Doc. 6-132)

Jacqueline Zadoc-Kahn (Courtesy of Etienne Eisenmann.)

They met and Einstein introduced him to several prominent physicists including Fritz Haber, Otto Hahn, Lise Meitner and Walter Gordon. It was an exciting time in Berlin. It was the beginning of the New Quantum Mechanics. As Bose describes in a letter to Jacqueline Zadoc-Kahn, a young student he met while attending Langevin's lectures at the Collège de France[10]

> *"Everybody (every physicist) seems to be quite excited in Berlin the ways things have been on with Physics. First on the 28th October Heisenberg spoke in the colloquium about his theory, then, in the last colloquium, there was a long lecture on the recent hypothesis of the spinning electron (perhaps you have heard of it). Everybody is quite bewildered and there is going to be very soon a discussion of Schrödinger's papers. Einstein seems quite excited about it; the other day coming from the colloquium, we found him jumping, in the same compartment, where we were, and forthwith he began to talk excitedly about the things we have just heard. He has to admit that it seems a tremendous thing, considering the lot of things which these new theories correlate and explain, but he is very much troubled by the unreasonableness of it all. We were all silent, but he talked almost all the time; unconscious of the interest and wonder that he is exciting in the mind of the passengers.*

According to Mehra and Rechenberg, Einstein proposed two problems to Bose to work on: first, the question whether the new statistics implied a novel type of interaction between the light quanta,[11] and second, as described by Bose himself: "He wanted me to try to see what the statistics of light-quanta and the transition probabilities of radiation would look like in the new theory." Apparently Bose made no progress on either of the two problems. It was probably because, as John Stachel says, "like many contemporaries who had mastered the old quantum theory, he [Bose] seems never to have been comfortable with the new quantum mechanics."[12]

Instead Bose spent sometime at the Kaiser Wilhelm Institute, Berlin-Dahlem in the laboratories of Hermann Mark. In an interview in 1974, Mark recalls his meeting Bose at a seminar Bose gave in Berlin in 1925–1926. "I don't recall whether he gave the seminar in English or German," says Mark. "But I think he gave it in German because he was fluent in German. Usually after the presentation, Laue would get up and say, 'alright fine, this was nice, and is there any question?' Then there would be a lengthy discussion on what the issue was. And in this case Laue said immediately, I am not sufficiently familiar with the topic and I would like to ask Prof. Einstein to comment. Then Einstein said, 'Well, I think this is one of the most worthwhile works of the last few years.' He talked about it and asked a few questions and Bose answered the questions and then Nernst asked a few questions. In other words it was a very sensational event on that day. The discussion on this specific seminar took a few hours! That was the first time I saw him and liked him. Then of course Karman, Leo Szilard, Eugene Wigner also asked questions. We were young and he was young — so we quickly became friends and we invited him to come to the Kaiser Wilhelm Institute in Dahlem and give us the same seminar to continue the discussion and so he came out a few weeks later and we had there a seminar."[13]

Bose liked the people and found the atmosphere at Kaiser Wilhelm Institute more agreeable and livelier than at the University of Berlin. He moved to a new apartment not far from the Institute and became a regular visitor. He became interested in Hermann and his collaborators' research on X-ray structure, study of crystals of metallic materials and the way they were formed. As Hermann recalls, "he always asked sharp questions and kept us thinking. You know, the most real impression we all had was when Bose was facing some new area of scientific endeavor or activity, he just sat there, listened for a while and he had a fantastic grasp over the fundamentals of the specific work and then he would ask questions and he would ask extremely relevant — not only intelligent — questions, he would ask questions which were piercing and pointing into future developments."[14]

The experimentalists had a great deal of fun listening to discussions about the new statistics between Bose, Eugene Wigner and Leo Szillard. The discussions would last "hours and hours," ending up finally in the small apartment of Mark and his wife Mia at the Institute for coffee or tea and refreshments. According to Mark, Bose was always good-humored, full of good jokes, loved singing German songs, and would frequently get up in parties and conduct. Before Bose was to depart for India, Mark invited him to Vienna for a visit with his parents and his son, and took him to the university to meet the theoretical physicists that included Erwin Schroedinger, Thirring and Flamm. Bose was invited to give a seminar at the university.

Bose also made frequent visits to 5 Heberland Strasse, the flat where Einstein lived. However, there is no record of their scientific conversations or of their differences concerning Bose's second and third papers. Bose never refers to these matters in his rare autobiographical sketches and articles on Einstein, although they seemed to have had extended discussions about matters outside physics. Bose, for instance, recalls an "interesting conversation" he had with him on one such occasion. It was the time in India when strong movements were going on to oust the British. The revolutionaries, on the one hand, were making futile attempts to drive out the British by violent armed struggle; on the other hand, the Congress party under the leadership of Mahatma Gandhi was advocating non-violent, non-co-operation ways to achieve the goal of independence. All this was confusing and strange to Einstein. Bose writes:[15]

> He confided to me "I think Englishmen are better than other Western colonial nations, and I feel that they are by far better than the French and the Dutch. You should not be surprised at a German like me praising the English (after the World War). Now, tell me do you really want that the British should quit your country? I said, "Of course, we all want to determine our own destinies." He was not quite convinced. He raised a hypothetical question. He said, "Suppose, there was a button near you and all the Englishmen would quit India if you were to press it. So, would you press that button?" I smiled and said, "If God were to grant me such an opportunity, I would not hesitate even for a moment to press it. "Really," he said and kept quiet for a while.

It was common knowledge that after the World War I, the Jews had an opportunity to establish a new State in the land of their ancestors and a new University was set up in Jerusalem. Bose knew that Einstein had taken the trouble to go and deliver an address at this new obscure university. The new State was under the British domination. Just as in India, British were playing divide and rule policy, playing the Jews against the Arabs. So Bose, known for his outspokeness, asked Einstein

> Well, why do you Jews then want to establish a new Israeli state? Even you seem to be fairly inclined in its favour. He said, "I have, of course, I now understand what you are saying — it is an emotional matter and cannot be explained by logic."

Professor in Dacca

In the middle of 1926, Bose returned from Germany to his university. He had completed the two years of study leave. While he was abroad, a professorship position was announced. He applied for the position with a letter of recommendation from Einstein that said

> The recent works of Mr. S.N. Bose, especially his theory of radiation equilibrium, signify in my opinion important and enduring progress of the physical theory. Also in personal discussion with Mr. Bose, I got the impression that he is a man of unusual gift and depth, from whom science has much to expect. He has also at his command an extensive knowledge and certain ability in our science. As a university teacher he will certainly develop a successful prosperous activity.

Einstein's letter and those of Langevin, Maurice de Broglie and Hermann Mark were not enough to make Bose the first choice for the position. He had no formal doctorate degree! Mostly as a formality, the position was offered to D.M. Bose, who was happily settled at the

Calcutta University. Bose had to wait a year until D.M. Bose declined the offer. He was then appointed as Professor and Head of the Department of Physics. He also became the Dean of Science the same year.

Since his study leave had been supported by the university, Bose felt an obligation to do something in return. His two-year stay in Europe had been in many ways a turning point in his career as a teacher. He had realized the importance of experimental research in the advancement of science and had spent a year in France getting first-hand experience working in the laboratories of de Broglie, Madame Curie, and Hermann Mark. So, on his return he occupied himself with experimental work, designing and building his own equipment to set up an X-ray crystallography laboratory. As Head of the Department, he also began to redirect the physics department towards experimental and theoretical research in other areas, setting up laboratories and library facilities. Soon it became the center of research in areas that included X-ray spectroscopy, X-ray diffraction, magnetic properties of matter, optical spectroscopy including Raman Spectra, and wireless.

Scientifically, he devoted himself almost exclusively to teaching and guiding research. "He would come to the Department around 7 a.m. and be there until late in the evenings except for a short lunch break," writes Partha Ghose. "Whenever he was not taking a class or attending a meeting, there would be a continuous stream of visitors to his room, and Satyendranath would get involved with their academic problems for hours. It could be any branch of Physics, Chemistry, History, hieroglyphics or indeed any other subject under the sun. He was engaged in creating a culture of intellectual discourse."[16]

It was a vibrant life at Dacca both scientifically and culturally. Noted scholars and writers visited the university regularly. The University had made great strides in providing higher education to Muslim students, whose number increased from 160 in 1921 to 600 in 1941. The residence halls for Hindu students that were originally run on caste lines, with separate dining halls for Brahmins and non-Brahmins were slowly being integrated. However, at the same time, during the years 1926–1941, the political situation in the country was beginning to disturb the academic atmosphere. The more peaceful Dacca of the early 1920s began to witness communal riots, famine and the war.

The partition of Bengal in 1905 had created a new Muslim-majority province of Eastern Bengal and Assam with Dacca as the provincial capital, and the Muslim population felt liberated from the political as well as the economic domination of Calcutta's Hindu landowners and lawyers. While this situation created a strong nationalist movement against the British, it also created a separatist Muslim political consciousness and demands for the safeguard of their rights as a minority community. It also played well into the hands of the British government to sow the seeds of division. With the British Government's support and encouragement, the *All India Muslim League* was formed in 1906. The Dacca University contributed in its own way to the development of an educated Muslim middle class that did not exist before. Gradually, this middle class, imbued with a spirit of Muslim cultural tradition, identified itself with the objective of the Muslim league, demanding the creation of a separate Muslim State. It became a seat of communal disharmony and violent disturbances.

As the Second World War drew close in 1945, the increased communal tensions, riots and violence, agonized Bose. It became difficult to pursue his scientific and intellectual pursuits. Earlier in 1938, he had received an offer of a position from Calcutta University that he had declined. But in 1945, due to the sudden death of Bidhu Bhusan Ray, the prestigious *Khaira* chair was lying vacant. And when he was offered this position, he accepted it willingly and returned to his alma mater.

Khaira Professor in Calcutta

During the quarter of century of Bose's absence, Calcutta University had undergone tremendous changes. When he had left for Dacca, the physics department was in its infancy with scarce resources for research that had led Saha to leave Calcutta and go to Allahabad University. He had returned to Calcutta in 1938 with his research interests focused on nuclear physics, hoping to establish an Institute for Nuclear Research. Sir C.V. Raman, India's first Nobel Laureate in physics, for his classic experiments on light scattering and the discovery that came to be known as Raman Effect[17] had left Calcutta for Bangalore, but his legacy had persisted. Work in optical scattering and Raman spectra had continued at the laboratories of the Indian Association for the Cultivation of Science (IACS). Raman's researches in acoustics were continued at the University Science College Campus. Researches in other fields had also flourished. D.M. Bose, who succeeded Raman as Palit Professor, was well known for his work in magnetism and atomic physics. S.K Mitra had pioneered work on radio and wireless. Bidhu Bhusan Ray, the preceding Khaira professor, whose untimely death had left the chair open for Bose, had set up an X-ray laboratory with an active group of research workers. In 1945, not only in the physics department, but in almost all departments, the faculty of the University College of Science and Technology had distinguished colleagues.

For Bose, it was a happy homecoming. The faculty and students of the University College extended him a warm welcome. He brought new life, his characteristic openness and informality with students, and his wide ranging interests. Soon a group of brilliant students formed research groups around him, working in the X-ray laboratory. Unfortunately for Bose, shortly thereafter, Calcutta became engulfed in communal violence on an unprecedented scale. Thousands of Hindus and Muslims were slaughtered in a three-day rampage in what has come to be known as "The great Calcutta killing of August 1946," an infamous, tragic episode in India's history. The University College of Science and Technology stood right in the midst of a sensitive area and was heavily affected. Some of its students living in nearby residence halls were brutally murdered. With British army personnel housed within the campus, academic life came to a standstill for sometime, although limited research activities continued. The following year, on 15 August, 1947, India gained its long awaited Independence, but with the country divided, Bengal and Punjab partitioned to create two disjointed East and West Pakistan. More bloodshed and an influx of refugees followed.

Bose at Dhaka University c.1930s.

National Professor S.N. Bose with Prime Minister Jawaharlal Nehru (l).

Bose in his 70s c.1960s.

A renaissance man, S.N Bose is seen here playing esraj. He also played flute, and even composed a raga. (Courtesy of Falguni Sarkar.)

S.N. Bose in 1953. (Courtesy of Etienne Eisenmann.)

Bose (seated 2nd from left) and his family c.1960s.

Aftermath, Independent India

All this had a profound effect on Bose. He felt betrayed and disillusioned. India was free — but mutilated. Part of Bengal where he had spent twenty-five best years of his life was now a foreign territory. He agonized, he did what he could to help anyone that needed help. It was another turning point in his life. He no longer could live isolated in the ivory tower of intellectual pursuits. He had to face the new challenge to face social realities. Independence also opened up new opportunities to serve the country at large.

Fortunately for India, the first Prime Minister after Independence, Jawaharlal Nehru, was a strong advocate of a strong national effort to utilize science and technology for the upliftment of Indian masses. "It was science alone," he wrote in Discovery of India, "that could solve the problems of hunger and poverty, of insanitation and illiteracy, of superstition and deadening custom and tradition, of vast resources running to waste, of a rich country inhabited by poor people." Long before Independence, in 1939, Nehru was instrumental in establishing a National Planning Committee under the auspices of the Indian National Congress. A committee for Science and Technology was a part of the National Planning Committee, which after Independence became the full pledged Council of Scientific and Industrial Research with the mission of establishing National Research Laboratories, Institutes of Technology and so on. Nehru called upon all the prominent scientists and intellectuals to participate in the new frontiers that Independence had opened. Bose, along with his colleagues, became an adviser to the Council and contributed his share in shaping the future of science in India.

During all these upheavals, work at the Khaira laboratory continued under Bose's guidance. Since the X-ray laboratory was not equipped for crystallography research, Bose and his students constructed a number of precision cameras such as a Weissenberg camera,[18] a differential thermal analyzer, a micro-focus X-ray tube, a special type of powder camera for clay analysis, and a fully automatic scanning spectrophotometer for thermo-luminescence studies. As in Dacca, he discouraged readymade, imported equipment, instead insisting every student design his or her own equipment, set up instruments from locally available materials and construct them with the help of local technicians in the laboratory. Soon new research areas opened up in many directions including both pure and applied fields. Professors and students in applied mathematics were frequent visitors with problems to be solved or to just have a stimulating conversation. His office was open to any one, to converse on any subject.

Bose was the President of the Indian Physical Society (1945–1948), and the President of the National Institute of Science (1948–1950). He was nominated a member of the Rajya Sabha, the upper chamber of the national parliament in 1954 and honored with the title Padma Vibhusan. He remained a member until he resigned in 1959. During the years 1947–1955, he was a host to a number of foreign visitors that included P.A.M. Dirac and J.D. Bernal. Dirac was apparently surprised Bose was not a FRS, a Fellow of the Royal Society of London. This honor came in 1958, twenty-four years after his revolutionary discovery.[19]

Travels Abroad

During the fifties, Bose made several trips to Europe, the first in 1951, twenty-six years after his first visit. He went to Paris at the invitation of UNESCO. Now that India was independent, he felt free to go to England and spent a week in London on his return trip to Calcutta. While there, he reconnected with many old acquaintances. In Paris he met Jacqueline, the young student he had met in 1924–1925 and had become close friends. She was married to Jacques Eisenmann, an engineer in the civil service and had a daughter and two sons. The war had interrupted her research career. She had volunteered to work in a laboratory of the Air Ministry till 1940. When Bose met her and her family, she was working in the documentation service of a professional organization called the *Institut Francais du Caoutchouc*.[20] It was a moving experience for both, as he writes from London on 22nd June, 1951,

> Dear Jacqueline,
>
> It was almost a mystic experience to see you, and hear your sweet voice and be near to you after so many years: It has filled me with bliss, and I am deeply grateful to my destiny.
>
> Years have elapsed with any news, and I had almost despaired of ever meeting you again in life: There was also a foolish fear, that under the relentless law of change with time it was hopeless to wish to recover in reality what we cherish in memory draped with cobweb of imagination. How all fears have melted away at your first sight!
>
> I am in London and shall stay here for about a week more, and then I go back to Calcutta:
>
> Hope you will now write again
>
> Yours Affectionately,
> Bose

In 1953, Bose was invited to attend the World Congress for General Disarmament and Peace at Budapest. The following year he represented India at the International Crystallography Conference in Paris. He returned to Paris again in 1955, this time at the invitation of the Council of National Scientific Research of France. It was the 50th anniversary of Einstein's "miraculous" year during which he published five papers that revolutionized the world of physics. A conference was to take place in July in Bern, Switzerland, the birthplace of Einstein's discoveries and Einstein himself was expected to attend. Bose, who had tried unsuccessfully to visit the United States and meet Einstein, was eager to attend the conference at Bern. He was also looking forward to the opportunity to discuss his own miraculous output of five papers on the Unified Theory that Einstein was occupied with. At least four of these papers were written within a span of one year (1953). Unfortunately, Einstein's death in April of that year led to the cancellation of that meeting. The following year Bose traveled to England for the Annual Meeting of the British Association for the Advancement of Science. He went again two years later to attend the Royal Society Meeting where he was elected a Fellow.

Vice Chancellor at Visva-Bharati

In 1956, at the age of sixty-two, Bose retired from the post of the Khaira professor and became the Vice Chancellor of Visva-Bharati, situated in Santiniketan, a small town, 180 kilometers north of Kolkata. It was here in Santiniketan (abode of peace), on an almost bare tract of land, Rabindranath Tagore had founded a school in 1901, and twenty years later Visva-Bharati, a university. In 1951, ten years after his death, the central government of India took over the institutions, which continues to run them. The school he founded, in rebellion against the mainstream strict, repetitive, parrot-like learning of his childhood, was based on the philosophy that learning becomes both effective and enjoyable when it is in natural environment without superficial barriers between teachers and students. The school that started with only five students, often referred to as Ashram system, soon became a renowned institution offering primary and secondary education. Sixteen years later, when relaxing on the beach in Santa Barbara, California, he conceived of a university, as he wrote to his son Rathindranath on 11 October 1911:[21]

> I have in mind to make Shantiniketan the connecting thread between India and the world. I have to found a world center for the study of humanity there. The days of petty nationalism are numbered — let the first step towards universal union occur in the fields of Bolpur. I want to make that place somewhere beyond the limits of nation and geography.

The foundation stone of his dream university to be called "Visva-Bharati" was laid on 22 December 1918 and exactly three years later it was inaugurated. He had traveled extensively in India, Europe and USA to raise funds and used his own Nobel Prize money to start the university. For Tagore, "Visva-Bharati" represented "India where she has her wealth of mind which is for all." He hoped Santiniketan would be a meeting place for western and Asian scholars, a place where the ancient learning would be rejuvenated through contact with modern thinking. Three departments formed the nucleus of the new university; the department of fine arts (Kala Bhavan), the department of music (Sangit Bhavan), and the department of Indology, founded for the study of Buddhist literature, Vedic and Classical Sanskrit, Pali, Prakrit and later on Tibetan and Chinese. Soon it became associated with scholars and visitors from all over the world and claimed Satyajit Ray, Indira Gandhi and Amartya Sen as some among its famous alumni.

Tagore founded the university, but had no definite plans for its future course. "I do not put faith in any new institutions, but in the individuals all over the world who think clearly, feel nobly and act rightly," he wrote at the very time he started the university. In response to a request to define his dream more clearly he replied, "I find it rather difficult to answer your question...In writing my stories I hardly ever have a distinct plot in my mind. I start with some general emotional motive which goes on creating its story form very often forgetting in the process its ...original boundaries. If I had in the commencement a definite outline which I was merely to fill in, it would certainly bore me — for I need the constant stimulation of surprises... The same thing happened with my Shantiniketan Institution. I merely started with this simple idea that education should never be disassociated from life."[22]

One thing that was certain in Tagore's mind was that Visva-Bharati was not to be a conventional degree peddling institution. But the high ideals of openness, open to new ideas, open to accept European renaissance through modern science became difficult to implement within the framework of a large institution. A steady stream of distinguished visitors came to visit Tagore and Shantiniketan. Most of them went away impressed with idea, but "unconverted with the actuality." One such distinguished visitor was, E.M. Forster, who came in 1945, four years after Tagore's death. He said to his hosts;

> I am not here to pass any verdict on India nor do I fortunately carry the white man's burden on my shoulder. All that I carry around my neck is the pleasant light burden of a lovely garland of marigold, which, I feel, is the symbol of your love and friendship.

But to an English friend he was more frank:

> We went over for a night to [the] shrine. It was less shriney than I expected, indeed there were some sensible remarks about the Passed Master. Much kindness, and my two companions (Muslims) were moved, as was I. I am afraid that the place does not cut much ice now, except as through the elderly and theosophic. The educational side of it is too casual.[23]

Indeed long before Bose became the Vice Chancellor, Tagore's ideal university had ceased to exit. For the first time, a full-time administrator, Bose went to Visva-Bharati full of new ideas. But what he soon discovered was a rigid way of life, adhering to narrow rituals and insular to change. Visva-Bharati till then did not have a basic infrastructure for scientific research. It offered science courses only up to the Intermediate level. Bose wanted to introduce advanced courses leading up to the graduate and post-graduate level. He wanted to build an Advanced Center of Science. He often spoke to the students about the role of science in every day life. He had such innovative ideas, for example, of building a small experimental plant which would use kitchen garbage as fuel and produce manure as well as methane gas for the classroom experiments.[24] This emphasis on science and breaking of some of the long established traditions of the university did not go well with the Executive Council. It harbored the false idea that "Gurudev Rabindranath Tagore", the founder of the institution, was anti-science. Science was at the root of materialism and contrary to the humanistic goals set by the founder. Bose failed to implement any of his ideas. Frustrated, he resigned from his position as the Vice Chancellor and returned to Calcutta at the end of 1958.

Professor Emeritus; National Professor

Calcutta University had appointed Bose, on his retirement, an Emeritus Professor and had allowed him to keep his former Khaira Professor office for holding seminars and discussions. He returned with the idea of devoting himself to research, particularly in the new developments in nuclear physics and complete some of his earlier work in organic chemistry. He had research laboratory facilities and a grant from the Indian Association for the Cultivation of Science. In 1959 he was appointed a National Professor, the highest honor which the nation could confer upon a scholar. He held it for three successive terms, each lasting five years.

For the rest of his life, Bose continued to live in Calcutta. He was one of the last survivors of the brilliant young men that had entered the University some sixty years ago. In the final phase of his career, Bose followed a distinctively different style in his scientific pursuits. His domain now went beyond basic research and beyond Khaira laboratory. The problems he became interested in were more in the applied areas of wider interest to the society at large. One such project was the extraction of helium from the hot springs of Bakreswar. The existence of helium in the radioactive water of hot springs was first detected by Syamadas Chatterjee at the University of Jadavpur. But it was through the efforts of Bose, who fully realized its strategic importance, that a field study unit was set up to collect the gas at the site and make a systematic analysis at the National Professor's laboratory. He made himself available, as in his academic life, to any group of young enthusiastic scholars undergoing training in diverse branches of science. "*Tempered by the years, with his luxuriant crop of white hair, he came to be regarded a sort of father confessor, giving counsel on any problem on which anyone sought his guidance. To him may be applied what Francis Bacon said about himself: I have taken all knowledge to be my province.*"[25]

Bose also became a strong advocate of modern science. Both Saha and Bose rose against a chauvinistic view that the Indian ancients knew it all — it was all there (including Einstein's relativity) in the *Vedas and the scriptures.* Bose was among the pioneers who realized the importance of science education in one's own mother tongue for its better foundation among young students. After his return to Calcutta from Dacca, he was instrumental in inaugurating the society called *Bangiya Bijnan Parishad.*[26] Under its auspices it began publishing a monthly periodical, called *Jnan O Bijnan*, a Bengali version of *Science and Culture*, the organ of the Science News Association started a decade earlier by Saha. The journal with provocative articles on science and religion, with popular articles on the latest discoveries in science and a section for children contributed significantly to a deeper and wider appreciation of science in Bengal.

In Calcutta, Bose's initiatives in teaching in Bengali did not go well. His lectures in Bengali in advanced courses were remembered, but met with ridicule and criticism from stalwarts who were not convinced that the Bengali language was a suitable medium for teaching highly technical ideas. For them, he found a good response. During his visit to Japan as the invited guest (1962) for a symposium on Science and Philosophy, he found that physics was taught in Japanese, research papers were being written in Japanese, even at seminars all discussions were conducted in Japanese. He returned from Japan firmly convinced that only the use of one's mother tongue can bring out the best in every individual. With the prestige of a National Professorship and plenty of freedom to think, write, and do as he pleased, Bose continued his crusade for the mother tongue along with his scientific research and projects. He also became involved with Indian Statistical Institute, as Vice President (1960–1967) and President from 1967 till the end of his life.

On his eightieth birthday, nationwide celebrations were organized. It was also the fiftieth anniversary of the discovery of Bose Statistics. An International Symposium in his honor took place in Calcutta. It was attended by prominent scientists from many countries and the proceedings of the symposium[27] record the impact Bose's work has had on so many diverse fields. Bose, responding to the occasion, spoke about his long years of struggle, and expressed his satisfaction at having seen his work appreciated and said "Now I feel I do not need to live any longer." It was a prophetic remark. The activities associated with the celebration, the receptions, seminars and exhibitions took a heavy toll on his health. After an attack of bronchial pneumonia, he died on the 4 February, 1974, only a month after he had completed eighty.

Conclusion

All his life Bose remained loyal and grateful to Einstein for his acceptance of his short paper and making it known in the scientific world. He had unbounded admiration for Einstein's scientific achievements. He called him the champion of the oppressed Jews and deplored anti-Semitism that drove him out from his native land. He wrote to him on 9 December, 1945,[28]

> As one who owes so much to you and your guidance in life, I have always been anxious to have news of yourself, especially under the new conditions when you are in a new country. Some of my friends who are more lucky than me, had chances to visit the States recently, and had the good fortune of seeing you in your new house. I am glad to know that you are well, and still take a lot of interest in India and Indians.

It is unfortunate he did not get a chance to see him again after 1925. Other than a few instances in informal discussions he kept his disagreements with Einstein to himself for the most part.[29]

Bose had a large family of five daughters and two sons (two other children died as infants). He was considerate and caring without being overly indulgent. They all grew up in a relaxed and loving family atmosphere. The only advice he gave them was to be honest and sincere. Ushabati, simple and unassuming, never made her presence felt. Bose, endearing and loving, did his best to make her life easy in the extended family. Unlike traditional husbands, he was very thoughtful of her in many small ways. "If she slept late, Satyendranath would take care of arranging breakfast for the children, prepare for himself a couple of toasts and tea and then rush off to college."[30] To the end he remained modest, friendly and easily accessible. His door was always open to anybody who cared to come in. He was completely at ease with people who had no idea of his achievements in science.

Judged against conventional standards, Bose's achievements in scientific research were not as sustained or as numerous as those of his contemporaries like C.V. Raman, M.N. Saha and K.S. Krishnan. Various interpretations could be made of this. Mostly self-taught under difficult circumstances, lacking books and journals, with responsibilities of heavy teaching both theory and experiments, Bose after great initial success fell into the trap of being a perfectionist, publishing only what he considered were worthwhile ideas. He never kept any notes, lecturing by memory. Many of his ideas ended up in students' theses.

Bose was a towering figure in many other ways. With his love for stimulating conversations, his wide ranging interests in all human endeavors and his intellectual fervor, he was a source of inspiration for many outside physics. A voracious reader with mastery in French, German and Sanskrit, he was also a connoisseur of classical music and played the instrument *Esraj*. More interested in absorbing new ideas than expressing them in print, his ability to tackle problems in entirely new fields was legendary. " ...his keen, perspicuous and versatile mind," says B.M. Udgaonkar, "trained in the method of physics and mathematics was allowing itself to range over a variety of fields, including chemistry and biology, soil science and mineralogy, philosophy and archeology, arts, literature and music."[31]

He was active politically throughout his life, serving on many committees to further the scientific and industrial development of India after Independence. Politically, the partition of India in 1947 had particularly disillusioned him. A free India — but mutilated! He was a fervent Bengali nationalist, and like Rabindranath Tagore, was a strong advocate and practitioner of writing and teaching in Bengali. To summarize, in the words of Partha Ghose, "Bose was indeed a unique confluence of intellectual brilliance, encyclopedic knowledge, selfless dedication to the motherland and unbounded compassion and love for fellow beings. He was a renaissance figure and a quintessential Bengali."[32]

References and Notes

1. The Early Phases of My Exploration in Science, S.N. Bose, (*The Man and His Work, Part II*, p. 246). [This article is included in this volume].
2. Ibid. p. 247.
3. The first two papers they wrote together were: On the Influence of Finite Volume of Molecules on the Equation of State, *Phil. Mag.*, Sr 6. **36**, 199, 1918; On the Equation of State*, Phil. Mag.*, Sr 6. **39**, 456, 1920. [Both these papers are included in this volume].
4. *The Principle of Relativity*, The Original Papers by A. Einstein and H. Minkowski, translated into English by M.N. Saha and S.N. Bose with a historical Introduction by P.C. Mahalanobis, University of Calcutta 1920.
5. There are conflicting accounts regarding the sequence of events that led Bose to send the manuscript of his first paper to Einstein. My account is based on those of W.A. Blanpied and John Stachel. See the articles, Satyendra Nath Bose: Co-Founder of Quantum Statistics by W.A. Blanpied and Einstein and Bose by John Stachel included in this volume.
6. *The Golden Age of Theoretical Physics*, Jagdish Mehra, (World Scientific, 2001) p. 514.
7. In the Company of Madame Curie, S.N. Bose (*The Man and His Work, Part II*, p. 290). [The article is included in this volume].

8. *Bose Statistics: A Historical Perspective.* Partha Ghose (*The Man and His Work, Part I*, pp. 35–71). [This article is included in this volume].
9. Private communication (Partha Ghose).
10. Sylvain Levy, the great French orientalist and a friend of Jacqueline's father Lèon Zadoc, knew Bose as a famous physicist. He introduced him to his friend's daughter, a budding young physics student.
11. *The Golden Age of Theoretical Physics*, Jagdish Mehra, (World Scientific, 2001) p. 526. See also *The Historical Development of Quantum Theory*, Jagdish Mehra and Helmut Rechenberg (Springer Verlag), p. 571.
12. Einstein and Bose, by John Stachel in *Einstein from 'B' to 'Z'* (Birkhäuser, 2002), p. 527. [This article is contained in this volume].
13. Satyen Bose in Berlin Purnima Sinha (*The Physicist, Bulletin of the Bangladesh Physical Society*, Jan–Feb 1995). [This article is contained in this volume].
14. Herman Mark expressed these views more formally in his recommendation letter, "... *His large and profound knowledge which stretches over the whole of physics as well as the wider territory of chemistry marks Mr. Bose out prominently. His most valuable quality which makes him of inestimable value to a collaborator is his deep and clear insight in the fundamentals of our science, his wealth of fruitful ideas and his capacity to combine theoretically important questions with experimentally feasible tests in an exceedingly happy manner...*".
15. Einstein (1965) S.N. Bose (*The Man and his Work, Part II*, p. 288).
16. S.N. Bose: The Man. Partha Ghose (Talk presented at a seminar on Albert Einstein, Dhaka, 30 May, 2005; Published in the *Proceedings of Bangaladesh Academy of Sciences*).
17. In 1928 Arnold Sommerfeld came to India. During his visit to Calcutta, Raman arranged a grand function at the Indian Association for the Cultivation of Science and gave his talk on the newly discovered phenomenon. Bose, who had come from Dacca to meet Sommerfeld, after listening to the talk and seeing the spectrum, had said to Raman, "Professor Raman, you have made a great discovery. It will be called the Raman Effect, and you will get the Nobel Prize."
18. This camera, the then most modern equipment of an X-ray crystallographer, was first fabricated by a mechanic in the laboratory workshop in Dacca University under the guidance of Bose.
19. At the request of Professor Asesh Das, Royal Society released the following information in 2008 concerning the names of the supporters of the nomination and election of Bose to a Fellowship of the Society in 1958 after the mandatory 50 years had passed.
    From general knowledge:
    Douglas Hartree, Max Born, Maurice Pryce, Cecil Powell, Frederick Frank, and Edward Guggenheim.
    From personal knowledge:
    Paul Dirac, Patrick Blackett, John Cockcroft, Alan Wilson, Rudolf Peierls, Homi Bhabha.
20. I am indebted to Etienne Eisenmann for this information regarding his mother and his family.
21. *Rabindranath Tagore: The Myriad-Minded Man.* Krishna Duttanand and Andrew Robinson (St. Martin's Press, New York, 1995) p. 205.
22. Ibid. p. 10.
23. Ibid. p. 11.
24. *Satyendra Nath Bose.* Santimay Chatterjee and Enakshi Chatterjee (National Book Trust. India, 1993) p. 60.
25. S.D. Chatterjee (*The Man and His Work, Part II*) p. 94.
26. Concerning the details of the resolution to form the society and its objectives, see (*The Man and His Work, Part II*) p. 96.

27. The Proceedings of this symposium were published under the title *Statistical Physics Symposium Celebrating Fifty Years of BOSE STATISTICS.* Edited by N. Mukunda, A.K. Rajagopal and K.P. Sinha, (Indian Institute of Science, Bangalore 1975).
28. Doc. 6-134, Einstein Archives, Hebrew U. of Jerusalem, Israel.
29. See *S.N. Bose by* Partha Ghose; Ref. 16.
30. *The Man and the Work, Part II*, p. 104.
31. The Unconventional Scientist, by Santimay Chatterjee and Enakshi Chatterjee, *Science Reporter*, January 1994, p. 19.
32. Partha Ghose; Ref. 16.

I. Selected Scientific Papers

A. Early Papers — Classical Physics

The papers in this section include the first published works of Bose, written jointly with Meghnad N. Saha. Both had just started their academic careers in the newly founded Calcutta University.

The equation of state for an ideal or perfect gas was derived from the kinetic theory assuming that the gas consisted of point particles with no attractive or repulsive forces. Experimental studies of real gases showed the perfect gas equation was not valid except in the case of highly rarified gases when extremely low pressures prevailed. It was recognized that a realistic equation of state demanded that the finite size of the molecules and a certain measure of intra-molecular forces be taken into account. The result was the celebrated Van der Waal's equation (1873) with two parameters, one to take into account the finiteness of the volume of each particle and the other to represent the forces of cohesion. While it was an important step forward, experimental results did not agree with the predictions of the theory if the parameters were regarded as absolute constants independent of volume and temperature. Various modifications including more parameters to fit the experiments were proposed by various authors. The derivation of Bose and Saha based on thermodynamic considerations and the theory of probability provided a new equation of state that had only two parameters and showed better agreements with experiments than other contemporary attempts.

Two other papers by Bose during the early years are mainly of mathematical nature. In the paper *The Stress Equations of Equilibrium*, Bose shows that the equations can be successfully integrated in the case of a semi-infinite body bounded by a plane. In the case of a sphere, he finds a transformed form of the equations and solutions for the stress coefficients in terms of an infinite series of spherical harmonics. The other paper concerns the property of a *Herpolhode*, the curve traced out by the end point of the angular velocity vector of a freely rotating rigid body. M. de Sparre and others had proved a theorem stating that the herpolhode had no point of inflexion. Bose offers an elegant alternative proof based on purely dynamical considerations of the motion, known as Poinsot motion.

It is also worth noting that Bose along with Saha translated Einstein's papers on Relativity from the original German. The book was published by the Calcutta University, in 1920. It was the first English version of Einstein's papers.

On the Influence of the Finite Volume of Molecules on the Equation of State.† *By*

MEGH NAD SAHA, *M. Sc., and* SATYENDRA NATH BASU, *M. Sc., Lecturers on Mathematical Physics, Calcutta University.*

It is wellknown that the departure of the actual behaviour of gases from the ideal state defined by the equation $p = \frac{NK\theta}{v}$ is due to two causes : (1) the finiteness of the volume of the molecules, (2) the influence of the forces of cohesion, i.e. the attractive forces amongst the molecules. van der Waals was the first to deduce an equation of state in which all these factors are taken into account; according to van der Waals, we have

$$p = \frac{NK\theta}{v-b} - \frac{a}{v^2} \quad \ldots \quad \ldots \quad \ldots \quad (1)$$

where $b = 4\times$ volume of the molecules, a defines the forces of cohesion.

In all subsequent modifications of this equation (Clausius, Dieterici, or D. Berthelot) the changes which have been proposed all relate to the influence of the cohesive forces; the part of the argument dealing with the finiteness of molecular volumes is generally left untouched.

But it has been found that the results of experiments do not agree with the predictions of theory if we regard a and b as absolute constants. Accordingly it has been proposed to regard both a and b as functions of volume and temperature.*

But before proceeding to these considerations, it is necessary to scrutinize whether the influence of finite molecular volumes is properly represented by the term b. From theoretical considerations, the conclusion has been reached that this is not the case. The argument is as follows :

According to Boltzmann's theory,

$$\text{the entropy } S = K \log W + C,$$

where K = Boltzmann's gas-constant, W = probability of the state. Let us now calculate the probability that a number N of molecules originally confined within the volume V_0 and possessing finite volumes, shall be contained in a volume. V. Neglecting the influence of internal forces, the probability for the first molecule is $\frac{V}{V_0}$, for the second molecule the probability is $\frac{V-\beta}{V_0-\beta}$, where $\beta = 8\times$ volume of a single molecule, for when

† Communicated by the Authors.

*Compare van der Waals, *Proc. Amst.* 1916 ; Van Laar, *Proc. Amst.* Vol. xvi, p. 44.

the first molecule is in position, the space enclosed by a concentric sphere of double the radius of the molecule will not be available for the second molecule. The available space is therefore $V-\beta$, whence the probability is $\frac{V-\beta}{V_0-\beta}$. Introducing similar considerations for the rest of the molecules, we have

$$W = \frac{V}{V_0} \cdot \frac{V-\beta}{V_0-\beta} \cdot \frac{V-2\beta}{V_0-2\beta} \cdots \frac{V-\overline{N-1}\beta}{V_0-\overline{N-1}\beta} \qquad \ldots \quad (2)$$

We are, of course, neglecting those cases in which partial overlapping of the regions occupied by two or more molecules occurs; for the number of such cases can at best be a small fraction of the total number. Even cases of actual association do not include these, for in that case, two discrete molecules become merged into one, without their outer surfaces being actually in contact.

From the relations $S = K \log W + C$

and

$$\left(\frac{\partial S}{\partial V}\right)_u = \frac{p}{\theta}$$

we can easily verify that

$$p = -\frac{K\theta}{\beta} \log \frac{V-n\beta}{V}$$

$$= -\frac{R\theta}{2b} \log \frac{V-2b}{V} \quad (R = NK) \qquad \ldots \quad (3)$$

As a first approximation, when b is small compared to v, we obtain $p = \frac{NK\theta}{v}$ (Boyle-Charles-Avogadro Law), and as a second approximation we obtain

$$p = \frac{NK\theta}{v-b} \text{ (van der Waals correction).}$$

We also note that

$$pV = NK\theta . \frac{x}{1-e^{-x}} \text{ where } x = \frac{\beta p}{K\theta} \qquad \ldots \quad (4)$$

To account for the influence of internal forces, we multiply, following the lead of Dieterici, the above expression (3) by $e^{-\frac{a}{NK\theta v}}$ having the same significance as before.

From this equation of state, we can easily verify the following results for the critical point :

Critical volume, $\qquad V_c = \frac{2c}{e-1} b = 3.166b,$

$$K = \frac{NK\theta}{p_c V_c} = 3.513.$$

The corresponding values of V_c from the van der Waals and the Dieterici equations are (3b, 2b) respectively, and of K are $\left(\frac{8}{3} = 2.66, \frac{e^2}{2}\ 3.695\right)$ respectively.

As a matter of fact, for the simpler gases, the value of 'K' obtained in this paper agrees better with the experimental results that the Dieterici value $\frac{e^2}{2}$ we have for oxygen* $K = 3.346$, for nitrogen† $K = 3.53$, for argon‡ $K = 3.424$, for xenon** $K = 3.605$. We need not consider the van-der-Waals value $\frac{8}{3}$, for it fails entirely.

The most serious drawback to Dieterici's equation is, according to Prof. Lewis (vide Lewis's Physical Chemistry, vol. ii. p. 117) that it makes b or the limiting volume $\frac{V_c}{2}$, while the limiting volume, obtained by the extrapolation of Cailletet-Mathias mean density line to the temperature $\theta = 0°$K is about $\frac{V_c}{4}$. The value of b obtained in this paper, viz., $\frac{V_c}{3{\cdot}16}$ therefore agrees better with this value.

It is yet premature to predict what influence this investigation will have on the speculations concerning the variability of the volume of molecules with temperature. A more detailed investigation dwelling upon this point, and the application of the formula (4) to Amagat's (pv, p) curves, will be communicated shortly. Meanwhile we point out that the factor $e^{-\frac{a}{NK\theta v}}$ has been introduced into the expression for 'p' only as a provisional measure, though it is considered that this step, though not quite exact, is one in the right direction. In the next paper an attempt will be made to introduce energy into probability calculations.

Sir T. N. Palit Laboratory of Science,
Calcutta.

Note added in proof—On consulting the literature on the subject, we noticed that in several papers in the Amsterdam Proceedings (*vide* vol xv, p. 240et *seq.*), Dr. Keesom of Leyden had also made attempts to deduce the equation of state from Boltzmann's entropy principle. But, in the expression (2) for W, he introduces, before differentiation, an approximation in which terms up to second order in $\frac{b}{v}$ are retained only. In this way, he arrives at the van der Waals' form $v-b$ for the influence of finite molecular volumes. In obtaining our present equation of state (4), no such approximation has been made.

(M. N. SAHA *and* S. N. BASU.)

*Mathias and K. Onnes, *Proc. Amst.* Feb. 1911.

†Berthelot, *Bull. de la Soc. France de Phys.* 167 (1901)

‡Mathias, Onnes, and Crommelin, *Proc. Amst.* 1913, p. 960, Vol. xv.

**Paterson, Cripps, Whytlaw-Gray, Proc. Roy. Soc. Lond. A. lxxvi, p. 579 (1912).

Reprinted from *Phil Mag* Ser 6, **36**, pp. 199-203, 1918 (Taylor and Francis, London).

On the Herpolhode

BY
SATYENDRANATH BASU

(*Read August 31st, 1919*)

M. de Sparre has shown that the horpolhode of Poinsot contains no point of inflexion; his proof was based upon the properties of elliptic functions. Various other proofs of the theorem have since been given by Mannheim, Saint-Germain[2], Routh[3], Le-Cornu[4] and others.

The following simple proof of the theorem is based solely on the dynamical considerations of the Poinsot motion; hence it might appear interesting and instructive.

Let A, B, C be the three principal moments of inertia of a body fixed at a point 0; let 0I and 0I′ be the two instantaneous axes, at time t and $t+dt$, I, I', etc., on the invariable plane trace out the horpolhode, length of 0I, 0I′, etc. being proportional to the resulting angular velocity at all times.

At a point of inflexion of the horpolhode, there will be a stationary tangent, i.e., the total change of the vector II' will be then along its own length.

Remarking that II′ has components proportional to $\dot{w}_1\,dt$, $\dot{w}_2\,dt$, $\dot{w}_3\,dt$, along the three moving principal axes, the condition reduces to

$$\frac{\ddot{w}_1-\dot{w}_2w_3+\dot{w}_3w_2}{\dot{w}_1}=\frac{\ddot{w}_2-\dot{w}_3w_1+\dot{w}_1w_3}{\dot{w}_2}=\frac{\ddot{w}_3-\dot{w}_1w_2+\dot{w}_2w_1}{\dot{w}_3}=\lambda \text{ say.} \quad \ldots \quad (1)$$

We have from Euler's equations

$$A\dot{w}_1=(B-C)w_2w_3,$$
$$B\dot{w}_2=(C-A)w_3w_1$$
$$C\dot{w}_3=(A-B)w_1w_2;$$

[1] An exactly similar proof of the theorem discussed in this paper has been given by Prof. W. Van der Woude in a paper entitled "Over de Herpoloide Van Poinsot", published in Nieuw Archief voor Wiskunde, Tweede Reeks, Deel XII, Eerste Stuk, pp. 94, 95 (1919). It appears however from an enquiry that this journal was not received in the library of the Society before the beginning of November, 1919 —S. K. B.

[2] Compte Rendu—1885

[3] Advanced Dynamics.

[4] Bulletin de Soc. Math. de France—1906.

from which

$$A\ddot{w}_1 = (B-C)(\dot{w}_2 w_3 + \dot{w}_3 w_2)$$

$$B\ddot{w}_2 = (C-A)(\dot{w}_3 w_1 + \dot{w}_1 w_3)$$

$$C\ddot{w}_3 = (A-B)(\dot{w}_1 w_2 + \dot{w}_2 w_1)$$

Substituting in (1) we have

$$\frac{\dot{w}_3}{w_3}(A+B+C) - \frac{\dot{w}_2}{w_2}(C+A-B) = \lambda(B-C)$$

and two similar equations.

So that

$$\frac{\dot{w}_1}{w_1}(B-C)(B+C-A) + \frac{\dot{w}_2}{w_2}(C-A)(C+A-B) + \frac{\dot{w}_3}{w_3}(A-B)(A+B-C) = 0$$

But remembering that

$$Aw_1 \dot{w}_1 + Bw_2 \dot{w}_2 + Cw_3 \dot{w}_3 = 0$$

$$A^2 w_1 \dot{w}_1 + B^2 w_2 \dot{w}_2 + C^2 w_3 \dot{w}_3 = 0$$

we have

$$\frac{w_1 \dot{w}_1}{BC(B-C)} = \frac{w_2 \dot{w}_2}{CA(C-A)} = \frac{w_3 \dot{w}_3}{AB(A-B)}$$

hence substituting, we get as a necessary condition for the point of inflexion

$$\frac{BC(B-C)^2(B+C-A)}{w_1^2} + \frac{CA(C-A)^2(C+A-B)}{w_2^2} + \frac{AB(A-B)^2(A+B-C)}{w_3^2} = 0$$

which is obviously impossible because $B+C-A$, $C+A-B$, $A+B-C$, are all positive, so that the left-hand side is an essentially positive quantity.

Reprinted from *Bull Cal Math Soc* **11**, pp. 21-22, 1919 (Calcutta Mathematical Society).

ON THE EQUATION OF STATE.

To the Editors of the Philosophical Magazine.

Sirs, —

In a paper published in the 'Philosophical Magazine' for August 1918 it has been shown that from Boltzmann's theory of entropy we can arrive at an equation of state

$$p = -\frac{R\theta_c}{2b} \log\left(1 - \frac{2b}{V}\right) = e^{-\frac{a}{NK\theta v}} \quad \ldots \quad (1)$$

Amongst other applications of this theorem it was shown that the value of the critical constant $K = \frac{R\theta_c}{p_c V_c}$ would be $\frac{2e}{e-1} = 3.53$, instead of van der Waals' $\frac{8}{3}$ and Dieterici's $\frac{e^2}{2} = 3.69$.

A table was given showing that in the case of the most of the elementary gases the value 3.53 corresponded better with experimental results than either $\frac{8}{3}$ or 3.69.

The list comprised He, N_2, O_2, Xe, but not Hydrogen, about which the experimental data were not satisfactory.

Recently the critical data for Hydrogen have been re-determined in the Laboratory of Prof. K. Onnes (vide Proc. K. Akad. Wetenschappen. Amsterdam, Vol. xx. 1917). It has been found that

$$\theta_c = 33°.18\text{K}, \quad d_c = .0310, \quad p_c = 12.8 \text{ atmospheres;}$$

from these data, $K = 3.35$.

It is superfluous to add that this value of K is in much better agreement with the value 3.53 than with either the value $\frac{8}{3} = 2.66$ or $\frac{e^2}{2} = 3.69$, thus corroborating the belief expressed in the aforesaid paper that the equation (1), though not final, marks a step in the right direction.

University College of Science,
Calcutta, India.

MEGH NAD SAHA.
SATYENDRA NATH BASU.

Reprinted from *Phil Mag* Ser 6, **39**, p. 456, 1920 (Taylor and Francis, London).

B. Quantum Mechanics and Quantum Statistics

Of the six papers in this section covering the period 1924 and 1943, two are on atomic physics. The remaining four (in reality only two, since two are translations from the published German versions) deal with quantum statistical mechanics. The best known among the latter is the one on Planck's Law and the Light Quantum Hypothesis, the paper Einstein translated himself into German and got it published in *Zeitscrift für Physik* in 1924. Bose had succeeded where others had failed — he had derived Planck's Law totally independent of classical mechanics, and in the process discovered a new quantum statistics for the light quanta or the photons. As Abraham Pais writes in his celebrated book on Einstein, Bose's *derivation of Planck's law was the fourth and the last of the revolutionary papers of the old quantum theory (the other three being by, respectively Planck, Einstein and Bohr).* Einstein extended the new statistics to mono-atomic gases and predicted the possibility of a new state of matter to be known later as Bose–Einstein condensation.

Bose's second paper *Thermal Equilibrium in the Radiation Field in the Presence of Matter* was also sent to Einstein, who translated it and got it published in *Zeitscrift für Physik* with a critical remark. After careful study of Einstein's objections, Bose wrote a third paper that he sent to Einstein. Unfortunately, this paper was never published and there is no trace of this paper in Einstein archives or elsewhere. Fortunately an analysis in depth is provided by Partha Ghose in his article, *Bose Statistics: a Historical Perspective* included in this volume.

In the first paper in this section, *On the Deduction of Rydberg's Law*, Bose uses Bohr–Sommerfeld quantization condition to generate Balmer series formula for any hydrogen-like atom with a potential consisting of a monopole and a dipole term. He develops approximation techniques to deal with Abelian integrals that enter into the calculations and derives an effective expression for the Rydberg constant in terms of the fundamental atomic constants. [Ed. Note: In the Introduction to this paper, E.C.G. Sudarshan notes that the formula was first derived by A. Hass, and Bose was probably unaware of it.]

In the paper *A Note on Dirac Equation and Zeeman Effect*, written along with K. Basu, the authors provide a general and elegant treatment of the problem dealt with by Hans Bethe and others. A novel feature is the use of *Sonine* polynomials.

*On the Deduction of Rydberg's Law from the Quantum Theory of Spectral Emission.** By SATYENDRA NATH BASU, *M. Sc., University Lecturer in Physics, University College of Science, Calcutta.*

It is well known that Rutherford's model of the atom has been fruitful in explaining many facts connected with atomic radiation. In the simplest case of hydrogen, with a nucleus consisting of a single positive charge, and an electron, Dr. Bohr has successfully applied the quantum theory to explain the Balmer series of hydrogen spectra. The mathematical problem of finding the spectral series for any atomic system has since been clearly formulated by Sommerfeld, and the quanta condition has been generalized in a form suitable for system with any number of degrees of freedom. If $q_1, q_2, q_3, \ldots q_n$ are co-ordinates to fix the position of the electron responsible for emission, and $p_1, p_2, p_3, \ldots, p_n$ are the corresponding generalized momenta, any statical path, according to Sommerfeld, is characterized by the conditions

$$\int p_1 dq_1 = n_1 h, \quad \int p_2 dq_2 = n_2 h, \quad \int p_n dq_n = n_n h$$

where n's are whole numbers and h is Planck's constant, the integral being extended generally over the complete orbit. The radiation is supposed to take place when the electron jumps from one statical path to another. The difference in energy, at the same time, flows away in the form of a homogeneous radiation of frequency ν, which can be calculated from the Bohr equation $h\nu = W_1 - W_2$. Sommerfeld has successfully applied this conception in explaining the fine structure of hydrogen lines. It is clear, however, that the problem of theoretically calculating the spectrum of any atom other than hydrogen is beset with difficulties of a formidable nature. It is exactly analogous to the dynamical problem of "n" bodies, where only in favourable cases we are able to find solutions. Nevertheless, from a purely experimental standpoint, we know that the visible radiation from any element can be classified in definite series. The frequency of any line in the series can be expressed as a difference of two terms, each of which has the form

$$\frac{N}{(m+\alpha+\beta/m^2)^2},$$

where m is a whole number and α and β are two constants depending upon the element and the nature of the series. So that if we are to explain the formation of the series

* Communicated by the Author.

+Bohr, Phil. Mag. July 1913.

**Sommerfeld, Ann. der Physik, li (1916).

from theoretical considerations following Bohr and Sommerfeld we must look upon each member multiplied by "h" as giving the energy of the atomic system when the radiating electron moves in a definite statical path. The complexity. of the inner atomic field under which the radiating electron moves is to be looked upon as bringing in the terms involving α and β. So it seems interesting to see what will be the corresponding expression for energy in a system by which the complex nature of the internal field may be approximately represented. In the case of any atom we have, in general, a condensed nuclear charge of $+ne$ (where n is the atomic number) surrounded by rings of electron at different distances. The number of electrons in total must be also equal to n in order to secure that the atom is electrically neutral in the ordinary state.

In X-ray emission the electron displaced comes from the inner rings; in the case of visible radiation, however, we have reasons to think that the displaced electron responsible for radiation comes from the outermost ring—the valency electrons, as they have been designated by Sommerfeld. When excited for radiation, we can suppose that the electron in the outermost ring is removed to a greater distance from the centre than the others, so that the force acting upon it may be regarded as the resultant of the various forces exerted by the central charge and the remaining electrons. The potential at any point can be regarded as given by

$$-\frac{ne^2}{r}+e^2\sum_{s=n-1}\frac{1}{r_s}$$

where r is the distance from the centre and r_s is the distance from the s-th electron. If we neglect the influence of the moving electron upon the arrangement of the others surrounding the nucleus, it is clear that the potential can be approximately represented as $-\frac{e^2}{r}+\frac{L\cos\theta}{r^2}$. The resultant field might be looked upon as due to a single positive charge together with a doublet of strength L in a certain fixed direction which we take as our Z-axis. If we neglect the disturbing effect of the outer electron, L may be taken to be approximately fixed in direction and in magnitude in the small interval of time during which the active emission takes place.

We may, therefore, take as our model a system consisting of a positive charge and a doublet of strength L. We proceed to calculate the energy in a statical path on the above simplified hypothesis.

The kinetic energy of the moving electron is obviously

$$T=\frac{1}{2}[m\dot{r}^2+mr^2\dot{\theta}^2+r^2\sin^2\theta\,\dot{\phi}^2];$$

the potential energy

$$V = -\frac{e^2}{r} + \frac{eL\cos\theta}{r^2}.$$

Two integrals can be at once written down :

$$mr^2 \sin^2\theta\dot{\phi} = c_1,$$

$$m[\dot{r}^2 + r^2\dot{\theta}^2 + r^2\sin^2\theta\dot{\phi}^2] - \frac{2e^2}{r} + \frac{2Le\cos\theta}{r^2} = -W.$$

To get another integral, we write

$$\frac{d}{dt}(mr^2\dot{\theta}) - mr^2\sin\theta\cos\theta\dot{\phi}^2 = \frac{eL\sin\theta}{r^2},$$

or

$$mr^2\dot{\theta}\frac{d}{d\theta}(mr^2\dot{\theta}) - \frac{c_1^2\cos\theta}{\sin^3\theta} - meL\sin\theta = 0.$$

Integrated, it gives

$$(mr^2\dot{\theta})^2 + \frac{c_1^2}{\sin^2\theta} + 2meL\cos\theta = c_2.$$

The expressions for three impulses $m\dot{r}$, $mr^2\dot{\theta}$ and $mr^2\sin^2\theta\dot{\phi}$ can now be written down : in terms of the constants of integration we have

$$mr^2\sin^2\theta\dot{\phi} = c_1$$

$$mr^2\dot{\theta} = \sqrt{c_2 - \frac{c_1^2}{\sin^2\theta} - 2meL\cos\theta}$$

$$m\dot{r} = \frac{1}{\sigma}\sqrt{-Wmr^2 + 2me^2r - c_2}.$$

The quanta conditions can be written down as

$$\int mr^2\sin^2\theta\dot{\phi}\,d\phi = n_1h, \qquad \ldots \quad (1)$$

$$\int mr^2\dot{\theta}\,d\theta = n_2h, \qquad \ldots \quad (2)$$

$$\int m\dot{r}dr = n_3h, \qquad \ldots \quad (3)$$

$-W$ being twice the total energy of the system. The integrals are to be extended over the whole range within which the expression within the square root remains positive.

The integration.

From (1) we have obviously

$$c_1 = \frac{n_1 h}{2\pi};$$

of the two remaining expressions, (3) can be integrated most easily : in fact,

$$n_3 h = \int \frac{1}{r} \sqrt{-Wmr^2+2me^2r-c_2}\ dr$$

gives after integration

$$n_3 h = 2\pi \left[e^2 \left(\frac{m}{W} \right)^{\frac{1}{2}} - c_2^{\frac{1}{2}} \right].$$

The second integral can be written as

$$\frac{n_2 h}{2} = I = \int \sqrt{(c_2-c_1^2)-c_2x^2-2meLx(1-x^2)}\ \frac{dx}{1-x^2}$$

by putting $\cos\theta = x$.

The right-hand side is to be integrated throughout the region, when the cubic remains positive. It cannot be integrated in finite terms; an approximation suitable for our purpose can, however, be made, assuming $2meL$ to be small compared with $(c_2-c_1^2) = A$. To see what this means we are to remember that $c_2-c_1^2$ is of the dimension of h^2; so that $2meL$ must be small compared with h^2, or L must be small compared with $\frac{h^2}{me}$. Now, if α, $-\beta$, and γ are taken as the three roots of the cubic, the cubic can be written down as

$$D(\gamma-x)(\alpha-x)(x+\beta),$$

where γ is the greatest of the positive roots and $D = 2meL$. The limits of the integral are obviously α and $-\beta$,

$$I = \int_{-\beta}^{\alpha} \sqrt{A-c_2x^2-Dx(1-x^2)}\ \frac{dx}{1-x^2}.$$

Hence

$$\frac{\partial I}{\partial D} = \frac{1}{2}\int_{-\beta}^{\alpha} \frac{-x dx}{\sqrt{D}[(\gamma-x)(\alpha-x)(x+\beta)]^{\frac{1}{2}}},$$

or

$$\frac{\partial I}{\partial D} = \frac{1}{2\sqrt{D}}\left[\int_{-\beta}^{\alpha} \frac{(\gamma-x)^{\frac{1}{2}}dx}{(\alpha-x)^{\frac{1}{2}}(\beta+x)^{\frac{1}{2}}} - \gamma \int_{-\beta}^{\alpha} \frac{dx}{(\gamma-x)^{\frac{1}{2}}(\alpha-x)^{\frac{1}{2}}(x+\beta)^{\frac{1}{2}}} \right].$$

Supposing

$$\alpha - x = (\alpha+\beta)\cos^2\theta,$$
$$\beta + x = (\alpha+\beta)\sin^2\theta,$$

we get

$$\frac{\partial I}{\partial D} = \frac{1}{2\sqrt{D}}\left[2\int_0^{\pi/2}[\gamma+\beta-(\alpha+\beta)\sin^2\theta]^{\frac{1}{2}}\,d\theta - 2\gamma\int_0^{\pi/2}\frac{d\theta}{[(\gamma+\beta)-(\alpha+\beta)\sin^2\theta]^{\frac{1}{2}}}\right]$$

$$= \frac{1}{2\sqrt{D}}\left[(\gamma+\beta)^{\frac{1}{2}}E\left(\pi/2, \sqrt{\frac{\alpha+\beta}{\gamma+\beta}}\right) - \frac{\gamma}{(\gamma+\beta)^{\frac{1}{2}}}K\left(\pi/2, \sqrt{\frac{\alpha+\beta}{\gamma+\beta}}\right)\right],$$

where E and K are the usual elliptic integrals, defined by

$$E(\pi/2, k) = \int_0^{\pi/2}\sqrt{1-k^2\sin^2\theta}\,d\theta$$

$$K(\pi/2, k) = \int_0^{\pi/2}\frac{d\theta}{\sqrt{1-k^2\sin^2\theta}}.$$

On the assumption that D is small, we have

$$\gamma = \frac{c_2}{D} + \frac{D(c_2-A)}{c_2^2},$$

$$\alpha = \sqrt{\frac{A}{c_2}} - \frac{D}{2c_2^2}(c_2-A),$$

$$\beta = \sqrt{\frac{A}{c_2}} + \frac{D}{2c_2^2}(c_2-A).$$

So that, expanding E and K and making necessary approximations, we have finally

$$\frac{\partial I}{\partial D} = \frac{\pi D}{4c_2^{5/2}}\left[c_2 - \frac{3A}{2}\right].$$

Also

$$I = \int\sqrt{(c_2-c_1^2)c_2x^2}\,\frac{dx}{1-x^2} = \pi\left[c_2^{\frac{1}{2}}-c_1\right].$$

So that we have

$$\frac{n_2 h}{2} = \pi(c_2^{\frac{1}{2}} - c_1) + \frac{\pi}{16}\frac{(2meL)^2}{c_2^{5/2}}(3c_1^2 - c_2).$$

Now, collecting all the quanta conditions, we have

$$c_1 = \frac{n_1 h}{2\pi},$$

$$c_2^{\frac{1}{2}} - c_1 + \frac{(meL)^2}{4c_2^{5/2}}(3c_1^2 - c_2) = \frac{n_2 h}{2\pi},$$

$$e^2\left(\frac{m}{W}\right)^{\frac{1}{2}} - c_2^{\frac{1}{2}} = \frac{n_3 h}{2\pi}.$$

So that we have

$$c_2^{\frac{1}{2}} + \frac{(meL)^2}{4c_2^{5/2}}\left(\frac{3n_1 h_1^2}{4\pi^2} - c_2\right) = \frac{(n_1 + n_2)h}{2\pi}$$

and

$$e^2\left(\frac{m}{W}\right)^{\frac{1}{2}} = \frac{n_3 h}{2\pi} + c_2^{\frac{1}{2}}.$$

Assuming

$$c_2^{\frac{1}{2}} = \frac{yh}{2\pi}$$

we have

$$\frac{yh}{2\pi} + \frac{(meL)^2(2\pi)^5}{4y^5h^5}\left[\frac{3n_1^2\,h^2}{4\pi^2} - \frac{y^2h^2}{4\pi^2}\right] = \frac{(n_1+n_2)}{2\pi}h$$

or

$$y + \left(\frac{2\pi^2\,meL}{h^2}\right)^2\frac{1}{y^5}(3n_1^2 - y^2) = n_1 + n_2.$$

Calling

$$\left(\frac{2\pi^2 meL}{h^2}\right)^2 = \alpha,$$

we have

$$e^2\left(\frac{m}{W}\right)^{\frac{1}{2}} = \frac{h}{2\pi}(n_3 + y),$$

where

$$y + \frac{\alpha}{y^5}(3n_1^2 - y^2) = n_1 + n_2\,;$$

we have approximately

$$y = (n_1+n_2) + \frac{\alpha}{(n_1+n_2)^3}\left[1-\frac{3n}{(n_1+n_2)^2}\right].$$

So that

$$W = \frac{4\pi^2e^4m}{h^2\left[n_1+n_2+n_3+\frac{\alpha}{(n_1+n_2)^3}\left(1-\frac{3n_1^2}{(n_1+n_2)^2}\right)\right]^2}.$$

So that the energy in the statical path

$$= -\frac{2\pi^2e^4m}{h^2\left[n_1+n_2+n_3+\frac{\alpha}{(n_1+n_2)^3}\left(1-\frac{3n}{(n_1+n_2)^2}\right)\right]^2}$$

if we suppose that the spectrum is due to ionized atoms in which the field can be approximately represented as a central charge of $E = pe$ and a doublet of different strength L'.

We have, by a similar reasoning, energy in a statical path

$$= -\frac{2\pi^2}{h^2}\frac{e^2E^2m}{\left\{n_1+n_2+n_3+\frac{\alpha'}{(n_1+n_2)^3}\left(1-\frac{3n_1^2}{(n_1+n_2)^2}\right)\right\}^2}$$

$$= \frac{-p^2Nh}{\left\{n_1+n_2+n_3+\frac{\alpha'}{(n_1+n_2)^3}\left(1-\frac{3n_1^2}{(n_1+n_2)^2}\right)\right\}^2}$$

where N = Rydberg number.

As a result of numerous investigations on the nature of the spectral series, it has been shown that for many elements the different series can be grouped according to the following schemes :

P-Series :

$$\left.\begin{aligned}\nu = (1.5,\ s)&-(m,\ p_1)\\ &-(m,\ p_2)\end{aligned}\right\} \qquad m = 2, 3, 4\ldots$$

II. Subordinate Series :

$$\nu = (2,\ p_1)-(m+\cdot 5,\ s)$$

$$m = 2, 3, 4\ldots$$

$$\nu = (2,\ p_2)-(m+\cdot 5,\ s)\cdot$$

1st Subordinate Series :

$$\nu = (2,\ p_1)-(m,\ d).$$

Companion :

$$\nu = (2,\ p_1)-(m,\ d')$$

$$m = 3, 4, 5.$$

$$\nu = (2,\ p_2)-(m,\ d'').$$

Symbol (m, f) stands for $\frac{N}{(m+f)^2}$ according to Rydberg, and $\frac{N}{\left(m+f+\frac{\phi}{m^2}\right)^2}$ according to Ritz.

The frequency of the lines emitted thus appears as the difference of two terms, each of which is to be regarded as corresponding to the energy in a definite statical path, on the Quantum theory. Sommerfeld* has recently given reasons for assuming that in any series of statical orbits corresponding to the different kinds of s, p, and d terms the azimuthal quanta generally preserve a certain definite value, whereas the radial quanta can have all values from 0 to ∞; he thus shows that in s, p and d terms the azimuthal quanta generally have values 1, 2, and 3 respectively.

Making the above assumptions in our formula, we see that the expression of the energy comes out as $-\frac{Nh}{(n_1+n_2+n_3+A)^2}$ in the same form as required by the Rydberg formula. The constants A, however, depends only upon n_1 and n_2; it diminishes for increasing values of the azimuthal quanta; so that they decrease progressively in the s, p, and d terms. Moreover, our form shows that A depends upon n_1 and n_2 separately, so that for the same value of n_1+n_2 we may have different values of the constant. Thus, if we suppose $n_1+n_2=1$ we have two values corresponding to the vaues 1, 0 and 0, 1; for $n_1+n_2=2$ we have three values; and so on. Thus we see, even on Sommerfeld's assumption, for the constancy of the azimuthal quanta we shall have two different s, three different p, four different d terms. At least two different values of p and three different values of d seem to be required by the series formula, which is essential for the explanation of doublets and triplets of constant frequency difference**. We thus see that our model serves at least as a qualitative explanation of the following facts :

(1) The progressive decrease of the characteristic numbers in the s, p, and d terms,

(2) The existence of different sets of s, p, and d terms for the same element.

It is clear, however, that our simplified assumption will not fit in any actual case exactly. The complex nature of the internal field can in no case be properly represented by a simple term, $\frac{L \cos \theta}{r^2}$, in the Potential. Moreover, we have reason to believe that the internal arrangement of the electrons itself will be influenced, in a large measure, by the motion of the outer electron, which we have neglected in our formula. In fact, Landé† has tried in a recent paper to take account of this disturbance in the

* Sommerfeld, Verh. d. Phys. Ges. May 1919.

** If we exclude the case $n_2 = 0$—*i.e.*, if we assume that the motion in a plane containing the axis of the doublet is excluded—we get the proper number of s, p, d terms as observed in the case of the alkali metals and the doublet system of alkaline earths.

† Landé, Phys. Zeit. 1909.

comparative simple case of the helium series. But at the same time, it is hoped that the calculation, in this comparatively simple case, will serve to illustrate at least some general principles at which we have arrived by an experimental study of the spectral series.

Summary

In this paper an attempt has been made to deduce the laws of regularity in the spectral series of elements on the basis of Bohr's quantum theory of spectral emission. Starting from Sommerfeld's assumption that the ordinary line-spectra of elements are due to the vibration of one outer electron (the valency electron), it has been shown that the field of the nucleus and the remaining $(n-1)$ electrons may be represented by the Potential $V=-\frac{e^2}{r}+\frac{eL\cos\theta}{r^2}$ i.e., the field due to a single charge plus a doublet of strength L. The axis of the doublet is variable, but the emission is supposed to take place so quickly that in that short time the axis does not appreciably change.

The quanta conditions have been applied according to Sommerfeld's rule, $nh=\int p_i\,dq_i$, and the energy of the system has been reduced to the quanta numbers. The energy comes out in the form

$$W=-\frac{Nh}{(n+n_3+z)^2}\,,\quad n+z=y$$

in the paper, where n_3 is the radial quantum, n is the azimuthal quantum, and z is given by an equation of the sixth degree, involving only the azimuthal quantum, and is a function of n only.

It has been next shown that if, in accordance with Sommerfeld's principle, we assume $n=1$ for the s-orbits, $n=2$ for the p-orbits, $n=3$ for the d-orbits, $n=4$ for the b-orbits, then, with a very simple assumption, we obtain a single value for the energy of the s orbits, a double value for the energy in the p-orbit, a treble value for the d-orbit. Then, applying Bohr's law $h\nu=W_n-W_{n'}$, we arrive at Rydberg's laws of the regularity in spectral series, in the case of the alkali metals.

Exact calculations are not tried on account of the uncertainty of the value of L; but it has been pointed out that the values of s, (p_1, p_2), (d_1, d_2, d_3) progressively decrease, as is actually the case.

If the value of L be supposed to vary with n_3 the radial quantum, then probably the above calculations would lead to Ritz's law.

Reprinted from Phil Mag **40**, *pp. 619-627, 1920 (Taylor and Francis, London).*

Plancks Gesetz und Lichtquantenhypothese.

Von **Bose** (Dacca-University, Indien).

(Eingegangen am 2. Juli 1924.)

Der Phasenraum eines Lichtquants in bezug auf ein gegebenes Volumen wird in „Zellen" von der Größe h^3 aufgeteilt. Die Zahl der möglichen Verteilungen der Lichtquanten einer makroskopisch definierten Strahlung unter diese Zellen liefert die Entropie und damit alle thermodynamischen Eigenschaften der Strahlung.

Plancks Formel für die Verteilung der Energie in der Strahlung des schwarzen Körpers bildet den Ausgangspunkt für die Quantentheorie, welche in den letzten 20 Jahren entwickelt worden ist und in allen Gebieten der Physik reiche Früchte getragen hat. Seit der Publikation im Jahre 1901 sind viele Arten der Ableitung dieses Gesetzes vorgeschlagen worden. Es ist anerkannt, daß die fundamentalen Voraussetzungen der Quantentheorie unvereinbar sind mit den Gesetzen der klassischen Elektrodynamik. Alle bisherigen Ableitungen machen Gebrauch von der Relation

$$\varrho_\nu d\nu = \frac{8\pi\nu^2 d\nu}{c^3} E,$$

d. h. von der Relation zwischen der Strahlungsdichte und der mittleren Energie eines Oszillators, und sie machen Annahmen über die Zahl der Freiheitsgrade des Äthers, wie sie in obige Gleichung eingeht (erster Faktor der rechten Seite). Dieser Faktor konnte jedoch nur aus der klassischen Theorie hergeleitet werden. Dies ist der unbefriedigende Punkt in allen Ableitungen, und es kann nicht wundernehmen, daß immer wieder Anstrengungen gemacht werden, eine Ableitung zu geben, die von diesem logischen Fehler frei ist.

Eine bemerkenswert elegante Ableitung ist von Einstein angegeben worden. Dieser hat den logischen Mangel aller bisherigen Ableitungen erkannt und versucht, die Formel unabhängig von der klassischen Theorie zu deduzieren. Von sehr einfachen Annahmen über den Energieaustausch zwischen Molekülen und Strahlungsfeld ausgehend, findet er die Relation

$$\varrho_\nu = \frac{\alpha_{mn}}{e^{\frac{\varepsilon_m - \varepsilon_n}{kT}} - 1}.$$

Indessen muß er, um diese Formel mit der Planckschen in Übereinstimmung zu bringen, von Wiens Verschiebungsgesetz und Bohrs Korrespondenzprinzip Gebrauch machen. Wiens Gesetz ist auf die klassische Theorie gegründet, und das Korrespondenzprinzip nimmt an, daß die Quantentheorie mit der klassischen Theorie in gewissen Grenzfällen übereinstimme.

In allen Fällen scheinen mir die Ableitungen nicht genügend logisch gerechtfertigt. Dagegen scheint mir die Lichtquantenhypothese in Verbindung mit der statistischen Mechanik (wie sie durch Planck den Bedürfnissen der Quantentheorie angepaßt worden ist) für die Ableitung des Gesetzes unabhängig von der klassischen Theorie hinreichend zu sein. Im folgenden will ich die Methode kurz skizzieren.

Die Strahlung sei in das Volumen V eingeschlossen und ihre Gesamtenergie E sei gegeben. Es seien verschiedene Sorten von Quanten von der jeweiligen Anzahl N_s und Energie $h\nu_s$ vorhanden ($s = 0$ bis $s = \infty$). Die Totalenergie E ist dann

$$E = \sum_s N_s h\nu_s = V\int \varrho_\nu d\nu. \qquad (1)$$

Die Lösung des Problems verlangt dann die Bestimmung der N_s, welche ϱ_ν bestimmen. Wenn wir die Wahrscheinlichkeit für jede durch beliebige N_s charakterisierte Verteilung angeben können, dann wird die Lösung durch die Bedingung bestimmt, daß diese Wahrscheinlichkeit bei Wahrung der Nebenbedingung (1) ein Maximum sein soll. Diese Wahrscheinlichkeit wollen wir nun aufsuchen.

Das Quantum hat ein Moment vom Betrage $\frac{h\nu_s}{c}$ in der Richtung seiner Fortbewegung. Der Momentanzustand des Quantums wird charakterisiert durch seine Koordinaten x, y, z und die zugehörigen Momente $p_x\, p_y\, p_z$; diese sechs Größen können als Punktkoordinaten in einem sechsdimensionalen Raum aufgefaßt werden, wobei wir die Beziehung haben

$$p_x^2 + p_y^2 + p_z^2 = \frac{h^2\nu^2}{c^2},$$

vermöge welcher der genannte Punkt auf einer durch die Frequenz des Quants bestimmten Zylinderfläche zu bleiben gezwungen ist. Zum

Frequenzbereich $d\nu_s$ gehört in diesem Sinne der Phasenraum

$$\int dx\, dy\, dz\, dp_x\, dp_y\, dp_z = V \cdot 4\pi \left(\frac{h\nu}{c}\right)^2 \frac{h\,d\nu}{c} = 4\pi \frac{h^3\nu^2}{c^3} V d\nu.$$

Wenn wir das gesamte Phasenvolumen in Zellen von der Größe h^3 einteilen, gehören zum Frequenzbereich $d\nu$ also $4\pi V \frac{\nu^2}{c^3} d\nu$ Zellen. In bezug auf die Art dieser Einteilung kann nichts Bestimmtes gesagt werden. Indessen muß die Gesamtzahl der Zellen als die Zahl der möglichen Anordnungen eines Quants in dem gegebenen Volumen angesehen werden. Um der Tatsache der Polarisation Rechnung zu tragen, erscheint es dagegen geboten, diese Zahl noch mit 2 zu multiplizieren, so daß wir für die Zahl der zu $d\nu$ gehörigen Zellen $8\pi V \frac{\nu^2 d\nu}{c^3}$ erhalten.

Nun ist es einfach, die thermodynamische Wahrscheinlichkeit eines (makroskopisch definierten) Zustandes zu berechnen. Es sei N^s die Zahl der zum Frequenzbereich $d\nu^s$ gehörigen Quanten. Auf wie viele Arten können diese auf die zu $d\nu^s$ gehörigen Zellen verteilt werden? Sei p_0^s die Zahl der vakanten Zellen, p_1^s die Zahl derer, die ein Quant enthalten, p_2^s die Zahl der Zellen, die zwei Quanten enthalten usf. Die Zahl der möglichen Verteilungen ist dann

$$\frac{A^s!}{p_0^s!\, p_1^s! \ldots}, \quad \text{wobei} \quad A^s = \frac{8\pi\nu^2}{c^3} d\nu^s,$$

und wobei

$$N^s = 0 \cdot p_0^s + 1 \cdot p_1^s + 2 p_2^s \ldots$$

die Zahl der zu $d\nu^s$ gehörigen Quanten ist.

Die Wahrscheinlichkeit des durch sämtliche p_r^s definierten Zustandes ist offenbar

$$\prod_s \frac{A^s!}{p_0^s!\, p_1^s! \ldots}.$$

Mit Rücksicht darauf, daß wir die p_r^s als große Zahlen betrachten können, haben wir

$$\lg W = \sum_s A^s \lg A^s - \sum_s \sum_r p_r^s \lg p_r^s,$$

wobei

$$A^s = \sum_r p_r^s .$$

Dieser Ausdruck soll ein Maximum sein unter der Nebenbedingung

$$E = \sum_s N^s h\nu^s; \quad N^s = \sum_r r p_r^s .$$

Die Durchführung der Variation liefert die Bedingungen

$$\sum_s \sum_r \delta p_r^s (1 + \lg p_r^s) = 0, \qquad \sum_s \delta N^s h\nu^s = 0$$

$$\sum_r \delta p_r^s = 0 \qquad \delta N^s = \sum_r r \delta p_r^s .$$

Hieraus folgt

$$\sum_s \sum_r \delta p_r^s (1 + \lg p_r^s + \lambda^s) + \frac{1}{\beta} \sum_s h\nu^s \sum_r r \delta p_r^s = 0.$$

Daraus folgt zunächst

$$p_r^s = B^s e^{-\frac{r h\nu^s}{\beta}} .$$

Da aber

$$A^s = \sum_r B^s e^{-\frac{r h\nu^s}{\beta}} = B^s \left(1 - e^{-\frac{h\nu^s}{\beta}}\right)^{-1},$$

so ist

$$B_s = A^s \left(1 - e^{-\frac{h\nu^s}{\beta}}\right).$$

Ferner hat man

$$N^s = \sum_r r p_r^s = \sum_r r A^s \left(1 - e^{-\frac{h\nu^s}{\beta}}\right) e^{-\frac{r h\nu^s}{\beta}}$$

$$= \frac{A^s e^{-\frac{h\nu^s}{\beta}}}{1 - e^{-\frac{h\nu^s}{\beta}}} .$$

Mit Rücksicht auf den oben gefundenen Wert von A^s ist also

$$E = \sum_s \frac{8\pi h {\nu^s}^3 d\nu^s}{c^3} V \frac{e^{-\frac{h\nu^s}{\beta}}}{1 - e^{-\frac{h\nu^s}{\beta}}} .$$

Mit Benutzung der bisherigen Resultate findet man ferner

$$S = k\left[\frac{E}{\beta} - \sum_s A^s \lg\left(1 - e^{\frac{h\nu^s}{\beta}}\right)\right],$$

woraus mit Rücksicht darauf, daß $\frac{\partial S}{\partial E} = \frac{1}{T}$, folgt, daß $\beta = kT$. Setzt man dies in obige Gleichung für E ein, so erhält man

$$E = \sum_s \frac{8\pi h \nu^{s^3}}{c^3} V \frac{1}{e^{\frac{h\nu^s}{kT}} - 1} d\nu^s,$$

welche Gleichung Plancks Formel äquivalent ist.

(Übersetzt von A. Einstein.)

Anmerkung des Übersetzers. Boses Ableitung der Planckschen Formel bedeutet nach meiner Meinung einen wichtigen Fortschritt. Die hier benutzte Methode liefert auch die Quantentheorie des idealen Gases, wie ich an anderer Stelle ausführen will.

Reprinted from *Z Physik* **26**, pp. 168-171, 1924 (Springer — Verlag, Heidelberg).

Planck's Law and the Light-Quantum Hypothesis

Bose (Dacca-University, India)
(Received 2 July, 1924)

The phase space of a light-quantum in a given volume is divided up in 'cells' of size h^3. The number of possible distributions of the light-quanta of a macroscopically defined radiation among the cells gives the entropy and with that all the thermodynamic properties of the radiation.

Planck's formula for the distribution of energy in the radiation of a blackbody forms the starting point for the quantum theory which has been developed in the last 20 years and has been very fruitful in all parts of physics. Since its publication in 1901 many methods for the derivation of this law has been proposed. It is recognized that the fundamental assumptions of the quantum theory are incompatible with the laws of classical electrodynamics. All derivations till now use the relation

$$\varrho_\nu \, d\nu = \frac{8\pi\nu^2 \, d\nu}{c^3} E,$$

that is, the relation between the density of radiation and the average energy of an oscillator, and they make assumptions about the number of degrees of freedom of the ether, which enters the above equation (the first factor on the right hand side). This factor could, however, be derived only from the classical theory. This is an unsatisfactory feature in all derivations, and it is not surprising that efforts are made again and again to give a derivation free from this logical flaw.

Einstein has given a remarkably elegant derivation. He has recognized the logical flaw in all previous derivations and has tried to deduce the formula independently of classical theory. Starting from very simple assumptions about the energy exchange between molecules and the radiation field, he finds the relation

$$\varrho_\nu = \frac{\alpha_{mn}}{e^{\frac{\varepsilon_m - \varepsilon_n}{kT}} - 1}.$$

In order to make this formula agree with that of Planck he has to use Wien's displacement law and Bohr's correspondence principle. Wien's law is based on classical theory and the correspondence principle assumes that the quantum theory agrees with the classical theory in certain limiting cases.

In all cases it appears to me that the derivations are not sufficiently justified from a logical point of view. On the other hand, the light-quantum hypothesis combined with statistical mechanics (as adapted by Planck to comform to the requirements of quantum theory) appears sufficient for the deduction of the law independent of classical theory. In the following I shall sketch the method briefly.

Let the radiation be enclosed in a volume V and its total energy be E. Suppose there are different types of quanta each having number N_s and energy $h\nu_s$ (s = o to $s = \infty$). The total energy is then

$$E = \sum_s N_s h \nu_s = V \int \varrho_\nu d\nu. \tag{1}$$

The solution of the problem then requires the determination of of N_s which in turn determine ρ_ν. If we can give the probability for each distribution characterized by arbitrary values of N_s, then the solution is determined by the condition that this probability is a maximum subject to the subsidiary condition (1). We now want to find this probability.

The quantum has a momentum of magnitude $\frac{h\nu_s}{c}$ in the direction of its motion. The instantaneous state of the quantum is characterised by its coordinates x, y, z and the corresponding momenta p_x, p_y, p_z. These six quantities can be considered to be the coordinates of a point in a six dimensional space, where we have the relation

$$p_x^2 + p_y^2 + p_z^2 = \frac{h^2 \nu^2}{c^2},$$

by virtue of which the point is forced to lie on the surface of a cylinder determined by the frequency of the quantum. The phase space belonging to the frequency interval $d\nu_s$ is.

$$\int dx\, dy\, dz\, dp_x\, dp_y\, dp_z = V \, . \, 4\pi \left(\frac{h\nu}{c}\right)^2 \frac{h\, d\nu}{c} = 4\pi \frac{h^3 \nu^2}{c^3} V d\nu.$$

If we divide the total phase space volume in cells of size h^3, then $4\pi V \frac{\nu^2}{c^3} d\nu$ cells will belong to the frequency interval $d\nu$. Nothing definite can be said about the method of this division. In any case, the total number of cells must be regarded as the number of possible arrangements of a quantum in the given volume. It seems, however, appropriate to multiply this number once again by 2 in order to take into account the fact of polarization, so that we obtain $8\pi V \frac{\nu^2\, d\nu}{c^3}$ as the number of cells belonging to $d\nu$.

Now it is easy to calculate the thermodynamic probability (macroscopically defined) of a state. Let N^s be the number of quanta belonging to the frequency range $d\nu^s$. In how many ways can these be distributed among the cells belonging to $d\nu^s$? Let p_0^s be the number of empty cells, p_1^s those which contain one quantum, p_2^s those which contain two quanta and so on. The number of possible distributions is then,

$$\frac{A^s!}{p_0^s!\, p_1^s! \ldots}, \quad \text{where} \quad A^s = \frac{8\pi\nu^2}{c^3} d\nu^s,$$

and

$$N^s = 0 \cdot p_0^s + 1 \cdot p_1^s + 2 p_2^s \ldots$$

is the number of quanta belong to the range $d\nu^s$.

The probability of the state defined by all p_r^s is clearly

$$\prod_s \frac{A^s!}{p_0^s!\, p_1^s! \ldots}.$$

Taking into account that we can consider p_r^s to be large numbers, we have

$$\lg W = \sum_s A^s \lg A^s - \sum_s \sum_r p_r^s \lg p_r^s,$$

where

$$A^s = \sum_r p_r^s.$$

This expression must be a maximum with the constraints

$$E = \sum_s N^s h\nu^s; \quad N^s = \sum_r r p_r^s.$$

Carrying out the variation, we obtain the conditions

$$\sum_s \sum_r \delta p_r^s (1 + \lg p_r^s) = 0, \qquad \sum_s \delta N^s h\nu^s = 0$$

$$\sum_r \delta p_r^s = 0 \qquad \delta N^s = \sum_r r\, \delta p_r^s.$$

From this it follows that

$$\sum_s \sum_r \delta p_r^s (1 + \lg p_r^s + \lambda^s) + \frac{1}{\beta} \sum_s h\nu^s \sum_r r\, \delta p_r^s = 0.$$

Next it follows from this that

$$p_r^s = B^s e^{-\frac{r h\nu^s}{\beta}}.$$

But since

$$A^s = \sum_r B^s e^{-\frac{r h\nu^s}{\beta}} = B^s \left(1 - e^{-\frac{h\nu^s}{\beta}}\right)^{-1},$$

therefore

$$B_s = A^s \left(1 - e^{-\frac{h\nu^s}{\beta}}\right).$$

We further have

$$N^s = \sum_r r p_r^s = \sum_r r A^s \left(1 - e^{-\frac{h\nu^s}{\beta}}\right) e^{-\frac{r h\nu^s}{\beta}}$$

$$= \frac{A^s e^{-\frac{h\nu^s}{\beta}}}{1 - e^{-\frac{h\nu^s}{\beta}}}.$$

Taking into account the value of A^s found above, we get

$$E = \sum_s \frac{8\pi h {\nu^s}^3 d\nu^s}{c^3} V \frac{e^{-\frac{h\nu^s}{\beta}}}{1 - e^{-\frac{h\nu^s}{\beta}}}.$$

Using the results found so far, one further finds

$$S = k\left[\frac{E}{\beta} - \sum_s A^s \lg\left(1 - e^{\frac{h\nu^s}{\beta}}\right)\right],$$

Whence, using the relation $\frac{\partial S}{\partial E} = \frac{1}{T}$, if follows that $\beta = kT$. Substituting this in the equation for E, we get

$$E = \sum_s \frac{8\pi h {\nu^s}^3}{c^3} V \frac{1}{e^{\frac{h\nu^s}{kT}} - 1} d\nu^s,$$

which is the same as Planck's formula.

(Translated by A. Einstein)

Translator's remarks :

In my opinion Bose's derivation signifies an important advance. The method used here gives the quantum theory of an ideal gas as I will work out elsewhere.

English translation of *Z Physik* **26**, pp. 168-171, 1924 (Springer — Verlag, Heidelberg).

Wärmegleichgewicht im Strahlungsfeld bei Anwesenheit von Materie.

Von **S. N. Bose** in Ramna (Indien).

(*Eingegangen am* 7, *Juli* 1924)

Die Wahrscheinlichkeit eines beliebigen Zustandes eines aus Strahlung und Molekeln bestehenden Systems wird berechnet, und es werden hieraus die Bedingungen fur das statistische Gleichgawicht abgeleitet. Im Anschluß hieran werden neue Ausdrücke für die statistische Wahrscheinlichkeit der Elementarvorgänge vorgeschlagen, welche der Wechselwirkung zwischen Strahlung and Materie entsprechen.

Debye[1] hat gezeigt, daß das Plancksche Gesetz mit Hilfe der statistischen Mechanik abgeleitet werden kann. Seine Ableitung ist jedoch insofern nicht völlig unbhängig von der klassischen Elektrodynamik, als er Gebrauch macht von dem Begriff der Eigenschwingungen des Äthers und annimmt, daß hinsichtlich der Energie das Spektralgebiet zwischen ν und $\nu + d\nu$ ersetzt werden kann durch $\frac{8\pi\nu^2}{c^3} V\, d\nu$ Resonatoren, deren Energie nur Vielfache von $h\nu$ betragen kann. Man kann aber zeigen, daß die Ableitung so abgeändert werden kann, daß man überhaupt keine Anleihen bei der klassischen Theorie zu machen braucht. $\frac{8\pi\nu^2}{c^3} V\, d\nu$ kann gedeutet werden als die Anzahl Elementargebiete des schesdimensionalen Phasenraumes für die Quanten. Die weitere Rechnung bleibt im wesentlishen dieselbe.

Einstein hat sich einer anderen Methode bedient. Er. betrachtet die Wechselwirkung zwischen den materiellen Telichen und dem Strahlungsfeld. Kennt man die Energieverteilung untr den materiellen Teilchen, so kann man das Gesetz der schwarzen Strahlung ermitteln aus der Bedingung der Stationarität dieser Verteilung beim Austausch von Energie zwischen Materie und Strahlung. Dieser Austausch sit weiterhin eine Folge gewisser Elementarprozesse. Eine geeignete Formulierung der Eigenschaften und Wahrscheinlichkeit des Elementar prozesses ermoglicht es, das Plancksche Gesetz abzuleiten, wenn man die Energieverteilung unter den Teilchen als gegeben ansieht. In seiner ersten Arbeit[2] sind die materiellen Gebilde Bohrsche Atome, die nur einer diskreten Reihe stationärer Zustände fähig sind. Energieaustausch findet statt infolge von Emissionsund Absroptionsprozessen und gleichzeitigen Übergängen der Atome von einem Zustand zu einem anderen. Er hat dargetan, daß das Plancksche Gesetz sich ergibt, wenn den Emissions- und Absorptionsprozessen gewisse Wahrscheinlichkeiten zugeschrieben werden. Neuerdings hat aber das Problem

(1) Debye, *Ann. d. Phys.* **33**, 1427, 1910.
(2) Einstein, *Phys. ZS.* **18**, 121, 1917.

des thermischen Gleichgewichts zwischen dem Strahlungsfeld und freien Elektronen Wichtigkeit erlangst durch die Arbeiten von Debye[3] Compton[4] und anderen, die sich mit der Zerstreuung der Strahlung durch Elektronen befassen, Pauli[5] hat dieses Problem in einer interessanten Arbeit diskutiert. Als Elementarproze*β* betrachtet er die Streuung, durch die ein Quant, das in einer bestimmten Richtung fliegt und innerhalb eines bestimmten Spektralintervalls. liegt, sich in ein anderes Quant eines anderen Spektralintervalls und von verschiedener Richtung verwandelt. Gleichzeitig andert ein Elektron, das sich mit einer gewissen Geschwindigkeit in einer bestimmten Richtung bewegt Gro*β*e und Richtung seiner Geschwindigkeit. Für alle diese Prozesse gilt das Erhaltungsgesetz von Impuls und Energie. Pauli hat gezeigt, da*β*, wenn die Wahrscheinlichkeit dieses Elementarprozesses $(A\rho+B\rho\rho')dt$ ist, wo ρ und ρ' die Strahlungsdichten der Frequenzen ν und ν', A und B unabhängig von q and q' aber abhänging von der Art des betrachteten Sto*β*es sind, dieses Wahrscheilichkeitsgesetz zum Planckschen Gesetz führt, falls für das Elektronengas das Maxwellsche Verteilungsgesetz gültig ist. Einstein unde Ehrenfest haben das Paulische Ergebnis noch verallgemeinert[6]. Sie betrachten einen Proze*β*, bei dem ein Molekül order ein Elektron Quanten $h\nu_1$, $h\nu_2$ usw. absorbiert und gleichzeitig $h\nu_1'$, $h\nu_2'$ ...emittiert. Es wird angenommen, da*β* die emittierten und absorbierten Quanten bestimmte Richtungen haben. Zugleich andert sich auch Gro*β*e und Richtung der Geschwindigkeit des betreffenden Moleküls. Energie und Impuls bleiben aber erhalten. Das Wahrscheinlichkeitsgesetz von Pauli wird folgenderma *β* on verallgemeinert :

$$dW_1 = \Pi b_1\rho_1\Pi(a_1'+b_1'\,\rho_1')dt,$$

$$dW_2 = \Pi(a_1+b_1\rho_1)\Pi(b_1'\,\rho_1')dt.$$

für den direkten und inverson Proze*β*.

Er zeigt, da*β* Gleichgewicht vorhanden ist, wenn gewisse Beziehungen zwischen den Koeffizienten bestehen. Dies ist sowohl eine Verallgemeinerung seines früheren Gesetzes fur Bohrsche Atom wide des Gesetzes von Pauli

Das Problem des Warmegleichgewichts in einem Strahlungsfeld in Anwesenheit von materiellen Gebilden kann aber nach den Methoden der statistischen Mechanik behandelt werden, unabhängig von jeder besonderen Hypothese über den Mechanismus der Elementarprozesse, auf denen der Energieaustausch beruht. Wir gelangen so zu einer allgemeinen Beziehung, die güetig ist für alle speziellen Hypothesen über die Elementarprozesse und ihre Wahrscheinlichkeiten. Wenn es moglich ist, die thermodynamische Wahrescheinlichkeit für irgend eine spezielle Verteilung der

(3) Debye, *Phys. ZS.* **24**, 161, 1923.
(4) Compton, *Phys. Rev.* **21**, 483, 1923.
(5) Pauli, *ZS. f. Phys.* **18**, 272, 1923.
(6) Einstein und Ehrenfest, *ZS. f. Phys.* 19, 301, 1923.

Quanten im Strahlungsfeld und die Wahrscheinlichkeit für irgend eine willkürliche Verteilung der Enerige unter den materiellen Teilchen niederzuschreiben, dann ist die thermodynamische Wahrscheinlichkeit des größeren Systems, das Materie uno Strahlung umfaßt, einfach das Produkt der bieden Wahrscheinlichketien. Die Gleichgewichtsbedingung lautet einfach, daß die resultierende Wahrscheinlichkeit ein Maximum sein muß. Das Gleichgewichts-problem kann somit in einer ganza ganz anderen Weise betrachtet werden. Statt geeignete Hypothesen uber Elementarprozesse, die das Plancksche Gesetz ergeben, aufzustellen, versuchen wir, die Relation abzuleiten, zu der alle Hypothesen fuhren mussen, wenn das Plancksche Gesetz fur das Strahlungsfeld und das Maxwellsche Gesetz für die materiellen Teilchen gelten sollen. Die Beziehung, die wir suchen, ist offenbar aquivalent mit der der zuvor aus der Bedingung, daß die thermodynamischche Wahrscheinlichkeit für das größ ere System ein Maximum ist, aabgeleiteten. Um das Problem in der angedeuteten Weise zu lüsen, mussen wir den Ausdruck für die thermodynamische Wahrscheinlichkeit für die Strahlung und für die materiellen Teilchen haben.

Thermodynamische Wahrscheinlichkeit für das Strahlungsfeld. Sie wurde in einer fruherent Arbeit[7] abgeleitet. Wenn die Zahl der Quanten im Spektralbereich ν und $\nu+d\nu$ gleich $N\,d\nu$ ist, so ist sie :

$$W = \prod_{\nu} \frac{(A_\nu+N_\nu d\nu)!}{A_\nu!\,N_\nu d\nu\,!},$$

wo $A_\nu = \frac{8\pi\nu^2}{c^3} V\,d\nu$ bedeutet. Es ist leicht zu sehen, daß dieses Wahrscheinlichkeitsgesetz zum Planckschen Gesetz führt, wenn wir die Bedingung aufstellen, daß W bei gegebener Energie ein Mximum sein muß.

Thermodynamische Wahrscheinlichkeit für materielle Teilchen. Man kann sic unschwer finden. Wir wollen unsere Voraussetzungen ein wenig verallgemeinern, um sowohl den Fall Bohrscher Atome mit diskreten Energieniveaus als auch den Fall der Verteilung der Translationsenergie bei materiellen Teilchen[8] zu umfassen. Wir nehmen an, der Phasenraum sie in Elementargebiete geteilt. Zu jeder Elementarzelle gehort eine bestimmte Zahl g, die die Wahrscheinlichkeit angibt, *daß* irgend ein Teilchen sich darin befindet. Die g sind im algemeinen einander gleich, ausgenommen den Fall Bohrscher Atome. Die thermodynamische Wahrscheinlichkeit irgend einer willkurlichen Verteillung n_1, n_2 use uber die verschiedenen Zellen ist :

$$\frac{N!\,g_1^{n_1} g_2^{n_2}\ \cdots}{n_1!\,n_2!\ \cdots}.$$

(7) Der Verfasser, erscheint im Phil. Mag. Siche auch Debye, *Ann. d. Phys.* **33**, 1427, 1910.

(8) Bohr, *ZS. f. Phys.* **18**, 117, 1923.

Die Bedingung des Wahrscheinlichkeitsmaximums bei gegebener Gesamtenergie und Zahl der Teilchen ist :

$$n_r = Cg_r\, e^{-\frac{E}{KT}}.$$

Jetzt sind wir in der Lage, die thermodynamische Wahrscheinlichkeit für das gesamte System zu berechnen. Sie ist offenbar :

$$W = \prod_s \frac{A_s+N_s!}{A_s!N_s!} \prod_T \frac{g_T^{n_T} N!}{n_T!},$$

wo

$$\Sigma\, N_s h\nu_s + \Sigma\, n_T E_T = E$$

und

$$\Sigma n_T = N.$$

Beim Gleichgewicht ist W ein Maximum, wobei die Gesamtenergie und die Zahl der Teilchen als gegeben anzusehen ist.

Wir betrachten den Elementarprozeß :

n_r verwandelt sich in n_r-1 und n_s in n_s+1; N_{ν_1}, N_{ν_2} ... vermindert sich um 1 und $N_{\nu_1'}$, $N_{\nu_2'}$ ... wächst um 1.

Die Bedingung der Stationaritat von W verlangt, daß W sich dabei nicht ändert.

Der betrachtete Elementarproze kann aufgefaßt werden als der Übergang eines Teilchens vom r-ten zum s-ten Zustand. Die Veränderung, die das Strahlungsfeld dabei erleidet, kann angesehen werden als das Ergebnis einer Zerstreuung nach einem Zusammenstoß.

Die gesuchte Bedingung lantet :

$$\frac{n_r}{g_r} \prod_\nu \frac{N_{\nu_r}}{N_{\nu_r}+A_{\nu_r}} = \frac{n_s}{g_s} \prod_{\nu'} \frac{N_{\nu'}}{N_{\nu'}+A_{\nu'}} \qquad \ldots \text{(I)}$$

wo $\qquad \Sigma h\nu' - \Sigma h\nu + E_s - E_r = 0.$

Diese Gleichung ist offenbar äquivalent mit

$$\frac{n_r}{g_r} \prod \frac{b_1\rho_{\nu_1}}{a_1+b_1\,\rho_\nu} = \frac{n_s}{g_s} \prod \frac{b'_1\rho_{\nu'}}{a'_1+b'_1\rho'_\nu},$$

wo

$$\frac{a_1}{b_1} = 8\pi \frac{\nu^2}{c^3} h\nu, \quad \frac{a_1'}{b_1'} = 8\pi \frac{\nu'^2}{c^3} h\nu'.$$

Dies kann auch geschrieben werden :

$$n_r \Pi b_1 \rho_{\nu_1} \Pi(a_1' + b_1' e'_\nu) = n_s \Pi b_1' \rho_\nu^1 \Pi(a_1 + b_1\rho_\nu),$$

wenn

$$g_r = g_s.$$

Das ist die verallegemeinerte Gleichung, die fur Elementarprozesse von der Art, wie sie Einstein und Ehrenfest[9] betrachetet haben, gilt. Nach der hier gegebenen Ableitung ergibt sich die Beziehung als eine Folgerung aus der Bedingung, daß beim Gleichgewicht die Wahrscheinlichkeit ein Maximum sein muß. Wenn aber der Mechanisms des Elementarprozesses uns gestattet, die Gleichung direkt niederzuchreiben fuhrt die Substitution des Wertes fur das Verhältnis $\frac{n_r}{n_s}$ zum Planckschen Gesetz.

Es is jedoch klar, daß Gleichung (I) fundamentaler ist als die transformierte Form von Einstein. Die nachfolgende Betrachtung wird zeigen, daß eine einfache Uberlegung den Wert des Wahrscheinlichkeitskoeffizienten in der gewünschten Form liefert.

Der Fall des Bohrschen Atoms.

Das ist der Elementarprozeß, den Einstein[10] in seiner fruheren Arbeit betrachtet hat.

Die fundamentale Gleichung (I) reduziert sich hier auf

$$\frac{n_r}{g_r} \frac{N_\nu}{A_\nu + N_\nu} = \frac{n_s}{g_s}.$$

Einsteins ursprüngliche Gleichung lautet :

$$n_r b_1 \rho_\nu dt = n_s(a_1' + b_1' \rho_\nu)dt \qquad \ldots \quad (1)$$

oder

$$\rho_\nu = \frac{a_1' n_s}{n_r b_1 - n_s b_1'}.$$

Nimmt man ferner an, daß

$$g_r b_1 = g_s b_1' \quad \text{und} \quad \frac{a_1'}{b_1'} = 8\pi \frac{h\nu^3}{c^3},$$

so folgt das Plancksche Gesetz.

Um seine Gleichung (I) abzuleiten, nimmt er an, daß Atome von höheren Energieniveaus zu Zuständen niederer Energie auf zwei Wegen übergehen konnen :

1. Eine Art von spontanem Übergang, wie bein einem radioactiven Prozeß, dessen Wahrscheinlickkeit unäbhangig vom Zustand des Strahlungsfeldes ist.

(9) Einstein und Ehrenfest, a. a. 0.

(10) Einstein, *Phys. ZS.* **18**, 121, 1917.

2. Ein durch die charakteristische Strahlung induzierter Übergang, dessen Wahrscheinlichkeit vom Zustand des Feldes abhängt.

Der Übergang von einem niederen zu einem höheren Energieniveau erfolgt durch Absorption, deren Wahrscheinlichkeit vom Felde abhängt.

Weiterhin werden gewisse Benziehungen zwischen den Koeffizienten postuliert, damit das Plancksche Gesetz herauskommt. Diese Beziehungen konnen nicht auf einfache Weise begründet werden.

Das gleiche Problem kann auf die folgende Weise behandelt werden, bei der viele von den vorkommenden willkürlichen Annahmen beseitigt werden.

Wir setzen auch voraus, daß der Übergang von niederen zu höheren Niveaus stets durch Absroption von Quanten einer charakteristischen Frequenz erfolgt. A. Der Übergang von höheren zu niederen Niveaus ist eine Art von spontaner Änderung, dessen Wahrscheinlichkeit ünabhangig vom Strahlungsfeld ist. Die zweite Hypothese von Einstein (negative Einstrahlung) ist nicht erforderlich. B. Wir wollen versuchen, den Wahrscheinlichkeitskoeffizienten für Übergang A zu berechnen.

Die Zahl der Elementargebiete, die den in Betracht kommenden Quanten entsprechen, ist A_ν. Die Gesamtzahl der vorhandenen Quanten ist $N_\nu d\nu$. Diese sind aber nicht gleichförming über den Phasenraum verteilt. Sei p_0 die Zahl der leeren Zellen, p_1 die *mit* 1 Quant, p_2 die *mit* 2 Quanten usw. besetzten. Damit die Strahlung und das materielle Teilchen in Wechselwirkung treten können, muß is dem Elementargebiet sein, das von einem Quant besetzt ist. Das besondere Elementargebiet, in dem sich das betrachtete Molekül befindet, wird, wenn wir es lange genug beobachten, alle moglichen Zustände durchlaufen : d.h. manchmal ist es leer, manchmal wird 1 Quant vorhanden sein, manchmal 2 usw. Die Dauer dieser Ereignisse wird Schleißlich proportional zu den Gleichgewichtswerten von $p_0, p_1 \ldots$ Wenn r Quanten und ein Teilchen zugleich vorhanden sind, dann gibt es entweder einen Energieaustausch oder gar nichts. Somit sind $r+1$ verschiedene Ereignisse möglich, nämlich Energieaustausch von 1 Quant, 2 Quanten usw. oder kein Austausch

Daher ist die Zahl der möglichen Fälle :

$$p_0+2p_1+3p_2+\ldots = A_s+N_s d\nu_s = \Sigma(r+1)p_r.$$

Die Zahl der Fälle, in denen Wechselwirkung eintritt, ist :

$$p_1+2p_2+\ldots = N_s d\nu_s = \Sigma r p_r.$$

Somit ist die Wahrscheinlichkeit einer Wechselwirkung :

$$\frac{\Sigma\, r p_r}{\Sigma(r+1)p_r} = \frac{N_s d\nu_s}{A_s+N_s d\nu_s}.$$

Nun mussen wir die sgezielle Art der Wechselwirkung in Rucksicht ziehen, d.h. die Absorption, die zu unterscheiden ist von Zerstreuung. Die Wahrscheinlichkeit hierfür sie β.

Die Wahrscheinlichkeit der Absorption ist :

$$\frac{\beta\Sigma r p_r}{\Sigma(r+1)p_r} = \beta\frac{N_\nu d\nu}{A_\nu + N_\nu d\nu} .$$

Der Wahrscheinlichkeitskoeffizient für einen Übergang vom Typus B ist offenbar irgend eine Konstante, die den Übergang von dem speziellen höheren Zustand zu dem speziellen niederen Zustand bestimmt. Diese Konstante sei α.

Die Gleichgewichtsbedingung kann dann geschrieben werden :

$$n_r\beta\,\frac{N_\nu d\nu}{A_\nu + N_\nu d\nu} = \alpha n_s$$

was mit der fundamentalen Gleichung (I) übereinstimmt, wenn wir annehmen, daβ

$$g_r\beta = g_s\alpha.$$

Es ist klar, daβ die Substitution von $\frac{n_r}{n_s}$ zum Planckschen Gesetz führt.

Der Fall von Pauli.

Er ist einfach zu behandeln. Die Wahrscheinlichkeit einer Wechselwirkung ist wire vorhin $\frac{\Sigma r p_r}{\Sigma(r+1)p_r}$. $\beta^s_{s'}$ sei der Wahrscheinlichkeitskoeffizient für Zerstreuung von ν_s in $\nu_{s'}$ falls eine Wechselwirkung statthat. Für die inverse Transformation seien die entsprechenden Koeffizienten $\frac{\Sigma r p'_r}{\Sigma(r+1)p'_r}$ und $\beta^{s'}_s$.

Beim Gleichgewicht ist also :

$$n_r\beta^s_{s'}\,\frac{N_s d\nu_s}{A_s + N_s d\nu_s} = n_o\beta^{s'}_s\,\frac{N_{s'} d\nu_{s'}}{A_{s'} + N_{s'} d\nu_{s'}} .$$

Nehmen wir ferner an, daβ $\beta^s_{s'} = \beta^{s'}_s$ so erhalten wir die fundamentale Gleichung (I).

Der Fall von Einstein und Ehrenfest.

Die Wahrscheinlichkeit einer gleichzeitigen Wechselwirkung ist das Produkt der Einzelwahrscheinlichkeiten, also gleich :

$$\Pi\,\frac{N_s d\nu_s}{A_s + N_s\,d\nu_s} .$$

Analog sei $\beta_{s'_1\ s'_2\ \ldots}^{s_1\ s_2\ \ldots}$ die Wahrscheinlichkeit für den speziellen Akt der Zerstreuung, und ein ähnlicher Ausdruck gelte für die inverse Streuung, woraus sich das Übrige leicht ergibt.

Bemerkungen. Aus unserer Rechung geht hervor, da β die Wahrscheinlichkeit einer Wechselwirkung eines Teilchens mit einem Quantum in einem Strahlungsfeld $\frac{N_\nu d\nu}{A_\nu + N_\nu d\nu}$ nicht einfach proportional zur Zahl der vorhandenen Quanten ist, wie man auf den ersten Blick vermuten könnte. Es ist leicht zu sehen, daβ diese Annahme zum Wienschen Gesetz führen würde, wie Pauli erkannt hat[11]. Pauli sah sich deshalb genotigt, eine andere Form für den Wahrscheinlichkeitskoeffizienten nzuunehmen. Die von Pauli angenommene und von Einstein und Ehrenfest verallgemeinerte Form erscheint aber als ganz willkurliche Hypothese, da man sich kein einfaches Bild von dem Zustandekommen eines solchen Ausdrucks machen kann. Die andere hier vorgeschlagene Form ist siemlich einfach und kann auf Grund elementarer Überlegungen gerechtfertigt werden. Es wird auch die Notwendigkeit vermieden, Felationen zwischen den Koeffizienten selbst annehman zu müssen. Bei der Ableitung des Wahrscheinlichkeitskoeffizienten für die Wechselwirkung (oder Kopplung, wie Bohr sagt) wurde angenommen, daβ selbst bei einem Zussmmenstoβ der Fall, daβ keine Wechselwirkung eintritt, gerade so wahrscheinlich ist, wie der Fal irgend einer spezielen Kopplung Diese Annahme ist ein fundamentaler Punkt in der hier gegebenen Ableitung. Nach der klassischen Theorie würde man erwarten, daβ irgend eine Wechselwirkung eintritt, wenn ein Quant mit einem materiellen Teilchen zusammentrifft. Eshandelt sich also um eine principielle Abweichung von der klassischen Theoire. Diese Hypothese ist aber (wie mir scheint) ganz ähnlich wie die Hypothese, die man gewohnlich über die Stabilität der stationaren Zustände macht. Die klassische Theorie lieβe uns erwarten, daβ keine stationären Zustände moglich sind, und daβ stets Wechselwirkung oder Kipplung (Bohr) zwischen Äther und erregtem Atom und damit verbundene Strahlung vorhanden ist. Auf der anderen Seite mussen wir, um die spektrale Emission zu erklaren, den Fall als moglich annehmen, daβ keine Wechselwirkung (Emission) eintritt. So kommen wir dauz, wegen der Persistenz der stationaren Zustande jedem beliebigen stationären Zustand einen Wahrscheinlickkeitskoeffizienten oder eine mittlere Lebensdaur zususchreiben.

Die oben benutzte Hypothese, daβ aich bei einem Zuszmmenstoβ keine Wechselwirkung eintreten kann, ist ganz analog der Annahme der Stabilität der stationären Zustände, die so grundlegend für die Bohrsche Theorie der Linienemission ist, und kann auf die gleiche Ursache zurückgefuhrt werden : Die den materiellen Teilchen innewohnende Wahrscheinlichkeit der Persistenz ihrer stationären Zustände. In

(11) Pauli, *Z.S f. Phys.*, a a 0).

diesem Zusammenhang ist es interessant, zu bemerken, daβ wir beim analogen Fall des Zusammenstoβes eines Elektrons und eines Atoms experimentell Fälle nachweisen konnen, wo bei Zusammenstoβen das Elektron durch das Atom hindurchgeht, ohne die intraatomare Bewegung oder seine eigene zu verandern.

Die Beziehung $g_r\beta = g_s\alpha$.

Die Koeffizienten, oder wie Bohr[12] sie nennt, Gwichte g werden eingefuhrt, wenn Valenzelektron im Atom eine entartet bedingt periodische Bewegung ausfuhrt. Diese numerischen Koeffizienten geben nach Bohr einfach an, auf wieviel verschiedene Weisen man von einer benachbarten nichtentarteten Bewegung zu der speziellen entarteten Bewegung als Grenzfall gelangen kann. Nehmen wir diese Koeffizienten einfach als proportional zur Zahl der Übergangsmöglichkeiten von einem bestimmten zu dem betrachteten Zustand an, so ist die obige Beziehung fast selbstverständlich. Es gibt g_s Wege, auf denen ein Übergang von irgend einem r-ten zu dem betrachteten s-ten Zustand durch Absorption möglich ist. Analog gibt es g_r Übergänge zu niederen Zuständen infolge von Emission. Der Wahrscheinlichkeitskoeffizient hat dann den Faktor g_s der Wahrscheinlichkeitskoeffizient der Emission den Faktor g_r. Wenn wir ferner annehmen, daβ die Wahrschelichkeit einer Transformation durch Strahlung auf einem bestimmten Wege die gleiche ist wie der Wahrscheinlichke-itskoeffizient der Emission, wenn das Atom in bestimmter Weise zu niederen Zuständen übergeht, so ergibt sich sofort die fragliche Relation[13].

Manindra Physikal Laboratorium, Dacca-Universität, 14 Juni 1924.

Ich halte Boses Hypothese uber die Wahrscheinlichkeit der Strahlungselementar-vorgange aus folgenden Grunden fur nicht zutreffend.

Für das statistische Gleichgewicht zwischen einem Bohrschen Zustande zu einem anderen gilt, wie Bose dargelegthat, die Beziehung

$$\frac{n_r}{g_r} \frac{N_\nu}{A_\nu+N_\nu} = \frac{n_s}{g_s}$$

Daraus folgt, daβ die Wahrscheinlichkeiten für die Übergange $r \rightarrow s$ und $s \rightarrow r$ der linken bzw. rechten Seite dieser Gleichung proportional sein müssen. Die Übergangswahrscheinlichkeiten für ein Molekül müssen sich also (wenn wir der Einfachheit halber die statistichen Gewichte beider Zustände gleich 1 setzen) wie $\frac{A_\nu}{A_\nu+N_\nu}$: 1 verhalten. Mehr kann aus der Kenntnis des thermodynamichen Gleichgewichts

(12) Bohr, *Z.S f. Phys.*, a.a.O.

(13) Vgl. in diesem Zusammenhang P. Hertz, *Repert. d. Phys.*, statistische Mechanik, 1. *Bd.*, Teil 2, S. 549.

nicht entonommen werden. Nach der von mir auftestellten Hypothese sollen diese Wahrscheinlichkeiten N_ν (d.h. der Strahlungsdichte) bzw. $A_\nu + N_\nu$ proportional sein, nach der Hypothese Boses $\frac{N_\nu}{A_\nu + N_\nu}$ *bzw.* 1.

Nach der letzteren Hypothese kann die aüßere Strahlung wohl einen Übergang von Zustande Z_r kleinerer Energie nach dem Zustande Z_s größerer Energie bewirken, nicht aber umgekhrt einen Übergang von Z_s nach Z_r. Dies widrespricht aber dem mit Recht allgemein anerkannten Grundsatz, daß die klassische Theorie einen Grenzfall der Quantentheorie darstellen müsse. Nach letzterer kann nämlich ein Strahlungsfeld auf einen Resonator sowohl positive als auch negative Energie übertragen (je nach der Phase), und zwar beides gleich wahrscheinlich. Die Wahrscheinlichkeiten beider Übergange müssen also von der Strahlungsdichte, d.h. von N_ν abhängen, im Gegensatz zu Boses Hypothese. Inwiefern die Qunatentheorie die klassische zum Grenzfall hat, hat Planck in der letzten Auflage seines Buches über Strahlungstheorie ausführlich erötert.

Zweitens mü te nach Boses Hypothese ein kalter Körper ein von der Strahlungsdichte abhängiges (mit ihr abnehmendes) Absorptionsvermögen besitzen. Die Körper sollten in Kaltem Zustande "nicht-Wiensche" Strahlung schwächer absorbieren als weniger intensive aus dem Gültigkeitsbereich der Wienschen Strahlungsformel. Dies wäre bei ultraroter Strahlung heißer Lichtquellen gewiß schon entdeckt worden, wenn es sich so verhielte.

A. Einsten

Reprinted from *Z Physik* **27**, pp. 384-393, 1924 (Springer — Verlag, Heidelberg).

Thermal Equilibrium in the Radiation Field in the Presence of Matter

S. N. Bose in Ramna (India)
(Received 7 July, 1924)

The probability of an arbitrary state of a system consisting of radiation and molecules is calculated and from it the conditions for statistical equilibrium are derived. Further, new expressions for the statistical probability of elementary processes, appropriate for the interaction between radiation and matter, are proposed.

Debye[1] has shown that Planck's law can be derived using statistical mechanics. His derivation is, however, not completely independent of classical electrodynamics, because he uses the concept of normal modes of the ether and assumes that for calculating the energy the spectral range between ν and $\nu + d\nu$ can be replaced by $\frac{8\pi\nu^2}{c^3} V d\nu$ resonators whose energy can be only multiples of $h\nu$. One can however show that the derivation can be so modified that one does not have to borrow anything from the classical theory. $\frac{8\pi\nu^2}{c^3} V d\nu$ can be interpreted as the number of elementary cells in the six dimensional phase space of the quanta. The further calculations remain essentially unchanged.

Einstein has used another method. He considers the interaction between material particles and the radiation field. If one knows the energy distribution among the particles, then one can derive the law of black-body radiation from the condition of stationarity of this distribution when there is an exchange of energy between matter and radiation. This exchange is moreover a result of certain elementary processes. With a suitable formulation of properties and the probability of these elementary processes one can derive Planck's law, if the energy distribution among the particles is assumed to be known. In his first paper[2] the material particles are Bohr atoms which can have only a series of discrete stationary states. The energy exchange occurs because of emission and absorption processes and the simultaneous transitions of atoms from one state to another. He has shown that Planck's law is obtained when the emission and absorption processes have certain probabilities. Recently the problem of thermal equilibrium between radiation and and free electrons has acquired added importance through the studies of Debye[3], Compton[4] and others, which deal with the scattering of radiation off electrons. Pauli[5] has discussed this problem in an interesting paper. As an elementary process he considers the scattering of an electron by a quantum which moves in a definite direction and has a frequency in a given spectral range and is converted into another quantum with a different frequency and moving in a different direction. Simultaneously an electron which moves with a certain velocity in a given direction changes the magnitude and the direction of its velocity. The laws of conservation of energy and momentum hold for all these processes. Pauli has shown that if the probability of this elementary process is $(A\rho + B\rho\rho')\,d\nu$ where ρ and ρ' are radiation densities for frequencies ν and ν', A and B are independent of ρ and ρ' but depend on the nature of collision, this probability law leads to Planck's law, provided Maxwell's distribution is valid for the electron gas. Einstein and Ehrenfest[6] have further generalized Pauli's result. They consider a process in which a molecule or an electron absorbs quanta $h\nu_1$, $h\nu_2$, etc. and simultaneously emits $h\nu_1'$, $h\nu_2'$ ... It is assumed that the emitted and absorbed quanta have definite directions. The magnitude and the direction of the velocity of the molecule under consideration also change at the same time. Energy and momentum are however conserved. The probability law of Pauli is generalized as follows,

$$dW_1 = \Pi b_1 \varrho_1 \, \Pi(a_1' + b_1' \varrho_1')\,dt,$$
$$dW_2 = \Pi(a_1 + b_1 \varrho_1)\, \Pi(b_1' \varrho_1')\,dt,$$

for the direct and the inverse processes. They show that equilibrium is obtained if certain relations between the coefficients exist. This is a generalization of Einstein's earlier result for Bohr atoms and also of Pauli's result.

The problem of thermodynamic equilibrium of radiation in the presence of material particles can, however, be studied using the methods of statistical mechanics, independently of any special assumption about the mechanism of the elementary processes on which the energy exchange depends. In this way we obtain a general relation which is valid for all special assumptions about the elementary processes and their probabilities. If it is possible to write down the thermodynamic probability for any special distribution of quanta of radiation and the probability for any arbitrary energy distribution among the particles, then the thermodynamic probability for the bigger system containing matter and radiation is simply the product of the two probabilities. The condition of equilibrium is simply that the resulting probability must be a maximum. The problem of equilibrium can thus be looked at in a completely different way. We try to derive the relation to which all assumptions must lead when Planck's law for radiation and Maxwell's law for particles hold instead of making assumptions about elementary processes adopted to give Planck's law. The relation which we are looking for is clearly equivalent to the previous one which is derived from the condition that the thermodynamic probability for the bigger system is a maximum. In order to solve the problem in the way indicated, we must have the expressions for the thermodynamic probability for radiation and the material particles.

Thermodynamic probability for radiation :

This has been derived earlier[7]. If $N_\nu\,d\nu$ is the number of quanta in the frequency range ν and $\nu + d\nu$, then the probability is,

$$W = \Pi_{\nu} \frac{(A_\nu + N_\nu d\nu)!}{A_\nu! \, N_\nu d\nu!},$$

where $A_\nu = \frac{8\pi\nu^2}{c^3} V d\nu$. It is easily seen that this probability law leads to Planck's law if we impose the condition that for a given energy W should be a maximum.

Thermodynamic probability for particles :

This can be easily found. We want to generalize our assumptions a little so that the case of the Bohr atom with discrete energy levels as well as the case of the distribution of translational energy of particles can be included. Let the phase space be divided in to cells. For every cell there is a probability g that a particle occupies it. The g's are in general equal except for the case of Bohr's atoms. The thermodynamic probability for any arbitrary distribution n_1, n_2 etc. among the different cell is,

$$\frac{N! \, g_1^{n_1} g_2^{n_2} \ldots}{n_1! \, n_2!}.$$

The condition that the probability be maximum for a given total energy and number of particles is,

$$n_r = C g_r c^{-\frac{E}{kT}}.$$

Now we are in a position to calculate the thermodynamic probability for the total system. It is clearly

$$W = \Pi_{s} \frac{A_s + N_s!}{A_s! \, N_s!} \Pi_{T} \frac{g_T^{n_T} N!}{n_T!},$$

where

$$\Sigma N_s h\nu_s + \Sigma n_T E_T = E$$

and

$$\Sigma n_T = N.$$

At equilibrium W is a maximum, the total energy and number of particles being given.

We consider the following elementary process : n_r changes to $n_r - 1$ and n_s to $n_s + 1$; N_{ν_1}, N_{ν_2}, ... reduce by 1 and $N_{\nu'_1}$, $N_{\nu'_2}$ increase by 1.

The condition of stationarity of W requires that W does not change on account of this process.

The elementary process under consideration can be considered as the transition of a particle from the rth to the sth state. The change which the radiation field undergoes can be looked upon as the result of a scattering after a collision. The required condition reads

$$\frac{n_r}{g_r} \prod_{\nu} \frac{N_{\nu_r}}{N_{\nu_r} + A_{\nu_r}} = \frac{n_s}{g_s} \prod_{\nu'} \frac{N_{\nu'}}{N_{\nu'} + A_{\nu'}}. \tag{I}$$

where

$$\Sigma h \nu' - \Sigma h \nu + E_s - E_r = 0.$$

The equation (I) is evidently equivalent to

$$\frac{n_r}{g_r} \Pi \frac{b_1 \varrho_{\nu_1}}{a_1 + b_1 \varrho_{\nu}} = \frac{n_s}{g_s} \Pi \frac{b_1' \varrho_{\nu}'}{a_1' + b_1' \varrho_{\nu}'}.$$

where

$$\frac{a_1}{b_1} = 8\pi \frac{\nu^2}{c^3} h\nu, \qquad \frac{a_1'}{b_1'} = 8\pi \frac{\nu'^2}{c^3} h\nu'.$$

This can also be written as

$$n_r \Pi b_1 \varrho_{\nu_1} \Pi (a_1' + b_1' \varrho_{\nu}') = n_s \Pi b_1' \varrho_{\nu}' \Pi (a_1 + b_1 \varrho_{\nu}),$$

if

$$g_r = g_s.$$

This is the generalized equation which holds for elementary processes of the type considered by Einstein and Ehrenfest[6]. According to the derivation given here the relation follows from the condition that for equilibrium the probability must be a maximum. However, if the mechanism of the elementary process allows us to write down the equation directly, the substitution of the value for the ratio $\frac{n_r}{n_s}$ gives Planck's law.

It is however clear that the equation (I) is more fundamental than the original one of Einstein. The following discussion will show that simple considerations give the value of the probability constants in their desired form.

The case of Bohr's atoms :

This is the elementary process which Einstein had considered in his earlier paper[2]. The fundamental equation (I) here reduces to

$$\frac{n_r}{g_r} \frac{N_{\nu}}{A_{\nu} + N_{\nu}} = \frac{n_s}{g_s}.$$

Einstein's original equation states

$$n_r b_1 \varrho_{\nu} dt = n_s (a_1' + b_1' \varrho_{\nu}) dt$$

or

$$\varrho_{\nu} = \frac{a_1' n_s}{n_r b_1 - n_s b_1'}.$$

If we assume further that

$$g_r b_1 = g_s b_1' \quad \text{and} \quad \frac{a_1'}{b_1'} = 8\pi \frac{h\nu^3}{c^3},$$

then Planck's law follows :

To derive his equation (I), Einstein assumes that atoms in higher energy levels make transitions to lower energy levels in two ways :

1. A type of spontaneous transition as in a radioactive process, whose probability is independent of the state of the radiation field.
2. A transition induced by the characteristic radiation, whose probability depends on the state of the radiation field.

The transition from a lower to a higher energy level is the result of an absorption whose probability depends on the radiation field. Further, certain relations between the coefficients are postulated so that Planck's law follows. These relations cannot be justified in a simple way.

The same problem can be treated in the following way removing many of the arbitrary assumptions.

We also presume that the transition from a lower to a higher state takes place through an absorption of a quantum of characteristic frequency.

A. The transition from higher to lower levels is a spontaneous change, whose probability is independent of the field of radiation. The second assumption of Einstein (negative radiation) is not necessary.

B. We shall try to calculate the probability coefficients for the transition A.

The number of cells corresponding to the quanta under consideration is A_ν. The total number of quanta present is $N_\nu\, d\nu$. They are however not distributed evenly in phase space. Let p_o be the number of empty calls, p_1 with 1 quantum, p_2 with two quanta etc. In order that the radiation and a particle may interact, it must be in a cell occupied by a quantum. The particular cell which the molecule under consideration occupies will pass through all possible states if we observe it long enough, that is, sometimes it is empty, sometimes occupied by 1 quantum sometimes by 2 quanta etc. The length of these events will be finally proportional to the equilibrium values of p_o, p_1. When r quanta and a particle are together, then either an exchange of energy takes place or nothing happens. Therefore $r + 1$ different events are possible, namely energy exchange of 1 quantum, 2 quanta etc. or no exchange. Therefore the number of possible cases is

$$p_0 + 2\,p_1 + 3\,p_2 + \cdots = A_s + N_s\, d\,\nu_s = \Sigma\,(r+1)\,p_r.$$

The number of cases in which interactions occur is,

$$p_1 + 2\,p_2 + \cdots = N_s\, d\,\nu_s = \Sigma\, r\,p_r.$$

Consequently the probability of an interaction is,

$$\frac{\Sigma\, r\,p_r}{\Sigma\,(r+1)\,p_r} = \frac{N_s\, d\,\nu_s}{A_s + N_s\, d\,\nu_s}.$$

Now we must take into consideration the special nature of the interaction, that is, absorption which is to be distinguished from scattering. Let the probability for this be β. The probability for absorption is

$$\frac{\beta \Sigma r p_r}{\Sigma (r+1) p_r} = \beta \frac{N_\nu d\nu}{A_\nu + N_\nu d\nu}.$$

The probability coefficient for a transition of type B is evidently any constant which determines the transition from the particular higher state to a particular lower state. Let this constant be α.

The equilibrium condition can then be written as

$$n_r \beta \frac{N_\nu d\nu}{A_\nu + N_\nu d\nu} = \alpha n_s,$$

which agrees with the fundamental equation (I) if we assume that

$$g_r \beta = g_s \alpha.$$

It is clear that substituting the value of $\frac{n_r}{n_s}$,.one gets Planck's law.

Pauli's case.

This can be easily discussed. The probability of an interaction is as before $\frac{\Sigma r p_r}{\Sigma (r+1) p_r}$. Let $\beta^{s}_{s'}$ be the probability coefficient for scattering from ν_s to $\nu_{s'}$ if there is an interaction. Let the corresponding coefficients for the inverse process $\frac{\Sigma r p_{r'}}{\Sigma (r+1) p_{r'}}$ and $\beta^{s'}_{s}$. Then at equilibrium

$$n_r \beta^{s}_{s'} \frac{N_s d\nu_s}{A_s + N_s d\nu_s} = n_s \beta^{s'}_{s} \frac{N_{s'} d\nu_{s'}}{A_{s'} + N_{s'} d\nu_{s'}}.$$

Let us assume further that $\beta^{s}_{s'} = \beta^{s'}_{s}$; then we obtain the fundamental equation (I).

Einstein and Ehrenfest's case.

The probability of a simultaneous interaction is the product of individual probabilities and therefore equal to

$$\Pi \frac{N_s d\nu_s}{A_s + N_{s'} d\nu_s}.$$

Analogously let $\beta^{s_1\, s_2 \ldots}_{s_1',\, s_2', \ldots}$ be the probability for scattering and let a similar expression hold for the inverse scattering. The rest then easily follows.

Remarks.

Our calculations show that the probability $\frac{N_\nu\, d\nu}{A_\nu + N_\nu\, d\nu}$ of an interaction between a particle and a quantum in a radiation field is not simply proportional to the number of the quanta present as one would presume at first sight. It is easy to see that this assumption leads to Wien's law, as realized by Pauli[5]. He found it necessary to assume another form for the probability coefficients. However the form assumed by Pauli and generalized by Einstein and Ehrenfest appears to be completely arbitrary because one cannot easily see how such an expression can be derived. The form suggested here is quite simple and can be justified on the basis of elementary considerations. The necessity of assuming relations between the coefficients themselves is also avoided. It was assumed in deriving the probability coefficients for the interaction (or coupling, as Bohr says) that even in a collision no interaction is as probable as the occurence of any special interaction. This assumption is a fundamental point in the derivation given here. From the classical theory one would expect that there is some interaction whenever a quantum and a material particle come together. Therefore here it is a question of a departure from classical theory. This hypothesis is (as it appears to me), however, very similar to the assumption one usually makes about the stability of stationary states. From classical theory one would expect that no stationary states are possible and that an interaction or coupling (Bohr) between ether and the excited atom and the radiation connected with it is always present. On the other hand, in order to explain spectral emission we must assume it to be possible that no interaction (emission) occurs. Because of the persistence of stationary states we are therefore led to assign to every stationary state a probability coefficient or a mean lifetime.

The assumption made above, that even in a collision no interaction need occur, is quite analogous to the assumption of the stability of stationary states which is so fundamental to Bohr's theory of line emission and can be traced back to the same origin—the probability of existence of the stationary states of the particles. It is interesting to remark in this connection that in the analogous case of a collision of an electron with an atom, experiments show that the electron goes through the atom without changing either the interatomic or its own motion.

The relation $g_r\ \beta = g_s\ \alpha$.

The coefficients or as Bohr calls them, the weights g, are introduced when the valence electron in the atom executes a finite degenerate periodic motion. These numerical coefficients give, according to Bohr, in how many ways from a neighbouring non-degenerate orbit one can arrive at the special degenerate orbit as the limiting case. Let us assume that these coefficients are simply proportional to the number of possibilities of transition from a definite state to the state under consideration ; then the above relation is almost obvious. There are g_s possible ways for a transition from an rth state to the given sth state through absorption. Similarly there are g_r transitions to the lower states as a result of emission. The probability coefficient β has then the factor g_s, the probability coefficient for emission the factor g_r. If we further assume that the probability of a transformation through radiation in a definite way is the same as the probability coefficient for emission when the atom makes transitions to lower states in a definite way, then the relation in question follows immediately[9].

Manindra Physical Laboratory, Dacca University, June 14, 1924.

I consider Bose's hypothesis about the probability of elementary radiation precesses as not appropriate for the following reasons.

For the statistical equilibrium between two states of a Bohr atom the following relation holds, as given by Bose :

$$\frac{n_r}{g_r}\frac{N_\nu}{A_\nu + N_\nu} = \frac{n_s}{g_s}.$$

It follows that the probabilities for the transitions $r \rightarrow s$ and $s \rightarrow r$ respectively on the left and the right hand side of the equation must be proportional to each other. The transition probabilities for a molecule must therefore (if for simplicity we put the statistical weights of both the states equal to 1) behave as $\frac{N_\nu}{A_\nu + N_\nu}$: 1. Nothing more can be learnt from the knowledge of the thermodynamic equilibrium. According to my assumptions these probabilities should be proportional to N_ν (i.e. the radiation density) and $A_\nu + N_\nu$ respectively, whereas according to Bose's assumption, $\frac{N_\nu}{A_\nu + N_\nu}$ and 1 respectively.

According to the latter assumption the external radiatian can indeed cause a transition from a state Z_r of lower energy to a state Z_s of higher energy, but not the reverse transition from Z_s to Z_r. This, however, contradicts the generally and rightly accepted fundamental principle that the classical theory should be a limiting case of the quantum theory. According to the latter a radiation field can transfer to a resonator positive as well as negative energy (depending on the phase) and indeed with equal probability. The probabilities of both the transitions must depend on the density of radiation, that is, on N_ν as oppsed to Bose's hypothesis. Planck has discussed in detail in the latest edition of his book on the theory of radiation to what extent the classical theory is the limiting case of the quantum theory.

Secondly, according to Bose's hypothesis a cold body should have an absorbing capacity dependent on the density of radiation (decreasing with it). Bodies in cold state should absorb "non- Wien" radiation to a weaker extent than less intense radiation as defined by the range of validity of Wien's radiation formula. If the behaviour had been such, then it would certainly have already been discovered in infra-red radiation from hot light sources.

A. Einstein

References

(rearranged sequentially)

1. Debye, *Ann. d. Phys.* **33**, 1427, 1910
2. Einstein, *Phys. ZS.* **18**, 121, 1917.
3. Debye, *Phys. Zs.* **24**, 161, 1923.
4. Compton, *Phys. Rev.* **21**, 483, 1923.

5. Pauli, *ZS. f. Phys.* **18**, 272, 1923.
6. Einstein and Ehrenfest, *ZS. f. Phys.* **19**, 301, 1923.
7. The author, to appear in Phil. Mag. See also Debye, *Ann. d. Phys.* **33**, 1427, (1910).
8. Bohr, *ZS. f. Phys.* **13**, 117, 1923.
9. See in this connection P. Hertz, *Repert. d. Phys., Statistiche Mechanik*, Vol. 1, Part 2, p. 549.

English translation of *Z Physik* 27, pp. 384-393, 1924 (Springer — Verlag, Heidelberg).

A NOTE ON DIRAC EQUATIONS AND THE ZEEMAN EFFECT

By S. N. BOSE
AND
K. BASU

(*Receieed for publication, Sept. 22, 1943*)

ABSTRACT. A new treatement has been given for solving Dirac's equations for hydrogenic atoms, and the radial functions are expressed in terms of a combination of Sonine's polynomials $T_n^{(\mu)}(\rho)$, $T_{n-1}^{(\mu)}(\rho)$ of only two consecutive degrees n, $n-1$; and the elementary properties of such polynomials have enabled us to tackle the Zeeman effect problem in general (homogeneous field) leading to the standard quadratic equation in energy for the effect.

1. With the help of the two-dimensional matrices s_x, s_y, s_z, of Pauli the wave-equations of Dirac can be put in the well-known matrix form :

$$\frac{\iota}{hc}\left[E+E_0+\frac{Ze^2}{r}\right]X+DY=0,\quad \frac{\iota}{hc}\left[E-E_0+\frac{Ze^2}{r}\right]Y+DX=0 \tag{1.1}$$

where D is the operator $\quad s_x\frac{\partial}{\partial x}+s_y\frac{\partial}{\partial y}+s_z\frac{\partial}{\partial z}.$

If, similarly $S=xs_x+ys_y+zs_z$, and $s=\frac{I}{r}(xs_x+ys_y+zs_y)$, then

$S.D.=r\frac{d}{dr}+L$, where $L=\iota\,(M_xs_x+M_ys_y+M_zs_z)$.

And the following commutation rules can be easily deduced :

$$s(L-1)+(L-I)s=0,\ D(L-1)+(L-1)D=0;$$

also $\quad s^2\equiv 1.$

Hence multiplying the equations on the left by S we have

$$\frac{\iota}{hc}\left[E+E_0+\frac{Ze^2}{r}\right]SX+\left[r\frac{d}{dr}+1\right]Y+(L-1)Y=0; \tag{1.2}$$

$$\frac{\iota}{hc}\left[E-E_0+\frac{Ze^2}{r}\right]SY+\left[r\frac{d}{dr}+1\right]X+(L-1)X=0.$$

Note on Dirac Equations and Zeeman Effect

Assuming $X \equiv \begin{vmatrix} u_1 \\ u_2 \end{vmatrix} = \frac{f(r)}{r}\Phi$; $Y \equiv \begin{vmatrix} u_3 \\ u_4 \end{vmatrix} = \frac{g(r)}{r}\Psi$ where Φ, ψ are similar one-column matrices and functions of (θ, ϕ), the equations can be rewritten as

$$\frac{\iota}{\hbar c} f(r)\left[E+E_0+\frac{Ze^2}{r}\right]s\Phi+\frac{dg(r)}{dr}\cdot\Psi+\frac{g(r)}{r}(L-1)\Psi = 0, \tag{1.3}$$

$$\frac{\iota}{\hbar c} g(r)\left[E-E_0+\frac{Ze^2}{r}\right]s\Psi+\frac{df(r)}{dr}\cdot\Phi+\frac{f(r)}{r}(L-1)\Phi = 0.$$

The equations become easily separable if the matrices Φ and Ψ are so chosen that at first $(L-1)\Psi = k\Psi$ and $s\Phi = \Psi$; and as $s^2 \equiv 1$, it follows therefore from commutation rules that

$$\Phi = s\Psi \text{ and } (L-1)s\Psi = -ks\Psi.$$

We observe at once that $L(L-1)\ \Psi = -\nabla_2\Psi = k(k+1)\Psi$, where

$$\nabla_2 \equiv \frac{1}{\sin\theta}\frac{\partial}{\partial\theta}\left(\sin\theta\frac{\partial}{\partial\theta}\right)+\frac{1}{\sin^2\theta}\frac{\partial^2}{\partial\phi^2};$$

and therefore k can be either a positive or a negative integer.

Secondly, if (Φ, Ψ) are the matrices for positive k, then (Ψ,Φ) are the matrices for negative k. Also remembering that the operator

$-\left[\iota\left(x\frac{\partial}{\partial y}-y\frac{\partial}{\partial x}\right)+\frac{1}{2}s_z\right]$ commutes with the equation-system, the angular matrices can be expressed in terms of spherical harmonics of order k and $k-1$ in the following form :

$$\Phi \equiv \begin{vmatrix} \sqrt{\frac{k+m}{2k-1}}\,Y^{\mu}_{k-1} \\ \sqrt{\frac{k-m-1}{2k-1}}\,Y^{\mu+1}_{k+1} \end{vmatrix},\ \Psi \equiv \begin{vmatrix} \sqrt{\frac{k-m}{2k+1}}\,Y^{\mu}_{k} \\ \sqrt{\frac{k+m+1}{2k+1}}\,Y^{\mu+1}_{k} \end{vmatrix} \quad (k\text{, a positive integer}) \tag{1.4}$$

where $Y^{\mu}_{k} \equiv \sqrt{\frac{2k+1}{2}\cdot\frac{(k-\mu)!}{(k+\mu)!}}\,\frac{1}{2^k k!}\,\frac{d^{k+\mu}}{dx^{k+\mu}}(x^2-1)^k\,\frac{e^{i\mu\phi}}{\sqrt{2\pi}}.\ (x=\cos\theta)$

The functions are interchanged for negative values of k. Writing $f(r) = \iota F(r)$ so as to remove the imaginary from the radial equations, we get

$$\left[\frac{dG}{dr}+k\frac{G}{r}\right]-\frac{1}{hc}\left[E+E_0+\frac{Ze^2}{r}\right]F=0,$$

$$\left[\frac{dF}{dr}-k\frac{F}{r}\right]+\frac{1}{hc}\left[E-E_0+\frac{Ze^2}{r}\right]G=0. \qquad (1.5)$$

2. To solve the equation we assume

$$F = f_0 e^{-\lambda r} r^\mu F_1(r),\ G = g_0 e^{-\lambda r} r^\mu G_1(r),\ \text{and}\ N = aE/(E_0^2-E^2)^{\frac{1}{2}} \qquad \ldots \quad (2.1)$$

where $Ze^2/hc = a$.

Substituting in (1.5), F_1 and G_1 are easily seen to satisfy the following equations :

$$\frac{dG_1}{dr}-\lambda G_1+\lambda F_1+\frac{k+\mu}{r}G_1+\frac{N}{E}(E_0-E)\frac{F_1}{r}=0, \qquad (2.2)$$

$$\frac{dF_1}{dr}-\lambda F_1+\lambda G_1-\frac{k-\mu}{r}F_1-\frac{N}{E}(E_0+E)\frac{G_1}{r}=0,$$

where $f_0/g_0 = -(E_0-E)/hc\lambda = \dfrac{\lambda hc}{E_0+E} = -\left(\dfrac{E_0-E}{E_0+E}\right)^{\frac{1}{2}}$ and $\lambda = \left(\dfrac{E_0^2-E^2}{hc}\right)^{\frac{1}{2}}$ (2.3)

If $F_1+G_1 = 2\chi^+$, $G_1-F_1 = 2\chi^-$, elimination leads easily to the following equations :

$$r\frac{d^2\chi^+}{dr^2}+(2\mu+1-2\lambda r)\frac{d\chi^+}{dr}+2\lambda(N-\mu)\chi^+=0, \qquad (2.4)$$

$$r\frac{d\chi^+}{dr}-(N-\mu)\chi^+=\left(N\frac{E_0}{E}-k\right)\chi^-, \qquad (2.5)$$

provided $\quad k^2-\mu^2 = N^2(E_0^2-E^2)/E^2 = \alpha^2.$

Taking the first equation it can be easily seen that it admits of polynomial solutions if $N-\mu = n$ (an integer). Writing ρ for $2\lambda r$, and writing a Sonine polynomial in the form

$$T_n^{(\mu)}(\rho) = B_n^{(\mu)}\left[\rho^n-\frac{n(n+2\mu)}{1\,!}\rho^{n-1}+\frac{n(n-1)(n+2\mu)(n+2\mu-1)}{2!}\rho^{n-2}-\ldots\right]$$

$$= B_n^{(\mu)}\cdot(-1)^n e^\rho \rho^{-2\mu}\frac{d^n}{d\rho^n}\left[e^{-\rho}\rho^{n+2\mu}\right] \qquad (2.6)$$

with constant $B_n^{(\mu)}$ so chosen that it is normalized according to

$$\int_0^\infty e^{-\rho}\rho^{2\mu}\, T_n^{(\mu)}(\rho)T_n^{(\mu)}(\rho)d\rho = 1$$

we get $B_n^{(\mu)} = 1/\{n\,!\;\Gamma(n+2\mu+1)\}^{\frac{1}{2}}$

Hence it easily follows that χ^+ and χ^- will have solutions as given below :

$$\chi^+ = \delta\left(N\frac{E_0}{E} - k\right)^{\frac{1}{2}} T_n^{(\mu)}(\rho) \tag{2.7}$$

$$\left[\, N = n+\mu = \frac{\alpha\epsilon}{\sqrt{1-\epsilon^2}}\;;\;\epsilon = \frac{E}{E_0}\,\right]$$

$$\chi^- = \delta\left(N\frac{E_0}{E} + k\right)^{\frac{1}{2}} T_{n-1}^{(\mu)}(\rho) \tag{2.8}$$

where the normalising factor δ is to be determined presently.

Our radial functions of eqs. (2.1) stand thus

$$F = -\delta\sqrt{\left(1-\frac{E}{E_0}\right)}\; e^{-\frac{\rho}{2}}\rho^{\mu}\left[\left(N\frac{E_0}{E}-k\right)^{\frac{1}{2}}T_n^{(\mu)}(\rho) - \left(N\frac{E_0}{E}+k\right)^{\frac{1}{2}}T_{n-1}^{(\mu)}(\rho)\right] \tag{2.9}$$

$$G = \delta\sqrt{\left(1+\frac{E_0}{E}\right)}\; e^{-\frac{\rho}{2}}\rho^{\mu}\left[\left(N\frac{E_0}{E}-k\right)^{\frac{1}{2}}T_n^{(\mu)}(\rho) + \left(N\frac{E_0}{E}+k\right)^{\frac{1}{2}}T_{n-1}^{(\mu)}(\rho)\right] \tag{2.10}$$

Normalisation requires

$$\int_0^\infty (F^2+G^2)dr = 1,$$

which gives $$\delta = \sqrt{\frac{\lambda}{2N}\frac{E}{E_0}}\,.$$

We can write the two solutions (A) and (B), corresponding to positive and negative values of k respectively.

$$\left[\begin{array}{ll} u_1 = \sqrt{\dfrac{k+\mu}{2k-1}}\,.\,\iota\,\dfrac{f_+(r)}{r}\,Y_{k-1}^{\mu}, & u_3 = \sqrt{\dfrac{k-\mu}{2k+1}}\,\dfrac{g_+(r)}{r}\,Y_k^{\mu} \\ u_2 = -\sqrt{\dfrac{k-\mu-1}{2k-1}}\,\iota\,\dfrac{f_+(r)}{r}\,Y_{k-1}^{\mu+1}, & u_4 = \sqrt{\dfrac{k+\mu+1}{2k+1}}\,\dfrac{g_+(r)}{r}\,Y_k^{\mu+1} \end{array}\right] \tag{A}$$

$$\begin{bmatrix} v_1 = \sqrt{\frac{k-\mu}{2k+1}}\,\iota\frac{f_-(r)}{r}Y_k^\mu & v_3 = \sqrt{\frac{k+\mu}{2k-1}}\,\frac{g_-(r)}{r}Y_{k-1}^\mu \\ v_2 = \sqrt{\frac{k+\mu+1}{2k+1}}\,\iota\frac{f_-(r)}{r}Y_k^{\mu+1}, & v_4 = -\sqrt{\frac{k-\mu-1}{2k-1}}\,\frac{g_-(r)}{r}Y_{k-1}^{\mu+1} \end{bmatrix} \quad \text{(B)}$$

wherein it is understood that f_- and g_- are obtained by changing the sign of k, in the expressions for F and G in (1.5) above.

3. When the atom is in an electromagnetic field defined by the vector potential $(A_x A_y A_z)$, the Dirac-equations become

$$\frac{\iota}{hc}\left[E+E_0+\frac{Ze^2}{r}\right]X+DY+\frac{e}{hc}AY=0 \quad (3.1)$$

$$\frac{\iota}{hc}\left[E-E_0+\frac{Ze^2}{r}\right]Y+DX+\frac{e}{hc}AX=0 \quad (3.2)$$

where $$A \equiv \iota[A_x s_x + A_y s_y + A_z s_z]$$

In the case of a constant magnetic field H, in the direction of Z-axis

$$A \equiv \tfrac{1}{2}Hr \begin{vmatrix} 0 & e^{-i\mu\phi}\sin\theta \\ -e^{i\mu\phi}\sin\theta & 0 \end{vmatrix} \equiv \tfrac{1}{2}Hra. \quad (3.3)$$

We observe in passing that in the absence of A, the wave function X is generally small compared with Y—the radial component of $F(r)$ has the factor $\sqrt{1-\frac{E}{E_0}}$, while $G(r)$ has $\sqrt{1+\frac{E}{E_0}}$; so that the perturbation effect of $\frac{e}{hc}AX$ is small as compared with effect due to the term $\frac{e}{hc}AY$ in eq. (3.1)

Remembering

$$e^{-i\phi}Y_k^{\mu+1}\sin\theta = -\sqrt{\frac{(k-\mu)(k-\mu+1)}{(2k+1)(2k+3)}}\,Y_{k+1}^\mu + \sqrt{\frac{(k+\mu)(k+\mu+1)}{(2k+1)(2k-1)}}\,Y_{k-1}^\mu\,;$$

$$e^{i\phi}Y_k^\mu\sin\theta = \sqrt{\frac{(k+\mu+1)(k+\mu+2)}{(2k+1)(2k+3)}}\,Y_{k+1}^{\mu+1} - \sqrt{\frac{(k-\mu)(k-\mu-1)}{(2k-1)(2k+1)}}\,Y_{k-1}^{\mu+1}\,;$$

we seek an approximate solution of the equation by choosing one set of angular function $(Y_k^\mu, Y_k^{\mu+1})$, and assuming the existence of both the sets $(Y_{k-1}^\mu, Y_{k-1}^{\mu+1})$ and $(Y_{k+1}^\mu, Y_{k+1}^{\mu+1})$ in X. This can be done by suitably combining the (A) and (B) types, thus

Take $u_j = C_l^+ u_j^+ + C_{-(l+1)}^- v_j^-$, $(j = 1, 2, 3, 4)$, where $(u_1^+, u_2^+, u_3^+, u_4^+)$ corresponds to $k = +l$ and $(v_1^-, v_2^-, v_3^-, v_4^-)$ corresponds to $k = -(l+1)$. More explicitly, their values are

$$u_1^+ = \sqrt{\frac{l+\mu}{2l-1}}\ \iota\frac{f_+(r)}{r} Y_{l-1}^\mu; \qquad v_1^- = \sqrt{\frac{l-\mu+1}{2l+3}}\ Y_{l+}^\mu\ \iota\frac{f_-(r)}{r};$$

$$u_1^+ = -\sqrt{\frac{l-\mu-1}{2l-1}}\ \iota\frac{f_+(r)}{r} Y_{l-1}^{\mu+1}; \quad v^- = \sqrt{\frac{l+\mu+2}{2l+3}}\ Y_{l+1}^{\mu+1}\iota\frac{f_-(r)}{r};$$

$$u_3^+ = \sqrt{\frac{l-\mu}{2l+1}}\ \frac{g_+(r)}{r} Y_l^\mu; \qquad v_3^- = \sqrt{\frac{l+\mu+1}{2l+1}}\ \frac{g_-(r)}{r} Y_l^\mu;$$

$$u_1^+ = \sqrt{\frac{l+\mu+1}{2l+1}}\ \frac{g_+(r)}{r} Y_l^{\mu+1}; \quad v_4^- = -\sqrt{\frac{l-\mu}{2l+1}}\ \frac{g_-(r)}{r}\ Y_l^{\mu+1}.$$

The constants λ, N, μ of the two types of solutions are different and are expressed by the following relations

$$\lambda_+ = (E_0^2 - E_+^2)^{\frac{1}{2}}/hc, \quad l^2 - \alpha^2 = \mu_+^2, \quad N_+ = \alpha E_+/[E_0^2 - E_+^2]^{\frac{1}{2}}, \tag{3.4}$$

and $\qquad N_+ = n_+ + \mu_+$[for (A)$-$type with $k = l$].

$$\lambda_- = (E_0^2 - E_-^2)^{\frac{1}{2}}/hc), \quad (l+1)^2 - a^2 = \mu_-^2, \quad N_- = aE_-/[E_0^2 - E_0^2]^{\frac{1}{2}}, \tag{3.5}$$

and $\qquad N_- = n_- + \mu_-$[for $(B)$$-$type with $k = -(l+1)$].

If E_+ and E_- differ slightly from one another, we have the following approximate relations

$$\lambda_+ = \lambda_-, \ \mu_- = \mu_+ + 1, \ N_+ = N_-, \ n_- = n_+ - 1, \text{ and } \frac{E_+}{E_0} \sim 1. \tag{3.6}$$

By following the usual method of perturbation we see easily that the characteristic equation for determining the Eigen value E would be

$$\begin{vmatrix} \frac{\iota}{h_c}\left[E - E_+\right] + pa_{11} & pa_{12} \\ pa_{21} & \frac{\iota}{h_c}\left[E - E_-\right] + pa_{22} \end{vmatrix} = 0 \tag{3.7}$$

where $p = eH/2hc$, and

$$a_{11} = \int [\bar{X}_+^* r\boldsymbol{a} Y_+ + Y_+^* r\boldsymbol{a} X_+] dV$$

$$a_{22} = \int [\bar{X}_-^* r\boldsymbol{a} Y_- + \bar{Y}_-^* r\boldsymbol{a} X_-] dV$$

$$a_{12} = \int [\bar{X}_+^* r\boldsymbol{a} Y_- + \bar{Y}_-^* r\boldsymbol{a} X_-] dV \tag{3.8}$$

$$a_{21} = \int [\bar{X}_-^* r\boldsymbol{a} Y_+ + \bar{Y}_-^* r\boldsymbol{a} X_+] dV$$

and $\bar{X}_+^*$ etc. are transposed conjugates of X_+ (the $\boldsymbol{a}$ is given in (3.3))

Carrying out integrations over the angle-variable-space we get

$$a_{11} = -\frac{2\iota(4\mu+2)l}{(2l-1)(2l+1)} \int_0^\infty rf_+g_+dr \ ;$$

$$a_{22} = \frac{2\iota(4\mu+2)(l+1)}{(2l+1)(2l+3)} \int_0^\infty rf_-g_-dr \ ;$$

$$a_{12} = a_{21} = \frac{\iota\sqrt{l(-\mu)(l+\mu+1)}}{2l+1} \int_0^\infty r(f_-g_+ + f_+g_-)\, dr.$$

4. To evaluate the integral {I} or {II}, we substitute the values of the corresponding f, g, and remembering $r = \rho/2\lambda$, we obtain

$$\{I\} = -\frac{\lambda_+}{2N_+}\frac{E_+}{E_0^2}\frac{(E_0^2-E_+^2)^{\frac{1}{2}}}{4\lambda^2}\left\{\left(N_+\frac{E_0}{E_+}-k\right)\int_0^\infty e^{-\rho}\rho^{2\mu+1}\left[T_n^{(\mu)}(\rho)\right]^2 d\rho\right.$$

$$\left.-\left(N_+\frac{E_0}{E_+}+k\right)\int_0^\infty e^{-\rho}\rho^{2\mu+1}\left[T_{n-1}^{(\mu)}(\rho)\right]^2 d\rho\right\}.$$

The integrals are evaluated quickly by repeated partial integration; thus

$$\int_0^\infty e^{-\rho}\rho^{2\mu+1}\Big]\, T_n^{(\mu)}(\rho)\,\Big]^2 d\rho = \frac{(-)^n}{\sqrt{n\,!\,\Gamma(n+2\mu+1)}}\int_0^\infty \rho T_n^{(\mu)}(\rho)\frac{d^n}{d\rho}\left[e^{-\rho}\rho^{2\mu+n}\right] d\rho$$

$$=\int_0^\infty \frac{e^{-\rho}}{n!}\frac{\rho^{2\mu+n}}{\Gamma(n+2\mu+1)}\,d\rho.\frac{d^n}{d\rho^n}\left[\rho\left(\rho^n-\frac{n(n+2\mu)}{1!}\rho^{n-1}+\dots\right)\right]$$

$$=(n+1)(n+2\mu+1)-n(n+2\mu) = 2N_+ + 1.$$

Similarly

Note on Dirac Equations and Zeeman Effect

$$\int_0^\infty e^{-\rho}\rho^{2\mu+1}\Big[\,T^{(\mu)}_{n-1}\,(\rho)\Big]^2 d\rho = 2N_+-1.$$

An easy substitution of the value of λ_+ in the foregoing leads to the following results :

$$\int_0^\infty rf_+g_+dr = -\frac{h_c}{4E_0}\left(1-2l\,\frac{E_0}{E_+}\right); \tag{4.1}$$

$$\int_0^\infty rf_-g_-dr = -\frac{h_c}{4E_0}\left(1+2(l+1)\,\frac{E_0}{E_-}\right). \tag{4.2}$$

The evaluation of the third integral cannot, however, be exactly expressed in a neat form ; for the simple reason that λ_+, N_+, E_+ as well as λ_-, N_-, E_- are different in the two solutions, as also $\rho_+ = 2\lambda_+ r$ and $\rho_- = 2\lambda_- r$. If however, we introduce approximations in the beginning, and take

$$\lambda_+ = \lambda_- = |\lambda|,\ E_+ = E_-,\ \ \mu_- = \mu_+ + 1,\ \ n_- = n_+ - 1,\ \ \rho_+ = \rho_-, \tag{4.3}$$

we see the third integral

$$\{\text{III}\} = -\frac{2|\lambda|}{2N}\,\frac{|E|}{E_0^2}\,\frac{(E_0-|E|^2)^{\frac{1}{2}}}{4|\lambda|^2}\Big\{\Big(N\frac{E_0}{E}-l\Big)^{\frac{1}{2}}\Big(N\frac{E_0}{E}+l+1\Big)^{\frac{1}{2}}\times$$

$$\int_0^\infty e^{-\rho}\rho^{2\mu+2}T^{(\mu)}_n(\rho)T^{(\mu+1)}_{n-1}(\rho)d\rho-\Big(N\frac{E_0}{E}+l\Big)^{\frac{1}{2}}\Big(N\frac{E_0}{E}-l-1\Big)^{\frac{1}{2}}$$

$$\times\int_0^\infty e^{-\rho}\rho^{2\mu+2}T^{(\mu)}_n(\rho)\,T^{(\mu+2)}_{n-2}(\rho)d\rho\Big\}.$$

The two definite integrals can be easily evaluated in the same way as before. We give the results below ;

$$\int_0^\infty e^{-\rho}\rho^{2\mu+2}\,T^{(\mu)}_n(\rho)\,T^{(\mu+1)}_{n-1}(\rho)d\rho = \int_0^\infty \frac{e^{-\rho}\rho^{n+2\mu}}{\sqrt{n-1!n\,!\;\Gamma(n+2\mu+1)\;\Gamma(n+2\mu+2)}} \tag{4.4}$$

$$\times\,\frac{d^n}{d\rho^n}\Big[\,\rho^{n+1}-(n-1)(n+2\mu+1)\rho^n+\dots\,\Big] = 2\sqrt{(N-l)(N+l+1)}.$$

$$\int_0^\infty e^{-\rho}\rho^{2\mu+2}T^{(\mu)}_{n-2}(\rho)T^{(\mu+1)}_{n-}\,d\rho = 2\sqrt{(N+l)(N-l-1)}. \tag{4.5}$$

Hence

$$\{III\} = -\frac{1}{2N}\frac{h_c}{E_0}\left[\sqrt{N\frac{E_0}{E}-l}\sqrt{N\frac{E_0}{E}+l+1}\sqrt{N-l}\sqrt{N+l+1}\right.$$

$$\left.-\sqrt{N\frac{E_0}{E}+l}\sqrt{N\frac{E_0}{E}-l-1}\sqrt{N+l}\sqrt{N-l-1}\right].$$

Observing $E_0/E \sim 1$, we see

$$\{III\} = -\frac{1}{2N}\cdot\frac{h_c}{E_0}2N = -\frac{h_c}{E_0}.$$

Finally, putting $p = eH/2h_c$, and $eh/2m_0c = \mu_0$ (Bohr-magneton) and substituting the values of the three integrals (and making $E_0/E \sim 1$) we see that the determinantal equation (3.7) takes the form :

$$\begin{vmatrix} E-E_+-\frac{2l}{2l+1}m\mu_0H & -\frac{\mu_0H}{2l+1}\sqrt{(l-m+\frac{1}{2})(l+m+\frac{1}{2})} \\ -\frac{\mu_0H}{2l+1}\sqrt{(l-m+\frac{1}{2})(l+m+\frac{1}{2})} & E-E_--\frac{2l+2}{2l+1}m\mu_0H \end{vmatrix} = 0 \qquad (4.6)$$

wherein we have put $\mu+\frac{1}{2} = m$ (magnetic quantum number).

The result (4.6) agrees completely with that quoted by Bethe (1933), supposed to have been worked out by him from Pauli's equations. Condon and Shortley (1935) obtained similar determinant from principles of quantum mechanics by applying two-fold perturbations (spin-orbit and magnetic) simultaneously. It may be noted that Darwin (1928) has many similar features with our mode of attack, and we can claim some elegance by our introduction of Sonines properties, which exhibit our solutions in a good perspective.

DACCA UNIVERSITY
RAMNA, DACCA.

CALCUTTA UNIVERSITY,
92, UPPER CIRCULAR ROAD, CALCUTTA.

REFERENCES

Bethe, H., 1933, *Handbuch der Physik*, **24** p 396.

Condon, E. U. and Shortley. G. H., 1935. *The Theory of Atomic Spectra* (Cambridge University Press), p. 233.

Darwin G. C., 1928, *Proc. Roy. Soc.*, London (A), **118**. p. 654-680.

Reprinted from *Ind J Phys* **17**, pp. 301-308, 1943 (Indian Association for the Cultivation of Science).

C. The Ionosphere — Mathematical Physics

Propagation of electromagnetic waves in the ionosphere encounters a complex refractive index due to the absorbing nature of the medium. Moreover, the refractive index is dependent upon not only the properties of the medium but also on the frequency and damping as the waves progress through the medium. Various theories were developed during the 1930s to explain the observed features.

One of the anomalous features was that the value of the dielectric constant of the ionized medium could exceed unity. Dispersion theory was shown to allow such an anomaly, that it could be greater or less or equal to unity depending upon the frequency and the degree of ionization in the medium. In the first paper in this section, Bose along with S.R. Khastgir, present a careful analysis of the conditions combined with a laboratory experiment of an artificial ionosphere created in a discharge tube. They show that dispersion theory alone cannot explain the experimental results.

In the second paper, Bose discusses the conditions for total reflection of the waves as they progress through the upper atmosphere. He finds that the proposed conditions — that the complex refractive index goes to zero or the group velocity vanish — ignore damping as a first approximation. However, damping is an essential factor in any realistic theoretical treatment, and Bose finds it extremely complicated in the usual treatment employing Maxwell equations with a complex refractive index and damping. He proposes an alternate scheme that utilizes the microscopic equations of Lorentz instead. The scheme is more general, adaptable for any absorbing medium as well as the ionosphere.

The next series of papers included in this section cover extended domains of applied and pure mathematics. The two papers on statistics deal with the mathematics involved in D^2-Statistic, a quantitative measure of the divergence between two statistical populations. P.C. Mahalanobis, a professor at the Presidency College, Calcutta, while working on some anthropometric data on statures of Anglo-Indians, had the idea of a "generalized distance" in statistics between two populations. He gave a precise definition in terms of the mean vectors of certain measured characteristics in the two groups. Bose, fascinated by the novel idea of a generalized distance in such context and certain aspects of its mathematical formulations, presents generalizations and exact formulas replacing approximations to calculate the moment coefficients.

The three other papers (1939, 1940 and 1945) concern various mathematical aspects of theoretical physics. The first deals with the decomposition of the Lorentz group into two commuting factors leading to the theory of spinors. The decomposition is generally shown to follow from an infinitesimal element approach of the Lorentz group SO (3, 1). Bose provides a purely algebraic finite element approach based on the properties of the matrix group O (4, C) in its defining representation and derives consequences on the physical Lorentz group. The second deals with the relativistically invariant solutions of the inhomogeneous Klein–Gordon equation. The authors, using complex contour integration methods of Herglotz and Sommerfeld, obtain general solutions by replacing the specific one provided in the earlier work by H.J.

Bhabha. The methods anticipate the later Stückelberg–Feynman techniques involving invariant Green's functions. The third concerns the solution of the Schrödinger equation for the Hydrogen atom in momentum space. Bose obtains complete solutions for both bound and scattering states with new interesting physical and mathematical insights into the quantum theory of the hydrogen atom.

These papers and those that follow in the next section reveal Bose's deep and abiding interest in the beauty and rigor of mathematical physics and his mastery over several branches of pure and applied mathematics.

Anomalous Dielectric Constant of Artificial Ionosphere

In a recent communication in *Nature*, Mitra and Roy[1] have pointed out an interesting feature in the ionospheric dispersion formula. They have shown that the formula can yield values of the dielectric constant of an ionized medium greater, equal to, or less than unity depending on the degree of ionization, collisional frequency and the exciting wave frequency. They therefore hold that the value of dielectric constant of an ionized gas greater than unity which has been recorded by many investigators for comparatively large ionizations is only an outcome of the complete dispersion formula. The object of this note is twofold. Taking the ionospheric dispersion formula we shall *first* deduce the conditions under which the dielectric constant may assume values greater or less than unity and *secondly* we shall show that in the experiments where anomalous values of the dielectric constant of an ionized gas have hitherto been obtained, the experimental conditions are such that the dispersion formula alone cannot explain the anomaly.

The dispersion formula is given by

$$\left(\mu - \frac{ick}{p}\right)^2 = 1 + \frac{1}{\alpha + i\beta},$$

where $\alpha = -\dfrac{mp^2}{4\pi Ne^2}$ and $\beta = \dfrac{mpv}{4\pi Ne^2}$

Separating the real and imaginary parts, we have

$$m^2 - \frac{c^2k^2}{p^2} = 1 + \frac{\alpha}{\alpha^2 + \beta^2} \qquad \text{..... (1)}$$

and

$$\frac{2ck}{p} \cdot \mu = \frac{\beta}{\alpha^2 + \beta^2} \qquad \text{..... (2)}$$

Thus $\dfrac{c^2k^2}{p^2} - \mu^2 = \dfrac{|\alpha|}{\alpha^2 + \beta^2} - 1$ so that

$$\mu^2 \underset{<}{\overset{>}{=}} 1 \text{ according as } \frac{c^2k^2}{p^2} \underset{<}{\overset{>}{=}} \frac{|\alpha|}{\alpha^2 + \beta^2}$$

which is equivalent to the condition :

$$\mu^2 \underset{<}{\overset{>}{=}} 1 \; if \; \frac{\mu^2c^2k^2}{p^2} \underset{<}{\overset{>}{=}} \frac{|\alpha|}{\alpha^2 + \beta^2},$$

but as $\dfrac{\mu^2c^2k^2}{p^2} = \dfrac{\beta^2}{4(\alpha^2 + \beta^2)^2}$ from (2)

the condition reduces to —

$$\mu^2 \underset{<}{\overset{>}{=}} 1 \text{ if } \frac{\beta^2}{4(\alpha^2 + \beta^2)^2} \underset{<}{\overset{>}{=}} \frac{|\alpha|}{\alpha^2 + \beta^2}$$

or

$$\frac{\beta^2}{\alpha^2 + \beta^2} \underset{<}{\overset{>}{=}} 4\,|\alpha|.$$

Now the left-hand side is always a proper fraction; therefore, if $4\,|\alpha| > 1$, *i.e.* $|\alpha| > \cdot 25$ we have then always

$$\frac{\beta^2}{\alpha^2 + \beta^2} < 4\,|\alpha| > 1, \; i.e. \; \mu^2 < 1 .$$

If however $|\alpha| < \cdot 25$, $\mu^2 > 1$ will require the further condition

$$\frac{\beta^2}{\alpha^2 + \beta^2} > 4\,|\alpha| \text{ or } \beta^2 > \frac{4|\alpha|\,\alpha^2}{1 - 4|\alpha|}.$$

Calling $\dfrac{\pi Ne^2}{m} = p_0^2$ or

$$p_0 = \sqrt{\frac{\pi e^2}{m}}\,\sqrt{N} = 2 \cdot 8 \times 10^4 \sqrt{N},$$

we have $4\,|\alpha| = \dfrac{p^2}{p_0^2}$ and we can say that if

$p > p_0 = 2{\cdot}8 \times 10^4 \sqrt{N}$, μ is always less than unity, whereas if $p < 2{\cdot}8 \times 10^4 \sqrt{N}$, or $|\alpha| < {\cdot}25$, for $\mu^2 > 1$, we must have

$$\nu^2 > \frac{p^4}{p_0^2 - p^2}$$

i..e., the collisional frequency must exceed the value

$$\frac{p^2}{\sqrt{p_0^2 - p^2}}$$

Thus the equations show that even when $p < 2{\cdot}8 \times 10^4 \sqrt{N}$, there exists a critical collisional frequency which has to be executed if μ is to be greater than unity.

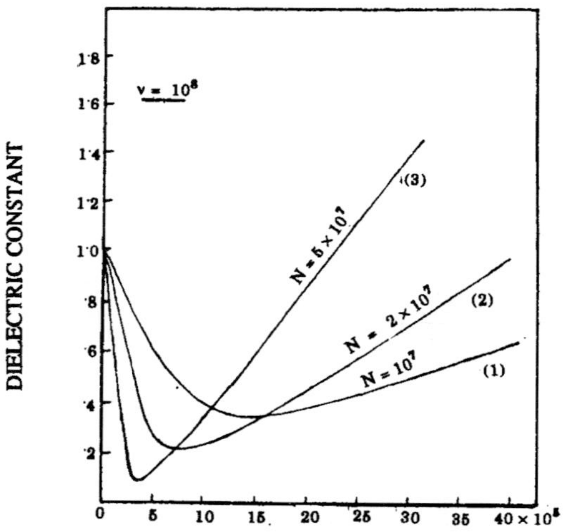

SQUARE OF WAVELENGTH (SQ. CM.)

Fig. I

In the experiments recently performed in this laboratory, we have obtained results similar to those of Appleton and Childs[2]. Using a wavelength of about 3·8 metres the dielectric constant of ionized air in a discharge tube has been found to be less than unity when the tube current is small. As the ionization is increased by increasing the tube current, the dielectric constant at first decreases and then increases gradually and even becomes greater than unity. The maximum electron density N in our discharge tube has been within 10^7. Taking $N=10^7$, $p=5 \times 10^8$, we get $|\alpha| = 7{\cdot}8$. This is much greater than ·25 so that the dielectric constant must be less than unity. Even when $N=10^8$, $|\alpha| = {\cdot}78$ and the dielectric constant should be theoretically less than unity.

In Fig. I are plotted the calculated values of the dielectric constants against λ^2 for some definite values of N and ν. Curve (2) which corresponds to Mitra and Banerjee's[3] experimental conditions shows a turning point at λ of the order 9m (λ^2 is of the order 8×10^5 sq.cm.) which is very different from the experimental value λ of the order 6m (λ^2 is of the order $3{\cdot}6 \times 10^5$ sq.cm.). In the experiments of Ali Imam and one[4] of us, which have been recently repeated by a different method, the turning point appears unmistakably at λ of the order 4m (λ^2 is of the order $1{\cdot}6 \times 10^5$ sq.cm.). Curve (1) which approximately corresponds to our experimental conditions shows however the turning point at λ of the order 12m (λ^2 is of the order 14×10^5 sq.cm.). It appears extremely doubtful if the disposition of the experimental apparatus can account for such wide discrepancy. It is likely that there must be some other explanation of these experimental results.

Physics Department,
Dacca University.
13.11.37

S.N. Bose,
S.R. Khastgir.

1. Mitra & Roy, *Nature, 140*, 586, 1937.
2. Appleton & Childs, *Phil, Mag. 10*, Dec. 1930.
3. Mitra & Banerjee, *Nature, 136*, 512, 1935.
4. Imam & Khastgir, *Phil. Mag.* May 1937 and *Ind. Jour. of Physics. 10*, Part I, Feb. 1937.

Reprinted from *Sci & Cult* 3, pp. 335-337, 1937 (The Indian Press Ltd, Calcutta).

ON THE TOTAL REFLECTION OF ELECTROMAGNETIC WAVES IN THE IONOSPHERE

BY S. N. BOSE,
DACCA UNIVERSITY.

(*Received for publication, April* 5, 1938)

The conditions under which the electromagnetic waves get totally reflected in the upper atmosphere appear still worth a closer investigation. The usual procedure has been to attribute to the medium a refractive index which is calculated according to the classical method of Lorentz from the mechanical equations of motion of electrons. In an absorbing medium, this refractive index is a complex quantity, and depends not only on the properties of the medium but also on the frequency of the waves : as progressive waves through the ionosphere are damped, the co-efficient of damping will also enter into the expression of the refractive index. The usual Appleton-Hartree condition for total reflexion has been deduced by neglecting the damping, as a first approximation, and by putting the refractive index equal to zero. Discussion of the conditions when the damping is not negligible, gets very complicated indeed. Certain workers in Allahabad have suggested the further condition that the group-velocity of the wave-train is zero when total reflection takes place. This is a reasonable hypothesis, but the method by which the group-velocity is calculated neglects damping, an essential factor in the physical process. It appears also that the calculation of the group-velocity of waves in an absorbing medium is not amenable to the ordinary methods unless the damping is negligible, an approximation which, as has been already pointed out, does not seem to agree closely with experience. It seems therefore necessary to change the method of analysis and to take, instead of the equations of Maxwell with a complex refractive-index, the microscopic equations of Lorentz as basis for calculation. The method sketched here is a general one suitable for problems of refractive-index of material media as well as for discussion of propagation of waves in the ionosphere. It seems further amenable to modifications to suit the requirements of modern wave mechanics.

We take the familiar equations of Lorentz

$$\frac{I}{c}\frac{\partial E}{\partial t} - \text{curl } H = -\frac{\rho V}{c}\, ; \; \frac{I}{c}\frac{\partial H}{\partial t} + \text{curl } E = 0.$$

$$\text{div } E = \rho, \quad \text{and} \quad \text{div } H = 0.$$

The density and the convection current, ρ and ρV, can be regarded as decomposable into two parts ρ_+, and ρ_- and $(\rho V)_+$ and $(\rho V)_-$, respectively, which thus takes account of the corpuscular distribution of positive and negative charges. A wave disturbance in the medium can be considered by putting

$$E = E_0 + Ee^S,\ H = H_0 + He^S;\quad \rho_+ = \rho_+^0 + q_+e^S,\ \rho_- = \rho_-^0 + q_-e^S,$$

$V_+ = V_+^0 + W_+e^S$ and $V_- = V_-^0 + W_-e^S$, where the elements with 0 indices correspond to the undisturbed state of the medium. The convection current, and the electric-density become on this assumption equal to

$$\overline{\rho V} = \rho_+V_+ + \rho_-V_- = (\overline{\rho V})_0 + \bar{\theta}e^S$$

where $\quad \bar{\theta} = [\rho_+^0 W_+ + q_+V_+^0 + \rho_-^0 W_- + q_-V_-^0)$ and $\theta_0 = q_+ + q_-$.

Thus the components $(\theta_x, \theta_y, \theta_z)$ of the vector $\bar{\theta}$ correspond to the amplitudes of the fluctuating part of the convection current, and θ_0, similarly to the fluctuating part of the electric density due to the disturbance.

The process is a familiar one : When S is a linear function of x, y, z, t, with a complex factor $2\pi\iota$, as is generally chosen for discussing monochromatic wave-propagation, $2\pi\iota\ S$ represents the phase of the disturbance and the surfaces for various values of S, represent the totality of all wave fronts as function of (x, y, z, t). In order to tackle the more general case of a damped wave-train we shall not at first put any limitation on the form of S but shall assume it to be any function whatever, whose differential co-efficients may also have complex values. Removing the contributions on the both side of the equation by the various quantities corresponding to the undisturbed state, we get

$$\frac{1}{c}\frac{\partial}{\partial t}(Ee^S) - \text{curl}\ (He^S) = -\frac{\theta}{c}\ e^S$$

$$\frac{1}{c}\frac{\partial}{\partial t}(He^S) + \text{curl}\ (Ee^S) = 0$$

$$\text{div}\ (Ee^S) = \theta_0 e^S, \text{ and div } (He^S) = 0.$$

We can also verify easily that

$$\text{div}\ [Ee^S] = [\text{div}\ E + (\text{grad}\ S.E)]e^S$$

$$\text{curl}\ [Ee^S] = [\text{curl}\ E + \text{grad}\ S \times E]e^S$$

$$\frac{\partial}{\partial t}\Big[\ Ee^S\ \Big] = \Big[\frac{dE}{dt} + E.\frac{\partial S}{\partial t}\ \Big]\ e^S$$

etc.

Under the usual conditions of propagation E, H, representing the amplitudes of the electric and magnetic waves, are slowly varying quantities whose differential co-efficients with regard to time and space co-ordinates can be neglected in comparison with the rapid variations of the function S which represents the phase of the disturbance. We shall therefore put the differential co-efficients of E, H equal to zero, and thus get finally the equations reduced to the following form :—

$$\frac{1}{c}\frac{\partial S}{\partial t}E-(\mathrm{grad}\ S\times H)=-\theta/c$$

$$\frac{1}{c}\frac{\partial S}{\partial t}H+(\mathrm{grad}\ S\times E)=0$$

$$(\mathrm{grad}\ S\cdot E)=\theta_0$$
$$(\mathrm{grad}\ S\cdot H)=0$$

These can be regarded as the characteristic set of the wave-equations of Lorentz, and since there are eight equations, if $\frac{\partial S}{\partial x}, \frac{\partial S}{\partial y}, \frac{\partial S}{\partial z}, \frac{\partial S}{\partial t}$ are taken as unknown, certain compatibility conditions have further to be satisfied which will give us the conditions under which wave-propagation can take place. We proceed to calculate these conditions by the usual method of vector-analysis. We have only to remember that, as the components of vectors may have complex-values, the geometrical interpretation of the various quantities may not be so immediate as in the usual cases : also a relation $A^2=0$, does not mean that the vector vanishes, as the components of the vector A_x, A_y, A_z, need not be zero, when $A_x^2+A_y^2+A_z^2=0$.

We shall call such vectors as singular, and in the problem that we have got before us, considerations of singular-vectors become important.

A few remarks regarding the geometrical interpretation of a singular vector may not be out-of-place; regarding the components of a singular vector A, as decomposable into real and imaginary parts, we can always choose a real direction (λ, μ, ν) which is perpendicular to a complex vector, *viz.*, $(A_x+A_x^{*}\iota, \ldots)$

i.e., such that $\lambda A_x+\mu A_y+\nu A_z=0$

$$\lambda A_x^{*}+\mu A_y^{*}+\nu A_z^{*}=0$$

i.e., an imaginary vector A, may be regarded as defining a real plane.

If co-ordinates are chosen so as to represent this plane as the plane of (xy), a singular vector A, i.e., whose $A^2=0$, can be represented as

$(A, \iota A, 0)$ where A, may be a complex quantity, i.e., A is of the form$[\iota\lambda+\iota\mu, -\mu+\iota\lambda, 0]$. Any vector perpendicular to a singular vector, will have in this co-ordinate system, the general form $(P, \iota P, Q)$ where P and Q may be complex quantities. A non-singular vector, C perpendicular to A, will have $C^2\neq 0$, *i.e.*, $Q\neq 0$; a singular vector perpendicular to A, will also necessarily have $Q=0$, *i.e.*, if A and B are two singular vectors, both conditions $(A\cdot B)=0$, and $(A\times B)=0$, may be satisfied at the same time and all singular vectors perpendicular to one another may be regarded as lying in one real plane.

If A and B are two perpendicular vectors of which one is singular while the other is not, any vector which is perpendicular to both A and B, must necessarily be singular, and will lie in the plane of A. This proposition follows immediately by writing out the general form of A and B in the manner explained above, and trying to write down the general form of C which satisfies the orthogonality condition. These preliminary remarks will be useful in discussing the singular case of total reflection, as we shall see later on. The other two relations in vector analysis which we shall have to use pretty frequently are

$$A.(B\times C) = B.(C\times A) = -B.(A\times C) = \text{etc.} \qquad \dots \quad \text{(I)}$$

and

$$A\times(B\times C) = B(AC)-C(AB). \qquad \dots \quad \text{(II)}$$

These have unrestricted validity even when the components of vectors have complex values, as they are formal identities, involving only re-arrangement of terms.

We shall at first deduce certain general conclusions from our equations

$$\frac{1}{c}\frac{\partial S}{\partial t}E-(\text{grad } S\times H) = -\theta/c \qquad \dots \quad (1)$$

$$\frac{1}{c}\frac{\partial S}{\partial t}H+(\text{grad } S\times E) = 0 \qquad \dots \quad (2)$$

$$(\text{grad } S\cdot E) = \theta_0 \qquad \dots \quad (3)$$

$$(\text{grad } S\cdot H) = 0. \qquad \dots \quad (4)$$

Scalar multiplication with grad S of (1) gives, taking account of (3),

$$\frac{1}{c}\left[\frac{\partial S}{\partial t}\theta_0+(\theta \text{ grad } S)\right]= 0, \qquad \dots \quad \text{(A)}$$

a relation which expresses the conservation of charge in the disturbed state of the medium.

Vector multiplication by grad S of (2) gives

$$\frac{1}{c}\frac{\partial S}{\partial t}(\text{grad } S\times H)+\text{grad } S\times(\text{grad } S\times E) = 0\,;$$

taking account of (1) we get

$$\left[\frac{1}{c^2}\left(\frac{\partial S}{\partial t}\right)^2-(\text{grad } S)^2\right]E+\frac{\theta}{c^2}\frac{\partial S}{\partial t}+\theta_0 \text{ grad } S=0. \qquad \ldots \quad (B)$$

Scalar multiplication of (2) with E gives $(EH)=0$ ... (C)

Scalar multiplication (1) with H gives $(\theta H)=0$ ... (D)

Relations (C), (D) and (4) show that the vectors E, θ, and grad S, are all in one plane perpendicular to H ; also (1) shows that grad S is perpendicular to $\frac{\partial S}{\partial t}E+\theta$. This vector can be regarded as representing the rate of change of induction, or $\frac{dD}{dt}$ of Maxwell's equation; thus H and D can be regarded as lying in the plane perpendicular to grad S while E and θ separately do not lie in it.

The arrangement of the various vectors thus agrees with our usual ideas about electromagnetic waves.

Energy and momentum relations in the wave field

Scalar multiplication of (1) and (2) by E and H respectively give

$$\left[\frac{1}{c}\frac{\partial S}{\partial t}E^2+\frac{(E\theta)}{c}\right]=E\cdot(\text{grad } S\times H)=-\text{grad } S.\ [E\times H] \qquad \ldots \quad (E)$$

$$\frac{1}{c}\frac{\partial S}{\partial t}H^2=-\text{ grad } S.\ [E\times H]\ ; \qquad \ldots \quad (F)$$

addition and subtraction of (E) and (F) give two further important relations

$$\frac{1}{c}\frac{\partial S}{\partial t}\frac{[E^2+H^2]}{2}+\frac{(E\theta)}{2c}=-\text{ grad } S.\ [E\times H] \qquad \ldots \quad (G)$$

$$\frac{1}{c}\frac{\partial S}{\partial t}(E^2-H^2)=-\frac{(E\theta)}{c}. \qquad \ldots \quad (H)$$

Also vector multiplication of (1), and (2) by H and E similarly gives

$$\frac{1}{c}\frac{\partial S}{\partial t}(E\times H)+H\times(\text{grad } S\times H)=-\frac{\theta\times H}{c}$$

or

$$\frac{1}{c}\frac{\partial S}{\partial t}(E\times H)+H^2\text{ grad } S=-\frac{\theta\times H}{c}\ ; \qquad \ldots \quad (K)$$

also $$\frac{1}{c}\frac{\partial S}{\partial t}(E\times H)+E^2\,\text{grad}\,S-E\,(\text{grad}\,S.E)=0$$

or $$\frac{1}{c}\frac{\partial S}{\partial t}\,(E\times H)+E^2\,\text{grad}\,S=\theta_0 E. \qquad \ldots \quad (L)$$

Subtracting (L) from (K) we have

$$\text{grad}\,S\,(H^2-E^2)=-\left[\theta_0 E+\frac{\theta\times H}{c}\right]. \qquad \ldots \quad (M)$$

Multiplying (K) and (L) by grad S we have the further relation

$$\frac{1}{c}\frac{\partial S}{\partial t}\,\text{grad}\,S.(E\times H)+\text{grad}\,S.\frac{(\theta\times H)}{c}=-H^2\,(\text{grad}\,S)^2$$

$$\frac{1}{c}\frac{\partial S}{\partial t}\,\text{grad}\,S.\,(E\times H)=\theta_0^2-(\text{grad}\,S)^2E^2;$$

but $$\text{grad}\,S.\,(\theta\times H)=-\,\theta.\,(\text{grad}\,S\times H)=-\,\frac{1}{c}\left(\theta^2+\frac{\partial S}{\partial t}\,(E\theta)\right)$$

or $$\frac{1}{c}\,\text{grad}\,S.\,(\theta\times H)=-\left[\frac{\theta^2}{c^2}+\frac{1}{c^2}\frac{\partial S}{\partial t}\,(E\theta)\right].$$

These two relations thus become

$$\frac{1}{c}\frac{\partial S}{\partial t}\,\text{grad}\,S.\,(E\times H)=\frac{\theta^2}{c^2}+\frac{1}{c^2}\frac{\partial S}{\partial t}\,(E\theta)-(\text{grad}\,S)^2H^2$$

$$\frac{1}{c}\frac{\partial S}{\partial t}\,\text{grad}\,S.\,(E\times H)=-(\text{grad}\,S)^2E^2+\theta_0^2.$$

From these we get

$$\frac{\theta^2}{c^2}-\theta_0^2+\frac{1}{c^2}\frac{\partial S}{\partial t}\,(E\theta)-(\text{grad}\,S)^2(H^2-E^2)=0,$$

or remembering (H) we see that

$$\left[\frac{1}{c^2}\left(\frac{\partial S}{\partial t}\right)^2-(\text{grad}\,S)^2\right](H^2-E^2)=S\,\theta_0^2-\frac{\theta^2}{c^2}. \qquad \ldots \quad (P)$$

Also $(\text{grad}\,S)^2H^2=(\text{grad}\,S.H)^2+(\text{grad}\,S\times H)^2$

$$=\frac{1}{c^2}\left[\left(\frac{\partial S}{\partial t}\right)^2E^2+2\,\frac{\partial S}{\partial t}\,(E\theta)+\theta^2\right];$$

we get

$$\frac{1}{c}\frac{\partial S}{\partial t}\text{ grad } S\,.\,(E\times H)=-\frac{1}{c^2}\frac{\partial S}{\partial t}\left[E^2\frac{\partial S}{\partial t}+(E\theta)\right]$$

$$=\theta_0^2-(\text{grad } S)^2E^2, \qquad \ldots \quad (Q)$$

which can be written in the form

$$\left[\frac{1}{c^2}\left(\frac{\partial S}{\partial t}\right)^2-(\text{grad } S)^2\right]E^2+\frac{(E\theta)}{c^2}\frac{\partial S}{\partial t}+\theta_0^2=0. \qquad \ldots \quad (Q)$$

This equation could have been immediately deduced from (B) by taking scalar multiplication with E.

Equation (B) :

$$\left[\frac{1}{c^2}\left(\frac{\partial S}{\partial t}\right)^2-(\text{grad } S)^2\right]E+\frac{\theta}{c^2}\frac{\partial S}{\partial t}+\text{grad } S(\text{grad } S\cdot E)=0$$

gives us generally the mutual disposition of E, θ and grad S when wave-propagation is possible, whereas the equation (Q) written in the form

$$\left[\frac{1}{c^2}\left(\frac{\partial S}{\partial t}\right)^2-(\text{grad } S)^2\right]+\frac{1}{c^2}\frac{\partial S}{\partial t}\frac{(E\theta)}{E^2}+\frac{(\text{grad } S\cdot E)}{E^2}=0 \qquad \ldots \quad (Q)$$

may be regarded as the equation for the refractive index, determining the velocity of propagation in any direction prescribed by grad S.

The Notion of the Refractive-index

A surface $S(x, y, z, t) = C$ may be regarded as moving perpendicular to itself with a velocity given by

$$W=\frac{\dfrac{\partial S}{\partial t}}{|\text{grad } S|}\;;$$

the refractive index can be defined as $\dfrac{c}{W}=\mu=\dfrac{|\text{grad } S|}{\dfrac{1}{c}\dfrac{\partial S}{\partial t}}$, agreeing with our usual definition, which however can be complex in the general case.

Introducing μ, equation (Q) can be written in the form

$$(1-\mu^2)+\frac{(E\theta)}{E^2}\left|\frac{\partial S}{\partial t}+\frac{c^2(\text{grad } S\cdot E)^2}{E^2}\right|\left(\frac{\partial S}{\partial t}\right)^2=0.$$

On Relation Between e and θ

Early in the paper we have expressed θ, in terms of motion of the positive and negative charges, namely,

$$\theta=[\rho_0^+W_0^+ + q^+V_0^+ + \rho_0^-W^- + q^-V_0^-].$$

These involve in general both the original distribution of the charges, their velocities as well as the amplitudes of velocities, and densities of positive and negative electricity due to the disturbance in the medium.

From general considerations we can express the relation of θ, and E in the following way.

We have seen that during propagation of disturbance both θ and E lie in the plane perpendicular to H. Therefore, θ can be decomposed into two components, one along E and another perpendicular to E, in this plane, *i.e.*, in the direction of $(E\times H)$.

θ thus can be put as equal to $\alpha E+\beta\,\frac{[E\times H]}{|E|}$

$$\because\quad (\theta E)=\alpha E^2,\quad \theta\times E=\frac{\beta E^2}{|E|}\,H;$$

also from equations of wave--propagation

$$\theta\times E=cH\,(\text{grad } S\cdot E)\ \text{ or }\ \beta^2/c^2=\theta_0^2/E^2.$$

Writing therefore

$$\theta=\alpha E+\beta\,\frac{[E\times H]}{|E|}$$

we see that

$$\alpha=\frac{(E\theta)}{E^2},\ \beta=\frac{C\theta_0}{|E|},$$

so that the refractive-index equation can be expressed in terms of α and β in the form

$$(1-\mu^2)+\frac{\alpha}{\partial S/\partial t}+\frac{\beta^2}{(\partial S/\partial t)^2}=0.$$

Further.

$$\theta^2 = \alpha^2E^2+\beta^2H^2.$$

As the vectors E and H are perpendicular to one another it follows that α and β are two co-efficients which stand in very simple relation to the total polarisation current; α and β however are not constants but may involve grad$S$ as well as $\frac{\partial S}{\partial t}$, *i.e.*, may be function of direction and frequency of the waves.

Condition for total Reflection

It is well-known that for electro-magnetic waves in vacuum we have $E^2-H^2 = 0$ as well as $(E \cdot H) = 0$. For waves in material media however $E^2-H^2 \neq 0$, but $(EH) = 0$ remains still valid. The ratio of the electric and magnetic vectors depends generally upon the properties of the medium as well as on the frequency of the wave and we can express this fact by the relations (H) and (M)

which are
$$\frac{1}{c}\frac{\partial S}{\partial t}(E^2-H^2) = -\frac{(E\theta)}{c}$$

$$\text{grad } S(H^2-E^2) = -F = -\left(E\theta_0+\frac{\theta\times H}{c}\right)$$

The ratio of the amplitudes of the vectors, H^2/E^2, can therefore be expressed by the following relations :

$$\frac{\partial S}{\partial t}+\frac{(E\theta)}{E^2} = \frac{H^2}{E^2}\frac{\partial S}{\partial t},$$

$$\text{grad } S-\frac{F}{E^2} = \frac{H^2}{E^2}\text{grad } S,$$

Depending upon the property of the medium and the direction of disposition of the vector E, $\frac{(E\theta)}{E^2}$ will generally vary with $\frac{\partial S}{\partial t}$ as well as with grad S. If, however, $\frac{\partial S}{\partial t}$ tends to such a value that $\frac{\partial S}{\partial t}+\frac{E\theta}{c^2}. = 0$, it follows that $H^2 = 0$ $(E \neq 0)$, independent of the amplitude of the wave traversing the medium; the disturbance in such a case loses its wave-character, so that we can say the wave cannot penetrate beyond that region and gets totally reflected. We shall therefore take this condition $\frac{H^2}{E^2} = 0$ or an equivalent condition as characterising the condition of total reflection and we can also correlate it with another physical idea.

We have also proved the following relation (F)

$$\frac{1}{c}\frac{\partial S}{\partial t}H^2 = -\text{grad } S.\ [E \times H]$$

and also

$$\frac{1}{c}\frac{\partial S}{\partial t}\left[\frac{E^2}{2}+\frac{H^2}{2}\right]+\frac{E\theta}{2c} = -\text{grad } S.\ (E\times H).$$

grad $S.(E\times H)$ may thus be regarded as proportional to the flow of energy across the surface S = constant, so that grad $S.(E\times H) = 0$ may be interpreted to mean that the flow of energy across the wave-front vanishes. This agrees with our usual idea of total-reflection at a boundary of two media, where the energy flows along the boundary interface in the limiting case.

We have also seen that

$$\frac{|\text{grad } S|}{\frac{1}{c}\frac{\partial S}{\partial t}} = \mu = \text{the refractive index of the medium.}$$

From (2) we have $\frac{1}{c^2}\left(\frac{\partial S}{\partial t}\right)^2 H^2 = (\text{grad } S\times E)^2 = E^2(\text{grad } S)^2-(\text{grad } S.E)^2$

$$= -\frac{1}{c}\frac{\partial S}{\partial t}\text{grad } S.[E\times H),$$

we have

$$\frac{H^2}{E^2} = \frac{(\text{grad } S)^2}{\frac{1}{c^2}\left(\frac{\partial S}{\partial t}\right)^2} - \frac{(\text{grad } S.E)^2}{E^2\frac{1}{c^2}\left(\frac{\partial S}{\partial t}\right)^2} = -\frac{\text{grad } S.[E\times H]}{\frac{1}{c}\frac{\partial S}{\partial t}E^2} = \frac{H.(\text{grad } S\times E]}{\frac{1}{c}\frac{\partial S}{\partial t}E^2}.$$

In case (grad S$\cdot E$) = 0, we see that the condition $\frac{H^2}{E^2} = 0$ is equivalent to the condition $\mu^2 = 0$, as has been assumed by Appleton to characterise the case of total reflection. If we now examine the condition (B) which gives the relation between the vectors E, θ and grad S, the condition $\theta = 0$ will mean that the wave is propagated under such conditions that the vector E is parallel to θ.

Multiplying (1) with E, we have

$$E\times\theta = cE\times(\text{grad } S\times H)$$

$$= -cH(\text{grad S}\cdot E),$$

so that (grad $S \cdot E$) = 0 will mean $E\times\theta = 0$, *i.e.*, E is paralled to θ.

We thus see that if the wave is such that E is $\|$ to θ the condition of total reflection $H = 0$, or grad $S.(E\times H) = 0$ is equivalent to the assumption $\mu = 0$; in the general case of wave propagation when E is not parallel to θ we would expect on the other hand the condition $H^2 = 0$ as giving a new result.

$H^2 = 0$, or grad $S.(E\times H) = 0$, may be regarded as equivalent to the assumption that the group-velocity vanishes. The condition of total-reflection postulated above is obviously suitable for a train of damped waves, because our analysis is general and the condition of non-penetrability has been deduced independent of all ideas of frequency.

We shall now apply the results of our analysis in the particular case of the ionosphere.

The contribution of the positive charges, as well as of the bound electrons in θ, may be easily seen to be negligible in comparison with the contribution of the free electrons.

We can assume for the ionosphere

$$\theta = [\rho_0^- W^- + q^- V_0^-]$$

where these quantities refer only to the free-electrons present. If we average over a volume containing a large number of electrons, but small compared with the wave-length of the disturbance,

$$\rho_0^- = Ne, \quad \theta = \overline{NeW}\ ;$$

the average value of the second term $q\bar{V}_0$ may be regarded as zero as the initial velocities of the charges may be regarded as distributed chaotically in all directions.

Writing the equation of motion of a free electron after Lorentz as

$$m\ddot{x} + gx = e\left[E_x + \frac{v_y h_z - v_z h_y}{c}\right], \text{ etc.}$$

where (h_x, h_y, h_z) may be regarded as the components of a steady magnetic force, we have, dividing by m and multiplying by $\rho = Ne$, and averaging, the following equation

$$\dot{\theta} + \frac{g}{m}\theta = \frac{Ne^2}{m}E + \left(\theta \times \frac{eh}{cm}\right)$$

where $\qquad (\theta_x = \overline{Nev_x}$, etc...).

Remembering that for an oscillatory disturbance $\theta = \theta \dfrac{\partial S}{\partial t} e^S$ as before also that the other terms in the equation varies at e^S, the equation can be written as

$$\left(\frac{\partial S}{\partial t}+\frac{g}{m}\right)\theta=\frac{Ne^2}{m}E+\left(\theta\times\frac{eh}{cm}\right) \qquad \ldots \text{ (Z)}$$

cancelling out e^S. $\frac{g}{m}=\nu$ is usually called the collision frequency of the electrons. We shall use this equation (Z) to calculate $\frac{(E\theta)}{E^2}$ or $\frac{\theta}{E}$.

Case I : This can be regarded as decomposable into two subcases.

(*a*) When $\frac{\theta\times eh}{cm}=0$, *i.e.*, θ is parallel to the external magnetic field, equation (Z) shows that θ is then parallel to E

and
$$\frac{E\theta}{E^2}=\frac{\theta}{E}=\frac{Ne^2}{m}\left(\frac{\partial S}{\partial t}+\nu\right).$$

The equation for determining the critical frequency for total reflection becomes

$$\frac{\partial S}{\partial t}+\frac{Ne^2}{m}\bigg/\left(\frac{\partial S}{\partial t}+\nu\right)=0$$

or
$$\left(\frac{\partial S}{\partial t}\right)^2+\nu\left(\frac{\partial S}{\partial t}\right)+\frac{Ne^2}{m}=0$$

or
$$\frac{\partial S}{\partial t}=-\frac{\nu}{2}\pm\iota\sqrt{\frac{Ne^2}{m}-\nu^2/4}$$

This determines the dependence of the critical frequency on the ionic density; the train totally reflected is therefore of the form

$$e^{-\nu/2t}e^{\pm\iota t\sqrt{Ne^2/m-\nu^2/4}}$$

or
$$w_c^2=Ne^2/m-\nu^2/4$$

and the damping co-efficient is $\nu/2$.

Case II : When $\theta \| E$ and $\theta^2=0$

that is the vector θ is singular, $\theta^2=0$ but $\theta\neq 0$

a vector $\left(\theta\times\frac{eh}{cm}\right)$ will represent a vector parallel to θ, when h is non-singular and perpendicular to θ according to our former remarks about singular vectors. Choosing the direction of h, as the direction of Z, we can write the relation

$$\left(\frac{\partial S}{\partial t}+\nu\right)\theta=\frac{Ne^2}{m}E+\left(\theta\times\frac{eh}{cm}\right)$$

in the form

$$\left(\frac{\partial S}{\partial t}+\nu\right)\theta_x=\frac{Ne^2}{m}E_x+\frac{eh}{cm}\theta_y$$

$$\left(\frac{\partial S}{\partial t}+\nu\right)\theta_y=\frac{Ne^2}{m}E_y-\frac{eh}{cm}\theta_x$$

or

$$\left(\frac{\partial S}{\partial t}+\nu\pm\iota\frac{eh}{cm}\right)(\theta_x+\iota\theta_y)=\frac{Ne^2}{m}(E_x\pm\iota E_y)$$

$\frac{\theta}{E}$ becomes equal to $\frac{Ne^2/m}{\left[\frac{\partial S}{\partial t}+\nu\pm\iota\frac{eh}{cm}\right]}$,

according as θ and E are of the form $(A, \mp\iota A, 0)$.

The condition of total reflection now becomes, from (B),

$$\frac{\partial S}{\partial t}\left(\frac{\partial S}{\partial t}+\nu\pm\iota\frac{eh}{cm}\right)+\frac{Ne^2}{m}=0;$$

calling

$$\frac{\partial S}{\partial t}+\nu/2\pm\iota\frac{eh}{2cm}=x$$

we have

$$x^2+\frac{Ne}{m}-\nu^2/4+\frac{e^2h^2}{4c^2m^2}\pm\frac{eh\iota\nu}{2cm}=0$$

Solving the above equation, we find that the two reflected waves are circularly polarised and the critical frequencies are

$$\omega_{c\pm}=\sqrt{\frac{Ne^2}{m}-\nu^2/4+\frac{e^2h^2}{4c^2m^2}}\mp\frac{eh}{2\,cm}$$

$$T_{c\pm}=\nu/2\left[1\pm\frac{eh}{2cm\,\omega'}\right]\text{ where }\omega'=\sqrt{\frac{Ne^2}{m}-\nu^2/4+\frac{e^2h^2}{4c^2m^2}}$$

So that when we have the critical frequency $=\frac{eh}{cm}$, *i.e.*, $\omega'=\frac{eh}{2cm}$, one of the circularly polarised wave will be suppressed, and the other will be reflected with undiminished intensity : as this will necessarily mean $\frac{Ne^2}{m}=\nu^2/4$ the reflection of plane-polarised wave, discussed above in case I, will not occur as the condition of total reflection gives $\omega_c = 0$, if $\frac{Ne^2}{m} = \nu^2/4$.

These two cases correspond to the cases discussed by Appleton by putting $\mu = 0$ and the result agrees with the usual formula, when the collision-frequency is neglected.
Case III. When $\theta \neq E$:

We shall now calculate the value of $\frac{(E\theta)}{E^2}$ in the general case.

We have

$$\left(\frac{\partial S}{\partial t}+\nu\right)\theta = \frac{Ne^2}{m}\,E+\left(\theta\times\frac{eh}{cm}\right).$$

Scalar multiplication with E gives

$$\left(\frac{\partial S}{\partial t}+\nu\right)(E\,\theta) = \frac{Ne^2}{m}\,E^2-\theta\left[E\times\frac{eh}{cm}\right];$$

also vector multiplication with E gives

$$\left(\frac{\partial S}{\partial t}+\nu\right)(E\times\theta) = \theta.\left(E\cdot\frac{eh}{cm}\right)-\frac{eh}{cm}\,(E\theta).$$

Scalar multiplication by $\frac{eh}{cm}$ of the above relation gives

$$\left(\frac{\partial S}{\partial t}+\nu\right)\frac{eh}{cm}\cdot(E\times\theta) = \left(\frac{eh}{cm}.\,\theta\right)\left(E.\,\frac{eh}{cm}\right)-\left(\frac{eh}{cm}\right)^2(E\theta).$$

Again, scalar multiplication of the original equation with $\frac{eh}{cm}$ gives

$$\left(\frac{\partial S}{\partial t}+\nu\right)\left(\frac{eh}{cm}\theta\right) = \frac{Ne^2}{m}\left(E\,\frac{eh}{cm}\right).$$

Combining (a), (b) and (c) we can easily deduce

$$\frac{(E\theta)\left[\left(\frac{\partial S}{\partial t}\right)+\nu\right]^2+\left(\frac{eh}{cm}\right)^2\Big]}{\left(\frac{\partial S}{\partial t}+\nu\right)} = \frac{\frac{Ne^2}{m}\,(E^2)\left[\left(\frac{\partial S}{\partial t}+\nu\right)^2+\frac{1}{E^2}\left(E.\,\frac{eh}{cm}\right)^2\right]}{\left(\frac{\partial S}{\partial t}+\nu\right)^2}$$

or $$\frac{(E\theta)}{E^2} = \frac{Ne^2}{m} \frac{\left[\left(\frac{\partial S}{\partial t} + \nu\right)^2 + \frac{1}{E^2}\left(E_1 \frac{eh}{cm}\right)^2\right]}{\left(\frac{\partial S}{\partial t} + \nu\right)\left[\left(\frac{\partial S}{\partial t} + \nu\right)^2 + \left(\frac{eh}{cm}\right)^2\right]}$$

when E and θ are non-singular vectors.

Now if we consider the case of total reflection as defined by grad $S \cdot [E \times H]$ $= 0$, [Q] gives $\frac{\partial S}{\partial t} = -\frac{(E\theta)}{E^2}$ as well as $\theta_0^2 - (\text{grad } S)^2 E = 0$.

Now also grad $S \cdot [E \times H]$ can be written in the form $H(\text{grad } S \times E)$, we see that grad $S \| E$

and $$\frac{1}{E^2}\left(E . \frac{eh}{cm}\right)^2 = \frac{1}{(\text{grad } S)^2}\left(\text{grad } S \frac{eh}{cm}\right)^2$$

$\frac{1}{E^2}\left(E_1 \frac{eh}{cm}\right)^2 = \left(\frac{eh}{cm}\right)^2 \cos^2\alpha$ if grad S is a vector whose components have real ratios.

The relation $\frac{\partial S}{\partial t} + \frac{(E\theta)}{E^2} = 0$ becomes

$$\frac{\partial S}{\partial t} + \frac{\frac{Ne^2}{m}\left[\left(\frac{\partial S}{\partial t} + \nu\right)^2 + \frac{1}{E^2}\left(E \frac{eh}{cm}\right)^2\right]}{\left(\frac{\partial S}{\partial t} + \nu\right)\left[\left(\frac{\partial S}{\partial t} + \nu\right)^2 + \left(\frac{eh}{cm}\right)^2\right]} = 0$$

If the disturbance is plane-polarised and the collision frequency is small we can put

$$\nu = 0, \quad \frac{1}{E^2}\left(E_1 \frac{eh}{cm}\right)^2 = \cos^2\alpha\left(\frac{eh}{cm}\right)^2$$

where α is the angle between the direction of propagation and the magnetic lines. The formula becomes, putting $\frac{\partial S}{\partial t} = \iota\, \omega_c$

$$\omega_c^2 = \frac{\frac{Ne^2}{m}\left[\omega_c^2 - \left(\frac{eh}{cm}\right)^2 \cos^2\alpha\right]}{\omega_c^2 - \left(\frac{eh}{cm}\right)^2}$$

agreeing with the formula supposed to be derived by putting group-velocity by Bajpai and Mathur.[1]

We shall have to remember however that E may be a complex-vecto general case, any complex vector $E = [E_x + \iota E_x$, etc.], may be regarded as p cular to a real direction, and choosing this direction as the direction of Z, we ca as components of the complex vector, $(E_x, \iota E_y, 0)$ where x, and y are certair directions and the external field is regarded as having components (h_x, h_y, h_z) special directions.

$$\left(E_1 \frac{eh}{cm}\right) \text{ then becomes equal to } \frac{e}{cm}(E_x h_x + \iota E_y h_y)$$

$$E^2 = E_x^2 - E_y^2 \neq 0$$

$$\frac{1}{E^2}\left(E_1 \frac{eh}{cm}\right)^2 = \frac{e^2}{c^2m^2}\left[\frac{E_x^2 h_x^2 - E_y^2 h_y^2}{E_x^2 - E_y^2} + 2\iota h_x h_y \frac{E_x E_y}{E_x^2 - E_y^2}\right]$$

The frequency equation for the elliptically polarised beams would be

$$0 = \frac{\partial S}{\partial t} + \frac{Ne^2}{m} \frac{\left[\left(\frac{\partial S}{\partial t} + \nu\right)^2 + \frac{e^2}{c^2m^2}\left[\frac{E_x^2 h_x^2 - E_y h_y^2}{E_x^2 - E_y^2} + \frac{2\iota E_x E_y}{E_x^2 - E_y^2} h_x h_y\right]\right]}{\left(\frac{\partial S}{\partial t} + \nu\right)\left[\left(\frac{\partial S}{\partial t} + \nu\right)^2 + \frac{e^2h^2}{c^2m^2}\right]};$$

when the collision frequency is neglected this similarly becomes

$$\omega^2 = \frac{Ne^2}{m} \frac{\left[\omega^2 - \frac{e^2}{c^2m^2}\left[\frac{E_x^2 h_x^2 - E_y^2 h_y^2}{E_x^2 - E_y^2}\right]\right]}{\omega^2 - e^2h^2/c^2m^2} \text{ when } \frac{E_x E_y}{E_x^2 - E_y^2} h_x h_y = 0.$$

This shows that the two cases will be either $h_x = 0$, or $h_y = 0$

$$\omega^2 = \frac{\frac{Ne^2}{m}\left[\omega^2 - \frac{e^2 h_x^2 E_x^2}{E_x^2 - E_y^2}\right]}{\omega^2 - \frac{e^2h^2}{c^2m^2}}, \text{ or } \frac{\frac{Ne^2}{m}\left[\omega^2 + \left[\frac{e^2}{c^2m^2} \frac{h_y^2 E_y^2}{(E_x^2 - E_y^2)}\right]\right]}{\omega^2 - \frac{e^2h^2}{c^2m^2}}.$$

We have therefore

or
$$\omega^2 = \frac{\frac{Ne^2}{m}\left[\omega^2 + \frac{e^2h^2}{c^2m^2} \frac{E_y^2}{E_x^2 - E_y^2} \cos^2 x\right]}{\omega^2 - \frac{e^2h^2}{c^2m^2}}$$

when h is $\perp$ to the major axis,

$$\omega^2 = \frac{\frac{Ne^2}{m}\left[\omega^2 - \frac{e^2h^2}{c^2m^2}\frac{E_x^2\cos^2\alpha}{E_x^2-E_y^2}\right]}{\omega^2 - \frac{e^2h^2}{c^2m^2}}$$

when h is $\perp$ to the minor axis.

The Propagation of the Disturbance

We shall analyse now the relation (B), *i.e.*,

$$\left[\frac{1}{c^2}\left(\frac{\partial S}{\partial t}\right)^2 - (\text{grad } S)^2\right]E + \frac{\theta}{c^2}\frac{\partial S}{\partial t}\theta_0 \text{ grad } S = 0$$

which may be regarded as prescribing the relation between the three vectors E, θ and grad S, when propagation takes place.

For this purpose, we shall have to consider the relation (A) which expresses the conservation of charge, *i.e.*,

$$\frac{1}{c}\left[\frac{\partial S}{\partial t}\theta_0^2 + (\theta \text{ grad } S)\right] = 0,$$

and the relation (Z), *i.e.*,

$$\left(\frac{\partial S}{\partial t} + \nu\right)\theta = \frac{Ne^2}{m}E + \left(\theta \times \frac{eh}{cm}\right).$$

We can utilise (A) and (Z) to eliminate E and θ_0 from (B) and thus get

$$\frac{\theta}{c^2}\frac{\partial S}{\partial t} + \frac{\left[\frac{1}{c^2}\left(\frac{\partial S}{\partial t}\right)^2 - \text{grad}^2 S\right]\left[\left(\frac{\partial S}{\partial \nu} + \nu\right)\theta - \left(\theta \times \frac{eh}{cm}\right)\right]}{\frac{Ne^2}{m}} = \frac{\text{grad } S\,(\theta.\ \text{grad } S)}{\frac{\partial S}{\partial t}}$$

or by re-arranging terms,

$$\left\{\left[\frac{1}{c^2}\left(\frac{\partial S}{\partial t}\right)^2 - \text{grad}^2 S\right]\frac{\partial S}{\partial t}\left(\frac{\partial S}{\partial t} + \nu\right) + \frac{Ne^2}{mc^2}\left(\frac{\partial S}{\partial t}\right)^2\right\}\theta$$

$$-\frac{\partial S}{\partial t}\left[\frac{1}{c^2}\left(\frac{\partial S}{\partial t}\right)^2 - (\text{grad}^2 S)\right]\left(\theta \times \frac{eh}{cm}\right) = \frac{Ne^2}{m}\text{grad } S\,(\theta.\ \text{grad } S). \quad \text{(R)}$$

This vector equation is equivalent to three homogeneous linear equations in $(\theta_x, \theta_y, \theta_z)$, whose co-efficients involve the differential co-efficients of S and the collision-frequency; when the relation is satisfied for non-zero values of $(\theta_x, \theta_y, \theta_z)$ the discriminant of the simultaneous equation must vanish. This will give a relation between the differential co-efficients of S alone, where the characteristic numbers of the ionospheric layer, *i.e.*, N, h and ν, will also be involved, and may be regarded as giving the equation of the characteristic surface, corresponding to the Ei-konal equation of Hamilton for a material medium

or equation $$\frac{1}{c^2}\left[\left(\frac{\partial S}{\partial t}\right)^2-(\text{grad } S)^2\right]=0 \qquad \text{for vacuum.}$$

It is better for perspicacity to introduce certain abbreviations at this stage; we shall put

$$g(S)=\frac{1}{c^2}\left(\frac{\partial S}{\partial t}\right)^2-\text{grad}^2 S$$

also $$\left[\frac{1}{c^2}\left(\frac{\partial S}{\partial t}\right)^2-\text{grad}^2 S\right]\frac{\partial S}{\partial t}\left(\frac{\partial S}{\partial t}+\nu\right)+\frac{Ne^2}{mc^2}\left(\frac{\partial S}{\partial t}\right)^2 \text{ is put}$$

$$=g(S)\frac{\partial S}{\partial t}\left(\frac{\partial S}{\partial t}+\nu\right)+\frac{Ne^2}{m}\left[g(S)+\text{grad}^2 S\right]=L(S)$$

$$=g(S)\,p(S)+\frac{Ne^2}{m}(\text{grad } S)^2=\frac{p(S)}{c^2}\left(\frac{\partial S}{\partial t}\right)^2-\frac{\partial S}{\partial t}\left(\frac{\partial S}{\partial t}+\nu\right)\text{grad}^2 S$$

where $$p(S)=\frac{\partial S}{\partial t}\left(\frac{\partial S}{\partial t}+\nu\right)+\frac{Ne^2}{m}.$$

The vector equation (R) is thus written

$$L(S)\,\theta-g(S)\frac{\partial S}{\partial t}\left(\theta\times\frac{eh}{cm}\right)=\frac{Ne^2}{m}\text{grad } S(\theta\cdot\text{grad } S).$$

Scalar multiplication with θ gives

$$L(S)\theta^2=\frac{Ne^2}{m}(\theta\cdot\text{grad } S)^2,$$

so that $(\text{grad } S\cdot\theta)=0$, either when $\theta^2=0$ or when $L(S)=0$, so that the vector equation becomes simplified, in both of these cases.

$E \parallel \theta$

Case I. When $L(S)=0,\quad \theta^2\neq 0$.

The vector equation (R) is satisfied only when

$$\left(\theta \times \frac{eh}{cm}\right) = 0, \text{ or } g(S) = 0, \text{ or } \frac{\partial S}{\partial t} = 0.$$

We reject $\frac{\partial S}{\partial t} = 0$, as we are considering wave propagation. We can easily see that $g(S) = 0$ is a singular case, which is not relevant to the problem before us, as $g(S) = 0$ will lead to the vanishing of the vector θ.

θ is parallel to h, the external magnetic field.

The characteristic equation thus becomes

$$\frac{p(S)}{c^2}\left(\frac{\partial S}{\partial t}\right)^2 - \left(\frac{\partial S}{\partial t} + \nu\right)\frac{\partial S}{\partial t}\,\text{grad}^2 S = 0 \qquad \ldots \quad (1)$$

or

$$\left(\frac{\partial S}{\partial x}\right)^2 + \left(\frac{\partial S}{\partial y}\right)^2 + \left(\frac{\partial S}{\partial z}\right)^2 = \frac{1}{c^2}\left(\frac{\partial S}{\partial t}\right)^2 \frac{\frac{\partial S}{\partial t}\left(\frac{\partial S}{\partial t} + \nu\right) + \frac{Ne^2}{m}}{\frac{\partial S}{\partial t}\left(\frac{\partial S}{\partial t} + \nu\right)}$$

or putting

$$\left(\frac{\partial S}{\partial x}\right)^2 + \left(\frac{\partial S}{\partial y}\right)^2 + \left(\frac{\partial S}{\partial z}\right)^2 = \frac{\mu}{c^2}\left(\frac{\partial S}{\partial t}\right)^2$$

$$\mu^2 = 1 + \frac{Ne^2}{m} \Big/ \frac{\partial S}{\partial t}\left(\frac{\partial S}{\partial t} + \nu\right)$$

which agrees with the ordinary expression for the refractive index for plane polarised ordinary ray. The singular solution of the equation (1) when $p(S)$ and $|\text{grad } S| = 0$ gives the case of total reflection.

Case II. When $\theta^2 = 0$

$$(\theta \text{ grad } S) = 0.$$

The vector equation (R) can be satisfied only when

$$L(S)\theta - g(S)\frac{\partial S}{\partial t}\left(\theta \times \frac{eh}{cm}\right) = 0$$

i.e., when $\left(\theta \times \frac{eh}{cm}\right)$ becomes parallel to θ.

This singular case, already discussed, means that θ is perpendicular to h, and choosing a new set of perpendicular axis, with h in the direction of Z, we see $L(S)$, $p(S)$, $g(S)$ are invariants for orthogonal transformations. The vector equation can be split up into two equations

$$L(S)\theta_x - g(S)\frac{\partial S}{\partial t}\,\theta_y\,\frac{eh}{cm} = 0$$

$$L(S)\theta_y + g(S)\,\frac{\partial S}{\partial t}\,\theta_x\frac{eh}{cm} = 0.$$

So that remembering $\theta_x^2+\theta_y^2 = 0$

we have $$L^2(S)+g^2(S)\left(\frac{\partial S}{\partial t}\right)^2\left(\frac{eh}{cm}\right)^2 = 0$$

which breaks up into two equations

$$g(S)\,p(S)+\frac{Ne^2}{m}\,(\text{grad } S)^2 \pm \iota g(S)\frac{\partial S}{\partial t}\,\frac{eh}{cm} = 0$$

or $$\frac{1}{c^2}\left(\frac{\partial S}{\partial t}\right)^2 - \text{grad}^2 S + \frac{\dfrac{Ne^2}{mc^2}\left(\dfrac{\partial S}{\partial t}\right)^2}{\dfrac{\partial S}{\partial t}\left(\dfrac{\partial S}{\partial t}+\nu \pm \iota\dfrac{eh}{cm}\right)} = 0$$

which can again be put in the form

$$\frac{\mu^2}{c^2}\left(\frac{\partial S}{\partial t}\right)^2 - (\text{grad } S)^2 = 0.$$

If $$\mu^2 = 1+\frac{Ne^2/m}{\dfrac{\partial S}{\partial t}\left(\dfrac{\partial S}{\partial t}+\nu\pm\iota\dfrac{eh}{cm}\right)};$$

total reflection is obtained as before by putting $\mu = 0$

$\text{grad}^2\,S$ can also become infinite, if $\dfrac{\partial S}{\partial t} = \pm\dfrac{\iota eh}{cm} - \nu$

i.e., for a frequency equal to the gyromagnetic frequency.

Case III.

Neither $L(S)$ nor $\theta^2 = 0$

i.e., when $(\text{grad } S\theta) \neq 0$

and θ is not parallel to E.

We have to proceed to further elimination to get the refractive index surface.

The elimination work becomes simplified if we choose a new set of axis, with Z axis, in the direction of h.

The vector equation is now equivalent to the three scalar equations

$$L(S)\theta_x - g(S)\frac{\partial S}{\partial t}\theta_y\frac{eh}{cm} = \frac{Ne^2}{m}S_x(S_x\theta_x+S_y\theta_y+S_z\theta_z) \quad \ldots \quad (i)$$

$$L(S)\theta_y + g(S)\frac{\partial S}{\partial t}\theta_x\frac{eh}{cm} = \frac{Ne^2}{m}S_y(S_x\theta_x+S_y\theta_y+S_z\theta_z) \quad \ldots \quad (ii)$$

$$L(S)\theta_z = \frac{Ne^2}{m}S_z(S_x\theta_x+S_y\theta_y+S_z\hat{\theta}_z) \quad \ldots \quad (iii)$$

where $\left(S_x = \frac{\partial S}{\partial x}, S_y = \frac{\partial S}{\partial y}, S_z = \frac{\partial S}{\partial z}\right).$

From (i) and (ii) we easily get the relations

$$\theta_x\left[L^2(S)+g^2(S)\left(\frac{\partial S}{\partial t}\right)^2\frac{e^2h^2}{c^2m^2}\right]$$

$$=\frac{Ne^2}{m}\left[L(S)S_x+S_yg(S)\frac{\partial S}{\partial t}\frac{eh}{cm}\right](S_x\theta_x+S_y\theta_y+S_z\theta_z)$$

$$\theta_y\left[L^2(S)+g^2(S)\left(\frac{\partial S}{\partial t}\right)^2\frac{e^2h^2}{c^2m^2}\right]$$

$$=\frac{Ne^2}{m}\left[L(S)S_y-S_xg(S)\frac{\partial S}{\partial t}\frac{eh}{cm}\right][S_x\theta_x+S_y\theta_y+S_z\theta_z]$$

and . $$L_z\theta_z = \frac{Ne^2}{m}S_z[S_x\theta_x+S_y\theta_y+S_z\theta_z].$$

$$\frac{\theta_x}{S_x\theta_x+S_y\theta_y+S_z\theta_z} = \frac{L(S)S_x+S_yg(S)\dfrac{\partial S}{\partial t}\dfrac{eh}{cm}}{L^2(S)+g^2(S)\left(\dfrac{\partial S}{\partial t}\right)^2\dfrac{e^2h^2}{c^2m^2}}\cdot\frac{Ne^2}{m} \quad \ldots \quad (iv)$$

$$\frac{\theta_y}{S_x\theta_x+S_y\theta_y+S_z\theta_z} = \frac{L(S)S_y-S_xg(S)\dfrac{\partial S}{\partial t}\dfrac{eh}{cm}}{L^2(S)+g^2(S)\left(\dfrac{\partial S}{\partial t}\right)^2\dfrac{e^2h^2}{c^2m^2}}\cdot\frac{Ne^2}{m} \quad \ldots \quad (v)$$

$$\frac{\theta_z}{S_x\theta_x+S_y\theta_y+S_z\theta_z} = \frac{\dfrac{Ne^2}{m}S_z}{L(S)}; \quad \ldots \quad (vi)$$

multiplying (iv), (v) and (vi) by S_x, S_y, S_z, we have

$$1 = \frac{Ne^2}{m}\left[\frac{L(S)(S_x^2+S_y^2)}{L^2(S)+g^2(S)\left(\frac{\partial S}{\partial t}\right)^2\left(\frac{eh}{cm}\right)^2}+\frac{S_z^2}{L(S)}\right]$$

or $$L(S)\left[L^2(S)+g^2(S)\left(\frac{\partial S}{\partial t}\right)^2\left(\frac{eh}{cm}\right)^2\right]$$

$$= \frac{Ne^2}{m}\left[L^2(S)(S_x^2+S_y^2+S_z^2+S_z^2g^2(S)\left(\frac{\partial S}{\partial t}\right)^2\left(\frac{eh}{cm}\right)^2\right].$$

Now writing $$S_z^2 = \text{grad}^2 S\cos^2\alpha$$

where $\alpha =$ the angle between the wavenormal and the magnetic field, we have

$$L^2(S)\left[L(S)-\frac{Ne^2}{m}\text{grad}^2 S\right]+g^2(S)\,L(S)\left(\frac{\partial S}{\partial t}\right)^2\left(\frac{eh}{cm}\right)^2$$

$$=\frac{Ne^2}{m}\left(\frac{eh}{cm}\right)^2 g^2(S)\left(\frac{\partial S}{\partial t}\right)^2\text{grad}^2 S\cos^2\alpha.$$

Remembering $$L(S)-\frac{Ne^2}{m}\text{grad}^2 S = p(S)g(S)$$

and rejecting the factor $g(S)$ the equation finally assumes the form

$$p(S)L^2(S)+g(S)\left(\frac{\partial S}{\partial t}\right)^2\left(\frac{eh}{cm}\right)^2 L(S)$$

$$-\frac{Ne^2}{m}\left(\frac{\partial S}{\partial t}\right)^2 g(S)\left(\frac{eh}{cm}\right)^2\text{grad}^2 S.\cos^2\alpha = 0.$$

When this equation is regarded as a quadratic in ($\text{grad}^2 S$) *i.e.*, when the value of $\frac{\partial S}{\partial t}$ is supposed to be prescribed, the two roots of grad²S, become, let us say, $f_1\left(\frac{\partial S}{\partial t}, \alpha\right)$ and $f_2\left\{\frac{\partial S}{\partial t}, \alpha\right\}$. So that we get two rays corresponding to any prescribed α.

Corresponding to the two refractive indices

$$\mu_1^2 = \frac{f_1\left\{\frac{\partial S}{\partial t}, \alpha\right\}}{\frac{1}{c^2}\left\{\frac{\partial S}{\partial t}\right\}^2}, \text{ and } \mu_2^2 = \frac{f_2\left\{\frac{\partial S}{\partial t}, \alpha\right\}}{\frac{1}{c^2}\left\{\frac{\partial S}{\partial t}\right\}^2}$$

When damping is neglected, $\frac{\partial S}{\partial t} = \iota\,\omega$.

We see from (*iv*), (*v*) and (*vi*), $\theta_x : \theta_y : \theta_z$

defines a complex-ratio. As $L(S)$ will become real, and $\frac{\partial S}{\partial t}$ pure imaginary, the two split rays will there be elliptically polarised, and the ratio of the axis, projected on a plane perpendicular of the magnetic field, is easily seen to be

$$\frac{g(S)\frac{\partial S}{\partial t}\frac{eh}{cm}}{L(S)}=\frac{g(S)\,\frac{\partial S}{\partial t}\left(\frac{eh}{cm}\right)}{\left[\frac{p(S)}{c^2}-\text{grad}^2S\right]}=\frac{\left[\frac{1}{c^2}\left(\frac{\partial S}{\partial t}\right)^2-\text{grad}^2S\right]\frac{eh}{cm}}{\frac{\partial S}{\partial t}\left[\frac{1}{c^2}\left(\frac{\partial S}{\partial t}\right)^2-\text{grad}^2S+\frac{Ne^2}{mc^2}\left(\frac{\partial S}{\partial t}\right)^2\right]};$$

for large values of grad^2S this tends to $\rightarrow\frac{eh}{cm\,\omega_c\iota}$.

Writing out the equation in extenso and putting $\text{grad}^2\,S=y$, we see the equation

$$0=p(S)\left[\frac{p(S)}{c^2}\left(\frac{\partial S}{\partial t}\right)^2-\frac{\partial S}{\partial t}\left(\frac{\partial S}{\partial t}+\nu\right)y\right]^2+\left[\frac{1}{c^2}\left(\frac{\partial S}{\partial t}\right)^2-y\right]\left(\frac{eh}{cm}\right)^2\left(\frac{\partial S}{\partial t}\right)^2$$

$$\left\{\frac{p(S)}{c^2}\left(\frac{\partial S}{\partial t}\right)^2-y\left(\frac{\partial S}{\partial t}+\nu\right)\frac{\partial S}{\partial t}+\frac{Ne^2}{m}\cos^2\alpha\right]\right\}.$$

$$\text{The co-efficient of } y^2=p(S)\frac{\partial S}{\partial t}\left(\frac{\partial S}{\partial t}+\nu\right)^2+\left(\frac{eh}{cm}\right)^2\left(\frac{\partial S}{\partial t}\right)^2\left[\frac{\partial S}{\partial t}\left(\frac{\partial S}{\partial t}+\nu\right)\right.$$

$$\left.+\frac{Ne^2}{m}\cos^2\alpha\right].$$

The absolute term on the other hand is equal to

$$\frac{p^3(S)}{c^4}\left(\frac{\partial S}{\partial t}\right)^4+\frac{1}{c^4}\left(\frac{\partial S}{\partial t}\right)^6p(S)\left(\frac{eh}{cm}\right)^2=\frac{p(S)}{c^4}\left(\frac{\partial S}{\partial t}\right)^4\left[p^2(S)+\left(\frac{\partial S}{\partial t}\right)^2\left(\frac{eh}{cm}\right)^2\right].$$

We get therefore

$$\frac{\partial S}{\partial t}\left(\frac{\partial S}{\partial t}+\nu\right)+\frac{\frac{Ne^2}{m}\left[\left(\frac{\partial S}{\partial t}+\nu\right)^2+\left(\frac{eh}{cm}\right)^2\cos^2\alpha\right]}{\left[\left(\frac{\partial S}{\partial t}+\nu\right)^2+\left(\frac{eh}{cm}\right)^2\right]}=0$$

as defining the critical frequencies for total reflection. This is in the general case when collision frequency is not neglected an equation of the fourth degree having two different conjugate complex roots corresponding to two elliptically polarised rays: or when collision frequency is neglected,

and $$\frac{\partial S}{\partial t}=\iota\omega,$$

the above relation reduces to

$$\omega_c^2 = \frac{\frac{Ne^2}{m}\left[\omega^2 - \left(\frac{eh}{cm}\right)^2 \cos^2\alpha\right]}{\omega^2 - \left(\frac{eh}{cm}\right)^2}$$

which agrees with the expression arrived at by Rai, Bajpai, and Mathur, and also with the result arrived at by another method.

From energy consideration, we have seen that when $E \parallel \theta$ the condition of total reflection grad $S.[E \times H) = 0$, is satisfied $(\text{grad } S)^2 = 0$: Here the propagation stops from penetrating further, as H vanishes, and the field beyond becomes electrostatic in character. When (grad S E) $\neq 0$, which will happen when the direction of propagation makes an angle with the magnetic field, the electric vector is not the wave front, and the total-reflection condition results when electric vector coincides with the wave normal. This happens as $(\text{grad } S)^2 \rightarrow \alpha$. The Poynting vector then lies in the wave front. The flow of energy across the surface stops, and the waves get totally reflected.

Conclusion

We have discussed the question of ionosphere at some length, but it will be easily seen, that the method can be applied with the same ease for discussion of propagation of light in a material medium. We have only to utilise the corresponding Lorentz-equation for motion of bound-electrons, which can be made to yield a relation similar to the equation (Z) deduced for the ionosphere. Readers interested in mathematical analysis will easily recognise the method as an application of the method of characteristics, used for discussions of wave-propagation by Hadamard, Debye, and others. So far as the writer is aware the method has not been applied to the microscopic equations of Lorentz, where however the extension does not seem to create any difficulty when the waves are looked upon as possible modes of oscillation of the stationary electro-magnetic field, generated by positive and negative charges, present in the medium.

Physical Laboratory,
Dacca.

Reference

1. Bajpai, R. R. and Mathur, K. B., *Ind. J. Phys.*, **11**, 165 (1937).

Reprinted from *Ind J Phys* **12**, pp. 121-144, 1938 (Indian Association for the Cultivation of Science).

ON THE COMPLETE MOMENT-COEFFICIENTS OF THE D^2-STATISTIC.

BY SATYENDRA NATH BOSE

PROFESSOR OF PHYSICS, UNIVERSITY OF DACCA.

The D^2-statistic was defined by Mahalanobis[1] to measure the divergence between two statistical populations, and the moment-coefficients were also calculated by him by approximative methods. Rajchandra Bose[2] has found the exact distribution of the D^2-statistic* which can be expressed in terms of Bessel functions. He has used the actual distribution function to obtain the moment-coefficients, and has shown that the results previously obtained by Mahalanobis are exact. In the present note a recurrence-formula for the D^2-statistic has been obtained directly without assuming the distribution, and certain properties of these moment-coefficients have been investigated.

2. Let $(a_1, a_2, \dots a_p)$ and $(a'_1, a'_2, \dots a'_p)$ be the observed mean values in two samples Σ_1 and Σ_2 of size n_1 and n_2 respectively drawn from two normal p-variate correlated populations with mean values $(\alpha_1, \alpha_2, \dots \alpha_p)$ and $(\alpha'_1, \alpha'_2, \dots \alpha'_p)$ respectively. The distribution of the differences in observed mean values $(a_1-a'_1, a_2-a'_2, \dots a_p-a'_p)$ can be written in the form :—

$$\text{Const.} \times e^{-\frac{n}{p\beta}\left[\beta_{11}\{(a_1-a'_1)-(\alpha_1-\alpha'_1)\}^2+ \dots 2\beta_{12}\{(a_1-a'_1)-(\alpha_1-\alpha'_1)\}\{(a_2-a'_2)-(\alpha_2-\alpha'_2)\}+\dots\right]}$$

$$\times d(a_1-a'_1), d(a_2-a'_2)\dots d(a_p-a'_p) \qquad \dots \quad (2.1)$$

where β and β_{ij} $(i, j = 1, 2, \dots p)$ are certain functions of the population variances and co-variances whose values have been given by Rajchandra Bose[3] and

$$\frac{2}{n} = \frac{1}{n_1} + \frac{1}{n_2} \qquad \dots \quad (2.2)$$

Then the D^2-statistic is defined by

$$D^2 = D_1^2 - \frac{2}{n} \qquad \dots \quad (2.31)$$

---

* In this paper p and n have been used in the place of P and $\bar{n}$ in the earlier papers of P. C. Mahalanobis and Raj Chandra Bose. The D^2-statistic was intended to be and was defined as a quantity determined entirely by the sample values of the variates. Raj Chandra Bose has investigated the exact distribution and S. N. Bose the moment-coefficients of a modified form of the D^2-statistic in which the population values of the variances and co-variances have been substituted for the corresponding sample estimates.—*Editor, Sankhyā.*

and $$D_1^2 = \frac{1}{p\beta} \mathop{S}_{i,j=1}^{i,j=p} [\beta_{ij}(a_i - a'_i)(a_j - a'_j)] \qquad \dots \quad (2.32)$$

The population value of D^2 is Δ^2 which is defined by

$$\Delta^2 = \frac{1}{p\beta} \mathop{S}_{i,j=1}^{i,j=p} [\beta_{ij}(\alpha_i - \alpha'_i)(\alpha_j - \alpha'_j)] \qquad \dots \quad (2.4)$$

3. If we put

$$(a_i - a'_i) = x_i, \text{ and } (\alpha_i - \alpha'_i) = \xi_i$$

we can write

$$D_1^2 = \frac{1}{p\beta} [\beta_{11}x_1^2 + \beta_{22}x_2^2 + \dots 2\beta_{12}x_1x_2 + \dots] \qquad \dots \quad (3.1)$$

$$\Delta^2 = \frac{1}{p\beta} [\beta_{11}\xi_1^2 + \beta_{22}\xi_2^2 + \dots 2\beta_{12}\xi_1\xi_2 + \dots] \qquad \dots \quad (3.2)$$

If we now use a linear transformation which changes the quadratic form

$$\frac{1}{2\beta} \mathop{S}_{i,j=1}^{i,j=p} (\beta_{ij}x_ix_j) \text{ to } (y_1^2 + y_2^2 + \dots y_p^2) \qquad \dots \quad (3.3)$$

then the same transformation will change

$$\frac{1}{2\beta} \mathop{S}_{i,j=1}^{i,j=p} (\beta_{ij}\xi_i\xi_j) \text{ to } (\eta_1^2 + \eta_2^2 + \dots \eta_p^2) \qquad \dots \quad (3.4)$$

where $(\eta_1, \eta_2, \dots \eta_p)$ are connected with $(\xi_1, \xi_2, \dots \xi_p)$ by the same linear transformation which connects $(y_1, y_2, \dots y_p)$ and $(x_1, x_2, \dots x_p)$

Also $\frac{1}{2} \mathop{S}_{i,j=1}^{i,j=p} (\beta_{ij}x_i\xi_j)$ will be transformed to $(y_1\eta_1 + y_2\eta_2 + \dots y_p\eta_p)$ ... (3.5)

We can then write

$$\tfrac{1}{2}p \cdot D_1^2 = y_1^2 + y_2^2 + \dots y_p^2 \qquad \dots \quad (3.6)$$

$$\tfrac{1}{2}p \cdot \Delta^2 = \eta_1^2 + \eta_2^2 + \dots \eta_p^2 \qquad \dots \quad (3.7)$$

Equation (2.1) can then be written in the form :—

$$\text{Const.} \times e^{-\frac{np}{4}[D_1^2 + \Delta^2 - 2D_1 . \Delta \cos\theta]} \times dV \qquad \dots \quad (3.8)$$

where $dV = dy_1, dy_2 \dots dy_p$

4. Let $\mu_k'(D_1^2)$ be the k-th raw moment of D_1^2, and let

$$M_k = \int_0^\infty D_1^{2k}.e^{-\frac{np}{4}[D_1^2+\Delta^2-2D_1.\Delta\cos\theta]}.dV \quad \dots \quad (4.1)$$

where the single sign of integration stands for the p-fold integration.

Then $$\mu_k'(D_1^2) = C.M_k \quad \dots \quad (4.2)$$

we also write $$np/4 = t \quad \dots \quad (4.3)$$

and $$M_k = \int_0^\infty D_1^{2k}.e^{-t(D_1^2+\Delta^2-2D_1.\Delta\cos\theta)}.dV \quad \dots \quad (4.11)$$

Then

$$\frac{\partial M_k}{\partial \Delta} = -2t \int D_1^{2k}(\Delta - D_1\cos\theta)e^{-(tD_1^2+\Delta^2-2D_1\Delta\cos\theta)}.dV \quad \dots \quad (4.4)$$

$$= -2t.\Delta.M_k + 2t\int D_1^{2k+1}\cdot\cos\theta\cdot e^{-t(D_1^2+\Delta^2-2D_1\Delta\cos 0)}.dV \quad \dots \quad (4.41)$$

Again

$$\frac{\partial M_k}{\partial t} = -\int D_1^{2k}(D_1^2+\Delta^2-2D_1\cdot\Delta\cos\theta)e^{-t(D^2+\Delta^2-2D_1\Delta\cos 0)}.dV \quad \dots \quad (4.5)$$

$$= -M_{k+1} - \Delta^2\cdot M_k + \frac{\Delta}{t}\left[\frac{\partial M_k}{\partial \Delta} + 2t\cdot\Delta\cdot M_k\right] \quad \dots \quad (4.51)$$

Therefore

$$M_{k+1} = \Delta^2\cdot M_k + \frac{\Delta}{t}\cdot\frac{\partial M_k}{\partial \Delta} - \frac{\partial M_k}{\partial t} \quad \dots \quad (4.6)$$

5. Now actually

$$M_0 = \int_0^\infty e^{-t[(y_1-\eta_1)^2+(y_2-\eta_2)^2+\dots(y_p-\eta_p)^2]}\cdot dy_1\, dy_2 \dots dy_p$$

$$= \left(\frac{\pi}{t}\right)^{p/2} \quad \dots \quad (5.1)$$

Therefore

$$\frac{\partial M_0}{\partial t} = -\frac{p(\pi)^{p/2}}{2(t)^{(p+2)/2}} = \frac{p\cdot(\pi)^{p/2}}{2(t)^{(p+2)/2}} = \frac{p\cdot M_0}{2t} \quad \dots \quad (5.2)$$

$$\frac{\partial M_0}{\partial \Delta} = 0 \quad \dots \quad \dots \quad \dots \quad \dots \quad (5.3)$$

Also $\mu_0'(D_1^2) = c \cdot M_0 = 1$ **and therefore**

$$c = \left(\frac{t}{\pi}\right)^{p/2} \quad \ldots \quad \ldots \quad \ldots \quad \ldots \quad (5.4)$$

We can use the above initial values to calculate $M_1, M_2, \ldots M_k$ etc., and hence with the help of equation (4.2) obtain the raw moments (about any arbitrary origin) of D^2 which can be written in the form :—

$$\mu_k'(D_1^2) = \left(\frac{t}{\pi}\right)^{p/2} M_k \quad \ldots \quad \ldots \quad (5.5)$$

6. It will be convenient to use new variables

$$u = \Delta^2, \quad \text{and} \quad s = 1/t \quad \ldots \quad \ldots \quad (6.1)$$

Equation (4.6) can now be written as

$$M_{k+1} = u \cdot M_k + 4\, u \cdot s \cdot \frac{\partial M_k}{\partial u} + 2s^2 \cdot \frac{\partial M_k}{\partial s} \quad \ldots \quad (6.2)$$

Remembering that $M_0 = (\pi \cdot s)^{p/2}$, we note from the form of equation (6.2) that M_k is a homogeneous function of order $(k+\frac{1}{2}p)$ in u and s.

Therefore

$$M_{k+1} = u \cdot M_k + 2u \cdot s \cdot \frac{\partial M_k}{\partial u} + 2s\left[u\frac{\partial M_k}{\partial u} + s\frac{\partial M_k}{\partial s}\right]$$

$$= u \cdot M_k + 2u \cdot s \cdot \frac{\partial M_k}{\partial u} + 2s(k+\tfrac{1}{2}p) \cdot M_k, \text{ by Euler's theorem} \quad \ldots \quad (6.3)$$

Thus

$$M_{k+1} = [s(2k+p)+u]M_k + 2u.s.\frac{\partial M_k}{\partial u} \quad \ldots \quad (6.4)$$

Writing briefly μ_k' for $\mu_k'(D_1^2)$, we have

$$\mu_k' = \frac{M_k}{(2s\pi)^{p/2}} \quad \ldots \quad (6.5)$$

Equation (6.4) may now be written in the form :—

$$\mu_{k+1}' = [s(2k+p)+u]\mu_k' + 2u \cdot s \cdot \frac{\partial \mu_k'}{\partial u} \quad \ldots \quad (6.6)$$

This is the fundamental recurrence formula. We notice that μ_k' is a homogeneous function of the k-th degree in s and u.

The general expression for μ_k' can be easily obtained.

$$\mu_k' = c_k \cdot u^k + c_{k-1} \cdot u^{k-1} \cdot s + \ldots c_0 \cdot s^k \qquad \ldots \quad (7.1)$$

Then from equation (6.6) we get

$$\mu_{k+1}' = C_k . u^{k+1} + [c_{k-1} + (p+4k)c_k] u^k . s + \ldots + [c_{k-2} + (p+4k-2)c_{k-1}] u^{k-1} . s^2 + \ldots$$
$$+ [c_0 + (p+2k+2)c_1] u . s^k + (p+2k)c_0 . s^{k+1} \qquad \ldots \quad (7.2)$$

From this equation we can immediately write down the coefficients occurring in μ_k' :—

μ_0'.........1

μ_1'.........1, p

μ_2'.........1, $2(p+2)$, $p(p+2)$,

μ_3'.........1, $3(p+4)$, $3(p+2)(p+4)$, $p(p+2)(p+4)$ etc.

The general coefficients can be easily proved by induction to be of the form :—

$$1 \, {}^kc_1[p+2(k-1)], \; {}^kc_2[\{p+2(k-1)\}\{p+2(k-2)\}], \ldots\ldots$$
$$\ldots\ldots {}^kc_r[\{p+2(k-1)\}\{p+2(k-2)\} \ldots\ldots \{p+2(k-r)\}],$$
$$\ldots {}^kc_k[\{p+2(k-1)\}\{p+2(k-2)\} \ldots p] \ldots\ldots \qquad \ldots \quad (7.4)$$

Remembering that $u = \Delta^2$, and $s = 1/2t = 2/(np)$ we finally obtain the general formula for the k-th raw moment-coefficient :—

$$\mu_k'(D_1^2) = \Delta^{2k} + \frac{2 . \, {}^kc_1(p+2k-2)}{np} . \Delta^{2k-2} + \frac{2^2 \cdot {}^kc_2(p+2k-2)(p+2k-4)}{n^2p^2} \cdot \Delta^{2k}$$

$$+ \ldots + \frac{2^k . (p+2k-2)(p+2k-4) \ldots (p+2)p}{n^k \cdot p^k} \qquad \ldots \quad (7.5)$$

$$= \underset{\alpha \to \infty}{\text{Lt.}} \; \Delta^{2k} \cdot F\left(-k; -k-\tfrac{1}{2}p+1; \alpha; \frac{4\alpha}{np\Delta^2}\right) \qquad \ldots \quad (7.6)$$

The two formulae (7.5) and (7.6) are in agreement with the results obtained by Rajchandra Bose[2] in equations (8.9) and (8.91) on *p*. 152 of *Sankhyā*, Vol. 2, Part 2.

8. It is clear from the preceding considerations that $\mu_k'(D^2)$'s, whose recurrence formulae have been calculated, belong to a Gaussian distribution in p-variables; it may be therefore interesting to deduce certain mathematical properties of these functions.

We start from the recurrence formula

$$\mu'_{k+1} = [s(2k+p)+u]\mu'_k + 2us\frac{\partial \mu'_k}{\partial s} \qquad \dots \quad (6.7)$$

Putting $\mu'_k = (2s)^k \cdot \psi_k(u/2s)$ and using $\chi = u/2s$ as the new variables, we have

$$\psi_{k+1}(x) = [(k+\tfrac{1}{2}p)+x]\psi_k(x)+x\psi'_k(x) \qquad \dots \quad (8.1)$$

Multiplying by $e^x. x^{k+\frac{1}{2}p-1}$, we have

$$e^x \cdot x^{k+\frac{1}{2}.-1}\psi_{k+1}(x) = \frac{d}{dx}[e^x \cdot x^{k+\frac{1}{2}p}\psi_k(x)],$$

$$\text{or} \quad e^x \cdot x^{k+\frac{1}{2}p+1}\psi_{k+1}(x) = x^2 \cdot \frac{d}{dx}[e^x. x^{k+\frac{1}{2}p}\psi_k(x)] \qquad \dots \quad (8.2)$$

By regular descent, since $\psi_0(x) = 1$, it follows easily that

$$e^x \cdot x^{k+\frac{1}{2}p}\psi_k(x) = \left(x^2\frac{d}{dx}\right)^k [e^x \cdot x^{\frac{1}{2}p}] \qquad \dots \quad (8.3)$$

which may be regarded as a solution of the recurrence formula.

If we put $t = 1/x = 2s/u$, we get,

$$\mu'_k = (2s)^k. \psi_k(x) = (2st)^k. e^{-1/t}. t^{\frac{1}{2}p}\left(-\frac{d}{dt}\right)^k [e^{1/t} \cdot t^{-\frac{1}{2}p}] \qquad \dots \quad (8.4)$$

which is identical with (8.6) in Rajchandra Bose's paper.[3]

Starting from the equation (8.3) and denoting by y, the expression $e^x x^{\frac{1}{2}p}$, and by D,. the operator $x^2 \dfrac{d}{dx}$, we easily find that y satisfies the differential equation

$$\frac{1}{x^2} Dy = \left(\tfrac{1}{2}p \cdot \frac{1}{x}+1\right)y.$$

Operating on both sides of this equation by D^{k+1} and remembering that $D^k y = \psi_k(x)$ one easily finds that $\psi_k(x)$ satisfies the differential equation

$$x\frac{d^2\psi_k}{dx^2}+(\tfrac{1}{2}p+x)\frac{d\psi_k}{dx} - k\psi_k = 0 \qquad \dots \quad (8.5)$$

If $$\psi_k(-x) = (-1)^k \cdot f_k(x),$$

$f_k(x)$ satisfies the differential equation

$$x\frac{d^2 f_k}{dx^2}+(\tfrac{1}{2}p-x)\frac{df_k}{dx}+k \cdot f_k = 0 \qquad \dots \quad (8.6)$$

The solutions are easily seen to be polynomials of order k, when k is an integer, and these in case of (8.6) have been studied by Sonine, Gegenbauer, and others. The differential equations have also been studied by Whittaker and others, and give rise to solutions of the confluent hypergeometric types.

9. There is an alternative method of arriving at the differential equation for μ', which is here appended as the method is capable of wide generalisation.

The Gaussian distribution function

$$F = \frac{1}{(2\pi t)^{\frac{1}{2}p}} \cdot e^{-\Sigma(x_r..y_r)^2/2t}, \text{ where } r = 1, 2, 3, \ldots p$$

can be easily seen to satisfy the differential equation

$$\frac{\partial^2 F}{\partial x_1^2} + \frac{\partial^2 F}{\partial x_2^2} + \ldots + \frac{\partial^2 F}{\partial x_p^2} = 2\frac{\partial F}{\partial t} \qquad \ldots \quad (9.1)$$

where instead of $(x_1, x_2, \ldots x_p)$, we may also take $(y_1, y_2 \ldots y_p)$ as variables.

The expression $\frac{1}{(2\pi t)^{\frac{1}{2}p}} \cdot e^{-\Sigma_r(x_r-y_r)^2/2t}$ denoted by F also satisfies the differential equation

$$\frac{\partial^2 F}{\partial y_1^2} + \frac{\partial^2 F}{\partial y_2^2} + \ldots\ldots + \frac{\partial^2 F}{\partial y_p} = 2\frac{\partial F}{\partial t} \qquad \ldots \quad (9.11)$$

Now, since this differential equation is linear which means that if $\psi_1, \psi_2, \ldots$ etc. are each a solution of the differential equation then $\psi_1+\psi_2+\ldots$ etc. is a solution of the same equation, and since an integral is nothing but a sum, therefore,

$$\psi(y_1, y_2, \ldots y_p, t) \equiv \int \phi(x_1, x_2, \ldots x_p) \cdot \frac{1}{(2\pi t)^{\frac{1}{2}p}} \cdot e^{-\Sigma(x_r-y_r)^2/2t} dx_1, dx_2 \ldots dx_p) \ \ldots \quad (9.2)$$

which consists of a sum of terms like $\frac{1}{(2\pi t)^{\frac{1}{2}p}} \cdot e^{-\Sigma(x_r-y_r)^2/2t}$ each satisfying the differential equation and each multiplied by a term like $\phi(x_1, x_2 \ldots x_p) \cdot dx_1\, dx_2 \ldots dx_p$ not depending upon the independent variables the y's, is easily seen to satisfy the equation (9.11).

Therefore

$$\mu_k'(y_1, y_2, \ldots y_p, t) = \int (x_1^2+x_2^2+\ldots x_p^2)^k \cdot \frac{1}{(2\pi t)^{\frac{1}{2}p}} \cdot e^{-\Sigma(x_r-y_r)^2/2t} \times dx_1 . dx_2 \ldots dx_p \quad (9.3)$$

also satisfies the same differential equation.

An orthogonal transformation of (x's) and (y's) will not change the integrand; also making a transformation

$$\left.\begin{array}{l} x_1' = \lambda x_1', \; x_2 = \lambda x_2', \quad \ldots\ldots\ldots\ldots t = \lambda^2 t' \\ y_1 = \lambda y_1', \; y_2 = \lambda y_2', \quad \ldots\ldots\ldots\ldots\ldots\ldots\ldots \end{array}\right\} \quad \ldots \quad (9.4)$$

multiplies the integral by λ^{2k}. Hence

$$\mu_k'(y_1, y_2, \ldots y_p, t) = \mu_k'[y_1^2+y_2^2+\ldots y_p^2, t] \quad \ldots \quad (9.41)$$

Putting
$$y_1^2+y_2^2+\ldots y_p^2 = R^2 \quad \ldots \quad (9.42)$$

it follows that
$$\mu_k'(y_1, y_2, \ldots y_p, t) = \mu_k'(R^2, t) \quad \ldots \quad (9.5)$$

also
$$\mu_k'(\lambda R^2, \lambda t) = \lambda^k \mu_k'(R^2, t) \quad \ldots \quad (9.51)$$

or μ_k' is a homogeneous function of (R^2, t) of the kth degree. Therefore μ_k' is a homogeneous solution of the kth degree of (R^2, t) alone of the equation (9.1). Making a change of variables to R and angle-coordinates in a p-dimensional space, μ_k' is seen to satisfy the differential equation

$$\frac{1}{R^{p-1}} \cdot \frac{\partial}{\partial R}\left[R^{p-1}\frac{\partial \mu_k'}{\partial R}\right] = 2\frac{\partial \mu_k'}{\partial t} \quad \ldots \quad (9.6)$$

This transformation is seen easily as $\square_p^2 \cdot F = \text{Div. grad } F$, and when F is a function of R alone, the expression follows as the hyper-spherical element perpendicular to R is proportional to R^{p-1}. Thus

$$\square_p^2 \cdot F = \frac{1}{R^{p-1}} \cdot \frac{\partial}{\partial R}\left[R^{p-1} \cdot \frac{\partial F}{\partial R}\right] = \frac{\partial^2 \mu_k'}{\partial R^2} + \frac{p-1}{R} \cdot \frac{\partial \mu_k'}{\partial R} = 2\frac{\partial \mu_k'}{\partial t} \quad \ldots \quad (9.7)$$

Finally choosing $R^2 = 2t \cdot x$, $\mu_k' = (2t)^k \cdot \psi_k(x)$ ... (9.8)

the differential equation of $\psi_k(x)$ is easily seen to be

$$x\frac{d^2\psi_k}{dx^2} + (\tfrac{1}{2}p+x)\frac{d\psi_k}{dx} - k\psi_k = 0 \quad \ldots \quad (9.9)$$

which is identical with the equation (8.5)

10. The method sketched above enables one to calculate the moment functions even when k is not an integer; actually however, unless k's are integers, μ_k''s will not be expressed by a polynomial but by an infinite series. The differential equation itself determines all the constants except one, which can be determined from simple considerations.

I will conclude my remarks by deducing certain properties of the general differential equation, which will be useful for determining the form of the general distribution function about which I wish to make some remarks in future.

Consider the equation

$$\Box_p^2 . F = \frac{\partial^2 F}{\partial x_1^2} + \frac{\partial^2 F}{\partial x_2^2} + \dots \frac{\partial^2 F}{\partial x_p^2} = 2 \frac{\partial F}{\partial t} \qquad \dots (10.1)$$

Suppose ψ is a solution of the differential equation, we then seek a solution of the equation of the form $F = \psi \, . \, \chi$.

Then

$$\psi \, . \, (\Box_p^2 . \chi) = \chi . (\Box_p^2 . \psi) + 2(D\psi, D\chi) = 2 \left[\psi \frac{\partial \chi}{\partial t} + \chi . \frac{\partial \psi}{\partial t} \right] \qquad \dots (10.2)$$

or
$$\psi(\Box_p^2 . \chi) + 2(D\psi, D\chi) = 2\psi \frac{\partial \chi}{\partial t} \qquad \dots (10.3)$$

Now if
$$\psi = \frac{1}{(2\pi t)^{\frac{1}{2}p}} . e^{-\Sigma x_r^2/2t}, \text{ then } \frac{\partial \psi}{\partial x_r} = -\frac{x_r}{t} . \psi \qquad \dots (10.4)$$

Therefore χ satisfies the differential equation

$$\Box_p^2 . \chi = -\frac{2}{t} \left[x_1 \frac{\partial \chi}{\partial x_1} + x_2 \frac{\partial \chi}{\partial x_2} + \dots + t \frac{\partial \chi}{\partial t} \right] \qquad \dots (10.5)$$

Suppose $F(x_1, x_2, \dots x_p, t)$ is a solution of the equation

$$\Box_p^2 . F = 2 \frac{\partial F}{\partial t} \qquad \dots (10.6)$$

Then
$$\psi(x_1, x_2, \dots x_p, t) = F(x_1/t, x_2/t, \dots x_p/t, -1/t) \qquad \dots (10.7)$$

will satisfy the differential equation (10.5), for putting $x_r' = x_r/t$, and $t' = -1/t$, it follows that

$$\frac{\partial \psi}{\partial x_r} = -t' \frac{\partial F}{\partial x_r'}, \qquad \dots (10.8)$$

$$t \frac{\partial \psi}{\partial t} + \Sigma x_r \frac{\partial \psi}{\partial x_r} = -t' \frac{\partial F}{\partial t'} \qquad \dots (10.81)$$

$$\square_p^2 . \psi - \frac{2}{t}\left[t\frac{\partial\psi}{\partial t} + \ldots\right] = t'^2\left[\square_p'^2 . F - 2\frac{\partial F}{\partial t'}\right] = 0 \quad \ldots \quad (10.82)$$

11. This theorem enables us to construct solutions of the fundamental differential equation of the form

$$\frac{1}{(2\pi t)^{\frac{1}{2}p}} . e^{-\Sigma x_r^2/2t} . H\left(\frac{x_1}{t}, \frac{x_2}{t}, \ldots \frac{x_p}{t}, -\frac{1}{t}\right) \quad \ldots \quad (11.1)$$

where $H(x_1, x_2, \ldots x_p, t)$ is a solution of the same differential equation (9.1).

We have seen that $\mu_k'(R^2, t)$ the moment functions, are solutions of the differential equation (9.1)

$$\mu_k' = (2t)^k . \psi_k(R^2/2t) = 2t^k . \psi_k(y) \quad \ldots \quad (11.2)$$

where $\psi_k(y)$ satisfies the differential equation (8.4). Therefore,

$$\mu_k'\left(\frac{x_1}{t}, \frac{x_2}{t}, \ldots \frac{x}{t}, -\frac{1}{t}\right) = (-1)^k . \frac{2^k}{t^k} . \psi_k\left[\frac{x_1^2 + x_2^2 + \ldots x^2}{2t^2(-1/t)}\right]$$

$$= (-1)^k . (2/t)^k . \psi_k(-y) \quad \ldots \quad (11.3)$$

where $(-1)^k . \psi_k(-y) = f(y)$ which satisfies the equation (8.5).

If $y = \Sigma x_1^2/2t$, and $(-1)^k . \psi_k(-y)$ are the polynomials of Sonine-Gegenbauer, it follows that

$$\frac{1}{(2\pi t)^{\frac{1}{2}p}} . e^{-y} . (-1)^k . (2/t)^k . \psi_k(-y) \qquad \text{are solutions of (9.1).}$$

12. A distribution function which satisfies the fundamental equation, and is of a certain type can be written in the form

$$\frac{1}{(2\pi t)^{\frac{1}{2}p}} . e^{-y} . \left[1 - C_1 . \frac{\psi_1(-y)}{t} + \ldots (-1)^k . C_k . \frac{\psi_k(-y)}{t^k} + \ldots\right] \text{where } y = \Sigma x_r^2/2t$$

Certain orthogonal properties of the Sonine polynomials will be useful to recall here. The functions $S_k(x) = (-1)^k . C_k . \psi_k(-x)$ satisfies, we have seen the differential equation

$$x\frac{d^2 S_k}{dx^2} + (\tfrac{1}{2}p - x)\frac{dS_k}{dx} + k . S_k = 0 \quad \ldots \quad (8.5)$$

which can be written in the self-adjoint form,

$$\frac{d}{dx}\left(x^{\frac{1}{2}p}.e^{-x}.\ \frac{dS_k}{dx} \right) + k.x^{\frac{1}{2}p-1}.\ e^{-x}S_k = 0 \qquad \dots \quad (12.2)$$

from which the orthogonal properties follow at once.

Taking two polynomials of orders k and k', we have

$$(k-k') \int_0^\infty e^{-x}.\ x^{\frac{1}{2}p-1},\ S_k.\ S_{k'}\, dx = 0 \qquad \dots \quad (12.3)$$

Hence if $$k \neq k',\ \int_0^\infty e^{-x}.\ x^{\frac{1}{2}p-1}.\ S_k.\ S_{k'}.\ dx = 0 \qquad \dots \quad (12.31)$$

also $$\int_0^\infty e^{-x}\cdot x^{\frac{1}{2}p-1}.\ S_k^2(x)dx \neq 0 = A = 1 \qquad \dots \quad (12.32)$$

since the arbitrary constant in the definition of the polynomial may be so adjusted as to make $A = 1$, which we may call the normalised polynomials.

Suppose the distribution functions to be $\phi(x_1, x_2 \dots x_p, t)$ then the distribution function in R^2 is obviously

$$\int_R^{R+dR} \phi(x_1, x_2 \dots x_p, t)dx_1 \cdot dx_2 \dots dx_p \qquad \dots \quad (12.4)$$

taken between the hyper-spheres of radii R and $(R+dR)$.

It will take the form

$$A \cdot \psi(R^2, t) \cdot R^{p-1} \cdot dR = \psi'(D, t) \cdot D^{\frac{1}{2}p-1} \cdot dD \text{ when } D = R^2. \qquad \dots \quad (12.41)$$

We shall call $\psi'(D, t)$ the distribution function, when the number of variables is p.

Then $$\int_0^\infty \psi'(D, t) \cdot D^{\frac{1}{2}p-1} \cdot dD = 1 \qquad \dots \quad (12.5)$$

$$\int_0^\infty D^k \cdot \psi'(D, t) \cdot D^{\frac{1}{2}p-1} \cdot dD = \mu_k(t) \qquad \dots \quad (12.51)$$

It is also clear that if $y = D/2t$, then substituting the expansion of $S_k\ (y)$ which can be easily obtained from the differential equation or from the expression for μ_k' found before, we get

$$\int_0^\infty (2t)^k \cdot S_k(y) \cdot \psi'(D,t) \cdot D^{\frac{1}{2}p-1} \cdot dD$$

$$= C_k \int_0^\infty \Big[D^k - k(\tfrac{1}{2}p+k-1) \cdot D^{k-1} \cdot (2t) + \frac{k(k-1)}{2!}(\tfrac{1}{2}p+k-1)(\tfrac{1}{2}p+k-2) \cdot D^{k-2} \cdot (2t)^2$$

$$+ \ldots \Big] \times \psi(D,t) \cdot D^{\frac{1}{2}p-1} \cdot dD \qquad \ldots \quad (12.6)$$

$$= C_k \Big[\mu_k' - \frac{k(\tfrac{1}{2}p+k-1)}{1!} \mu_{k-1}' \cdot (2t) + \text{etc} \ldots\Big] \qquad \ldots \quad (12.61)$$

This expression is thus calculable in terms of the moments of the different orders up to k.

Assuming now

$$\psi(D,t) = \frac{1}{(2\pi t)^{\frac{1}{2}p}} \cdot e^{-y}\Big[1 + \frac{A_1}{t} S_1(y) + \frac{A_2}{t^2} S_2(y) + \ldots + \frac{A_k}{t^k} \cdot S_k(y) + \ldots\Big] \qquad \ldots \quad (12.7)$$

and remembering that the above series is so chosen that each term, and therefore the whole series, if convergent, satisfies the differential equation $\Box_p^2 F = 2\partial F/\partial t$, we have:

$$\int \psi(D,t) \cdot D^{\frac{1}{2}p-1} \cdot (2t)^k S_k(y) \cdot dD$$

$$= \int \frac{1}{(2\pi t)^{\frac{1}{2}p}} \cdot e^{-y}\Big[1 + \frac{A}{t} S_1(y) + \ldots \frac{A_k}{t_k} \cdot S_k(y) + \ldots\Big](2t)^k S_k(y)(2t)^p \cdot y^{\frac{1}{2}p-1} \cdot dy$$

$$= \frac{1}{(2\pi t)^{\frac{1}{2}p}} \cdot 2^k A_k$$

$$= C_k\Big[\mu_k' - \frac{k(\tfrac{1}{2}p+k-1)}{1!} \mu_{k-1}'(2t) + \frac{k(k-1)}{2!}(\tfrac{1}{2}p+k-1)(\tfrac{1}{2}p+k-2)\mu_{k-2}'(2t)^2 + \ldots\Big] \quad (12.8)$$

From (12.6) since $y = D/2t$, we get finally

$$A_k = \frac{(2\pi)^{\frac{1}{2}p}}{2^k} \cdot C_k \cdot \Big[\mu_k' - \frac{k(\tfrac{1}{2}p+k-1)}{1!} \mu_{k-1}' + \ldots\Big] \qquad \ldots \quad (12.9)$$

which determines the constants A_k in the distribution formula in terms of the moments. Hence the distribution function can be calculated in terms of the moments. Rajchandra Bose's distribution function obviously can be put in the form indicated above, the actual calculation would give interesting integral relations involving the Bessel functions.

REFERENCES

1. P. C. Mahalanobis: On Tests and Measures of Divergence: *Jour. Asiat. Soc. Beng.* XXVI (1930), 541–588.

P. C. Mahalanobis: On the Generalized Distance in Statistics: *Proc. Nat. Inst. Sc. India*, Vol. II. No. 1 (1936), 49–55.

2. Rajchandra Bose: On the Exact Distribution and Moment-coefficients of the D^2-Statistic: *Sankhyā*, Vol. 2, Part 2 (1936), 143–154.

3. Rajchandra Bose: On the Distribution of Differences of Mean Values in Two Samples and the Definition of the D^2-Statistic: *Sankhyā*, Vol. 2, Part 4 (1936), 379–384.

1st June, 1936.
Ramna, Dacca.

Reprinted from *Sankhyā — Ind J Stat* **2**, pp. 385-396, 1936 (Statistical Publishing Society, Calcutta).

SANKHYĀ

THE INDIAN JOURNAL OF STATISTICS

Edited by: P. C. MAHALANOBIS

VOL. 3. PART 2. 1937.

ON THE MOMENT-COEFFICIENTS OF THE D^2-STATISTIC AND CERTAIN INTEGRAL AND DIFFERENTIAL EQUATIONS CONNECTED WITH THE MULTIVARIATE NORMAL POPULATION

By S. N. BOSE,

UNIVERSITY OF DACCA.

CONTENTS.

1. Algebraic Identities deduced from the Polynomial Form of the Moment-coefficients of the D^2-statistic 106
2 Algebraic Identities deduced from the Differential Form of the Moment-Coefficients ... 107
3. The Integral Equation connected with the Moment-coefficients and its Solutions ... 110
4 The Spherical Form of the Fundamental Differential Equation connected with the Multivariate Normal Population 113
5. The Differential Equation in the Parabolic Form 113
6. The Differential Equation in the Product Form 114
7. Further Reduction of the Parabolic Form to the Polar Form 115
8. Solutions of the Differential Equation in different cases 116
9. On the Series Solution of the Differential Equation 120

INTRODUCTION.

In a previous paper[1] I have considered certain recursion formulæ and relations between the moment-coefficients of the D^2-statistic. Starting with certain further algebraic identities between the moment-coefficients of different orders of the D^2-statistic, I reach a number of fundamental integral and differential equations connected with the multivariate normal distribution. A number of solutions of these equations are discussed in the present paper.

[1] "On the Complete Moment-Coefficients of the D^2-statistic." *Sankhyā*, Vol. 2 (4), 1936, 385-396.

Algebraic Identities deduced from the Polynomial Form of the Moment-Coefficients of the D^2-statistic.

1. I start with the identities that exist between the moment-functions of the different orders of the D^2-statistic, which can be regarded as the square modulus statistic for a Gaussian distribution in p-variates.

It has been already proved[2] that

$$\mu_k'(l^2) = \int (x_1^2+x_2^2+\ldots+x_p^2)^k \cdot \frac{e^{-\Sigma_r(x_r-y_r)^2/2t}}{(2\pi t)^{p/2}} \cdot dV \qquad \ldots \quad (1.1)$$

where $\qquad x_1^2+x_2^2+\ldots x_p^2 = l^2 \qquad \ldots \quad (1.11)$

and $\qquad y_1^2+y_2^2+\ldots y_p^2 = \Delta^2 \quad (\text{or } u) \qquad \ldots \quad (1.12)$

and μ_k is a homogeneous function of $(\Delta^2, 2t)$ of the k-th degree of the form[3] :

$$\mu_k(l^2) = \beta^k + k(k+p/2-1)\cdot\beta^{k-1}(2t)+\ldots$$
$$+\frac{\Gamma(k+1)\cdot\Gamma(k+p/2)\cdot\beta^{k-r}(2t)^r}{\Gamma(r+1)\cdot\Gamma(k+1-r)\cdot\Gamma(k+p/2-r)}+\ldots+\frac{\Gamma(k+p/2)}{\Gamma(p/2)}\cdot(2t)^k \qquad \ldots \quad (1.2)$$

It should be noted here that in (1·2) and in subsequent equations we are writing μ_k for μ_k' and β for Δ^2 for convenience of printing.

It is easily seen by direct algebraical methods that

$$\beta^k = \mu_k - k(k+p/2-1)\mu_{k-1}\cdot(2t)+\ldots+(-1)^r\cdot\mu_{k-r}\cdot(2t)^r\frac{\Gamma(k+1)\Gamma(k+p/2)}{\Gamma(r+1)\,\Gamma(k+1-r)\,\Gamma(k+p/2-r)}$$
$$+\ldots+(-1)^k\cdot\frac{\Gamma(k+p/2)}{\Gamma(p/2)}\cdot(2t)^k \equiv B_k \qquad \ldots \quad (1.3)$$

So that the various identical relations between μ's involving p and t, which will be called the identities of the first type, can be written in the form $(B_k)^{k'} = (B_{k'})^k$, or symbolically, $(k\ k') = 0$. These identities can be written down by taking any two expressions B_k and $B_{k'}$ and their existence is self-evident from their form.

Also since B_k's are polynomials in p or in $(2t)$, by taking any three expressions $\beta^a = B_a$, $\beta^b = B_b$, and $\beta^c = B_c$, we can eliminate (β, p) or (β, t) and thus arrive at identities involving alternatively either μ's and t's or μ's and p's. These identities can be written as

$$[abc]_p = 0, \quad [abc]_t = 0 \text{ etc.} \qquad \ldots \quad (1.4)$$

where the suffix denotes the variable which has been eliminated in the process.

---

[2] *Sankhyā*, 2(4), 1936, p. 392, equation (9).

[3] *Sankhyā*, 2(4), 1936, p. 389, equation (7.5).

Finally by taking any four of the expressions we can eliminate β, p and t, arrive at relations between μ's alone, which can be written as

$$\{a\, b\, c\, d\} = 0 \qquad \dots \quad (1.5)$$

These idential relations are interesting, and may be regarded as characteristic of the modulus distribution of the Gaussian type.

Direct algebraical computation will easily verify the following typical relations :—

$$2pt^2 - 4t\mu_1 + (\mu_2 - \mu_1^2) = 0 \qquad \dots \quad (1.51)$$

$$[123]_p = 8\mu_1 t + 8t(\mu_1^2 - \mu_2) + (\mu_3 - 3\mu_1\mu_2 + 2\mu_1^3) \qquad \dots \quad (1.52)$$

The expression (1234) = 0 is more complicated; by taking B_1, B_2, B_3, B_4, and putting

$$\left.\begin{aligned} &A_2 = \mu_1^2 - \mu_2, \quad A_3 = \mu_3 - 3\mu_1\mu_2 + 2\mu_1^3 \\ \text{and} \quad &A_4 = \mu_4 - 4\mu_2\mu_3 - 3\mu_2^2 + 12\mu_1^2\mu_2 - 6\mu_1^4 \end{aligned}\right\} \qquad \dots \quad (1.61)$$

the existence of the following two quadratic equations can be proved,

$$\left.\begin{aligned} &8\mu_1 t^2 + 8A_2 t + A_3 = 0 \\ \text{and} \quad &A_2 t^2 + A_3 t - A_4/12 = 0 \end{aligned}\right\} \qquad \dots \quad (1.62)$$

from which the bracket (1234) = 0 follows as the t-eliminant of the equations.

Algebraic identities Deduced from the Differential Form of the Moment-coefficient.

2. It is satisfactory to note here that the identities of the above type, which can be arrived at by laborious algebraical methods, can be deduced much more simply by the following analytical procedure, which furnishes the *raison d'être* of identities of the above types, and especially of the existence of the quadratic equations. This method also allows us to investigate the inverse problem, namely, the nature of the distribution function for the modulus when the identities of the above kind are postulated for the moment-functions.

We start from the differential form of the moment-coefficient deduced before[4] for the D^2-statistic :—

$$\mu_k = (2st)^k \cdot e^{-1/t} \cdot t^{p/2} \cdot \left(-\frac{d}{dt}\right)^k (e^{1/t} \cdot t^{-p/2}) \qquad \dots \quad (2.1)$$

[4] *Sankhyā*, 2(4), 1936, p. 390, equation (8.4).

where $t = 2s/u$ according to the present notation.

Then it follows from the fact that μ_k is a homogeneous function of $(2s, u)$ of the k-th degree, that

$$\frac{\mu_k}{(2st)^k} = \frac{\mu_k}{(ut^2)^k} = \frac{\mu_k}{u^k t^{2k}} = \psi_k(t) = e^{-1/t}\cdot t^{p/2}\cdot \left(\frac{-d}{dt}\right)^k (e^{1/t}\cdot t^{-p/2}) \quad \ldots \quad (2.2)$$

So that $\psi(t)$ is a polynomial of $-k$-th degree in t alone. Also the differential form of $\psi(t)$ can be transformed into an integral form. Assuming that

$$\frac{e^{1/t}}{t^{p/2}} = Z(t) = \int_0^\infty e^{-xt}\cdot \phi(x)\cdot dx \quad \ldots \quad (2.3)$$

where $\phi(x)$ is a function whose properties we shall investigate later, it follows that

$$\psi_k(t) = \frac{1}{Z(t)}\cdot \left(-\frac{d}{dt}\right)^k \int_0^\infty e^{-xt}\cdot \phi(x)\cdot dx \quad \ldots \quad (2.31)$$

which can be written in the form :—

$$\psi_k = \frac{\int_0^\infty x^k\cdot e^{-xt}\cdot \phi(x)\cdot dx}{\int_0^\infty e^{-xt}\cdot \phi(x)\cdot dx} \quad \ldots \quad (2.32)$$

From this it follows that

$$\psi_k(t)\cdot \int_0^\infty e^{-xt}\cdot \phi(x)\cdot dx = \int_0^\infty x^k\cdot e^{-xt}\cdot \phi(x)\cdot dx \quad \ldots \quad (2.41)$$

Hence by differentiation with regard to t, we have

$$\frac{d\psi_k}{dt}\cdot \int_0^\infty e^{-xt}\cdot \phi(x)\cdot dx - \psi_k\cdot \int_0^\infty xe^{-xt}\cdot \phi(x)\cdot dx = -\int_0^\infty x^{k+1}\cdot e^{-xt}\cdot \phi(x)\cdot dx \ \ldots \quad (2.42)$$

Remembering the general form of ψ_k in (2.31) it follows that

$$\frac{d\psi_k}{dt} = \psi_1\cdot \psi_k - \psi_{k+1} \quad \ldots \quad (2.43)$$

This is fundamental relation which allows us to calculate the various differential coefficients of ψ's in terms of ψ's of different orders. Thus

$$\left.\begin{aligned} \frac{d\psi_1}{dt} &= \psi_1^2 - \psi_2, \quad \frac{d^2\psi_1}{dt^2} = 2\psi_1^3 - 3\psi_1\psi_2 + \psi_3 \\ \frac{d^3\psi_1}{dt^3} &= 6\psi_1^4 - 12\psi_1^2\psi_2 + 3\psi_2^2 + 4\psi_1\psi_3 - \psi_4 \end{aligned}\right\} \quad \ldots \quad (2.51)$$

Since $(\psi_1 t^2)$ is a linear function of t it follows that the following differential relations are true.

$$\frac{d^2(\psi_1 t^2)}{dt^2} = 0, \quad \frac{d^3(\psi_1 t^2)}{dt^3} = 0, \quad \frac{d^4(\psi_1 t^2)}{dt^4} = 0 \quad \ldots \quad (2.52)$$

$$\left.\begin{aligned} t^2 \cdot \frac{d^2\psi_1}{dt^2} + 4t \cdot \frac{d\psi_1}{dt} + 2\psi_1 = 0 \\ t^2 \cdot \frac{d^3\psi_1}{dt^3} + 6t \cdot \frac{d^2\psi_1}{dt^2} + 3\frac{d\psi_1}{dt} = 0 \end{aligned}\right\} \quad \ldots \quad (2.61)$$

and generally $t^2 \cdot \dfrac{d^k\psi_1}{dt^k} + 2kt \cdot \dfrac{d^{k-1}\psi_1}{dt^{k-1}} + k(k-1) \cdot \dfrac{d^{k-2}\psi_1}{dt^{k-2}} = 0 \quad \ldots \quad (2.62)$

If we now remember that the differential coefficients of ψ_1's are expressible in ψ^k's of different orders, and that ψ's are only μ's divided by powers of t, the existence of the quadratic equations of the type found by algebraical method is evident. This at the same time allows us to write down the equations with much less labour.

Using equations (2.51) and equation (2.2), we have

$$\left.\begin{aligned} t^2(2\psi_1^3 - 3\psi_1\psi_2 + \psi_3) + 4t(\psi_1^2 - \psi_2) + 2\psi_1 = 0 \\ t^2(6\psi_1^4 - 12\psi_1^2\psi_2 + 3\psi_2^2 + 4\psi_1\psi_3 - \psi_4) + 6t(2\psi_1^3 - 3\psi_1\psi_2 + \psi_3) + 3(\psi_1^2 - \psi_2) = 0 \end{aligned}\right\} \quad \ldots \quad (2.63)$$

Now substituting the values of ψ's in terms of μ's the above two equations reduce respectively to (2.65) and (2.66) given below

$$\frac{t^2(2\mu_1^3 - 3\mu_1\mu_2 + \mu_3)}{u^3 t^6} + \frac{4t(\mu_1^2 - \mu_2)}{u^2 t^4} + \frac{2\mu_1}{ut^2} = 0 \quad \ldots \quad (2.64)$$

which can be written, by substituting $2s/u$ for t, in the form

$$8\mu_1 s^2 + 8(\mu_1^2 - \mu_2)s + (2\mu_1^2 - 3\mu_1\mu_2 + \mu_3) = 0 \qquad \ldots \quad (2.65)$$

and $(\mu_1^2 - \mu_2^2)s^2 + (2\mu_1^3 - 3\mu_1\mu_2 + \mu_3)s + (1/12)(6\mu_1^4 - 12\mu_1^2\mu_3 + 3\mu_2^2 + 4\mu_1\mu_3 - \mu_4)$ .. (2.66)

These are the quadratic equations arrived at by direct elimination and referred to in (1.62)

The fundamental { } expression alluded to before in (1.5) can also be written in a form involving the differential co-efficients of ψ_1 's of different orders.

Thus writing

$$\psi_1^k \equiv \frac{d^k \psi_1}{dt^k} \qquad \ldots \quad (2.7)$$

we have also two relations like

$$\frac{d^k (\psi_1 t^2)}{dt} = 0, \quad \frac{d^\sigma (\psi_1 t^2)}{dt^\sigma} = 0 \qquad \ldots \quad (2.81)$$

which can be written in the form

$$\left.\begin{aligned} \psi_1^k \cdot t^2 + 2k \cdot \psi_1^{k-1} \cdot t + k(k-1)\psi_1^{k-1} &= 0 \\ \psi_1^\sigma \cdot t^2 + 2\sigma \cdot \psi_1^{\sigma-1} \cdot t + \sigma(\sigma-1)\psi_1^{\sigma-2} &= 0 \end{aligned}\right\} \qquad \ldots \quad (2.82)$$

Hence the following relation which is equivalent to (1.5) can be arrived at as the t-eliminant of the above two quadratic equations :—

$$[k(k-1)\psi_1^\sigma \cdot \psi_1^{k-2} - \sigma(\sigma-1)\psi_1^{\sigma-2} \cdot \psi_1^k]^2$$

$$= 4[k\sigma(\sigma-1)\psi_1^{k-1} \cdot \psi_1^{\sigma-2} - \sigma k(\sigma-1)\psi_1^{\sigma-1} \cdot \psi_1^{k-2}][\sigma \cdot \psi_1^k \psi_1^{\sigma-1} - k\psi_1^\sigma \cdot \psi_1^{k-1}] \ldots \quad (2.9)$$

There is however one fundamental relation. The other relations deduced should therefore reduce to the fundamental one calculated above, where the different relations between μ's of different orders are taken into account.

The Integral Equation connected with the Moment-coefficients and its Solutions.

3. I shall now take up the integral form of $\psi_k(t)$

$$\psi_k(t) = \frac{\int_0^\infty x^k \cdot e^{-xt} \cdot \phi(x) \cdot dx}{\int_0^\infty e^{-xt} \cdot \phi(x) \cdot dx} \qquad \ldots \quad (2.32)$$

The form shows that $\psi_k(t)$'s are moment-functions of a certain x-distribution and a knowledge of $\phi(x)$ will determine the distribution function to which $\psi(t)$'s can be related. In other words $\phi(x)$ will determine the μ_k's with which $\psi_k(t)$'s are simply connected. The existence of the algebraic identities between the μ's follows from two separate hypotheses—(1) the existence of the above integral relation, and (2) the linear form of $\psi_1(t)\cdot(t)^2$. We can therefore see what form of $\phi(x)$ (which for the present discussion can be regarded as arbitrary at first) is determined by the hypothesis that $\psi_1(t)\cdot t^2$ is a linear function of t.

Since

$$\psi_1(t)=\frac{\int_0^\infty x\cdot e^{-xt}\cdot\phi(x)\cdot dx}{\int_0^\infty e^{-xt}\cdot\phi(x)\cdot dx} \quad \ldots \quad (3.10)$$

That is,

$$\psi_1(t)\cdot\int_0^\infty e^{-xt}\cdot\phi(x)\cdot dx=\int_0^\infty x\cdot e^{-xt}\cdot\phi(x)\cdot dx \quad \ldots \quad (3.11)$$

The assumption of the linear form of $\psi_1(t)\cdot t^2$ implies that

$$(At+B)\int_0^\infty e^{-xt}\cdot\phi(x)\cdot dx=\int_0^\infty x\cdot t^2\cdot e^{-tx}\cdot\phi(x)\cdot dx \quad \ldots \quad (3.12)$$

or

$$\int_0^\infty e^{-xt}\,[x\cdot\phi(x)\cdot t^2-A\cdot t\cdot\phi(x)-B\cdot\phi(x)]\cdot dx=0 \quad \ldots \quad (3.13)$$

The above relation can be transformed by integration by parts. Thus

$$\int_0^\infty e^{-xt}\cdot x\cdot\phi(x)\cdot t^2\cdot dx=\Big[-t\cdot x\cdot\phi(x)\cdot e^{-xt}-\frac{d}{dx}\{x\phi(x)\}\cdot e^{-xt}\Big]_0^\infty+\int\frac{d^2}{dx^2}\{x\phi(x)\}\cdot e^{-xt}\cdot dx \quad \ldots \quad (3.21)$$

Similarly

$$\int_0^\infty t\cdot e^{-xt}\cdot\phi(x)\cdot dx=\big[-\phi(x)\cdot e^{-xt}\big]_0^\infty+\int_0^\infty e^{-xt}\cdot\frac{d\phi}{dx}\cdot dx \quad \ldots \quad (3.22)$$

The relation (3.13) implies

$$\Big[-t\cdot x\cdot\phi(x)\cdot e^{-xt}\cdot-\frac{d}{dx}\{x\phi(x)\}\cdot e^{-xt}+A\cdot\phi(x)\cdot e^{-xt}\Big]_0^\infty$$
$$+\int_0^\infty e^{-xt}\Big\{\frac{d^2}{dx^2}\{x\phi(x)\}-A\cdot\frac{d\phi(x)}{dx}-B\cdot\phi(x)\Big\}\,dx=0 \quad \ldots \quad (3.23)$$

Thus

$$\left\{e^{-xt}\left\{-x\frac{d\phi}{dx}-(tx+1-A)\phi(x)\right\}\right\}_0^\infty+\int_0^\infty e^{-xt}\left\{\frac{d^2}{dx^2}\{x\phi(x)\}-A\cdot\frac{d\phi}{dx}-B\phi\right\}=0 \qquad \ldots \quad (3.24)$$

This shows that $\phi(x)$ is a regular solution existing up to infinity of the following differential equation :

$$\frac{d^2}{dx^2}[x\cdot\phi(x)]-A\cdot\frac{d\phi(x)}{dx}-B\cdot\phi(x)=0 \qquad \ldots \quad (3.25)$$

if $\left(\frac{d\phi}{dx}\right)$ is finite and either $A=1$, or $\phi(x)=0$, when $x=0$.

The differential equation

$$\frac{d^2}{dx^2}(x\cdot\phi)-A\cdot\frac{d\phi}{dx}-B\phi\cdot(x)=0 \qquad \ldots \quad (3.25)$$

can be transformed by the substitution

$$x=R^2/\rho \qquad \ldots \quad (3.26)$$

So that

$$dx=\frac{2R\cdot dR}{\rho}, \text{ and } \frac{d}{dx}=\frac{\rho}{2R}\cdot\frac{d}{dR} \qquad \ldots \quad (3.27)$$

Hence the equation becomes

$$\frac{d^2\phi}{dR^2}-\frac{(2A-3)}{R}\cdot\frac{d\phi}{dR}-\frac{4B\cdot\phi}{\rho}=0 \qquad \ldots \quad (3.28)$$

The different cases for different positive and negative values of A and B can be easily seen to be solvable in terms of the functions of the type $J_\lambda(r)$ or $J_\lambda(ir)$.

When $A=p/2$, $B=1$, the case reduces to that of the D^2-statistic. The differential equation then becomes

$$\frac{d^2\phi}{dR^2}-\frac{(p-3)}{R}\cdot\frac{d\phi}{dR}-\phi=0, \text{ when } \rho=4 \qquad \ldots \quad (3.31)$$

The solution can be easily seen to be of the form $A\cdot R^{(p-2)/2}\cdot I_{\frac{1}{2}(p-2)}(R)$

or

$$\phi(x)=A\cdot(2\sqrt{x})^{(p-2)/2}\cdot I_{\frac{1}{2}(p-2)}(2\sqrt{x}) \qquad \ldots \quad (3.32)$$

Therefore if A is so chosen that

$$\frac{e^{1/t}}{t^{p/2}} = A \cdot \int_0^\infty (2\sqrt{x})^{(p-2)/2} \cdot I_{\frac{1}{2}(p-2)}(2\sqrt{x}) \cdot e^{-xt} \cdot dx \qquad \ldots \quad (3.33)$$

then

$$\psi_k(t) = \frac{A \cdot \int_0^\infty x_k \cdot (2\sqrt{x})^{(p-2)/2} \cdot I_{\frac{1}{2}(p-2)}(2\sqrt{x}) \cdot e^{-xt} \cdot dx}{e^{1/t}\ t^{-p/2}} \qquad \ldots \quad (3.34)$$

or

$$\psi_k(t) = A \cdot \int_0^\infty x^k \cdot e^{(-1/t-x/t)} \cdot t^{p/2} \cdot (2\sqrt{x})^{(p-2)/2} \cdot I_{\frac{1}{2}(p-2)}(2\sqrt{x}) \cdot dx \qquad \ldots \quad (3.35)$$

Remembering that $\quad \psi_k(t) = \frac{\mu_k}{(2st)^k}$ $\qquad \ldots \quad (3.36)$

We have $\psi_k = A \cdot \int_0^\infty (2stx)^k \cdot e^{(-1/t-x/t)} \cdot t^{p/2}(2\sqrt{x})^{(p-2)/2} \cdot I_{\frac{1}{2}(p-2)}(2\sqrt{x}) \cdot dx \ldots$ (3.37)

Assuming $(2stx) = r^2$ as a new variable, then

$$\psi_k = A \cdot \int_0^\infty r^{2k} \cdot e^{-(1/t+r^2/2s)}\ t^{p/2} \cdot \{2\sqrt{(r^2/2st)}\}^{(p-2)/2} \cdot I_{\frac{1}{2}(p-2)}\{2\sqrt{(r^2/2st)}\} \cdot \frac{d(r^2)}{2st} \qquad (3.38)$$

Remerbering that

$$t = \frac{2s}{\Delta^2}, \quad \frac{r^2}{2st} = \frac{r^2\Delta^2}{4s^2} \quad \text{or} \quad 2\sqrt{\frac{r^2}{2st}} = \frac{r\Delta}{s} \qquad \ldots \quad (3.39)$$

$$\text{we have } \mu_k = A \cdot \int_0^\infty \frac{r^{2k}}{s} \cdot e^{-(\Delta^2+r^2)/2s} \cdot 2^{(p-1)/2} \cdot \frac{r^{p/2}dr}{\Delta^{(p-2)/2}} \cdot I_{\frac{1}{2}(p-2)}\left(\frac{r\Delta}{s}\right) \qquad \ldots \quad (3.41)$$

$$= 2^{(p-1)/2} \cdot A \cdot \int_0^\infty \frac{r^{2k} \cdot e^{-(\Delta^2+r^2)2s}}{s \cdot \Delta^{(p-2)/2}} \cdot r^{p/2}dr \cdot I_{\frac{1}{2}(p-2)}\left(\frac{r\Delta}{s}\right) \qquad \ldots \quad (3.42)$$

Now putting $s = 1/n$, we have

$$\mu_k = 2^{(p-1)/2} \cdot A \cdot \int_0^\infty \frac{r^{2k} \cdot n}{\Delta^{(p-2)/2}} \cdot e^{-\frac{1}{2}n(\Delta^2+r^2)} \cdot r^{p/2} \cdot I_{\frac{1}{2}(p-2)}(nr\Delta) \cdot dr \ \ldots \quad (3.43)$$

Thus the μ's can be regarded as the moment functions for the distribution

$$dF = \text{const.}\ n \cdot \frac{e^{-\frac{1}{2}n(\Delta^2+r^2)}}{\Delta^{(p-2/2)}} \cdot r^{p/2} \cdot I_{\frac{1}{2}(p-2)}(nr\Delta) \cdot dr \qquad \ldots \quad (3.44)$$

which agrees with R. C. Bose's formula.[5]

[5] R. C. Bose : "On the Exact Distribution and Moment co-efficients of the D2-Statistic". *Sankhyā*, 2(2), 1936, p. 149, equation (6.4).

Spherical Form of the Fundamental Differential Equation Connected with the Multivariate Normal Population.

4. I shall now discuss the differential equation

$$\frac{\partial^2 F}{\partial x_1^2}+\frac{\partial^2 F}{\partial x_2^2}+\ldots\ldots+\frac{\partial^2 F}{\partial R_p^2}=2\cdot\frac{\partial F}{\partial t} \quad \ldots \quad (4.1)$$

which may be regarded as the fundamental equation for the distribution function of the multivariate normal population in p-variates.

If F is a function of (R, t) where

$$R^2=x_1^2+x_2^2+x_3^2+\ldots+x_p^2 \quad \ldots \quad (4.2)$$

then F satisfies the differential equation

$$\frac{\partial^2 F}{\partial R^2}+\frac{p-1}{R}\quad\frac{\partial F}{\partial R}=2\cdot\frac{\partial F}{\partial t} \quad \ldots \quad (4.3)$$

The Differential Equation in Parabolic Form.

5. When F is a function of the form

$$F(x_1^2+x_2^2+\ldots+x_p^2,\ a_1x_1+\ldots+a_px_p,\ t)$$

by suitably choosing one axis say y_1, perpendicular to the hyperplane

$$a_1x_1+a_2x_2+\ldots\ a_px_p=0 \quad \ldots \quad (5.1)$$

and similar orthogonal axis-system as before, F takes the form

$$F(R, y_1, t)\ \text{or}\ F(y_1, y_1^2+\ldots+y_p^2, t)$$

The equation (4.1) can be transformed to the form

$$\frac{\partial^2 F}{\partial y_1^2}+\frac{1}{r^{p-2}}\cdot\frac{\partial}{\partial r}\left(r^{p-2}\cdot\frac{\partial F}{\partial r}\right)=2\cdot\frac{\partial F}{\partial t} \quad \ldots \quad (5.2)$$

where $$r^2=y_2^2+y_3^2+\ldots+y_p^2 \quad \ldots \quad (5.2)$$

Further decomposition can be carried on in the same way.

If F is of the form $F(x_1, x_2, R, t)$; the differential equation can be written in the form

$$\frac{\partial^2 F}{\partial x_1^2}+\frac{\partial^2 F}{\partial x_2^2}+\frac{1}{\omega^{p-3}}\cdot\frac{\partial}{\partial\omega}\left(\omega^{p-3}.\frac{\partial F}{\partial\omega}\right)=2\cdot\frac{\partial F}{\partial t} \quad \ldots \quad (5.4)$$

where $$\omega^2=x_2^2+x_3^2+\ldots+x_p^2,\ \text{and so on.} \quad \ldots \quad (5.5)$$

The Differential Equation in the Product from.

6. If $F(x_1, x_2, \ldots x_r, t)$ can be written as the product of two functions

$$\phi_1(x_1, x_2, \ldots x_r, t) \cdot \phi_2 (x_{r+1}, x_{r+2}, \ldots x_p, t) \qquad \ldots \quad (6.1)$$

then the fundamental differential equation (4.1) breaks up at once to the form

$$\frac{1}{\phi_1}\left[\Box_r^2 \phi_1 - 2\cdot \frac{\partial \phi_1}{\partial t}\right] + \frac{1}{\phi_2}\left[\Box_{r+1\cdots p}\phi_2 - 2\cdot \frac{\partial \phi_2}{\partial t}\right] = 0 \qquad \ldots \quad (6.2)$$

so that the functions ϕ_1 and ϕ_2 satisfy respectively equations of the form

$$\Box_{1\cdot 2\cdots r}^2 \phi_1 - 2\cdot \frac{\partial \phi_1}{\partial t} = F(t, a)\phi_1$$

and

$$\Box_{r+1\cdots p}^2 \phi_2 - 2\cdot \frac{\partial \phi_2}{\partial t} = - F(t, a)\phi_2 \qquad \ldots \quad (6.3)$$

Now these equations are transformable to the forms

$$\Box_r^2 \phi_1 \cdot e^{-\frac{1}{2}\int F(t, a)\, dt} = 2\frac{d}{dt}\left[\phi_1 \cdot e^{-\frac{1}{2}\int F(t, a)\, dt}\right]$$

$$\Box_{r+1\cdots p}^2 \varphi_2 \cdot e^{\frac{1}{2}\int F(t, a)\, td} = 2\frac{d}{dt}\left[\phi_2 \cdot e^{\frac{1}{2}\int F(t, a)\, dt}\right] \qquad \ldots \quad (6.4)$$

Therefore if L_1 and L_2 are solutions of the equations (6.4), their product $L_1 \cdot L_2$ is a solution of the original equation (4.1). It is easily seen that

$$\frac{e^{-x^2/2t}}{t^{1/2}}$$

satisfies the differential equation

$$\frac{\partial^2 F}{\partial x_1^2} = 2\cdot \frac{\partial F}{\partial t} \qquad \ldots \quad (6.5)$$

Hence

$$\frac{e^{-\Sigma_r (x_r^2/2t)}}{t^{p/2}}$$

is a solution of the equation in p-variables *i.e.* of the equation

$$\Box_p^2 F = 2\cdot \frac{\partial F}{\partial t} \qquad \ldots \quad (6.6)$$

FURTHER REDUCTION OF THE PARABOLIC FORM TO POLAR FORM

7. The equation (5.2) can be written in the form

$$\frac{\partial^2 F}{\partial x_1^2}+\frac{\partial^2 F}{\partial r^2}+\frac{p-2}{r}\cdot\frac{\partial F}{\partial r}=2\cdot\frac{\partial F}{\partial t} \qquad \ldots \quad (7.1)$$

Since

$$R^2 = r^2+x_1^2 \qquad \ldots \quad (7.11)$$

we can introduce in place of r and x_1, variables R and θ defined by the equations

$$r = R\sin\theta \quad \text{and} \quad x_1 = R\cos\theta \qquad \ldots \quad (7.2)$$

Then

$$\frac{\partial^2 F}{\partial x^2}+\frac{\partial^2 F}{\partial p^2}=\frac{\partial^2 F}{\partial R^2}+\frac{1}{R}\cdot\frac{\partial F}{\partial R}+\frac{1}{R^2}\cdot\frac{\partial^2 F}{\partial \theta^2} \qquad \ldots \quad (7.31)$$

and

$$\frac{\partial F}{\partial r}=\sin\theta\cdot\frac{\partial F}{\partial R}+\frac{\cos\theta}{R}\cdot\frac{\partial F}{\partial \theta} \qquad \ldots \quad (7.32)$$

The transformed equation in R, θ becomes

$$\frac{\partial^2 F}{\partial R^2}+\frac{1}{R}\cdot\frac{\partial F}{\partial R}+\frac{1}{R^2}\cdot\frac{\partial^2 F}{\partial \theta^2}+\frac{(p-2)}{R\sin\theta}\cdot\left[\sin\theta\cdot\frac{\partial F}{\partial R}+\frac{\cos\theta}{R}\cdot\frac{\partial F}{\partial R}\right]=\cdot 2\,\frac{\partial F}{\partial t} \ldots \quad (7.33)$$

or

$$\frac{\partial^2 F}{\partial R^2}+\frac{(p-1)}{R}\cdot\frac{\partial F}{\partial R}+\frac{(p-2)}{R}\cdot\cot\theta\cdot\frac{\partial F}{\partial \theta}+\frac{1}{R^2}\cdot\frac{\partial^2 F}{\partial \theta^2}=2\cdot\frac{\partial F}{\partial t} \qquad \ldots \quad (7.34)$$

Since

$$x_1^2+x_2^2+\omega^2=R^2 \qquad \ldots \quad (7.41)$$

the equation

$$\frac{\partial^2 F}{\partial x_1^2}+\frac{\partial^2 F}{\partial x_2^2}+\frac{\partial^2 F}{\partial \omega^2}+\frac{(p-3)}{\omega}\cdot\frac{\partial F}{\partial \omega}=2\cdot\frac{\partial F}{\partial t} \qquad \ldots \quad (7.42)$$

can be transformed by the substitutions

$$\omega = R\cos\theta, \quad x_1 = R\sin\theta\cos\phi, \quad x_2 = R\sin\theta\sin\phi \qquad \ldots \quad (7.51)$$

to

$$\frac{1}{R^2}\frac{\partial}{\partial R}\left(R^2\cdot\frac{\partial F}{\partial R}\right)+\frac{1}{R^2\sin\theta}\cdot\frac{\partial}{\partial\theta}\left(\sin\theta\cdot\frac{\partial F}{\partial\theta}\right)+\frac{1}{R^2\sin^2\theta}\cdot\frac{\partial^2 F}{\partial\phi^2}$$

$$+\frac{(p-3)}{R\sin\theta}\left[\frac{\partial F}{\partial R}\cdot\cos\theta-\frac{\sin\theta}{R}\cdot\frac{\partial F}{\partial R}\right]=2\,\frac{\partial F}{\partial t} \qquad \ldots \quad (7.52)$$

Equation (5.2) could be investigated by assuming a solution of the form

$$F = f((R, t) \cdot P(\theta) \qquad \ldots \quad (7.61)$$

so that the equation breaks up in the form

$$\frac{\partial^2 f}{\partial R^2} + \frac{(p-1)}{R} \cdot \frac{\partial f}{\partial R} - \frac{\lambda f}{R^2} = 2\frac{\partial f}{\partial t} \text{ and } \frac{\partial^2 P}{\partial \theta^2} + (p-2) \cdot \cot \theta \cdot \frac{\partial P}{\partial \theta} + \lambda P = 0 \quad \ldots \quad (7.62)$$

The equation (5.4) could be similarly investigated by assuming a solution of the form

$$F = f(R, t) \cdot P(\theta) \cdot Q(\phi) \qquad \ldots \quad (7.7)$$

so that the equation (7.42) breaks up into the three equations

$$\left.\begin{aligned} \frac{\partial^2 f}{\partial R^2} + \frac{(p-1)}{R} \cdot \frac{\partial f}{\partial R} - \frac{\lambda}{R^2} \cdot f &= 2 \cdot \frac{\partial f}{\partial t} \\ \frac{\partial^2 Q}{\partial \phi^2} + n^2 \cdot Q &= 0 \\ \frac{1}{\sin \theta} \cdot \frac{\partial}{\partial \theta}\left(\sin \theta \cdot \frac{\partial Q}{\partial \theta} \right) + \frac{n^2}{\sin^2 \theta} \cdot \frac{\partial Q}{\partial \theta} - (p-3) \cdot \tan \theta \cdot \frac{\partial Q}{\partial \theta} + \lambda Q &= 0 \end{aligned}\right\} \quad \ldots \quad (7.8)$$

Solutions of the Differential Equation in Different Cases.

8. We shall first consider the equation of the type (4.1). The existence of a simple solution of the type

$$\frac{e^{-\Sigma_r (x_r^2/2t)}}{(2\pi t)^{p/2}} = F_1 \qquad \ldots \quad (8.11)$$

is verified by direct substitution.

We shall have occasion to make use of the theorem of Appell and Brill, which has been proved in the case of p-variables earlier[6], that if $H(x_1, x_2, \ldots x , t)$ is a solution of the differential equation (4.1), then

$$H'(x_1, x_2, \ldots \cdot \ldots x_p, t) = \frac{e^{-\Sigma_r (x_r^2/2t)}}{(2\pi)^{p/2}} \cdot H\left(\frac{x_1}{t}, \frac{x_2}{t}, \ldots \frac{x_p}{t}, -\frac{1}{t} \right) \ldots \quad (8.12)$$

is also a solution.

Therefore in the case of a spherically symmetric solution of the type

$$H(R^2, t) = H(x_1, x_2, \ldots x_p, t) \qquad \ldots \quad (8.13)$$

[6] *Sankhyā*, 2(4), 1936, p. 393, equation (11.1).

$$H'(x_1, x_2, \ldots, x_p, t) = \frac{e^{-\Sigma_r(x_r^2/2t)}}{(2\pi\ t)^{p/2}} \cdot H\left(\frac{R^2}{t^2}, -\frac{1}{t}\right) \qquad \ldots \quad (8.14)$$

is also a solution.

We therefore transform the equation

$$\frac{\partial^2 F}{\partial R^2} + \frac{p-1}{R} \cdot \frac{\partial F}{\partial R} = 2 \cdot \frac{\partial F}{\partial t} \qquad \ldots \quad (4.3)$$

by the substitution $R^2 = y$, and the equation becomes

$$2y \cdot \frac{\partial^2 F}{\partial y^2} + p \cdot \frac{\partial F}{\partial y} = \frac{\partial F}{\partial t} \qquad \ldots \quad (8.16)$$

Further putting $2t = T$, it becomes ... (8.17)

$$y \cdot \frac{\partial^2 F}{\partial y^2} + \frac{p}{2} \cdot \frac{\partial F}{\partial y} = \frac{\partial F}{\partial T} \qquad \ldots \quad (8.18)$$

First consider the equation (4.3). If we assume

$$F = e^{\lambda t} \cdot \psi(R) \qquad \ldots \quad (8.21)$$

$\psi(R)$ satisfies the differential equation

$$\frac{\partial^2 \psi}{\partial R^2} + \frac{p-1}{R} \frac{\partial \psi}{\partial R} = 2\lambda\psi \qquad \ldots \quad (8.22)$$

whose solution can be written in either of the forms

$$\frac{e^{\lambda t}}{R^{(p-2)/2}} \cdot I_{\frac{1}{2}(p-2)}(R\sqrt{2\lambda}) \qquad \ldots \quad (8.23)$$

$$\frac{e^{-\lambda t}}{R^{(p-2)/2}} \cdot J_{\frac{1}{2}(p-2)}(R\sqrt{2\lambda}) \qquad \ldots \quad (8.24)$$

Therefore, from (8.13) and (8.14) we get two further types of solutions. Carrying out the transformation suggested it is seen that

$$\frac{e^{-\Sigma(x_r^2/2t)}}{(2\pi t)^{p/2}} \cdot \frac{e^{-\lambda/t}}{R^{(p-2)/2}} \cdot t^{(p-2)/2} \cdot I_{\frac{1}{2}(p-2)}\{\sqrt{(2\lambda)} \cdot R/t\} \qquad \ldots \quad (8.25)$$

and

$$\frac{e^{-\Sigma(x_r^2/2t)}}{(2\pi t)^{p/2}} \cdot \frac{e^{\lambda/t}}{R^{(p-2)/2}} \cdot t^{(p-2)/2} \cdot J_{\frac{1}{2}(p-2)}\{\sqrt{(2\lambda)} \cdot R/t\} \qquad \ldots \quad (8.26)$$

are also solutions of the equation (4.3).

Hence writing $\lambda = \xi^2/2$, we have as solutions

$$\frac{e^{-(\xi^2+R^2)/2t}}{t\cdot(2\pi)^{p/2}\cdot R^{(p-2)/2}} I_{\frac{1}{2}(p-2)}(R\xi/t) \quad \ldots \quad (8.27)$$

and

$$\frac{e^{-(R^2-\xi^2)/2t}}{t\cdot(2\pi)^{p/2}\cdot R^{(p-2)/2}} J_{\frac{1}{2}(p-2)}(R\xi/t) \quad \ldots \quad (8.28)$$

We shall now seek polynomial types of solution. Stating from the formula (4.3) and putting

$$y = R^2,\ T = 2t \quad \ldots \quad (8.31)$$

we get

$$y\cdot\frac{\partial^2 F}{\partial y^2} + \frac{p}{2}\cdot\frac{\partial F}{\partial y} = \frac{\partial F}{\partial T} \quad \ldots \quad (8.32)$$

Let us put

$$F = T^k\cdot\phi(y/T) \quad \ldots \quad (8.33)$$

Since

$$\frac{\partial F}{\partial y} = T^{k-1}\cdot\phi',\ \frac{\partial^2 F}{\partial y^2} = T^{k-2}\cdot\phi'',\ \frac{\partial^2\phi}{\partial T^2} = k\cdot T^{k-1}\cdot\phi - y\cdot T^{k-2}\cdot\phi \quad \ldots \quad (8.34)$$

the equation (8.32) becomes

$$y\cdot T^{k-2}\phi'' + (p/2)\cdot T^{k-1}\cdot\phi' = kT^{k-1}\cdot\phi - y\cdot T^{k-2}\cdot\phi' \quad \ldots \quad (8.35)$$

or dividing by T^{k-1} and putting $y/T = s$ ... (8.36)

$$s\cdot\phi_k'' + \{(p/2)+s\}\cdot\phi_k' - k\cdot\phi_k = 0 \quad \ldots \quad (8.37)$$

This equation has been already discussed in a previous paper.[7]

It is easily seen that with integer values of k, it gives polynomial solutions. These solutions are denoted by $\mu_k(y, T)$ here. They are identical with the moment-functions discussed at some length in the earlier paper.[8] The solutions can be written in the form

$$T^k\cdot\mu_k(y/T) \quad \ldots \quad (8.41)$$

The application now of (8.12), allows us to deduce solutions of the type

$$\frac{e^{-R^2/2t}}{(t)^{p/2}}\cdot\frac{1}{T^k}\cdot S(s),\ \text{where } s = y/T = R^2/2t \quad \ldots \quad (8.42)$$

[7] *Sankhyā*, 2(4), 1936, p. 390, equation (8.5).
[8] *Sankhyā*, 2(4), 1936 pp. 387—390.

and $$s \cdot S_k'' + \{(p/2)-s\} \cdot S_k' + k \cdot S_k = 0 \qquad \ldots \quad (8.43)$$

We therefore see that

$$\frac{e^{-R^2/2t}}{(2\pi t)^{p/2}} \quad \frac{1}{T^{2k}} \cdot S_k(y, T) \qquad \ldots \quad (8.44)$$

is also a solution where $S_k(y, T)$ is a homogeneous polynomial in (y, T).

We shall now seek relations between solutions of type (4.3) and (5.3) with type (8.44). We note that since (8.44) is a solution of the fundamental equation (4.3), writing

$$R^2 = y,\ 2t = T \qquad \ldots \quad (8.45)$$

we see that

$$L = \frac{e^{-y/T}}{T^{p/2+k}} \cdot S_k(y/T) \text{ satisfies the equation (8.32)} \qquad \ldots \quad (8.46)$$

that is, $$y \cdot \frac{\partial^2 L}{\partial y^2} + \frac{p}{2} \cdot \frac{\partial L}{\partial y} = \frac{\partial L}{\partial T} \qquad \ldots \quad (8.47)$$

This ıesult will be useful later on.

Consider the fundamental equation (8.32) in $(y\ T)$.

$$y \cdot \frac{\partial^2 \psi}{\partial y^2} + \frac{p}{2} \cdot \frac{\partial \psi}{\partial y} = \frac{\partial \psi}{\partial T} \qquad \ldots \quad (8.32)$$

If we write $$\psi = e^{\lambda T} \cdot F(y) \qquad \ldots \quad (8.51)$$

we can see that

$$y \cdot \frac{\partial^2 F}{\partial y^2} + \frac{p}{2} \cdot \frac{\partial F}{\partial y} = \lambda \cdot F \qquad \ldots \quad (8.52)$$

The form of the equation shows that

$$\psi = e^{\lambda T} \cdot F(\lambda y) \qquad \ldots \quad (8.53)$$

is a solution of (8.52) where $F(y)$ satisfies the equation

$$\frac{\partial^2 F}{\partial y^2} + \frac{p}{2y} \cdot \frac{\partial F}{\partial y} = \frac{F}{y} \qquad \ldots \quad (8.54)$$

Appell's theorem applied to this form gives us a solution of the form

$$t^{-p/2} \cdot e^{-R^2/2t} \cdot e^{-\lambda/T} \cdot F(\lambda y/T) \qquad \ldots \quad (8.55)$$

or $$t^{-p/2} \cdot e^{-R^2/t^2} \cdot F(\lambda R^2/T) \qquad \ldots \quad (8.56)$$

Remembering that $T = 2t$, and putting $\lambda = \Delta^2$ ... (8.57)

the solution reduces to

$$t^{-p/2} \cdot e^{-(R^2+\Delta^2)/2t} \cdot F \cdot (\Delta^2 R^2/4t) \qquad \ldots \quad (8.54)$$

Remembering again that $T = 2t$, we see that the solution of (8.54) is of the form

$$t^{-p/2} \cdot e^{-R^2/2t} \cdot e^{-\Delta^2/2t} \cdot \psi(\Delta R/t) \qquad \ldots \quad (8.61)$$

Now a solution of the above type gives rise to a distribution

$$dF = H \cdot dx_1 \cdot dx_2 \ldots dx_p \qquad \ldots \quad (8.62)$$

where H is a solution of the equation. Therefore

$$dF = A \cdot e^{-(R^2+\Delta^2)/2t} \cdot t^{-p/2} \cdot \psi(\Delta R/t) \cdot R^{(p-2)/2} \qquad \ldots \quad (8.63)$$

If we compare the result with that in the earlier paper, it is easily seen that the function

$$\psi(\Delta R/t) = A \cdot (n\lambda l)^{-(p-2)/2} \cdot I_{\frac{1}{2}(p-2)} \cdot (nl\lambda) \qquad \ldots \quad (8.64)$$

using the notation of R. C. Bose's paper[9].

If we remember that $n = 1/t$, $l = R$, $\lambda = \Delta$ the identity of the solution can also be easily seen.

By transforming the equation

$$Z \cdot \frac{\partial^2 F}{\partial Z} + \frac{p}{2} \cdot \frac{\partial F}{\partial Z} = F \qquad \ldots \quad (8.72)$$

by the substitution $Z = x^2/4$, we have

$$\frac{\partial^2 F}{\partial x^2} + \frac{(p-1)}{x} \cdot \frac{\partial F}{\partial x} = F \qquad \ldots \quad (8.73)$$

Since I_n/x satisfies the equation

$$\frac{\partial^2 F}{\partial x^2} + \frac{(2n+1)}{x} \cdot \frac{\partial F}{\partial x} = F \qquad \ldots \quad (8.74)$$

the solution of the above is easily seen to be

$$A \cdot x^{-(p-2)/2} \cdot I_{\frac{1}{2}(p-2)}(x) \qquad \ldots \quad (8.75)$$

On the Series Solution of the Differential Equation.

9. I have assumed in an earlier paper the possibility of expansion of solutions or the equation (4.1) as series of the type[10]

[9] Raj Chandra Bose: "On the Exact Distribution and Moment-coefficients of the D^2-statistic" *Sankhyā*, 2(2), 1936, p. 148, equation (6.4). [10] *Sankhyā*, 2(4), 1936, p. 396, equation (12.7).

$$\Sigma_k A_k \cdot \frac{e^{-R^2/2t}}{t^{p/2}} \cdot \frac{1}{t^k} \cdot S_k(R^2/2t)$$

where Σ_k denotes a summation over all values of k from 1 to ∞, and S_k denotes a Sonine polynomial of the k-th order

The justification for the statement will be given here, and *I* shall, by actually developing the distribution function, arrive at the same result as before. A distribution function of the type

$$t^{-p/2} \cdot e^{-(\xi^2+R^2)/2t} \cdot F(\xi^2R^2/4T^2)$$

where $R^2 = x_1^2 + x_2^2 + \ldots + x_p^2$, necessarily implies that

$$\int t^{-p/2} \cdot e^{-(\xi^2+R^2)/2t} \cdot F(\xi^2R^2/4t^2) \cdot R^{p-1} \cdot dR = 1 \quad \ldots \quad (9.12)$$

Assume now that $$R^2/2t = Z, \quad \text{and} \quad \xi^2/2t = a \quad \ldots \quad (9.13)$$

The integral then takes the form

$$2^{(p-2)/2} \cdot \int_0^\infty t^{-p/2} \cdot e^{-(a+z)} \cdot F(az) \cdot z^{(p-2)/2} \cdot t^{p/2} \cdot dZ = 1 \quad \ldots \quad (9.14)$$

or $$2^{(p-2)/2} \cdot e^{-a} \int_0^\infty e^{-z} \cdot F(az) \cdot z^{(p-2)/2} \cdot dz = 1 \quad \ldots \quad (9.15)$$

Therefore so long as a remains finite

$$\int_0^\infty e^{-z} \cdot z^{(p-2)/2} \cdot F(az) \cdot dz = 2^{-(p-2)/2} \cdot e^a = \text{finite} \quad \ldots \quad (9.16)$$

Therefore a constant c can be found such that

$$\int_0^\infty e^{-z} \cdot z^{(p-2)/2} \cdot [F(az)-c] \cdot dz = 0 \quad \ldots \quad (9.17)$$

so that if we consider the function $$F(az)-c \equiv F_1(z) \quad \ldots \quad (9.18)$$

we have $$\int_0^\infty e^{-z} \cdot z^{(p-2)/2} \cdot F_1(z) \cdot dz = 0 \quad \ldots \quad (9.19)$$

Now the Sonine polynomials $S_k(z)$ satisfy the differential equation

$$z \cdot \frac{d^2S_k}{dz^2} + \left(\frac{p}{2} - z\right) \cdot \frac{dS_k}{dz} + k \cdot S_k = 0 \quad \ldots \quad (9.21)$$

or written in the self-adjoint form

$$\frac{d}{dz} \cdot \left(z^{p/2} \cdot e^{-z} \cdot \frac{dS_k}{dz} \cdot \right) + k \cdot S_k \cdot e^{-z} \cdot z^{(p-2)/2} = 0 \quad \ldots \quad (9.22)$$

We see that $e^{-z} \cdot z^{(p-2)/2}$ is the density function for the orthogonal set of polynomials. Hence follows the possibility of expansion of a function $F_1(z)$ which satisfies the equation

$$\int_0^\infty e^{-z} \cdot z^{(p-2)/2} \cdot F_1(z) \cdot dz = 0 \quad \ldots \quad (9.23)$$

Assuming therefore that $F(az) = \Sigma_k\, A_k \cdot S_k(z)$ ... (9.24)

the co-efficients A_k will be (because of the orthogonality of the polynomial functions S_k mentioned earlier) found to be given by

$$A_k = \frac{\int_0^\infty F(az) \cdot S_k(z) \cdot e^{-z} \cdot z^{(p-2)/2} \cdot dz}{\int_0^\infty e^{-z} \cdot z^{(p-2)/2} \cdot S_k(z) \cdot dz}$$

Let us suppose that the functions are normalized such that

$$\int_0^\infty e^{-k} \cdot z^{(p-2)/2} \cdot S_k^2(z) \cdot dz = 1 \quad \ldots \quad (9.26)$$

Then

$$A_k = \int_0^\infty F(az) \cdot S_k(z) \cdot e^{-z} \cdot z^{(p-2)/2} \cdot dz \quad \ldots \quad (9.27)$$

To calculate A_k we at first write $az = x$; then ... (9.28)

$$A_k = \int_0^\infty F(x) \cdot S_k(z/a) \cdot a^{-p/2} \cdot e^{-x/a} \cdot x^{(p-2)/2} \cdot dx \quad \ldots \quad (9.31)$$

As noted in *Sankhyā*, 2(4), 1836, pp. 394, $L = a^{-(p+2k)/2} \cdot e^{-x/a} \cdot S_k(x/a)$ ... (9.32) satisfies the differential equation

$$x \cdot \frac{\partial^2 L}{\partial x^2} + \frac{p}{2} \cdot \frac{\partial L}{\partial x} = \frac{\partial L}{\partial a} \quad \ldots \quad (9.33)$$

$$A_k/a^k = \int_0^\infty F(x) \cdot x^{(p-2)/2} \cdot L \cdot dx \quad \ldots \quad (9.41)$$

or
$$\frac{\partial}{\partial a}\left(\frac{A_k}{a_k}\right) = \int_0^\infty x^{(p-2)/2} \cdot F(x) \cdot \frac{\partial L}{\partial a} \cdot dx \quad \ldots \quad (9.42)$$

$$= \int_0^\infty x^{(p-2)/2} \cdot F(x) \cdot \left[x \cdot \frac{\partial^2 L}{\partial x^2} + \frac{p}{2} \cdot \frac{\partial L}{\partial x}\right] \cdot dx \quad \ldots \quad (9.43)$$

Now by successive integration by parts, because the quantities outside the sign fo integration vanish at both limits, it easily follows that

$$\frac{\partial}{\partial a}\left(\frac{A_k}{a_k}\right) = \int_0^\infty L \cdot \left[\frac{\partial^2}{\partial x^2} \cdot (x^{p/2} \cdot F) - \frac{p}{2} \cdot \frac{\partial}{\partial x}(x^{\frac{1}{2}p-1} \cdot F)\right] \cdot dx \ldots \quad (9.44)$$

$$= \int_0^\infty L \cdot x^{p/2} \cdot \left[\frac{\partial^2 F}{\partial x^2} + \frac{p}{2x} \cdot \frac{\partial F}{\partial x}\right] \cdot dx \quad \ldots \quad (9.45)$$

As
$$\frac{\partial^2 F}{\partial x^2} + \frac{p}{2x} \cdot \frac{\partial F}{\partial x} = \frac{F(2x)}{x} \quad \ldots \quad (9.46)$$

we have now

$$\frac{\partial}{\partial \alpha}\left(\frac{A_k}{a_k}\right) = \int_0^\infty L \cdot x^{(p-2)/2} \cdot F_{(x)} \cdot dx = \int_0^\infty a^{-(p+2k)/2} \cdot e^{-x/a} \cdot S_k(x/a) \cdot x^{(p-2)/2} F_{(x)} \cdot dx$$
$$\ldots \quad (9.51)$$

Writing $x = az$ as before, we have

$$\frac{\partial}{\partial a}\left(\frac{A_k}{a_k}\right)\frac{1}{a^k} = \frac{1}{a^k}\int_0^\infty e^{-z} \cdot S_k(z) \cdot z^{(p-2)/2} \cdot F(az) \cdot dx \quad \ldots \quad (9.52)$$

Therefore we have
$$\frac{\partial}{\partial a} \cdot \left(\frac{A_k}{a^k}\right) = \frac{A_k}{a^k} \quad \ldots \quad (9.61)$$

or
$$A_k = C \cdot a^k \cdot e^a \quad \ldots \quad (9\ 62)$$

To detrmine the constant C in the expression for A_k we observe that

$$\int_0^\infty F \cdot (az) \cdot e^{-z} \cdot z^{(p-2)/2} \cdot S_k(z) \cdot dz = C \cdot a^k \cdot e^a \quad \ldots \quad (9.63)$$

But when a is very small $C \cdot a^k \cdot e^a = C \cdot a^k \cdot (1+a+a^2/2+\ldots) = C \cdot a^k \ldots \quad (9.64)$

Therefore
$$C \cdot a^k = \int_0^\infty B_k \cdot a^k \cdot z^k \cdot e^{-z} \cdot z^{(p-2)/2} \cdot S_k(z) \cdot dz \quad \ldots \quad (9.65)$$

where B_k is the coefficient of z^k in the expansion of $F(z)$ as a power series, as terms of lower order in $F(a\,z)$ below z^k, will contribute zero values because of the orthogonal properties of $s_k(z)$'s

$$\text{Now as} \int_0^\infty z^k \cdot e^{-z} \cdot z^{(p-2)/2} \cdot S_k(z) \cdot dz = \frac{1}{\gamma_k}\int_0^\infty e^{-z} \cdot z^{(p-2)/2} \cdot S_k^2 \cdot (z) \cdot dz = \frac{1}{\gamma_k} \quad \ldots \quad (9.66)$$

where γ_k is the coefficient of the highest term in the series of normalised Sonnine polynomials, we have

$$C \cdot a^k = \frac{B_k \cdot a^k}{r_k}, \text{ or } \quad C = \frac{B^k}{\gamma^k} \quad \ldots \quad (9.71)$$

To calculate therefore the coefficients of B_k and γ_k we consider the differential equation which $F(z)$ satisfies, that is,

$$z \cdot \frac{d^2F}{\partial z^2} + \frac{p}{2} \cdot \frac{\partial F}{\partial z} = F(z) \quad \ldots \quad (9.81)$$

Assuming a series solution $\qquad F = \Sigma B_k \cdot z^k \qquad \ldots \quad (9.82)$

is easily seen that $\qquad k \cdot (k-1) \cdot B_k + \frac{1}{2}p \cdot k \cdot B_k = B_{k-1} \qquad \ldots \quad (9.83)$

or $\qquad k \cdot (k+\frac{1}{2}p-1) \cdot B_k = B_{k-1} \qquad \ldots \quad (9.84)$

or $\qquad B_k = \dfrac{B\Gamma\ (p/2)}{\Gamma(k+1) \cdot \Gamma(k+\frac{1}{2}p)} \qquad \ldots \quad (9.85)$

Therefore from (9.24), (9.62), (9.66) and (9.85) we have

$$F(az) = B_0\,\Sigma \frac{\Gamma(p/2)}{\Gamma(k+1)\,\Gamma(k+\frac{1}{2}p)} \cdot a^k\,e^a \cdot S_k'\,(z) \quad \ldots \quad (9.86)$$

where $\qquad S'(z) = S_k(z)/\gamma_k \qquad \ldots \quad (9.87)$

Putting now $\qquad a = \xi^2/2t, \text{ and } \quad z = R^2/2t \qquad \ldots \quad (9.88)$

we see that

$$F\left(\frac{\xi^2R^2}{4t^2}\right) = F(0)\,\Sigma \frac{\Gamma(p/2)}{\Gamma(k+1)\,.\,\Gamma(k+\frac{1}{2}p)} \cdot e^{\xi^2/2t} \cdot (\xi^2/2t)^k \cdot S_k(R^2/2t) \quad \ldots \quad (9.89)$$

or $\qquad F(x) = F(0)\Sigma \dfrac{\Gamma(p/2)}{\Gamma(k+1) \cdot \Gamma(k+\frac{1}{2}p)} \cdot e^{\xi^2/2t} \cdot (\xi^2/2t)^k \cdot S_k \cdot \left(\dfrac{\xi^2}{2t} \cdot x\right) \quad \ldots \quad (9.91)$

Therefore the type of solution is of the form

$$\frac{e^{-(\xi^2+R^2)/2t}}{t^{p/2}} F(\xi^2 R^2/4t^2)$$

$$= F(0)\Sigma \frac{\Gamma(p/2)}{\Gamma(k+1)\cdot\Gamma(k+\frac{1}{2}p)} \cdot \frac{e^{-R^2/2t}}{t^{p/2}} \cdot \frac{\xi^{2k}}{(2t)^k} \cdot S_k \,.\, (R^2/2t) \qquad \ldots \quad (9.92)$$

which agrees with the previous result[11], if we remember the expression

$$\Delta^k = \xi^{2k} = \mu_k - k(k+\tfrac{1}{2}p-1)\cdot\mu_{k-1}(2t) + \ldots \qquad \ldots \quad (9.93)$$

It follows from the above analysis that the general form of the distribution function in (R, t) can be assumed to be of the form

$$\Sigma \frac{e^{-R^2/2t}}{(2\pi t)^{p/2}} \left\{ A_0 + \frac{A_1}{t^2} \cdot\cdot S_1(R^2/2t) + \frac{A_2}{t^4} S_2(R^2/2t) + \ldots \right\} \qquad \ldots \quad (9.94)$$

where $S_1, S_2, \ldots S_k$ are what can be called Sonine polynomials. The coefficients $A_0, A_1, A_2 \ldots$ etc. can be calculated in any particular case, when the moments of the modulus distributions are known. The distribution-function in *Sankhya*[12] may thus be seen to fall in the general class, as I have shown by a direct expansion of the expression, and also by identification of the two results.

It is interesting to stydy the general types of solutions of the equation $\Box_p^2 F = 2\frac{\partial F}{dt}$ for different types of complexity; certain results already arrived at will be published in due course.

[11] *Sankhyā*, 2(4), pp. 395—396.
[12] *Sankhyā*, 2(4), 1936, pp. 395-396.

Reprinted from *Sankhyā — Ind J Stat* 3, pp. 105-124, 1937 (Statistical Publishing Society, Calcutta).

Studies in Lorentz Group

By
S. N. Bose

1. The decomposition of any Lorentz transformation into two commutable factors is a well-known result, and forms a convenient starting point for the theory of spinors. Einstein arrived at the decomposition from consideration of the infinitesimal elements of the Lorentz group. The result is here obtained from purely algebraic consideration, which gives at the same time a slightly more general result inasmuch as the factorisation can be shown to hold for elements of the general orthogonal group C_4, with det $C_4 = +1$. The Lorentz group L is really a sub-group of C_4 inasmuch as the co-efficients are restricted by certain reality conditions, for every matrix C in L has C_{rs} real $(r, s = 1, 2, 3)$, while $C_{4r}, C_{r4}, r = 1, 2, 3$ are pure imaginaries, and C_{44} real.

If C is a matrix, the transposed matrix will be denoted by $\bar{C}$; a symmetric matrix for which $\bar{C} = C$ will be denoted by S. If $\bar{C}+C = 0$, the matrix is anti-symmetric and will be denoted by A.

A non-symmetric matrix $C = S+A$ while $\bar{C} = S-A$. We shall assume here the elements of a matrix to be complex numbers.

2. Prop. I. An anti-symmetric matrix A having six non-zero elements can be further decomposed into two component matrices P and Q, which have only three non-zero elements such that

$$\left.\begin{array}{lll} P_{12} = P_{34}, & P_{23} = P_{14}, & P_{31} = P_{24} \\ Q_{12} = -Q_{34}, & Q_{23} = -Q_{14}, & Q_{31} = -Q_{24} \end{array}\right\}. \quad \dots (1)$$

If

$$A = P+Q,$$

$$A_{rs} = P_{rs}+Q_{rs}$$

where

$$\left.\begin{array}{lll} P_{12} = \tfrac{1}{2}(A_{12}+A_{34}), & P_{23} = \tfrac{1}{2}(A_{23}+A_{14}), & P_{31} = \tfrac{1}{2}(A_{31}+A_{24}) \\ Q_{12} = \tfrac{1}{2}(A_{12}-A_{34}), & Q_{23} = \tfrac{1}{2}(A_{23}-A_{14}), & Q_{31} = \tfrac{1}{2}(A_{31}-A_{24}) \end{array}\right\}. \quad \dots (2)$$

If further we introduce certain basic matrices, defined by

$$P_1 = \begin{vmatrix} 0 & 0 & 0 & -1 \\ 0 & 0 & -1 & 0 \\ 0 & 1 & 0 & 0 \\ 1 & 0 & 0 & 0 \end{vmatrix}, \quad P_2 = \begin{vmatrix} 0 & 0 & 1 & 0 \\ 0 & 0 & 0 & -1 \\ -1 & 0 & 0 & 0 \\ 0 & 1 & 0 & 0 \end{vmatrix}, \quad P_3 = \begin{vmatrix} 0 & -1 & 0 & 0 \\ 1 & 0 & 0 & 0 \\ 0 & 0 & 0 & -1 \\ 0 & 0 & 1 & 0 \end{vmatrix},$$

and

$$Q_1 = \begin{vmatrix} 0 & 0 & 0 & 1 \\ 0 & 0 & -1 & 0 \\ 0 & 1 & 0 & 0 \\ -1 & 0 & 0 & 0 \end{vmatrix}, \quad Q_2 = \begin{vmatrix} 0 & 0 & 1 & 0 \\ 0 & 0 & 0 & 1 \\ -1 & 0 & 0 & 0 \\ 0 & -1 & 0 & 0 \end{vmatrix}, \quad Q_3 = \begin{vmatrix} 0 & -1 & 0 & 0 \\ 1 & 0 & 0 & 0 \\ 0 & 0 & 0 & 1 \\ 0 & 0 & -1 & 0 \end{vmatrix},$$

in which the first set P have the properties of the above P matrices in (1), and the second set Q those of the Q matrices. We note that

$$P_1^2 = P_2^2 = P_3^2 = Q_1^2 = Q_2^2 = Q_3^2 = -E$$
$$P_1 = P_2P_3 = -P_3P_2, \quad P_2 = P_3P_1 = -P_1P_3, \quad P_3 = P_1P_2 = -P_2P_1$$
$$Q_1 = Q_2Q_3 = -Q_3Q_2, \quad Q_2 = Q_3Q_1 = -Q_1Q_3, \quad Q_3 = Q_1Q_2 = -Q_2Q_1,$$
$$P_r\, Q_s = Q_s\, P_r. \qquad \dots \quad (3)$$

Any P matrix can then be represented as

$$P = a_1P_1 + a_2P_2 + a_3P_3, \qquad \dots \quad (4)$$

and any Q matrix as

$$Q = b_1Q_1 + b_2Q_2 + b_3Q_3. \qquad \dots \quad (5)$$

Hence

$$A = a_1P_1 + a_2P_2 + a_3P_3 + b_1Q_1 + b_2Q_2 + b_3Q_3 \qquad \dots \quad (6)$$

Carrying out the multiplication of all P_i's and Q_i's ($i = 1, 2, 3$), we get in all sixteen matrices which we can utilise as basic matrices for the representation of any arbitrary matrix C thus (the indices of Q will be raised in what follows when Q's appear with summations)

$$C = C_0^0\, E + \Sigma C_0^r\, P_r + \Sigma C_r^0\, Q^r + \Sigma C^{\lambda}_{\mu}\, P_{\lambda}\, Q^{\mu} \qquad \dots \quad (7)$$

$$\text{spur of } C = 4C_0^0.$$

Any symmetric matrix

$$S = S_0E + \sum_{r,\,s} S_s^r P_r Q^s. \qquad \dots \quad (7')$$

Since $P_i^2 = -E$, we have generally

$$P^2 = -(a_1^2 + a_2^2 + a_3^2)E. \qquad \dots \quad (8)$$

Prop. II. If S is any orthogonal matrix, *i.e.*, $S^2 = E$, det $S \neq 0$, then, either (*i*) det $S = -1$, spur $S \neq 0$, or (*ii*) det $S = +1$, spur $S = 0$.

We take a symmetric matrix

$$S = \begin{vmatrix} a & h & g & x \\ h & b & f & y \\ g & f & c & z \\ x & y & z & d \end{vmatrix}$$

Since $S^2 = E$, $S = S^{-1}$, and as det $S \neq 0$, the inverse can be written down by utilising the minors of the det S from which the following results follow easily.

If $s =$ det S, and A, B, C, etc. are minors of the determinant

$$D = \begin{vmatrix} a & h & g \\ h & b & f \\ g & f & c \end{vmatrix}$$

then

$$\begin{array}{llll} s(ad-x^2) = A & s(fd-yz) = F & BC-F^2 = aD & \\ s(bd-y^2) = B & s(gd-zx) = G & \text{etc.} & \text{etc.} \\ s(cd-z^2) = C & s(hd-xy) = H & ,, & ,, \end{array}$$

We also remember the usual orthogonality relations of the type

$$ax+hy+gz+xd = 0, \quad \text{etc.}$$
$$a^2+h^2+g^2+x^2 = 1, \quad \text{etc.}$$

So we easily deduce

$$Ax+Hy+Gz+sx = 0$$
$$Hx+By+Fz+sy = 0$$
$$Gx+Fy+Cz+sz = 0,$$

so that, if x, y, z be not all zero, we have by elimination

$$s^3+s^2(A+B+C)+sD(a+b+c)+D^2 = 0,$$

or, as $s^2 \neq 0$, and $D = sd$

$$(s+A+B+C) = -d(a+b+c+d).$$

Again
$$A+B+C = s[d(a+b+c)-x^2-y^2-z^2]$$
$$= s[d(a+b+c+d)-1];$$

so that
$$(A+B+C+s) = sd(a+b+c+d).$$

Hence either $s = -1$, or $a+b+c+d = 0$, or $d = 0$.

The special cases

$$d = 0 \quad \text{and} \quad x = y = z = 0$$

can be treated in the same way, and we will have in every case $s = -1$ or spur $S = 0$, which proves our general result.

Prop. III. If $S^2 = E$ and spur $S = 0$ (so that $s = +1$), then S is always decomposable into two commutative anti-symmetric factors.

As spur $S = 0$, we can write by (7)

$$S = \Sigma S^{\lambda}_{\mu} P_{\lambda} Q^{\mu}.$$

Now putting

$$S^{\lambda}_{\mu} P_{\lambda} = \Delta_{\mu} \qquad \ldots \quad (9)$$

where Δ_{μ} are certain matrices of the P class, we have

$$S = \Delta_1 Q_1 + \Delta_2 Q_2 + \Delta_3 Q_3. \qquad \ldots \quad (10)$$

Again as

$$S^2 = E = \sum_r \Delta_r^2 + Q_1(\Delta_2\Delta_3 - \Delta_3\Delta_2) + Q_2(\Delta_3\Delta_1 - \Delta_1\Delta_3) + Q_3(\Delta_1\Delta_2 - \Delta_2\Delta_1),$$

we must have

$$\begin{aligned} \Delta_2\Delta_3 &= \Delta_3\Delta_2 \\ \Delta_3\Delta_1 &= \Delta_1\Delta_3 \\ \Delta_1\Delta_2 &= \Delta_2\Delta_1. \end{aligned} \qquad \ldots \quad (11)$$

Hence the Δ matrices are *commutable.*

It follows easily, as the Δ's are all of the same P class,

$$\Delta_1 : \Delta_2 : \Delta_3 = k_1 : k_2 : k_3$$

where k_1, k_2, k_3 are certain numbers:

hence

$$\frac{\Delta_1}{k_1} = \frac{\Delta_2}{k_2} = \frac{\Delta_3}{k_3} = (l_1 P_1 + l_2 P_2 + l_3 P_3) \qquad \ldots \quad (12)$$

so that

$$S = (l_1 P_1 + l_2 P_2 + l_2 P_3)(k_1 Q_1 + k_2 Q_2 + k_3 Q_3) \qquad \ldots \quad (13)$$

which puts the symmetric orthogonal matrix with zero spur into two commutative anti-symmetric factors.

Also

$$\det S = |\det(l_1 P_1 + l_2 P_2 + l_3 P_3)| \; |\det(k_1 Q_1 + k_2 Q_2 + k_3 Q_3)|$$

whence

$$1 = (l_1^2+l_2^2+l_3^2)^2(k_1^2+k_2^2+k_3^2)^2. \qquad \dots \quad (14)$$

When the elements of S are given, we can determine the elements of the factor matrices in the following way.

As

$$S = \begin{vmatrix} a & h & g & x \\ h & b & f & y \\ g & f & c & z \\ x & y & z & d \end{vmatrix}, \text{ and } \quad a+b+c+d = 0,$$

we can put

$$S = \Delta_1 Q_1+\Delta_2 Q_2+\Delta_3 Q_3.$$

Then

$$\left.\begin{aligned} SQ_1 &= -\Delta_1+(\Delta_3 Q_2-\Delta_2 Q_3) \\ SQ_2 &= -\Delta_2+(\Delta_1 Q_3-\Delta_3 Q_1), \text{ etc.} \end{aligned}\right\} \qquad \dots \quad (15)$$

Hence Δ_1, Δ_2, Δ_3 are obtained by separating the anti-symmetric parts of SQ_1, SQ_2 and SQ_3. These are expressible as follows :

$$\begin{aligned} \Delta_1 &= \tfrac{1}{2}[(a+d)P_1+(h-z)P_2+(g+y)P_3] \\ \Delta_2 &= \tfrac{1}{2}[(h+z)P_1+(b+d)P_2+(f-x)P_3] \\ \Delta_3 &= \tfrac{1}{2}[(g-y)P_1+(f+x)P_2+(c+d)P_3]. \end{aligned} \qquad \dots \quad (16)$$

as

$$\frac{a+d}{2k_1} = \frac{h+z}{2k_2} = \frac{g-y}{2k_3} = l_1$$

$$\frac{h-z}{2k_1} = \frac{b+d}{2k_2} = \frac{f+x}{2k_3} = l_2$$

$$\frac{g+y}{2k_1} = \frac{f-x}{2k_2} = \frac{c+d}{2k_3} = l_3,$$

hence

$$a+d = 2k_1 l_1, \quad b+d = 2k_2 l_2, \quad c+d = 2k_3 l_3.$$

Also

$$a+b+c+d = 0,$$

so

$$\begin{aligned} d &= k_1 l_1+k_2 l_2+k_3 l_3 \\ a &= k_1 l_1-k_2 l_2-k_3 l_3, \text{ etc.} \\ h &= k_1 l_2+k_2 l_1, \\ z &= k_2 l_1-k_1 l_2. \end{aligned}$$

Hence,

$$S = \left| \begin{array}{llll} k_1l_1-k_2l_2-k_3l_3, & k_1l_2+k_2l_1, & k_1l_3+k_3l_1, & k_3l_2-k_2l_3 \\ \quad - \quad , & k_2l_2-k_1l_1-k_3l_3, & k_2l_3+k_3l_2, & k_1l_3-k_3l_1 \\ \quad - & \quad - \quad , & k_3l_3-k_1l_1-k_2l_2, & k_2l_1-k_1l_2 \\ \quad - & \quad - & \quad - \quad , & k_1l_1+k_2l_2+k_3l_3 \end{array} \right.$$

$$= (k_1P_1+k_2P_2+k_3P_3)(l_1Q_1+l_2Q_2+l_3Q_3)\,; \qquad \ldots \quad (17)$$

where also

$$k_1^2+k_2^2+k_3^2 = +1,$$
$$l_1^2+l_2^2+l_3^2 = +1.$$

3. *Decomposition of Non-symmetric Matrices*

A general orthogonal matrix can be written as

$$C = S+A \qquad \ldots \quad (18)$$

$$C\overline{C} = \overline{C}C = E.$$

where

From this follows

$$SA-AS = 0, \quad S^2-A^2 = E. \qquad \ldots \quad (19)$$

Decomposing

$$A = P+Q,$$

where P and Q are linear combinations of P and Q type matrices, we have

$$S^2-P^2-Q^2-2PQ = E$$

i.e.,

$$S^2 = (1+P^2+Q^2)E+2PQ. \qquad \ldots \quad (20)$$

Now

$$\text{spur } C = \text{spur } S = \mu \text{ (say)},$$

hence by (7),

$$S = \mu E+\lambda PQ.$$

Again as

$$S^2 = \mu^2+\lambda^2P^2Q^2+2\mu\lambda PQ. \qquad \ldots \quad (21)$$

we have comparing (20) and (21), firstly

$$\lambda = \frac{1}{\mu}. \qquad \ldots \quad (22)$$

Thus

$$C = \mu E+P+Q+\frac{PQ}{\mu},$$

or $$\mu C = [\mu+Q][\mu+P]. \qquad \dots (23)$$

Hence if
$$\mu = k_1k_2$$
$$\mu C = [k_1k_2+Q][k_1k_2+P]$$
$$= k_1k_2\left[k_2+\frac{Q}{k_1}\right]\left[k_1+\frac{P}{k_2}\right].$$

Identifying (20) with (21) we put
$$k_2^2-\frac{Q^2}{k_1^2} = 1, \quad k_1^2-\frac{P^2}{k_2^2} = 1,$$
$$\mu^2-Q^2 = k_1^2, \quad \mu^2-P^2 = k_2^2.$$

Hence
$$(\mu^2-Q^2)(\mu^2-P^2) = \mu^2$$
which relation is equivalent to $$\mu^2+\frac{Q^2P^2}{\mu^2} = 1+P^2+Q^2.$$

The numbers k_1 and k_2 are thus determined. Thus (23) which is the factorisation aimed at is completely established.

The general form of C can also be calculated.

If
$$C = AB$$
where
$$\left.\begin{aligned} A &= A_0+A_1P_1+A_2P_2+A_3P_3 \\ B &= B_0+B_1Q_1+B_2Q_2+B_3Q_3 \end{aligned}\right\} \qquad \dots (24)$$
and
$$A_0^2+A_1^2+A_2^2+A_3^2 = 1$$
$$B_0^2+B_1^2+B_2^2+B_3^2 = 1.$$

Explicitly,
$$A = \begin{vmatrix} A_0 & -A_3 & A_2 & -A_1 \\ A_3 & A_0 & -A_1 & -A_2 \\ -A_2 & A_1 & A_0 & -A_3 \\ A_1 & A_2 & A_3 & A_0 \end{vmatrix} \quad B = \begin{vmatrix} B_0 & -B_3 & B_2 & +B_1 \\ B_3 & B_0 & -B_1 & +B_2 \\ -B_2 & B_1 & B_0 & +B_3 \\ -B_1 & -B_2 & -B_3 & B_0 \end{vmatrix} \qquad \dots (25)$$

Multiplying we have

$$C = \begin{vmatrix} \begin{array}{c}(A_0B_0+A_1B_1)\\ -(A_2B_2+A_3B_3)\end{array} & \begin{array}{c}-(A_3B_0+A_0B_3)\\ +(A_1B_2+A_2B_1)\end{array} & \begin{array}{c}+(A_2B_0+A_0B_2)\\ +(A_1B_3+A_3B_1)\end{array} & \begin{array}{c}-(A_1B_0-A_0B_1)\\ +(A_2B_3-A_3B_2)\end{array} \\ \\ \begin{array}{c}(A_3B_0+A_0B_3)\\ +(A_1B_2+A_2B_1)\end{array} & \begin{array}{c}(A_0B_0+A_2B_2)\\ -(A_3B_3+A_1B_1)\end{array} & \begin{array}{c}-(A_0B_2+A_1B_0)\\ +(A_2B_3+A_3B_2)\end{array} & \begin{array}{c}-(A_2B_0-A_0B_2)\\ +(A_3B_1-A_1B_3)\end{array} \\ \\ \begin{array}{c}-(A_2B_0+A_0B_2)\\ +(A_1B_3+A_3B_1)\end{array} & \begin{array}{c}+(A_1B_0+A_0B_1)\\ (A_2B_3+A_3B_2)\end{array} & \begin{array}{c}(A_0B_0+A_3B_3)\\ -(A_1B_1+A_2B_2)\end{array} & \begin{array}{c}-(A_3B_0-A_0B_3)\\ +(A_1B_2-A_2B_1)\end{array} \\ \\ \begin{array}{c}(A_1B_0-A_0B_1)\\ +(A_2B_3-A_3B_2)\end{array} & \begin{array}{c}+(A_2B_0-A_0B_2)\\ +(A_3B_1-A_1B_3)\end{array} & \begin{array}{c}(A_3B_0-A_0B_3)\\ +(A_1B_2-A_2B_1)\end{array} & \begin{array}{c}+(A_1B_1+A_2B_2)\\ +(A_3B_3+A_0B_0)\end{array} \end{vmatrix} \quad \ldots \quad (26)$$

Now

$$\det C = (A_0^2+A_1^2+A_2^2+A_3^2)^2(B_0^2+B_1^2+B_2^2+B_3^2)^2 = 1$$

and

$$\text{spur } C = 4A_0B_0. \qquad \ldots \quad (27)$$

Two special forms of C can be at once noticed :

If C is a matric of the proper Lorentz group, then

C_{14}, C_{24}, C_{34}, C_{41}, C_{42}, C_{43} *are pure imaginaries,*

so that

$$(A_2B_3-A_3B_2)+(A_1B_0-B_1A_0) = i\lambda.$$

$$(A_2B_3-A_3B_2)-(A_1B_0-B_1A_0) = i\lambda',$$

from which follows

$$(A_2B_3-A_3B_2) \text{ as well as } (A_1B_0-B_1A_0),$$

etc, are all pure imaginaries.

A_0B_0, A_1B_1, A_2B_2, A_3B_3 are real quantities as well as $(A_2B_3+A_3B_2)$, etc. and $(A_3B_0+A_0B_3)$, etc. are real.

We have then

$$\frac{A_0}{B_0^*} = \frac{A_1}{B_1^*} = \frac{A_2}{B_2^*} = \frac{A_3}{B_3^*} = k, \text{ real} \qquad \ldots \quad (28)$$

as $A_0^2+A_1^2+A_2^2+A_3^2 = B_0^2+B_1^2+B_2^2+B_3^2 = +1$ (B_0^* representing the conjugate of B_0, etc.).

We thus see that the sets (A_0, A_1, A_2, A_3) and (B_0, B_1, B_2, B_3) *are conjugate to each other*.

There are now two cases possible.

Case I. When (A_0, A_1, A_2, A_3) are real, then

$$A_0 = B_0, \quad A_1 = B_1, \quad A_2 = B_2, \quad A_3 = B_3$$

and from (26) further

$$C_{14} = C_{24} = C_{34} = C_{41} = C_{42} = C_{43} = 0$$

$$C_{44} = 1.$$

This represents a three dimensional rotation.

The axis of rotation is easily seen to be

$$\frac{l}{A_1} = \frac{m}{A_2} = \frac{n}{A_3} \qquad \dots (29)$$

The angle of rotation is given by

$$\tan \psi = \frac{2A_0G}{A_0^2 - G^2}, \text{ where } G^2 = A_1^2 + A_2^2 + A_3^2.$$

Case II. A_1, A_2, A_3 are pure imaginaries, A_0 real.

The resultant transformation is here seen to be an Einstein transformation, with translation along the line

$$\frac{l}{A_1} = \frac{m}{A_2} = \frac{n}{A_3}$$

and velocity given by

$$\frac{V}{c} = \frac{2A_0G}{A_0^2 + G^2}$$

where

$$A_0^2 - G^2 = 1, \quad \text{and} \quad G^2 = A_1^2 + A_2^2 + A_3^2.$$

4.

We shall now consider certain general matrices of the form

$$\left.\begin{aligned} A &= \mu E + aP_1 + bP_2 + cP_3 \\ \text{and} \qquad B &= \mu E + aQ_1 + bQ_2 + cQ_3. \end{aligned}\right\} \qquad \dots (30)$$

We can easily see that

$$|\det A| = [\mu^2 + a^2 + b^2 + c^2]^2$$

$$|\det B| = [\mu^2 + a^2 + b^2 + c^2]^2.$$

So when a, b, c are complex quantities and $|\det A|$ and $|\det B|$ real and positive we must have

$$\mu^2+a^2+b^2+c^2 = \mu^{*2}+a^{*2}+b^{*2}+c^{*2}$$

where μ^*, etc. are conjugates of μ, etc.

We now consider A and B to be factors of decomposition of C, and take the determinant of each of the factors to be $+1$, so that

$$\mu^2+a^2+b^2+c^2 = \pm 1.$$

The matrices of A and B types are orthogonal matrices; but since their elements are generally complex, they do not belong to the Lorentz sub-group of Group C_4.

We will now discuss certain general characteristics of the A and B matrices, which will bring into light also the intimate relation that exists between the decompositions as carried out here and the spinor theory.

Prop. All matrices of A and B types are reducible :

$$A = \left|\begin{array}{cc:cc} p_0 & -p_3 & p_2 & -p_1 \\ p_3 & p_0 & -p_1 & -p_2 \\ \hdashline -p_2 & p_1 & p_0 & -p_3 \\ p_1 & p_2 & p_3 & p_0 \end{array}\right| = \left|\begin{array}{cc} S & -Q \\ Q & S \end{array}\right| \qquad \ldots \quad (31)$$

where S and Q are certain matrices of the second order. That such matrices are reducible is a well-known result.

Let

$$\psi = A\phi$$

define a transformation of a four-component ϕ to a four component ψ.

We have

$$\begin{aligned} \psi_1 &= p_0\phi_1 - p_3\phi_2 + p_2\phi_3 - p_1\phi_4 \\ \psi_2 &= p_3\phi_1 + p_0\phi_2 - p_1\phi_3 - p_2\phi_4 \\ \psi_3 &= -p_2\phi_1 + p_1\phi_2 + p_0\phi_3 - p_3\phi_4 \\ \psi_4 &= p_1\phi_1 + p_2\phi_2 + p_3\phi_3 + p_0\phi_4. \end{aligned} \qquad \ldots \quad (32)$$

From this it can be easily deduced that

$$\left|\begin{array}{c} \psi_1+i\psi_2 \\ \psi_3-i\psi_4 \end{array}\right| = \left|\begin{array}{cc} p_0+ip_3 & p_2-ip_1 \\ -p_2-ip_1 & p_0-ip_3 \end{array}\right| \left|\begin{array}{c} \phi_1+i\phi_2 \\ \phi_3-i\phi_4 \end{array}\right| \qquad \ldots \quad (33)$$

and there is also the corresponding conjugate relation obtained by putting $-i$ for i, (33) can also be written as

$$\begin{vmatrix} \psi_3+i\psi_4 \\ -\psi_1+i\psi_2 \end{vmatrix} = \begin{vmatrix} p_0+ip_3 & p_2-ip_1 \\ -p_2-ip_1 & p_0-ip_3 \end{vmatrix} \begin{vmatrix} \phi_3+i\phi_4 \\ i\phi_2-\phi_1 \end{vmatrix} \qquad \ldots \text{(34)}$$

Combining (33) and (34) we obtain

$$\begin{vmatrix} \psi_3+i\psi_4 & \psi_1+i\psi_2 \\ i\psi_2-\psi_1 & \psi_3-i\psi_4 \end{vmatrix} = \begin{vmatrix} p_0+ip_3 & p_2-ip_1 \\ -p_2-ip_1 & p_0-ip_3 \end{vmatrix} \begin{vmatrix} \phi_3+i\phi_4 & \phi_1+i\phi_2 \\ i\phi_2-\phi_1 & \phi_3-i\phi_4 \end{vmatrix} \qquad \ldots \text{(35)}$$

which is a well-known spinor transformation.

Again, choosing two sets of variables as (x_1, x_2, x_3, x_4) and (y_1, y_2, y_3, y_4) the corresponding transformation formula for B matrices can be written as

$$\begin{vmatrix} y_3+iy_4 & -y_1+iy_2 \\ y_1+iy_2 & y_3-iy_4 \end{vmatrix} = \begin{vmatrix} q_0-iq_3 & -q_2-iq_1 \\ q_2-iq_1 & q_0+iq_3 \end{vmatrix} \begin{vmatrix} x_3+ix_4 & -x_1+ix_2 \\ x_1+ix_2 & x_3-ix_4 \end{vmatrix} \qquad \ldots \text{(36)}$$

which can be also written in the form

$$\begin{vmatrix} y_3+iy_4 & y_1+iy_2 \\ -y_1+iy_2 & y_3-iy_4 \end{vmatrix} = \begin{vmatrix} x_3+ix_4 & x_1+ix_2 \\ -x_1+ix_2 & x_3-ix_4 \end{vmatrix} \begin{vmatrix} q_0-iq_3 & q_2-iq_1 \\ -q_2-iq_1 & q_0+iq_3 \end{vmatrix}. \qquad \ldots \text{(37)}$$

(35) can also be written in this notation with the transformation matrix to the left on the right-hand side.

Thus A can be regarded as inducing a front transformation while B induces a back transformation, or *vice-versa*, which shows the commutative nature of the transformation at once.

The general reducibility of the A and B types of matrices shows that the Lorentz Group can be studied advantageously in the two-dimensional representation of the A and B class of transformation.

DEPARTMENT OF PHYSICS,
DACCA UNIVERSITY.

Reprinted from *Bull Cal Math Soc* **31**, pp. 137-147, 1939 (Calcutta Mathematical Society).

THE COMPLETE SOLUTION OF THE EQUATION:

$$\nabla^2\phi - \frac{\partial^2\phi}{c^2\,\partial t^2} - k^2\phi = -4\pi\rho(xyzt).$$

By S. N. Bose, *Dacca University, and* S. C. Kar, *Bangabasi College, Calcutta.*

(*Received November 17, 1940.*)

1. In what follows we desire to present what appears to us to be a complete Kirchhoff-like solution of the equation above and indeed by two distinct methods. The one is a method of complex integration and the other an extension of a method adopted by Love * for ordinary retarded potentials. The equation itself is one which seems to have acquired some importance in view of a recent work of J. H. Bhabha † on the mesotron, who has stated a solution of the equation, in which the part involving a surface integral does not appear. To the best of our knowledge and belief the solution at which we arrive and which exhibits the surface integral is original. As the whole work, however, was started by a successful derivation, by an adaptation of the well-known method of Herglotz and Sommerfeld ‡, of the potentials of a moving mesotron, such as Bhabha uses, we let this derivation precede the main investigation, since this way of arriving at the potentials appears to us to be also new and of sufficient interest.

2. To start with, we seek a solution of the equation

$$\nabla^2\psi - \frac{\partial^2\psi}{c^2\,\partial\theta^2} - k^2\psi = 0 \quad \cdots \quad (1)$$

of the form: $\psi = \psi(s)$,

where
$$s^2 = c^2(\theta - t)^2 - \Sigma(\xi - x)^2 = c^2(\theta - t)^2 - r^2 \quad \cdots \quad (2)$$

and x, y, z, t are parameters. The transformation of eq. (1) to the single variable s leads to the equation

$$\frac{d^2\psi}{ds^2} + \frac{3}{s}\frac{d\psi}{ds} + k^2\psi = 0$$

or
$$\frac{d^2}{ds^2}(s\psi) + \frac{1}{s}\frac{d}{ds}(s\psi) + \left(k^2 - \frac{1}{s^2}\right)(s\psi) = 0. \quad \cdots \quad (3)$$

Hence we infer: $\psi = \dfrac{J_1(ks)}{s}$ or $\dfrac{Y_1(ks)}{s}$, where $J_1(ks)$ and $Y_1(ks)$ are Bessel functions of the first and second kinds respectively.

3. *Retarded potentials of a moving mesotron.*

We put $\chi = \dfrac{kY_1(ks)}{s}$

$$= \frac{kJ_1(ks)}{s} ln(ks) - \frac{k}{s}\sum_{m=0}^{\infty}(-1)^m \frac{2m+1}{m(m+1)} J_{2m+1}(ks) - \frac{J_0(ks)}{s^2}\,\S \qquad .. \quad (4)$$

and choose for the four potentials (ϕ, A_1, A_2, A_3) of the mesotron solutions of eq. (1), stated as an equation in x, y, z, t, with ξ, η, ζ, θ as parameters, in the forms

$$\phi = \int_{\theta_1}^{\theta_2} d\theta f_4(\theta)\chi, \quad A_1 = \int_{\theta_1}^{\theta_2} d\theta f_1(\theta)\chi, \text{ etc.} \qquad .. \quad (5)$$

The equation of restraint on the four potentials becomes

$$0 = \frac{\partial\phi}{c\,\partial t} + \frac{\partial A_1}{\partial x} + \frac{\partial A_2}{\partial y} + \frac{\partial A_3}{\partial z}$$

$$= -\int_{\theta_1}^{\theta_2} d\theta\left[f_4\frac{\partial}{c\,\partial\theta} + f_1\frac{\partial}{\partial\xi} + f_2\frac{\partial}{\partial\eta} + f_3\frac{\partial}{\partial\zeta}\right]\chi$$

$$= -B\int_{\theta_1}^{\theta_2} d\theta\left[\frac{\partial}{\partial\theta} + \frac{d\xi}{d\theta}\frac{\partial}{\partial\xi} + \frac{d\eta}{d\theta}\frac{\partial}{\partial\eta} + \frac{d\zeta}{d\theta}\frac{\partial}{\partial\zeta}\right]\chi$$

$$= -B\int_{\theta_1}^{\theta_2} d\theta\frac{d\chi}{d\theta} \quad .. \quad .. \quad .. \quad .. \quad .. \quad .. \quad (6)$$

if, after Herglotz and Sommerfeld, we treat ξ, η, ζ as functions of θ and put

$$(f_4, f_1, f_2, f_3) = B\left(c, \frac{d\xi}{d\theta}, \frac{d\eta}{d\theta}, \frac{d\zeta}{d\theta}\right), \qquad .. \quad .. \quad (7)$$

where B is an absolute constant.

We assume now θ to have been complex of the form $\mu + i\nu$ and proceed to perform the integrations in eqs. (5) along a path on the complex plane shown in the diagram below.

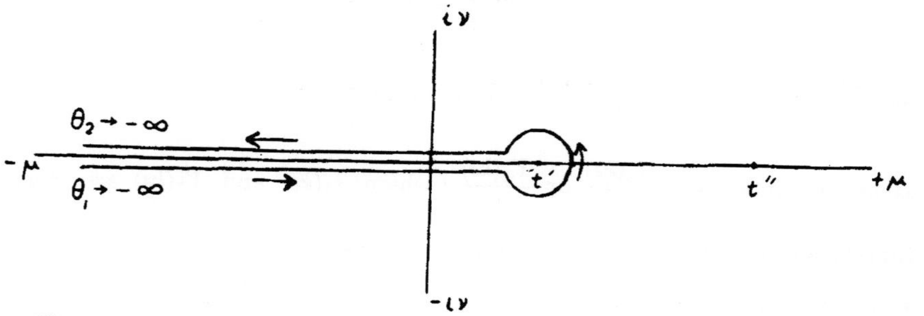

5B

$$\nabla^2\phi - \frac{\partial^2\phi}{c^2\,\partial t^2} - k^2\phi = -4\pi\rho(xyzt). \qquad 95$$

The following considerations govern the choice of this path. Since, from eq. (2),

$$s^2 = c^2(\theta - t')(\theta - t''), \qquad \ldots \quad \ldots \quad \ldots \quad (8)$$

where $t' = t - \frac{r}{c}$ and $t'' = t + \frac{r}{c}$, the only singularities of χ are those at $\theta = t'$ and $\theta = t''$ and since we are seeking, for the moment, the retarded potentials we let the path embrace just the one singularity at $\theta = t'$. Since, further, $ln s$ in χ is multiple-valued we cannot satisfy the equation of constraint (6) by just having a closed contour about t' but have to let the path run along the real axis from $-\infty$ on one Riemann surface round t' to $-\infty$ on a second, as, with $\theta \to -\infty$, $\chi \to 0$.

Retaining only such terms in χ as involve a singularity we may write eqs. (5) as

$$\phi = Bc \int_{\theta_1}^{\theta_2} d\theta \left[\frac{kJ_1(ks)}{s} ln s - \frac{1}{s^2}\right],$$

$$A_1 = B \int_{\theta_1}^{\theta_2} d\theta \left[\frac{kJ_1(ks)}{s} ln s - \frac{1}{s^2}\right] \frac{d\xi}{d\theta},$$

etc.,

and evaluate the integrals as follows:

$$\int_{\theta_1}^{\theta_2} d\theta \frac{1}{s^2}\frac{d\xi}{d\theta} = \oint d\theta \frac{1}{s^2}\frac{d\xi}{d\theta} = \frac{2\pi i}{\left[\frac{d(s^2)}{d\theta}\right]_{t'}} \cdot \frac{dx'}{dt'} = -\frac{\pi i}{cr\left(1 - \frac{v_r}{c}\right)} v_x, \text{ etc.,}$$

$$\int_{\theta_1}^{\theta_2} d\theta \frac{J_1(ks)}{s}\frac{d\xi}{d\theta} ln s = \frac{1}{2}\int_{\theta_1}^{\theta_2} d\theta \frac{J_1(ks)}{s}\frac{d\xi}{d\theta} ln(\theta - t')$$

$$= -\pi i \int_{-\infty}^{t'} d\theta \frac{J_1(ks)}{s}\frac{d\xi}{d\theta}, \text{ etc.}$$

We identify now $B\pi i$ with the charge e of the mesotron and obtain as its potentials

$$\left.\begin{aligned} \phi &= \frac{e}{r\left(1 - \frac{v_r}{c}\right)} - eck\int_{-\infty}^{t'} d\theta \frac{J_1(ks)}{s}, \\ (A_1, A_2, A_3) &= \frac{e\vec{v}}{cr\left(1 - \frac{v_r}{c}\right)} - ek\int_{-\infty}^{t'} d\theta \frac{J_1(ks)}{s}\left(\frac{d\xi}{d\theta}, \frac{d\eta}{d\theta}, \frac{d\zeta}{d\theta}\right), \end{aligned}\right\} \ldots \quad (9)$$

which are the forms used by Bhabha.

In concluding this section we remark that the advanced potentials of the mesotron, also used by Bhabha, may be obtained in like manner by the choice of a path of integration made to run along the real axis from $+\infty$ on one Riemann surface round the other singularity at $\theta = t''$ back to $+\infty$ on a second.

4. *Solution of the proposed equation for retarded values of t : First method.* We write

$$\nabla^2\phi - \frac{\partial^2\phi}{c^2\,\partial\theta^2} - k^2\phi = -4\pi\rho(\xi\eta\zeta\theta) \qquad \ldots \qquad \ldots \qquad (10)$$

and

$$\nabla^2\chi - \frac{\partial^2\chi}{c^2\,\partial\theta^2} - k^2\chi = 0,$$

where χ is given by (4). Hence we obtain firstly,

$$\sum \frac{\partial}{\partial\xi}\left(\chi\frac{\partial\phi}{\partial\xi} - \phi\frac{\partial\chi}{\partial\xi}\right) - \frac{\partial}{c\,\partial\theta}\left(\chi\frac{\partial\phi}{c\,\partial\theta} - \phi\frac{\partial\chi}{c\,\partial\theta}\right) = -4\pi\chi\rho(\xi\eta\zeta\theta)$$

and then, upon integration with respect to θ,

$$\int_{\theta_1}^{\theta_2} d\theta \sum \frac{\partial}{\partial\xi}\left(\chi\frac{\partial\phi}{\partial\xi} - \phi\frac{\partial\chi}{\partial\xi}\right) - \frac{1}{c}\left[\chi\frac{\partial\phi}{c\,\partial\theta} - \phi\frac{\partial\chi}{c\,\partial\theta}\right]_{\theta_1}^{\theta_2} = -4\pi\int_{\theta_1}^{\theta_2} d\theta\,\chi\rho(\xi\eta\zeta\theta).$$

We assume now θ to have been complex of the form $\mu+i\nu$ and proceed to perform the integrations along the same path on the complex plane as we have used in the previous section. Since, with $\theta \to -\infty$, both χ and $\frac{\partial\chi}{\partial\theta} \to 0$, we have

$$\int_{\theta_1}^{\theta_2} d\theta \sum \frac{\partial}{\partial\xi}\left(\chi\frac{\partial\phi}{\partial\xi} - \phi\frac{\partial\chi}{\partial\xi}\right) = -4\pi\int_{\theta_1}^{\theta_2} d\theta\,\chi\rho(\xi\eta\zeta\theta).$$

This we may write again in the form

$$\sum \frac{\partial}{\partial\xi}\int_{\theta_1}^{\theta_2} d\theta\left(\chi\frac{\partial\phi}{\partial\xi} - \phi\frac{\partial\chi}{\partial\xi}\right) = -4\pi\int_{\theta_1}^{\theta_2} d\theta\,\chi\rho(\xi\eta\zeta\theta), \qquad \ldots \qquad (11)$$

if we observe that we may treat the path of integration as independent of ξ, η, ζ, since the path about the singularity at t' may always be slightly varied to accommodate the shift of t' involved in the operations, $\frac{\partial}{\partial\xi}$, etc.

We set down now in brief the necessary calculations.

(i) $$\int_{\theta_1}^{\theta_2} d\theta\,\chi\rho(\xi\eta\zeta\theta) = \int_{\theta_1}^{\theta_2} d\theta\left[\frac{kJ_1(ks)}{s}\ln s - \frac{1}{s^2}\right]\rho(\xi\eta\zeta\theta)$$

$$= -\pi ik\int_{-\infty}^{t'} d\theta\,\frac{J_1(ks)}{s}\rho(\xi\eta\zeta\theta) + \frac{\pi i}{cr}\rho(\xi\eta\zeta t').$$

$$\nabla^2\phi - \frac{\partial^2\phi}{c^2\,\partial t^2} - k^2\phi = -4\pi\rho(xyzt).$$

(ii) $$\int_{\theta_1}^{\theta_2} d\theta\,\chi\frac{\partial\phi}{\partial\xi} = \int_{\theta_1}^{\theta_2} d\theta\left[\frac{kJ_1(ks)}{s}\,lns - \frac{1}{s^2}\right]\frac{\partial\phi}{\partial\xi}$$

$$= -\pi ik\int_{-\infty}^{t'} d\theta\,\frac{J_1(ks)}{s}\frac{\partial\phi}{\partial\xi} + \frac{\pi i}{cr}\left(\frac{\partial\phi}{\partial\xi}\right)_{t'}.$$

(iii) $$\int_{\theta_1}^{\theta_2} d\theta\,\phi\frac{\partial\chi}{\partial\xi} = -(\xi - x)\int_{\theta_1}^{\theta_2} d\theta\,\phi\frac{1}{s}\frac{d\chi}{ds}$$

$$= -(\xi - x)\int_{\theta_1}^{\theta_2} d\theta\,\phi\frac{1}{s}\frac{d}{ds}\left[\frac{kJ_1(ks)}{s}\,lns - \frac{J_0(ks)}{s^2}\right]$$

$$= -(\xi - x)\int_{\theta_1}^{\theta_2} d\theta\,\phi\left[k\frac{d}{ds}\left(\frac{J_1(ks)}{s}\right)\frac{lns}{s} + \frac{2kJ_1(ks)}{s^3} + \frac{2J_0(ks)}{s^4}\right]$$

$$= -(\xi - x)\left[k\int_{\theta_1}^{\theta_2} d\theta\,\phi\frac{d}{ds}\left(\frac{J_1(ks)}{s}\right)\frac{lns}{s} + \frac{k^2}{2}\int_{\theta_1}^{\theta_2} d\theta\,\frac{\phi}{s^2}\right.$$

$$\left. + 2\int_{\theta_1}^{\theta_2} d\theta\,\frac{\phi}{s^4}\right]$$

Replacing $\frac{1}{s^4}$ by $\frac{1}{c^4(t''-t')^2}\left[\frac{1}{(\theta-t')^2} - \frac{2}{(\theta-t')(\theta-t'')} + \frac{1}{(\theta-t'')^2}\right]$

we have

$$\int_{\theta_1}^{\theta_2} d\theta\,\phi\frac{\partial\chi}{\partial\xi}$$

$$= -(\xi - x)\left[-\pi ik\int_{-\infty}^{t'} d\theta\,\phi\frac{d}{ds}\left(\frac{J_1(ks)}{s}\right)\frac{1}{s} - \frac{k^2\pi i}{2cr}(\phi)_{t'} + \frac{\pi i}{cr^2}\left(\frac{\partial\phi}{c\,\partial\theta}\right)_{t'} + \frac{\pi i}{cr^3}(\phi)_{t'}\right]$$

$$= -\pi ik\int_{-\infty}^{t'} d\theta\,\phi\frac{\partial}{\partial\xi}\left(\frac{J_1(ks)}{s}\right) + \frac{k^2\pi i}{2c}(\phi)_{t'}\frac{\partial r}{\partial\xi} - \frac{\pi i}{cr}\frac{\partial r}{\partial\xi}\left(\frac{\partial\phi}{c\,\partial\theta}\right)_{t'} + \frac{\pi i}{c}(\phi)_{t'}\frac{\partial}{\partial\xi}\left(\frac{1}{r}\right).$$

Substituting these results in eq. (11) and dropping the factor $\frac{\pi i}{c}$ we get

$$-4\pi\left[\frac{\rho(\xi\eta\zeta t')}{r}-ck\int_{-\infty}^{t'}d\theta\,\frac{J_1(ks)}{s}\,\rho(\xi\eta\zeta\theta)\right]$$

$$=\sum\frac{\partial}{\partial\xi}\left[\frac{1}{r}\left(\frac{\partial\phi}{\partial\xi}\right)_{t'}+\frac{1}{r}\frac{\partial r}{\partial\xi}\left(\frac{\partial\phi}{c\,\partial\theta}\right)_{t'}-(\phi)_{t'}\frac{\partial}{\partial\xi}\left(\frac{1}{r}\right)-ck\int_{-\infty}^{t'}d\theta\,\frac{J_1(ks)}{s}\frac{\partial\phi}{\partial\xi}\right.$$

$$\left.+ck\int_{-\infty}^{t'}d\theta\,\phi\,\frac{\partial}{\partial\xi}\left(\frac{J_1(ks)}{s}\right)-\frac{k^2}{2}(\phi)_{t'}\frac{\partial r}{\partial\xi}\right]. \quad .. \quad (12)$$

We integrate this equation now further through a closed volume about the point P (x, y, z) but exclude the volume of a small sphere about that point. We obtain thus

$$-4\pi\iiint d\xi\,d\eta\,d\zeta\left[\frac{\rho(\xi\eta\zeta t')}{r}-ck\int_{-\infty}^{t'}d\theta\,\frac{J_1(ks)}{s}\,\rho(\xi\eta\zeta\theta)\right]$$

$$=\iint dS\left[\frac{1}{r}\left(\frac{\partial\phi}{\partial n}\right)_{t'}+\frac{1}{r}\frac{\partial r}{\partial n}\left(\frac{\partial\phi}{c\,\partial\theta}\right)_{t'}-(\phi)_{t'}\frac{\partial}{\partial n}\left(\frac{1}{r}\right)\right.$$

$$\left.-ck\int_{-\infty}^{t'}d\theta\,\frac{J_1(ks)}{s}\frac{\partial\phi}{\partial n}+ck\int_{-\infty}^{t'}d\theta\,\phi\,\frac{\partial}{\partial n}\left(\frac{J_1(ks)}{s}\right)-\frac{k^2}{2}(\phi)_{t'}\frac{\partial r}{\partial n}\right]$$

$$+\iint dS'\left[\frac{1}{r}\left(\frac{\partial\phi}{\partial n}\right)_{t'}+\frac{1}{r}\frac{\partial r}{\partial n}\left(\frac{\partial\phi}{c\,\partial\theta}\right)_{t'}-(\phi)_{t'}\frac{\partial}{\partial n}\left(\frac{1}{r}\right)\right.$$

$$\left.-ck\int_{-\infty}^{t'}d\theta\,\frac{J_1(ks)}{s}\frac{\partial\phi}{\partial n}+ck\int_{-\infty}^{t'}d\theta\,\phi\,\frac{\partial}{\partial n}\left(\frac{J_1(ks)}{s}\right)-\frac{k^2}{2}(\phi)_{t'}\frac{\partial r}{\partial n}\right],$$

where the second surface integral is over the surface of the small sphere about P. Its limiting value, with $r\to 0$, is seen to be

$$-\phi_{P,t}\iint dS'\,\frac{\partial}{\partial n}\left(\frac{1}{r}\right)=-4\pi\phi_{P,t}.$$

$$\nabla^2\phi - \frac{\partial^2\phi}{c^2\,\partial t^2} - k^2\phi = -4\pi\rho(xyzt).$$

We arrive hence at the complete Kirchhoff-like solution for retarded values of t in the form

$$\phi(xyzt) = \iiint d\xi\, d\eta\, d\zeta \left[\frac{\rho(\xi\eta\zeta t')}{r} - ck\int_{-\infty}^{t'} d\theta\, \frac{J_1(ks)}{s}\,\rho(\xi\eta\zeta\theta)\right]$$

$$+\frac{1}{4\pi}\iint dS\left[\frac{1}{r}\left(\frac{\partial\phi}{\partial n}\right)_{t'} + \frac{1}{r}\frac{\partial r}{\partial n}\left(\frac{\partial\phi}{c\,\partial\theta}\right)_{t'} - (\phi)_{t'}\frac{\partial}{\partial n}\left(\frac{1}{r}\right)\right.$$

$$\left. - ck\int_{-\infty}^{t'} d\theta\left\{\frac{J_1(ks)}{s}\frac{\partial\phi}{\partial n} - \phi\frac{\partial}{\partial n}\left(\frac{J_1(ks)}{s}\right)\right\} - \frac{k^2}{2}(\phi)_{t'}\frac{\partial r}{\partial n}\right]. \quad \ldots \quad (13)$$

In closing this section we remark that a like procedure with the second path of integration pointed out in the previous section yields the complete Kirchhoff-like solution for advanced values of t in the form

$$\phi(xyzt) = \iiint d\xi\, d\eta\, d\zeta \left[\frac{\rho(\xi\eta\zeta t'')}{r} - ck\int_{t''}^{\infty} d\theta\, \frac{J_1(ks)}{s}\,\rho(\xi\eta\zeta\theta)\right]$$

$$+\frac{1}{4\pi}\iint dS\left[\frac{1}{r}\left(\frac{\partial\phi}{\partial n}\right)_{t''} - \frac{1}{r}\frac{\partial r}{\partial n}\left(\frac{\partial\phi}{c\,\partial\theta}\right)_{t''} - (\phi)_{t''}\frac{\partial}{\partial n}\left(\frac{1}{r}\right)\right.$$

$$\left. - ck\int_{t''}^{\infty} d\theta\left\{\frac{J_1(ks)}{s}\frac{\partial\phi}{\partial n} - \phi\frac{\partial}{\partial n}\left(\frac{J_1(ks)}{s}\right)\right\} - \frac{k^2}{2}(\phi)_{t''}\frac{\partial r}{\partial n}\right]. \quad \ldots \quad (14)$$

5. *Solution of the proposed equation for retarded values of t : Second method.*

To begin with, we replace θ by $t-\frac{r}{c}(=t')$ in eq. (10). With the notation

$$\left.\begin{aligned} \rho\left(\xi,\eta,\zeta,t-\frac{r}{c}\right) &= [\rho], \\ \phi\left(\xi,\eta,\zeta,t-\frac{r}{c}\right) &= [\phi], \end{aligned}\right\} \quad \ldots \quad \ldots \quad \ldots \quad (15)$$

etc.,

equation (10) takes then the form

$$[\nabla^2\phi] - \left[\frac{\partial^2\phi}{c^2\,\partial\theta^2}\right] - k^2[\phi] = -4\pi[\rho]. \quad \ldots \quad \ldots \quad (16)$$

Now

$$\left.\begin{aligned}\frac{\partial}{\partial\xi}[\phi] &= \left[\frac{\partial\phi}{\partial\xi}\right]-\left[\frac{\partial\phi}{\partial\theta}\right]\frac{1}{c}\frac{\partial r}{\partial\xi},\\ \frac{\partial}{\partial t}[\phi] &= \left[\frac{\partial\phi}{\partial\theta}\right],\\ \frac{\partial^2}{\partial\xi^2}[\phi] &= \left[\frac{\partial^2\phi}{\partial\xi^2}\right]-2\left[\frac{\partial^2\phi}{\partial\xi\,\partial\theta}\right]\frac{1}{c}\frac{\partial r}{\partial\xi}\\ &\quad+\left[\frac{\partial^2\phi}{\partial\theta^2}\right]\left(\frac{1}{c}\frac{\partial r}{\partial\xi}\right)^2-\left[\frac{\partial\phi}{\partial\theta}\right]\frac{1}{c}\frac{\partial^2 r}{\partial\xi^2},\\ \frac{\partial^2}{\partial\xi\,\partial t}[\phi] &= \left[\frac{\partial^2\phi}{\partial\xi\,\partial\theta}\right]-\left[\frac{\partial^2\phi}{\partial\theta^2}\right]\frac{1}{c}\frac{\partial r}{\partial\xi}.\end{aligned}\right\}\quad\text{.. (17)}$$

Hence we deduce firstly

$$\frac{\partial^2}{\partial\xi^2}[\phi]+\frac{2}{c}\frac{\partial r}{\partial\xi}\frac{\partial^2}{\partial\xi\,\partial t}[\phi]+\frac{1}{c}\frac{\partial^2 r}{\partial\xi^2}\frac{\partial}{\partial t}[\phi] = \left[\frac{\partial^2\phi}{\partial\xi^2}\right]-\left(\frac{1}{c}\frac{\partial r}{\partial\xi}\right)^2\left[\frac{\partial^2\phi}{\partial\theta^2}\right]$$

and then

$$\nabla^2[\phi]+\frac{2}{c}\sum\frac{\partial r}{\partial\xi}\frac{\partial^2}{\partial\xi\,\partial t}[\phi]+\frac{2}{cr}\frac{\partial}{\partial t}[\phi]-k^2[\phi] = [\nabla^2\phi]-\left[\frac{\partial^2\phi}{c^2\,\partial\theta^2}\right]-k^2[\phi]$$
$$= -4\pi[\rho]\quad\text{..}\quad\text{..}\quad\text{.. (18)}$$

in view of eq. (16). Dividing both sides of eq. (18) by r, we have, since $\nabla^2\frac{1}{r}=0$,

$$\sum\frac{\partial}{\partial\xi}\left\{\frac{1}{r}\frac{\partial}{\partial\xi}[\phi]-[\phi]\frac{\partial}{\partial\xi}\left(\frac{1}{r}\right)+\frac{2}{c}\frac{1}{r}\frac{\partial r}{\partial\xi}\frac{\partial}{\partial t}[\phi]\right\}-\frac{k^2[\phi]}{r} = -4\pi\frac{[\rho]}{r}.\quad\text{.. }(A)$$

We next proceed to combine eq. (10) with

$$\nabla^2 Q-\frac{\partial^2 Q}{c^2\,\partial\theta^2}-k^2 Q = 0,$$

where

$$Q = \frac{J_1(ks)}{s},$$

and obtain firstly

$$\sum\frac{\partial}{\partial\xi}\left(Q\frac{\partial\phi}{\partial\xi}-\phi\frac{\partial Q}{\partial\xi}\right)-\frac{\partial}{c\,\partial\theta}\left(Q\frac{\partial\phi}{c\,\partial\theta}-\phi\frac{\partial Q}{c\,\partial\theta}\right) = -4\pi Q\rho(\xi\eta\zeta\theta)$$

and then, since both Q and $\frac{\partial Q}{\partial\theta}\to 0$ with $\theta\to-\infty$,

$$\sum\int_{-\infty}^{t'}d\theta\frac{\partial}{\partial\xi}\left(Q\frac{\partial\phi}{\partial\xi}-\phi\frac{\partial Q}{\partial\xi}\right)-\frac{1}{c}\left[Q\frac{\partial\phi}{c\,\partial\theta}-\phi\frac{\partial Q}{c\,\partial\theta}\right] = -4\pi\int_{-\infty}^{t'}d\theta\,Q\rho(\xi\eta\zeta\theta)$$

$$\nabla^2\phi - \frac{\partial^2\phi}{c^2\,\partial t^2} - k^2\phi = -4\pi\rho(xyzt).$$

or $$\sum\frac{\partial}{\partial\xi}\int_{-\infty}^{t'} d\theta\left(Q\frac{\partial\phi}{\partial\xi}-\phi\frac{\partial Q}{\partial\xi}\right)+\sum\left[Q\frac{\partial\phi}{\partial\xi}-\phi\frac{\partial Q}{\partial\xi}\right]\frac{1}{c}\frac{\partial r}{\partial\xi}-\frac{1}{c}\left[Q\frac{\partial\phi}{c\,\partial\theta}-\phi\frac{\partial Q}{c\,\partial\theta}\right]$$

$$= -4\pi\int_{-\infty}^{t'} d\theta\, Q\rho(\xi\eta\zeta\theta).$$

Rearranging the terms in this equation we write

$$\sum\frac{\partial}{\partial\xi}\int_{-\infty}^{t'} d\theta\left(Q\frac{\partial\phi}{\partial\xi}-\phi\frac{\partial Q}{\partial\xi}\right)-\frac{[\phi]}{c}\left[\sum\frac{\partial Q}{\partial\xi}\frac{\partial r}{\partial\xi}-\frac{\partial Q}{c\,\partial\theta}\right]+\frac{[Q]}{c}\left[\sum\frac{\partial\phi}{\partial\xi}\frac{\partial r}{\partial\xi}-\frac{\partial\phi}{c\,\partial\theta}\right]$$

$$= -4\pi\int_{-\infty}^{t'} d\theta\, Q\rho(\xi\eta\zeta\theta) \quad .. \quad .. \quad .. \quad .. \quad (19)$$

and observe that

$$\left[\frac{\partial Q}{\partial\theta}\right] = 0 = \left[\frac{\partial Q}{\partial\xi}\right] = \left[\frac{\partial Q}{\partial\eta}\right] = \left[\frac{\partial Q}{\partial\zeta}\right],$$

$$[Q] = \frac{k}{2},$$

and, in view of eqs. (17),

$$\left[\sum\frac{\partial\phi}{\partial\xi}\frac{\partial r}{\partial\xi}-\frac{\partial\phi}{c\,\partial\theta}\right] = \sum\frac{\partial r}{\partial\xi}\frac{\partial}{\partial\xi}[\phi]$$

$$= \sum\frac{\partial}{\partial\xi}\left(\frac{\partial r}{\partial\xi}[\phi]\right)-\frac{2}{r}[\phi].$$

Equation (19) now takes the form

$$\sum\frac{\partial}{\partial\xi}\left\{\int_{-\infty}^{t'} d\theta\left(Q\frac{\partial\phi}{\partial\xi}-\phi\frac{\partial Q}{\partial\xi}\right)+\frac{k}{2c}\frac{\partial r}{\partial\xi}[\phi]\right\}-\frac{k}{c}\frac{[\phi]}{r}$$

$$= -4\pi\int_{-\infty}^{t'} d\theta\, Q\rho(\xi\eta\zeta\theta). \quad .. \quad .. \quad .. \quad (B)$$

Combining now equations (A) and (B) we get

$$\sum\frac{\partial}{\partial\xi}\left\{\frac{1}{r}\frac{\partial}{\partial\xi}[\phi]-[\phi]\frac{\partial}{\partial\xi}\left(\frac{1}{r}\right)+\frac{2}{c}\frac{1}{r}\frac{\partial r}{\partial\xi}\frac{\partial}{\partial t}[\phi]-ck\int_{-\infty}^{t'} d\theta\left(Q\frac{\partial\phi}{\partial\xi}-\phi\frac{\partial Q}{\partial\xi}\right)\right.$$

$$\left.-\frac{k^2}{2}\frac{\partial r}{\partial\xi}[\phi]\right\}$$

$$= -4\pi\left\{\frac{[\rho]}{r}-ck\int_{-\infty}^{t'} d\theta\, Q\rho(\xi\eta\zeta\theta)\right\}$$

or, in view of eqs. (17),

$$\sum \frac{\partial}{\partial \xi}\left\{\frac{1}{r}\left[\frac{\partial \phi}{\partial \xi}\right]+\frac{1}{r}\frac{\partial r}{\partial \xi}\left[\frac{\partial \phi}{c\,\partial \theta}\right]-[\phi]\frac{\partial}{\partial \xi}\left(\frac{1}{r}\right)-ck\int_{-\infty}^{t'} d\theta\left(Q\frac{\partial \phi}{\partial \xi}-\phi\frac{\partial Q}{\partial \xi}\right)\right.$$

$$\left.-\frac{k^2}{2}\frac{\partial r}{\partial \xi}[\phi]\right\}$$

$$=-4\pi\left\{\frac{[\rho]}{r}-ck\int_{-\infty}^{t'} d\theta\, Q\rho(\xi\eta\zeta\theta)\right\}. \qquad \ldots \quad \ldots \quad (20)$$

This equation (20) is easily seen to be identical with eq. (12) and subsequent integration through a closed volume, as in the last section, will now obviously lead to the complete solution of the proposed equation in the form (13).

It is evident that we may in like manner obtain the complete solution for advanced values of t if we replace θ in eq. (10) at the outset by $t+\frac{r}{c}$ $(=t'')$ and later perform integrations with respect to θ from t'' to $+\infty$.

6. In closing this communication we like to make a few observations on the methods here employed. Firstly, the method of complex integration is obviously also available for the Kirchhoff solution of the equation

$$\nabla^2\phi-\frac{\partial^2\phi}{c^2\,\partial\theta^2}=-4\pi\rho(\xi\eta\zeta\theta).$$

We may have this solution, of course, immediately out of the solutions (13) and (14) with $k\to 0$; we may have it, too, by a like procedure repeated with a solving function $\chi(s)=\frac{1}{s^2}$. Secondly, both methods may be adapted without difficulty for solution of the equation

$$\nabla^2\phi-\frac{\partial^2\phi}{c^2\,\partial\theta^2}+k^2\phi=-4\pi\rho(\xi\eta\zeta\theta)$$

and the only difference should be that real Bessel functions of imaginary arguments would take the place of those we have used above. Lastly, underlying the method of complex integration there is obviously the assumption that ρ and ϕ are analytic functions of θ over a region of the complex plane wide enough to embrace the path of integration—an assumption from which the second method is happily free.

References.

* Love: Mathematical Theory of Elasticity, 4th Ed. (1927), pp. 302-3 or *Lond. Math. Soc. Proc.*, Ser 2, Vol. 1, (1904).

† J. H. Bhabha: *Proc. Roy. Soc.*, Vol. 172, p. 384, (1939).

‡ Herglotz and Sommerfeld: *Frenkel's Lehrbuch der Elektrodynamik*, Bd. 1, p. 177, (1926).

§ Forsyth: A Treatise on Differential Equations, 2nd Ed., p. 167.

Reprinted from *Proc Nat Inst Sc India* 7, pp. 93-102, 1941 (Indian National Science Academy).

ON AN INTEGRAL EQUATION ASSOCIATED WITH THE EQUATION FOR HYDROGEN ATOM

By

S. N. Bose

(*Received on May* 15, 1945)

I

The Schrödinger functions 'ϕ' characterising the stationary states of hydrogen atom are now very familiar things in analysis as also the differential equation which they satisfy, namely,

$$\nabla^2\phi+\frac{8\pi^2m}{h^2}\left(E+\frac{e^2}{r}\right)\phi=0. \qquad \ldots \quad (1,1)$$

The associated functions M, defined by the relation,

$$\phi=\int M\exp.\,2\pi\iota(lx+my+nz)dl\,dm\,dn, \qquad \ldots \quad (1,2)$$

can be utilised for defining the probability in momentum space if the momentum variables are introduced by the relation $p_x=hl$, $p_y=hm$, $p_z=hn$.

When inversion is possible, (1,2) implies also

$$M(l,m,n)=\int\phi\exp.-2\pi\iota(lx+my+nz)dx\,dy\,dz \qquad \ldots \quad (1,3)$$

and solutions of (1,1) can be utilised to calculate M's. Elsasser (1933) has followed this method and arrived at fairly complicated formula. Another alternative would be to set up an appropriate equation for them, and investigate its solutions. This is an integral equation, whose complete solution is presented here. The analysis presents several interesting features, and leads to expressions of M's, which can be immediately utilised to study their properties or to apply them to physical problems.

II

If we use the semi-convergent integral

$$\frac{1}{R}=\frac{1}{\pi}\int\frac{dl\,dm\,dn}{l^2+m^2+n^2}\exp.\,2\pi\iota(lx+my+nz),\quad R^2=x^2+y^2+z^2, \qquad \ldots \quad (2,1)$$

then combination with (1,2) leads to the following sex-tuple integral for ϕ/R, after a change of variable and order of integration :

$$\frac{\phi}{R}=\frac{1}{\pi}\int\frac{M(\lambda-l,\mu-m,\nu-n)}{l^2+m^2+n^2}\exp.\,2\pi\iota(\lambda x+\mu y+\nu z)dl\,dm\,dn\,d\lambda\,d\mu\,d\nu.$$

M and

$$\frac{1}{\pi}\int\frac{M(\lambda-l,\ \mu-m,\ \nu-n)dl\ dm\ dn}{l^2+m^2+n^2}$$

can thus be regarded as Fourier transforms of ϕ and ϕ/R.

Using then in the Schrödinger equation (1,1) we deduce the following integral equation

$$(l^2+m^2+n^2-k^2)M = \lambda\int\frac{M(l-x,\ m-y,\ n-z)}{x^2+y^2+z^2}\,dx\,dy\,dz, \qquad \ldots\ \text{(A)}$$

where

$$k^2 = 2mE/h^2. \quad \text{and} \quad \lambda = 2me^2/\pi h^2.$$

M is here assumed to be finite and single-valued throughout the domain of integration, as also

$$\int |M|^2\,dx\,dy\,dz = 1.$$

(A) is the characteristic integral equation for M-functions of hydrogen; for $k^2 < 0$, it leads to the discrete spectrum, while $k^2 > 0$: yields the continuous spectrum.

It is easy to transform (A) to the Fredholm type, when $k^2 < 0 = -a^2$. We first transform the origin, *i.e.*, put $l-x = x'$, etc, in the right side. (A) becomes

$$(l^2+m^2+n^2-k^2)M = -\int\frac{\lambda M(x'y'z')dx'\,dy'\,dz'}{[(l-x')^2+(m-y')^2+(n-z')^2]}\ ; \qquad \ldots\ (2{,}32)$$

when $k^2 = -a^2$, we make a similarity transformation

$$l = l'a,\ x' = ax'',\ etc.,$$

and put

$$M(l, m, n) = M'(l', m', n')$$

whereby

$$dx'\,dy'\,dz' = a^3dx''\,dy''\,dz''$$

and

$$(l^2+m^2+n^2+a^2) = a^2(l'^2+m'^2+n'^2+1), \text{ etc.}$$

The relation becomes

$$(l^2+m^2+n^2+1)M = -\frac{\lambda}{a}\int\frac{M'(x'y'z')dx'dy'dz'}{(l-x')^2+(m-y')^2+(n-z')^2.}$$

If further

$$\sqrt{(1+l^2+m^2+n^2)}M = \phi,$$

then

$$\phi(A) = \kappa\int\frac{\phi(P)dv_P}{[(l-x)^2+(m-y)^2+(n-z)^2]\sqrt{(1+l^2+m^2+n^2)}\sqrt{(1+x^2+y^2+z^2)}}\ , \qquad \ldots\ (2{,}33)$$

where

$$\kappa = \frac{e^2}{\pi h}\left(\frac{2m}{-E}\right)^{\frac{1}{2}}.$$

(2,33) is thus seen of the standard form

$$\phi(A) = -\lambda \int \phi(P)K(A, P)dv_p,$$

where the kernel is symmetrical in A and P. Such a transformation will however make the kernel imaginary within the domain, if $k^2 > 0$; to have a uniform treatment to cover both the cases, we will not utilise the transformation mentioned above.

III

We require the following simple result in our subsequent calculations. If r_1 and r_2 are the distances of a point $P(x, y, z)$ from two fixed points $A(a, b, c)$ and $F(f, g, h)$, the integral

$$I = \int \frac{dx\,dy\,dz}{r_1^2 r_2^2}$$

extended over the whole domain easily transforms to

$$\frac{2\pi}{c}\int_1^\infty \int_{-1}^1 \frac{dX dY}{X^2 - Y^2}$$

in bipolar co-ordinates with $AF = 2c$,

$$dx\,dy\,dz = c^3(\cosh^2\xi - \cos^2\eta)\sinh\xi \sin\eta\, d\xi d\eta d\phi,$$

and

$$X = \cosh\xi \quad \text{and} \quad Y = \cos\eta,$$

whence, by easy integration,

$$I = \frac{\pi^3}{2c} = \frac{\pi^3}{AF}. \qquad \dots \quad (3,1)$$

Again, if

$$A(a, b, c) = \lambda \int \frac{M(a-x, b-y, c-z)dx\,dy\,dz}{x^2+y^2+z^2} = -\lambda \int \frac{M(x'y'z')dx'dy'dz'}{(a-x')^2+(b-y')^2+(c-z')^2} \qquad \dots \quad (3,2)$$

we have, by multiplying the equation with

$$\frac{da\,db\,dc}{(f-a)^2+(g-b)^2+(h-c)^2}$$

and integrating over the whole domain,

$$\int \frac{A(a\,b\,c)da\,db\,dc}{[(f-a)^2+(g-b)^2+(h-c)^2]} = -\lambda \int \int \frac{M(x'y'z')dx'\,dy'dz'\,da\,db\,dc}{[(a-x')^2+\dots][(f-a)^2+\dots]}.$$

Changing the order of integration and utilising (3,1) we have

$$\int \frac{M(x'y'z')dx'dy'dz'}{[(f-x')^2+(g-y')^2+(h-z')^2]^{\frac{1}{2}}} = -\frac{1}{\pi^3\lambda}\int \frac{A(abc)da\,db\,dc}{[(f-a)^2+(g-b)^2+(h-c)^2]}$$

$$= \frac{1}{\pi^3\lambda}\int \frac{A(f-a',\ldots)da'db'dc'}{a'^2+b'^2+c'^2}\ ;$$

hence, operating with $\nabla^2_{(fgh)}$ on both sides, as

$$\nabla^2_{(fgh)}\int \frac{M(x'y'z')dx'dy'dz'}{[(f-x')^2+\ldots]^{\frac{1}{2}}} = -4\pi M(f,g,h)$$

from potential theory, and

$$\nabla^2_{(fgh)}\{A(f-a,g-b,h-c)\} = (\nabla^2 A)(f-a,g-b,h-c),$$

we have finally

$$M(f,g,h) = -\frac{1}{4\pi^4\lambda}\int \frac{(\nabla^2 A)(f-a,g-b,h-c)\,da\,db\,dc}{a^2+b^2+c^2}$$

$$= \frac{1}{4\pi^4\lambda}\int \frac{(\nabla^2 A')dx'dy'dz'}{[(f-x')^2+(g-y')^2+(h-z')^2]} \qquad \ldots \quad (3,3)$$

as a solution of (3,2).

This important result furnishing a solution of integral equation (3,2) with obvious restrictions about the nature of the function A, enables us to tackle our present problem.

IV

According to a well-known theorem due to Hobson, if an operator $S\left(\frac{d}{dx},\frac{d}{dy},\frac{d}{dz}\right)$ is constructed by substituting $\frac{d}{dx}$, …, etc. for x, y, z in a solid harmonic S_n of degree n (a positive integer),

$$S_n\left(\frac{d}{dx},\frac{d}{dy},\frac{d}{dz}\right)F(r^2) = 2^n S_n(x,y,z)\left(\frac{d}{dr^2}\right)^n F. \qquad \ldots \quad (4,1)$$

Let us assume, with regard to the integral equation (A), that

$$M(x,y,z) = S_n(x,y,z)f(r)$$

or

$$M(x,y,z) = S_n\left(\frac{d}{dx},\frac{d}{dy},\frac{d}{dz}\right)A(r^2) \qquad \ldots \quad (4,2)$$

in view of (4,1) ; also $(r^2-k^2)M(x,y,z)$ can be written as

$$S_n\left(\frac{d}{dx}, \frac{d}{dy}, \frac{d}{dz}\right) B(r^2) \qquad \ldots \quad (4,3)$$

with the same operàtor, as the same surface-harmonic will occur in both. Also as

$$S_n\left(\frac{d}{dx}, \frac{d}{dy}, \frac{d}{dz}\right) A = 2^n S_n(x, y, z) \left(\frac{d}{dr^2}\right)^n A = M,$$

and

$$S_n\left(\frac{d}{dx}, \frac{d}{dy}, \frac{d}{dz}\right) B = 2^n S_n(x, y, z) \left(\frac{d}{dr^2}\right)^n B = (r^2-k^2)M,$$

we have

$$\left(\frac{d}{dr^2}\right)^n B = (r^2-k^2) \left(\frac{d}{dr^2}\right)^n A = \left(\frac{d}{dr^2}\right)^{n-1} [A(r^2-k^2)-nA],$$

implying

$$\frac{dB}{dr} = \frac{d}{dr}[A(r^2-k^2)]-2nAr. \qquad \ldots \quad (4,4)$$

We have also from (A)

$$S_n\left[\frac{d}{da}, \ \ldots\right] B = \lambda \int \frac{S_n\left[\frac{d}{d(a-x)}, \ \ldots\right] A \, dx\, dy\, dz}{x^2+y^2+z^2}$$

$$= \lambda S_n\left(\frac{d}{da}, \ \ldots\right) \int \frac{A[(a-x)^2+\ldots]dx\,dy\,dz}{x^2+y^2+z^2}.$$

Hence removing the operator S_n

$$B = \lambda \int \frac{A[a-x)^2+\ldots]dx\, dy\, dz,}{x^2+y^2+z^2}, \qquad \ldots \quad (4,51)$$

and hence on account of (3.3)

$$A = -\frac{1}{4\pi^4\lambda} \int \frac{(\nabla^2 B)(a-x, b-y, c-z)\, dx\, dy\, dz}{x^2+y^2+z^2}. \qquad \ldots \quad (4,52)$$

We now perform the integration ; assuming

$$(a-x)^2+(b-y)^2+(c-z)^2 = f^2 = r^2+\rho^2-2r\rho\cos\theta$$

according to the accompanying figure

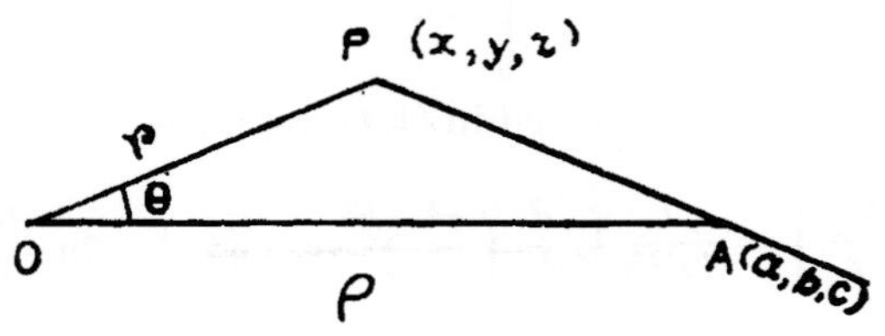

we have, as $fdf = r\rho \sin \theta$,

$$\rho B(\rho^2) = 2\pi\lambda \int_0^\infty \frac{dr}{r} \int_{f^2=(\rho-r)^2}^{f^2=(\rho+r)^2} \tfrac{1}{2} A(f^2)\, df^2$$

$$= 2\pi\lambda \int_0^\infty \frac{dx}{x}[G_1(\rho+x)-G_1(\rho-x)],$$

where

$$\frac{1}{x}\frac{dG}{dx} = A(x^2)$$

and G_1, an even function of x. Similarly from (4,52)

$$\rho A = -\frac{1}{2\pi^3\lambda}\int_0^\infty \frac{dx}{x}[L(\rho+x)-L(\rho-x)],$$

where

$$\frac{dL}{xdx} = \nabla^2 B(x^2);$$

hence integrating and putting

$$\frac{dQ}{dx} = L, \text{ or } \frac{1}{x}\frac{d^2Q}{dx^2} = \nabla^2 B$$

we have

$$G_1(\rho) = -\frac{1}{2\pi^3\lambda}\int \frac{dx}{x}[Q(\rho+x)-Q(\rho-x)],$$

where

$$\frac{1}{x}\frac{d^2Q}{dx^2} = \nabla^2 B = \frac{1}{x}\frac{d^2}{dx^2}(xB),$$

$$\text{or} \quad Q = xB,$$

and therefore an odd function of x.

Putting finally

$$\frac{xB(x^2)}{2\pi^2\lambda} = G_2 = \frac{Q}{2\pi^2\lambda},$$

we have the following relations between G_1 and G_2 in the skew-reciprocal form :

$$G_2(\rho) = \frac{1}{\pi}\int_0^\infty \frac{G_1(\rho+x)-G_1(\rho-x)}{x}\,dx, \qquad \ldots (4,53)$$

$$G_1(\rho) = -\frac{1}{\pi}\int_0^\infty \frac{G_2(\rho+x)-G_2(\rho-x)}{x}\,dx. \qquad \ldots (4,54)$$

If the functions G_1 and G_2 are introduced in (4,4) we have

$$\frac{d}{dr}\left[\frac{G_2}{r}\right] = \frac{1}{2\pi^2\lambda}\left[\frac{d}{dr}\left(\frac{r^2-k^2}{r}\cdot\frac{dG_1}{dr}\right)-2n\frac{dG_1}{dr}\right],$$

or

$$G_2 = \frac{1}{2\pi^2\lambda}\left[(x^2-k^2)\frac{dG_1}{dx} - 2nxG_1 \right],$$

or

$$\frac{G_2}{(x^2-k^2)^{n+1}} = \frac{1}{2\pi^2\lambda}\frac{d}{dx}\left[\frac{G_1}{(x^2-k^2)^n}\right]. \quad \dots \quad (4{,}55)$$

The skew-reciprocal relations (4,53) and (4,54) then at once suggest that

$$\frac{G_1}{(x^2-k^2)^{n+1}} = -\frac{1}{2\pi^2\lambda}\frac{d}{dx}\left[\frac{G_2}{(x^2-k^2)^n}\right] \quad \dots \quad (4{,}56)$$

is true at the same time.

(4,55) and (4,56), lead at once, by integration, to

$$G_1 = \mathsf{A}(x^2-k^2)^n \cos\left(\frac{2\pi^2\lambda}{2k}\log\left|\frac{x-k}{x+k}\right|\right),$$

$$G_2 = -\mathsf{A}(x^2-k^2)^n \sin\left(\frac{2\pi^2\lambda}{2k}\log\left|\frac{x-k}{x+k}\right|\right), \quad (4{,}57)$$

when $k^2 > 0$; when $k^2 = -a^2$, on the other hand,

$$G_1 = \mathsf{B}\,(x^2+a^2)^n \cos\left(\frac{2\pi^2\lambda}{a}\tan^{-1}\frac{x}{a}\right),$$

$$G_2 = -\mathsf{B}(x^2+a^2)^n \sin\left(\frac{2\pi^2\lambda}{a}\tan^{-1}\frac{x}{a}\right). \quad \dots \quad (4{,}58)$$

The two equivalent forms of M are

$$M(x,y,z) = \mathsf{A}\,S_n(x,y,z)\left(\frac{1}{r}\frac{d}{dr}\right)^{n+1}\left[(r^2-k^2)^n\cos\left(\frac{2\pi^2\lambda}{2k}\log\left|\frac{r-k}{r+k}\right|\right)\right]$$

or

$$= -2\pi^2\lambda\,\frac{\mathsf{A}S_n(x,y,z)}{r^2-k^2}\left(\frac{1}{r}\frac{d}{dr}\right)^n\left[\frac{(r^2-k^2)^n}{r}\sin\left(\frac{2\pi^2\lambda}{2k}\log\left|\frac{r-k}{r+k}\right|\right)\right] \quad \dots \quad (4{,}591)$$

for the case : $k^2 > 0$; while, if $k^2 = -a^2$,

$$M(x,y,z) = \mathsf{B}S_n(x,y,z)\left(\frac{1}{r}\frac{d}{dr}\right)^{n+1}\left[(r^2+a^2)^n\cos\left(\frac{2\pi^2\lambda}{a}\tan^{-1}\frac{r}{a}\right)\right]$$

or

$$= -2\pi^2\lambda\,\mathsf{B}\frac{S_n(x,y,z)}{r^2+a^2}\left(\frac{1}{r}\frac{d}{dr}\right)^n\left[\frac{(r^2+a^2)^n}{r}\sin\left(\frac{2\pi^2\lambda}{r}\tan^{-1}\frac{r}{a}\right)\right]. \quad \dots \quad (4.592)$$

V

As M is to be single-valued, we see, in the case of $k^2\ (<0) = -a^2$, that a restriction comes in regarding the choice of constants of the problem.

$\tan^{-1}(r/a)$ is a multiple-valued function of the form $\theta+N\pi$; if therefore $\cos\left(\frac{2\pi^2\lambda}{a}\tan^{-1}\frac{r}{a}\right)$ and $\sin\left(\frac{2\pi^2\lambda}{a}\tan^{-1}\frac{r}{a}\right)$ are single-valued, it follows that $\pi^2\lambda/a$ must be an integer $= N$. This gives discrete energy-values :

$$k^2 = \frac{2mE}{h^2} = -\frac{\pi^4\lambda^2}{N^2},$$

or

$$E = -\frac{2\pi^2 e^4 m}{h^2 N^2}$$

agreeing with the well-known result. Also as

$$2\tan^{-1}\frac{r}{a} = \cos^{-1}\frac{a^2-r^2}{a^2+r^2},$$

$$\cos\left(N\cos^{-1}\frac{a^2-r^2}{a^2+r^2}\right) = \frac{f_N(r^2)}{(a^2+r^2)^N}$$

we see that $M = 0$ unless $N \geqslant n+1$.

Again starting from the series

$$\frac{1}{1-2h\cos\theta+h^2} = \sum_1 h^{N-1}\frac{\sin N\theta}{\sin\theta},\quad \frac{1-h\cos\theta}{1-2h\cos\theta+h^2} = \sum_0 h^N\cos N\theta, \tag{5,1}$$

and writing

$$\cos\theta = \frac{a^2-r^2}{a^2+r^2} \text{ and } \sin\theta = \frac{2ar}{a^2+r^2}$$

we have

$$\frac{2a(a^2+r^2)}{\{(1-h)^2a^2+(1+h)^2r^2\}} = \Sigma\, h^{N-1}\frac{(a^2+r^2)}{r}\sin N\theta, \qquad \dots \tag{5,2}$$

$$\frac{a^2(1-h)+r^2(1+h)}{\{(1-h)^2a^2+(1+h)^2r^2\}} = \Sigma h^N\cos N\theta. \qquad \dots \tag{5,3}$$

Multiplying (5,2) with $(a^2+r^2)^{n-1}$ and differentiating n times with regard to r, we have

$$\frac{1}{[(1-h)^2a^2+(1+h)^2r^2]^{n+1}}$$

$$= \frac{(-1)^n}{n!2^{2n+1}a^{2n+1}} \Sigma\, h^{N-n-1} \left(\frac{d}{dr^2}\right)^n \left[\frac{(a^2+r^2)^n}{r} \sin\left(N\cos^{-1}\left(\frac{a^2-r^2}{a^2+r^2}\right)\right)\right]; \quad \ldots \quad (5,4)$$

from (5,3) we deduce similarly

$$\left(\frac{1-h^2}{[(1-h)^2a^2+(1+h)^2r^2]^{n+2}}\right)$$

$$= \frac{(-1)^{n+1}}{2^{2n+1}a^{2n+2}} \cdot \frac{1}{(n+1)!} \Sigma\, h^{N-n-1}\left(\frac{d}{dr^2}\right)^{n+1}\left[(a^2+r^2)^n \cos\left(N\cos^{-1}\frac{a^2-r^2}{a^2+r^2}\right)\right]. \quad (5,5)$$

Remembering the Gegenbauer expansion

$$\frac{1}{(\alpha^2-2\alpha h\cos\theta+h^2)^\nu} = \frac{1}{\alpha^{2\nu}} + \sum_\mu \frac{(2h)^\mu}{\alpha^{2\nu+\mu}} F^\nu_\mu(\cos\theta)$$

we can easily deduce, if $\nu = n+1$, $\alpha = 1$ and

$$\cos\theta = \frac{a^2-r^2}{a^2+r^2},$$

the following results :

$$\frac{1}{[a^2(1-h)^2+r^2(1+h)^2]^{n+1}} = \frac{1}{(a^2+r^2)^{n+1}}\left[1+\sum_1 (2h)^\lambda F^{n+1}_\lambda\left(\frac{a^2-r^2}{a^2+r^2}\right)\right], \quad \ldots \quad (5,6)$$

$$\frac{1-h^2}{[a^2(1-h)^2+r^2(1+h)^2]^{n+2}} = \frac{1}{(a^2+r^2)^{n+2}}\left[1+\sum_\lambda \frac{(n+1+\lambda)}{n+1}(2h)^\lambda F^{n+1}_\lambda\left(\frac{a^2-r^2}{a^2+r^2}\right)\right]. \quad (5,7)$$

Comparing (5.4) and (5.5) with (5.6) and (5.7) we have

$$\frac{1}{(a^2+r^2)}\left(\frac{d}{dr^2}\right)^n\left[\frac{(a^2+r^2)^n}{r}\sin N\theta\right] = \frac{(-1)^n 2^{N+n}a^{2n+1}n!}{(a^2+r^2)^{n+2}} F^{n+1}_{N-n-1}\left(\frac{a^2-r^2}{a^2+r^2}\right),$$

and

$$\left(\frac{d}{dr^2}\right)^{n+1}\left[(a^2+r^2)^n\cos N\theta\right] = \frac{(-1)^{n+1}2^{N+n}a^{2n+2}N\cdot n!}{(a^2+r^2)^{n+2}} F^{n+1}_{N-n-1}\left(\frac{a^2-r^2}{a^2+r^2}\right).$$

Hence

$$M(x,y,z) = \mathbf{B}S_n(x,y,z)\left(\frac{1}{r}\frac{d}{dr}\right)^{n+1}\left[(r^2+a^2)^n\cos\left(N\cos^{-1}\frac{a^2-r^2}{a^2+r^2}\right)\right]$$

or

$$= -2Na\frac{\mathbf{B}S_n(x,y,z)}{a^2+r^2}\left(\frac{d}{rdr}\right)^n\left[\frac{(a^2+r^2)^n}{r}\sin\left(N\cos^{-1}\frac{a^2-r^2}{a^2+r^2}\right)\right]$$

$$= CS_n(x,y,z)\,F^{n+1}_{N-n-1}\left(\frac{a^2-r^2}{a^2+r^2}\right)$$

VI

When $k^2<0$, the problem can be transformed to a homogeneous integral equation of the Fredholm type, with a symmetrical kernel

$$\phi(A)=\lambda\int\phi(P)K(A,P)dv_P.$$

The results that we have obtained above enable us also to say that

$$\phi = C\sqrt{1+x^2+y^2+z^2}\,S_n(x,y,z)\left(\frac{1}{r}\frac{d}{dr}\right)^{n+1}\left[(1+x^2+y^2+z^2)^n\cos\left(N\cos^{-1}\frac{1-r^2}{1+r^2}\right)\right]$$

$$= D\frac{S_n(x,y,z)}{(1+x^2+y^2+z^2)^{n+3/2}}F^{n+1}_{\lambda}\left(\frac{1-r^2}{1+r^2}\right) \qquad \dots (6,1)$$

are eigen-functions corresponding to the eigen-value $-(n+1+\lambda)/\pi^2$ for the kernel

$$K(A,P)=\frac{1}{(x-x')^2+(y-y')^2+(z-z')^2\sqrt{(1+x^2+y^2+z^2)}\sqrt{(1+x'^2+y'^2+z'^2)}}.$$

Writing

$$M(x,y,z)=\frac{\mathbf{C}S_n(x,y,z)}{(a^2+r^2)^{n+2}}F^{n+1}_{\lambda}\left(\frac{a^2-r^2}{a^2+r^2}\right)=\mathbf{C}Y_n(\theta,\phi)\frac{r^n}{(a^2+r^2)^{n+2}}F^{n+1}_{\lambda}\left(\frac{a^2-r^2}{a^2+r^2}\right);$$

also

$$\int M^2dxdydz=1=\mathbf{C}^2\int Y_n^2(\theta,\phi)d\omega\int\frac{r^{2n+2}}{(a^2+r^2)^{2n+4}}\left[F^{n+1}_{\lambda}\left(\frac{a^2-r^2}{a^2+r^2}\right)\right]^2dr.$$

To determine the radial integral, we use (5,7) and write

$$\mathrm{I}=\int\frac{(1-h^2)(1-t^2)r^{2n+2}dr}{[a^2(1-h)^2+r^2(1+h)^2]^{n+2}[a^2(1-t)^2+r^2(1+t)^2]^{n+2}}$$

$$=\sum_{\lambda}\sum_{\sigma}\frac{(n+1+\lambda)(n+1+\sigma)}{(n+1)^2}(2h)^{\lambda}(2t)^{\sigma}\int\int\frac{F^{n+1}_{\lambda}F^{n+1}_{\sigma}r^{2n+2}}{(a^2+r^2)^{2n+4}}dr,$$

which reduces to

$$I=\frac{(-1)^{n+1}}{(n+1)!}\cdot\frac{(1-h^2)(1-t^2)}{(1+h)^{2n+4}(1+t)^{2n+4}}\cdot\left\{\frac{d}{d(p+q)}\right\}^{n+1}\int_0^{\infty}\frac{dr}{[p+r^2][q+r^2]},$$

where

$$p = \frac{a^2(1-h)^2}{(1+h)^2},\ q = \frac{a^2(1-t)^2}{(1+t)^2},$$

when, by easy calculation,

$$I = \frac{\pi}{2^{4n+5}a^{2n+5}} \cdot \frac{(2n+1)\,!}{n!(n+1)!} \frac{(1+h)(1+t)}{(1-ht)^{2n+3}}. \qquad \dots\ (6.2)$$

Equating coefficient of $(ht)^\lambda$

$$\int \frac{r^{2n+2}}{(a^2+r^2)^{2n+4}} \left(F_\lambda^{n+1} \right)^2 dr = \frac{\pi(2n+1+\lambda)!}{2^{4n+2\lambda+5}\, a^{2n+5}(n+1+\lambda)\lambda!\{n!\}^2};$$

hence

$$\overline{M}\ \text{(normalised)} = 2^{2n+\lambda+3}a^{n+3}\left(\frac{N}{2\pi a} \right)^{\frac{1}{2}} n\,! \left\{ \frac{\lambda\,!}{(N+n)!} \right\}^{\frac{1}{2}} \frac{F_\lambda^{n+1}\left(\frac{a^2-r^2}{a^2+r^2} \right)}{(a^2+r^2)^{n+2}} \overline{Y}_n(\theta, \phi), \qquad \dots\ (6.3]$$

where $\overline{Y}_n(\theta, \phi)$ is taken to be a normalised spherical harmonic.

We can normalise the eigen-functions of the homogeneous integral equation, for which similarly the value of

$$\int \frac{r^{2n+2}}{(1+r^2)^{2n+3}} \left(F_\lambda^{n+1} \right)^2 dr$$

will be necessary; for it, we utilise again (5.7) and write

$$I = \int_0^\infty \frac{(1-h^2]r^{2n+2}dr}{[(1-h)^2+(1+h)^2r^2]^{n+2}[(1-t)^2+(1+t)^2r^2]^{n+1}}$$

$$= \sum_\lambda \sum_\sigma (2h)^\lambda(2t)^\sigma \frac{(n+1+\lambda)}{(n+1)} \int \frac{F_\lambda^{n+1}\, F_\sigma^{n+1}\, r^{2n+2}\, dr}{(1+r^2)^{2n+3}};$$

after easy integration

$$I = \frac{\pi}{2^{4n+4}} \cdot \frac{(2n+1)!}{n!\,(n+1)\,!} \cdot \frac{1}{(1-ht)^{2n+2}}. \qquad \dots\ (6.4)$$

Equating coefficient we have

$$\int \frac{\{F_\lambda^{n+1}\}^2\, r^{2n+2}\, dr}{(1+r^2)^{2n+3}} = \frac{\pi}{2^{4n+2\lambda+4}(n!)^2(n+1+\lambda)} \cdot \frac{(2n+1+\lambda)!}{\lambda!}.$$

The normalised eigen-functions of the homogeneous equation is

$$\psi = \frac{2^{2n+\lambda+2}n!\,(n+1+\lambda)^{\frac{1}{2}}}{\sqrt{\pi}} \cdot \left(\frac{\lambda!}{(2n+1+\lambda)!} \right)^{\frac{1}{2}} \frac{Y_n(\theta, \phi)r^n}{(1+r^2)^{n+3/2}} F_\lambda^{n+1}\left(\frac{1-r^2}{1+r^2} \right). \dots\ (6.5)$$

The completeness of the eigen-functions series will mean the equality

$$K(x, y) = \sum_{n} \frac{\pi^2 \psi_r^n(x)\, \psi_r^n(y)}{(n+1+\lambda)}$$

(multiple summations as there are repeated roots) leading to the relation

$$\frac{1}{[(x-x')^2+(y-y')^2+(z-z')^2][1+r^2]^{\frac{1}{2}}[1+r'^2]^{\frac{1}{2}}}$$

$$= \sum_{1} \frac{\pi^2}{N} \sum_{n=0}^{n=N} \frac{2^{2N+2n+2}}{\pi} (n!)^2 N \left[\frac{(N-n-1)!}{(N+n)!} \right] \frac{r^n}{(1+r^2)^{3/2}} \cdot \frac{r'^n}{(1+r'^2)^{3/2}} F_\lambda^{n+1} F_\lambda'^{n+1}$$

$$\times \frac{(2n+1)}{4\pi} \left[\sum_{1}^{n} 2P_n^m(\mu) P_n^m(\mu') \frac{(n-m)!}{(n+m)!} \cos m(\phi-\phi') + P_n(\mu)\, P_n(\mu') \right]$$

and leading to an expansion of

$$\frac{1}{(x-x')^2+(y-y')^2+(z-z')^2} = \sum_{N} 2^{2N} \sum_{0}^{N-1} 2^{2n} F_{N-n-1}^{n+1} \left(\frac{1-r^2}{1+r^2} \right) F_{N-n-1}^{n+1}$$

$$\left(\frac{1-r'^2}{1+r'^2} \right) \frac{r^n r'^n}{(1+r^2)^{n+1}(1+r'^2)^{n+1}} \left[\frac{(N-n-1!)}{(N+n)!} \right] (n!)^2 \Big[P_n(\mu) P_n(\mu')$$

$$+2 \sum_{1}^{n} \frac{(n-m)!}{(n+m)!} P_n^m(\mu) P_n^m(\mu') \cos m(\phi-\phi') \Big]; \qquad \ldots \quad (6.7)$$

by a change of axis making $(x'\ y'\ z')$ on the z axis

$$\frac{1}{r^2+r'^2-2rr'\mu} = \sum_{N} 2^{2N} \sum_{n} 2^{2n} F_{N-n-1}^{n+1} \left(\frac{1-r^2}{1+r^2} \right) \cdot F_{N-n-1}^{n+1} \left(\frac{1-r'^2}{1+r'^2} \right) \frac{r^n r'^n}{(1+r^2)^{n+1}\,(1+r'^2)^{n+1}}$$

$$\times (n!)^2 \left[\frac{(N-n-1)\,!}{(N+n)\,!} \right] (2n+1)\ P_n(\mu) \qquad (6.8)$$

from the well-known addition theorem of $P_n(\mu)$.

To verify the correctness of the result we note that

$$\frac{2}{2[r^2+r'^2-2rr'\mu]}=\frac{2}{(1+r^2)(r'^2+1)\left[1-\frac{1-r^2}{1+r^2}\cdot\frac{1-r'^2}{1+r'^2}-\frac{4rr'}{(1+r^2)(1+r'^2)}\cos\phi\right]};$$

hence writing

$$\cos\theta=\frac{1-r^2}{1+r^2},\quad \cos\theta'=\frac{1-r'^2}{1+r'^2},$$

we have

$$\frac{1}{r+r'^2-2rr'\mu^2}=\frac{2}{(1+r^2)(1+r'^2)[1-\cos\theta\cos\theta'-\sin\theta\sin\theta'\cos\phi]}$$

$$=\frac{4}{(1+r^2)(1+r'^2)}\sum_1^N\frac{\sin N\chi}{\sin\chi}, \tag{6·81}$$

if $\pi/2>\chi>0$ from (5.1), and $\cos\chi=\cos\theta\cos\theta'+\sin\theta\sin\theta'\cos\phi$; hence

$$\frac{\sin N\chi}{\sin\chi}=\sum_{n=0}^{N-1}2^{2(N-1)}(2n+1)(n!)^2\left[\frac{(N-n-1)\,!}{(N+n)\,!}\right]$$

$$\times\sin^n\theta\sin^n\theta' F^{n+1}_{N-n-1}(\cos\theta)\cdot F^{n+1}_{N-n-1}(\cos\theta')\cdot P_n(\cos\phi), \tag{6.9}$$

a result which is a particular case of a general class of identity deduced by Gegenbauer (Whittaker, 1927).

PHYSICAL LABORATORY,
DACCA UNIVERSITY.

REFERENCES

Elsasser, (1933), *Zeit. für. Physik.*, **81**, 332.

Whittaker, (1927), *Modern Analysis*, 335, Ex. 42.

Reprinted from *Bull Cal Math Soc* **37**, pp. 51-61, 1945 (Calcutta Mathematical Society).

D. Unified Field Theory

The most intriguing feature of these papers is that Bose wrote them within a span of just over a year. He had never before worked so intensely on one topic and published in such rapid succession. He was probably inspired by what Einstein announced in popular press as *A New Unified Theory*, and published the material as an appendix of his book *Meaning of Relativity*.

Bose, without referring to this specific theory, works out the mathematical details concerning divergence identities derived from a variance principle and the determination of the affine coefficients in the theory. Finally, he attempts to find a solution to the tensor equation of the unified theory. In reply to the letter sent presumably along with a copy of his paper, Einstein wrote:

THE INSTITUTE FOR ADVANCED STUDY

PRINCETON, NEW JERSEY

October 4th, 1952

SCHOOL OF MATHEMATICS

Professor S.N.Bose
University College of Science
92, Upper Circular Road
Calcutta,9, India.

Dear Professor Bose:

Thank you for your letter of September 20th. I am glad to see that you are interested in this theory and that you have devoted so much work and penetration to the solution of the equation

$$g_{\underset{+}{i}\underset{-}{k};l}=0$$

I believe, to be sure, that the solution of those equations is not of great help toward the answer of the question: Do the singularity-free solutions of the equation system have physical meaning? Are there at all singularity-free solutions which correspond to the atomistic character of matter and radiation? It seems to me that the mathematical methods available at present are not powerful enough to answer this question.

However, I am now firmly convinced that the equation-system

$$g_{\underset{+}{i}\underset{-}{k};l}=0 \qquad \Gamma_i=0$$

$$R_{\underline{ik}}=0 \qquad R_{\underset{\vee}{ik},l}+R_{\underset{\vee}{kl},i}+R_{\underset{\vee}{li},k}=0$$

represents the only formally natural generalization of the relativistic law of gravitation.

With kind regards,

sincerely yours,

Albert Einstein.

It is worth noting from this letter that Einstein knew by then his dream of finding a singularity-free pure field theory to describe the discrete nature of matter and radiation was far from realization.

The Identities of Divergence in the New Unitary Theory

Note of Mr. S. N. Bose, presented by Mr. Louis de Broglie

Einstein([1]) deduced the four identities of divergence from his new theory of Bianchi identities, clearing the ambiguity of covariant differentiation.

The method of Hilbert([2]), when slightly modified, also immediately gives the result. Perhaps it is interesting to indicate the mode of calculation.

The relation between the coefficients of affine connection $\Gamma^{\alpha}_{\mu\nu}$ and $g^{\mu\nu}$ is not simple in the new theory. Thus the method of Hilbert regarding the variation of an invariant function of $g^{\mu\nu}$ and its derivatives cannot be directly applied here. In the new theory, this relationship between Γ and g is initially not assumed and the variation gives rise to the equations as well as the equations of connection of Γ and g. The proposed new demonstration takes this into account. The method will be understood according to the following indications.

Let us form the tensor $R_{\mu\nu}$, contracted from the generalized Riemann tensor. Its covariant character is evident from the law of transformation of $\Gamma^{\alpha}_{\mu\nu}$ as in the symmetric theory. $g^{\mu\nu} R_{\mu\nu}$ is thus an invariant under changes of the coordinates, as in the expression $|g|^{1/2} dx_1\, dx_2\, dx_3\, dx_4 = |g|^{1/2} dv_4$.

If $R_{\mu\nu} = 0$ the integral

(1) $$I \doteq \int R_{\mu\nu}\, \delta g^{\mu\nu} \sqrt{g}\, dv_4 = 0$$

for any variation of $g^{\mu\nu}$ alone ; $R_{\mu\nu}$ being a function of the coefficients of affine connection is not changed in the variation. Let us observe with Hilbert ([2]) that

(2) $$p^{\mu\nu} = \frac{\delta g^{\mu\nu}}{\delta x^{\sigma}} \varphi^{\sigma} - g^{\mu\sigma} \varphi^{\nu}_{\sigma} - g^{\sigma\nu} \varphi^{\mu}_{\sigma}$$

obtained from an arbitrary contravariant vector φ^{σ} can be taken for $\delta g^{\mu\nu}$ in (1). Hence

(3) $$0 = \int R_{\mu\nu} \sqrt{|g|} \left(\frac{\partial g^{\mu\nu}}{\partial x^{\sigma}} \varphi^{\sigma} - g^{\mu\sigma} \varphi^{\nu}_{\sigma} - g^{\sigma\nu} \varphi^{\mu}_{\sigma} \right) dv_4.$$

The integration is extended over a manifold of four dimensions limited by a manifold of three dimensions and one has everywhere $R_{\mu\nu} = 0$.

We observe that

([1]) *Canad. J. Math.*, **2**, 1950, p. 120 et seq.

([2]) *Math. Ann.* **92**, 1924, p. 1 et seq.

$$\frac{d}{dx^\sigma}[R_{\mu\nu}\, g^{\mu\nu}\, \varphi^\sigma \sqrt{|g|}] = \frac{\partial R_{\mu\nu}}{\partial x^\sigma}\, g^{\mu\nu} \sqrt{|g|}\, \varphi^\sigma + \tag{4}$$

$$R_{\mu\nu} \frac{\partial g^{\mu\nu}}{\partial x^\sigma} \sqrt{|g|}\, \varphi^\sigma + R_{\mu\nu}\, g^{\mu\nu} \left[\sqrt{g} \frac{\partial \varphi^\sigma}{\partial x^\sigma} + \frac{\partial \sqrt{g}}{\partial x^\sigma} \varphi^\sigma \right]$$

If we consider $\varphi^\sigma \equiv \Delta x^\sigma$as a suitable small and arbitrary variation of the coordinates which vanishes on the boundary, we observe that

$$\left.\begin{array}{l} \dfrac{\partial \phi^\sigma}{\partial x^\sigma}\, dv_4 = \Delta dv_4 \text{ and } \dfrac{\partial \sqrt{g}}{\partial x^\sigma}\, \varphi^\sigma = \Delta \sqrt{g}. \\ \text{and} \\ \left[\sqrt{g}\, \dfrac{\delta \varphi^\sigma}{\delta x^\sigma} + \dfrac{\delta \sqrt{g}}{\delta x^\sigma}\, \varphi^\sigma \right] dv_4 = \Delta(\sqrt{g} dv_4) \end{array}\right\} \tag{5}$$

and consequently = 0, since $\sqrt{g} d\upsilon_4$ is invariant. By multiplying (4) with $d\upsilon_4$ and integrating on the manifold with four dimensions we have

$$\int \left[\frac{\partial R_{\mu\nu}}{\partial x^\sigma}\, g^{\mu\nu} \sqrt{|g|}\, \varphi^\sigma + R_{\mu\nu} \frac{\partial g^{\mu\nu}}{\partial x^\sigma} \sqrt{|g|}\, \varphi^\sigma \right] dv_4 = \int \frac{\partial}{\partial x^\sigma} [R_{\mu\nu}\, g^{\mu\nu}\, \varphi^\sigma \sqrt{|g|}] dv_4, \tag{6}$$

= a surface integral = 0, as per the hypothesis of variation.
We immediately obtain from (3) and (6)

$$\int \left[R_{\mu\nu} (g^{\mu\sigma}\, \varphi^\nu_\sigma + g^{\sigma\nu}\, \varphi^\mu_\sigma) + \frac{\partial R^{\mu\nu}}{\partial x^\sigma}\, \varphi^\sigma \right] \sqrt{g} dv_4 = 0$$

A partial integration for eliminating the derivatives of φ^σ immediately gives the identities

$$g^{\mu\nu} \sqrt{g} \frac{\partial R_{\mu\nu}}{\partial x^\sigma} - \frac{d}{dx^\nu} [g^{\mu\nu}\, R_{\mu\sigma} + g^{\nu\mu}\, R_{\sigma\mu}] \sqrt{g} = 0$$

identical to the result of Einstein ([3]). Nevertheless, in a certain way our result is slightly more general. We can say that the reasoning of Einstein is an algebraic method where the connections between the affine coefficients and the $g^{\mu\nu}$ as well as the conditions of integrability and a particular hypothesis $\Gamma^\alpha_{\alpha\mu} - \Gamma^\alpha_{\mu\alpha} = 0$ are assumed *a priori*.

Our demonstration is independent of these restrictive hypotheses. In fact, while trying to deduce the general equations $R_{\mu\nu} = 0$ and the equations of connection from the principle of variation $\delta \int g^{\mu\nu} R_{\mu\nu} \sqrt{g}\; d\upsilon_4 = 0$, we are led to the connection of Einstein only if $\Gamma^\alpha_{\sigma\alpha} - \Gamma^\alpha_{\alpha\sigma} = 0$.

On the other hand, by assuming *a priori* the connection of Einstein, the compatibility of the variation with the initial hypothesis leads to $\Gamma_\mu = \Gamma^\alpha_{\mu\alpha} - \Gamma^\alpha_{\alpha\mu} = 0$.

Since the demonstration given here does not use the equations of connection, it

([3]) *Canad. J. Math.*, **2**, 1950, p. 125, equation (176).

clearly shows that the above identities are independent of the equations of connection. In fact, the $R_{\mu\nu}$ are functions of Γ alone, and the demonstration of the identities of divergence only assumes the existence of the first derivatives of $g^{\mu\nu}$. It survives even when due to singularities of the field of $g^{\mu\nu}$, there are no second derivatives or integrable equations of connection.

English translation of *Comptes rendus de l'Academie des Sciences* 236, pp. 1333-1335, 1953 (Gauthier — Villars Publishers, Paris).

A Unitary Field Theory with $\Gamma_\mu \neq 0$

By S. N. BOSE,
University of Calcutta

Summary — The equations of the unitary field are established by varying an integral which satisfies the postulate of hermiticity. One obtains an absolute equivalence with the old theories either with $\Gamma_\mu = 0$ or with $\| a_{\lambda\mu} \| = 0$ when $\Phi_\lambda = 0$; one can interpret Φ_λ as the Lorentz force if $a_{\lambda\mu}$ are taken as the components of the electromagnetic field.

The approximate equations with antisymmetric components take the Maxwellian form for a certain value of the constants. Nevertheless, for another value, it is possible to have a field without any singularity.

1. Introduction

The characteristic equations of Einstein's field theory can be classified into three groups :

I. The field equations which involve only the connection coefficients $\Gamma^\lambda_{\mu\nu}$ and their derivatives.

II. The connection equations which link the tensor g to the affine connection coefficients.

III. The four relations of restrictive character

$$\Gamma^\lambda_{\mu\lambda} - \Gamma^\lambda_{\lambda\mu} = 2\Gamma_\mu = 0.$$

Note that the relations in (III) resemble the 24 equations,

$$\Gamma^\lambda_{\mu\nu} - \Gamma^\lambda_{\nu\mu} = 0.$$

which, in the relativistic theory of gravitation, are consequences of the postulate of symmetry of the affine connections.

In unitary field theory as well as in the theory of gravitation the equations of the field (I) and the equations of connection (II) can be simultaneously deduced from the principle of variation where the arbitrary variation of an integral is equal to zero, the variation being subjected to restrictions of type (III).

Nevertheless, the arbitrary conditions are much less numerous in the unitary theory than in the theory of gravitation.

In this article we are trying to eliminate all the restrictive conditions and establish a system of fundamental equations by relying only on the principle of variation.

We have used the postulate of hermiticity recently enunciated by Einstein to construct the integral, the variation of which leads to this new generalization.

The specific results given below will show that the additional terms containing Γ_μ occur as well in the field equations as in the equations of connection.

The equations of connection which concern the tensor g give rise to certain remarks. Here they take the form

$$g^{\mu\nu}_{\lambda} + g^{\mu\alpha}\,\Gamma^{\nu}_{\lambda\alpha} + g^{\alpha\nu}\,\Gamma^{\mu}_{\alpha\lambda} = 3g^{\mu\nu}\,\Phi_{\lambda},$$

whereas in the two proceeding theories the first member of this equality is zero.

Here Φ_λ is defined by the following equation

$$\Phi_{\lambda} = \frac{a_{\lambda\beta}\,k\beta}{\sqrt{|g|}},\ \text{with}\ a_{\lambda\beta} = \frac{1}{2}\,(g_{\lambda\beta} - g_{\beta\lambda})$$

and

$$k^{\beta} = \frac{1}{3}\,\theta(g^{\beta\nu} + g^{\nu\beta})\Gamma_{\nu}.$$

If $\Gamma_\nu = 0$, $k^\beta = 0$ and consequently if we also have $\Phi = 0$, the equation again takes the old form.

However, there exits another possibility : that $a_{\alpha\beta}$ being the antisymmetric part of the covariant tensor, is considered as linked to a vector with the six components (**E**, **H**) of the electromagnetic theory.

If the determinant

$$\| a_{\lambda\beta} \| = (a_{12}a_{34} + a_{31}a_{24} + a_{23}a_{14})^2 = 0,$$

we can have $\Phi_\lambda = 0$ even if k^μ and Γ_μ are not zero.

Note that since we always suppose

$$\| g^{\beta\nu} + g^{\nu\beta} \| \neq 0,$$

k^μ and Γ_μ are simultaneously zero.

The condition $\| a_{\lambda\beta} \| = 0$ combined with the usual rule of correlation just mentioned immediately leads.to (**E H**) = 0, a classical property of the electromagnetic field.

One is also tempted to put a correlation on the one hand between k^μ and the four dimensional charge current vector (because $k^{\mu}_{,\mu} = 0$ comes immediately from the general equations) and on the other hand, between Φ_λ and the ponderomotive force of Lorentz, due to the classical identifications

$$(a_{23}a_{31}\,a_{12}) \rightarrow \vec{H}, \qquad (a_{41}\,a_{42}\,a_{43}) \rightarrow i\vec{E}.$$

We can interpret $\Phi_\lambda = 0$ as determining a stationary distribution, where the Lorentz force is zero. This brings back the equation of connection to the form which was found in previous theories.

In the field equations the symmetric element $P^{\lambda}_{\mu\nu}$ and the antisymmetric element $G^{\lambda}_{\mu\nu}$ have totally different roles.

By means of hypotheses on the orders of relative magnitude of the various terms we could bring back the fundamental equations to a simple form.

Physically, these hypotheses imply that it is possible to neglect gravitation in a study of relations between electromagnetic quantities.

The equations thus transformed have an interesting form and we can interpret them as simple consequences of the Maxwell theory if we attribute particular values to the arbitrary constants.

But it is also possible to obtain a field without any singularities for other particular values. The field is then not Maxwellian.

These encouraging aspects from the new system of equations have made me think that it would be appropriate to publish the first findings in order to provoke criticisms and observations.

2. The new equations

Following Einstein, all the relations of unitary field theory must satisfy the postulate of hermiticity, i.e. remain unchanged when $g^{\mu\nu}$, $g_{\mu\nu}$, $\Gamma^{\lambda}_{\mu\nu}$ are changed to $g^{\nu\mu}$, $g_{\nu\mu}$, $\Gamma^{\lambda}_{\nu\mu}$ all together through permutations of the indices μ and ν.

We can easily see that in this transformation, the Einstein tensor **E** of components

$$E_{\mu\nu} = \Gamma^{\lambda}_{\mu\nu,\lambda} - \Gamma^{\lambda}_{\mu\lambda,\nu} + \Gamma^{\xi}_{\mu\nu}\Gamma^{\lambda}_{\xi\lambda} - \Gamma^{\xi}_{\lambda\mu}\Gamma^{\lambda}_{\xi\nu}$$

become the tensor **H** of components

$$H_{\nu\mu} = \Gamma^{\lambda}_{\nu\mu,\lambda} - \Gamma^{\lambda}_{\lambda\mu,\nu} + \Gamma^{\epsilon}_{\nu\mu}\Gamma^{\lambda}_{\lambda\epsilon} - \Gamma^{\epsilon}_{\lambda\mu}\Gamma^{\lambda}_{\nu\xi}.$$

Instead of the expression

$$I = g^{\mu\nu} E_{\mu\nu}\sqrt{|g|} = g'^{\mu\nu} E_{\mu\nu}$$

which occurs in the integral which is subjected to variation in order to obtain the fundamental equations, we adopt a form which is more general

$$I' = \frac{1}{2}(g'^{\mu\nu} E_{\mu\nu} + g'^{\nu\mu} H_{\nu\mu}) + a\gamma^{\mu\nu}\Gamma_{\mu}\Gamma_{\nu} + bA^{\mu\nu}(\Gamma_{\mu,\nu} - \Gamma_{\nu,\mu})$$

where

$$\gamma^{\mu\nu} = \frac{1}{2}(g'^{\mu\nu} + g'^{\nu\mu}) = \frac{1}{2}\sqrt{|\bar{g}|}\,(g^{\mu\nu} + g^{\nu\mu})$$

$$A^{\mu\nu} = \frac{1}{2}(g'^{\mu\nu} - g'^{\nu\mu}) = \frac{1}{2}\sqrt{|\bar{g}|}\,(g^{\mu\nu} - g^{\nu\mu}),$$

and

$$|\bar{g}| = |g|,$$

since a permutation of μ and ν does not modify the determinant. The additional terms also satisfy the condition of hermiticity since Γ_{μ} $A^{\mu\nu}$ changes sign and $\gamma^{\mu\nu}$ retains its sign with permutation. Let us separate the symmetric part $P^{\lambda}_{\mu\nu}$ and the antisymmetric part $V^{\lambda}_{\mu\nu}$ in the affine connection coefficients $\Gamma^{\lambda}_{\mu\nu}$ and let us also put

$$V_{\mu\nu} = G^{\lambda}_{\mu\nu} - \frac{1}{3}\delta^{\lambda}_{\mu}\Gamma_{\nu} + \frac{1}{3}\delta^{\lambda}_{\nu}\Gamma_{\mu},$$

with

$$\Gamma_{\mu} = V^{\lambda}_{\mu\lambda},$$

one has

$$G^{\lambda}_{\mu\lambda} = 0.$$

Through substitution in I′, all the terms that are products of the symmetric factor and the antisymmetric factor disappear in the summation, and we finally have

$$I' = \gamma^{\mu\nu}(\bar{E}_{\mu\nu} - G^{\lambda}_{\mu\varepsilon} G^{\varepsilon}_{\lambda\nu} + x\Gamma_{\mu}\Gamma_{\nu}) + A^{\mu\nu}[G^{\lambda}_{\mu\nu;\lambda} - y(\Gamma_{\mu,\nu} - \Gamma_{\nu,\mu})]$$

where x, y are arbitrary constants.

$$\bar{E}_{\mu\nu} = P^{\lambda}_{\mu\nu,\lambda} - \frac{1}{2}(P^{\lambda}_{\mu\lambda,\nu} + P^{\lambda}_{\nu\lambda,\mu}) + P^{\varepsilon}_{\mu\nu}P^{\lambda}_{\varepsilon\lambda} - P^{\varepsilon}_{\mu\lambda}P^{\lambda}_{\varepsilon\nu}$$

is the symmetrized component of the Einstein tensor with symmetric affine connection and

$$G^{\lambda}_{\mu\nu;\lambda} = G^{\lambda}_{\mu\nu,\lambda} - G^{\lambda}_{\mu\varepsilon}P^{\varepsilon}_{\lambda\nu} - G^{\lambda}_{\varepsilon\nu}P^{\varepsilon}_{\mu\lambda} + G^{\varepsilon}_{\mu\nu}P^{\lambda}_{\varepsilon\lambda}$$

is the classical covariant derivative, also calculated with symmetric connection.

During the course of arbitrary variation of

$$L = \int I' dv_4, \quad dv_4 = dx^1 dx^2 dx^3 dx^4.$$

$\gamma^{\mu\nu}$, $A^{\mu\nu}$ automatically retain their symmetry character. The variations of $\gamma^{\mu\nu}$ and $A^{\mu\nu}$ can thus be considered as arbitrary, which immediately gives rise to the field equations

$$\frac{\partial I'}{\partial \gamma^{\mu\nu}} = \bar{E}_{\mu\nu} - G^{\varepsilon}_{\mu\lambda}G^{\lambda}_{\varepsilon\nu} + x\Gamma_{\mu}\Gamma_{\nu} = 0,$$

$$\frac{\partial I'}{\partial A^{\mu\nu}} = G^{\lambda}_{\mu\nu;\lambda} - y(\Gamma_{\mu,\nu} - \Gamma_{\nu,\mu}).$$

In order to obtain the equations of connection, let us first make I′ equal to a divergence in four dimensions (H) and let us consider only H since the divergence transforms itself to an integral over a three dimensional domain where all the coefficients of arbitrary variations cancel each other.

Let us recall the fact that in the variation of the elements Γ_{μ}, $P^{\lambda}_{\mu\nu}$, $G^{\lambda}_{\mu\nu}$ the four relations $G^{\lambda}_{\mu\lambda} = 0$ always remain valid ; the 24 component of $G^{\lambda}_{\mu\nu}$ are not arbitrary. We can thus apply the usual method of employing undermined multipliers k^{μ} and varying the function

$$H - 2k^{\mu}G^{\lambda}_{\mu\lambda} = H'.$$

These coefficients k^{μ} will be determined in the end from the final equation.

Thus through easy calculation we obtain a system of equations

$$\gamma^{\mu\nu}_{,\lambda} + \gamma^{\mu\alpha}P^{\nu}_{\lambda\alpha} + \gamma^{\alpha\nu}P^{\mu}_{\alpha\lambda} - \gamma^{\mu\nu}P^{\alpha}_{\lambda\alpha} = -[A^{\mu\alpha}G^{\nu}_{\lambda\alpha} + A^{\alpha\nu}G^{\mu}_{\alpha\lambda}],$$

$$A^{\mu\nu}_{,\lambda} + A^{\mu\alpha}P^{\nu}_{\lambda\alpha} + A^{\alpha\nu}P^{\mu}_{\alpha\lambda} - A^{\mu\nu}P^{\alpha}_{\lambda\alpha} - k^{\mu}\delta^{\nu}_{\lambda} + k^{\nu}\delta^{\mu}_{\alpha} = -[\gamma^{\mu\alpha}G^{\nu}_{\lambda\alpha} + \gamma^{\alpha\nu}G^{\mu}_{\alpha\lambda}],$$

$$yA^{\mu\nu}_{\nu} + x\gamma^{\mu\nu}\Gamma_{\nu} = 0,$$

and consequently

$$A^{\mu\nu}_{,\nu} = 3k^{\mu},$$

$$\gamma^{\mu a}_{,a} + \gamma^{a\beta} P^{\mu}_{a\beta} + A^{a\beta} G^{\mu}_{a\beta} = 0.$$

From where we immediately obtain

$$k^{\mu}_{,\mu} = 0$$

and

$$k^{\mu} = \theta\gamma^{\mu\nu}\Gamma_{\nu}, \quad \text{avec } \theta = -\frac{x}{3y} \cdot$$

We can now give a suitable form to the equations of connection : through addition and regrouping we have

$$g'_{\lambda}{}^{\mu\nu} + g'^{\mu a} L^{\nu}_{\lambda a} + g'^{a\nu} L^{\mu}_{\lambda a} - g'^{\mu\nu} L^{a}_{\lambda a} = k^{\mu}\delta^{\nu}_{\lambda} - k^{\nu}\delta^{\mu}_{\lambda},$$

where

$$L^{\nu}_{\lambda a} = G^{\nu}_{\lambda a} + P^{\nu}_{\lambda a},$$

by multiplying with the coefficient $g'_{\mu\nu}$ defined by

$$g'^{\mu\nu} g'_{\mu\lambda} = \delta^{\nu}_{\lambda}, \quad g'^{\mu\nu} g'_{\lambda\nu} = \delta^{\mu}_{\lambda}.$$

we first obtain

$$L'^{a}_{\lambda a} = \frac{1}{2}\ \frac{c,\lambda}{c} + a'_{\lambda\beta} k^{\beta},$$

$$c = |g|$$

and

$$a'_{\lambda\beta} = \frac{1}{2}\ (g'_{\lambda\beta} - g'_{\beta\lambda}),$$

then, by replacing $L^{a}_{\lambda a}$ by its value and dividing it by $\sqrt{c}$, we get an easy transformation

$$g^{\mu\nu}_{\lambda} + g^{\mu a} L^{\nu}_{\lambda a} + g^{a\nu} L^{\mu}_{a\lambda} - \frac{g^{\mu\nu} a_{\lambda\beta} k^{\beta}}{\sqrt{c}} = \frac{k^{\mu}\delta^{\nu}_{\lambda} - k^{\nu}\delta^{\mu}_{\lambda}}{\sqrt{c}} \cdot$$

By observing that

$$g^{\mu a} g_{\beta a} k^{\beta} = k^{\mu}, \quad g^{a\nu} g_{a\beta} k^{\beta} = k^{\nu}$$

and by regrouping and adding certain terms, we finally obtain

$$g^{\mu\nu}_{,\lambda} + g^{\mu a}\, \Gamma'^{\nu}_{\lambda a} + g^{a\nu}\, \Gamma'^{\mu}_{a\lambda} = 3g^{\mu\nu} \frac{a_{\lambda\beta} k^{\beta}}{\sqrt{c}} = 3g^{\mu\nu}\Phi_{\lambda} \ ;$$

where

$$\Gamma'^{\nu}_{\lambda a} = L^{\nu}_{\lambda a} - g_{\beta a} \frac{k\beta}{\sqrt{c}}\ \delta^{\nu}_{\lambda} + g_{\lambda\beta} k^{\beta} \delta^{\nu}_{a},$$

$$\Gamma'^{\mu}_{a\lambda} = L^{\mu}_{a\lambda} - g_{\beta\lambda} \frac{k^{\beta}}{\sqrt{c}}\ \delta^{\mu}_{a} + g_{a\beta} \frac{k^{\beta}}{\sqrt{c}}\ \delta^{\mu}_{\lambda}.$$

The new affine coefficients are

$$\Gamma'^{\lambda}_{\mu\nu} = Q^{\lambda}_{\mu\nu} + U_{\mu\nu}$$

where

$$Q^{\lambda}_{\mu\nu} = P^{\lambda}_{\mu\nu} + \Phi_{\mu}\delta^{\lambda}_{\nu} + \Phi_{\nu}\delta^{\lambda}_{\mu}$$

and

$$U^{\lambda}_{\mu\nu} = G^{\lambda}_{\mu\nu} - \Psi_{\nu}\delta^{\lambda}_{\mu} + \Psi_{\mu}\delta^{\lambda}_{\nu} ;$$

$$\Psi_{\nu} = -\frac{s_{\nu\beta}k^{\beta}}{\sqrt{c}} , \phi_{\nu} = \alpha_{\nu\beta}\frac{k^{\beta}}{\sqrt{c}} ;$$

By introducing these new affine coefficients in the fundamental equation, we can write

$$\bar{E}'_{\mu\nu} + \frac{3}{2}[\Phi_{\mu;\nu} + \Phi_{\nu;\mu}] + 3\Phi_{\mu}\Phi_{\nu} + \frac{x}{\theta^2} b_{\mu\alpha} b_{\nu\beta}\frac{k^{\alpha}k^{\beta}}{c} - G^{\lambda}_{\mu\varepsilon} G^{\varepsilon}_{\lambda\nu} = 0,$$

$$(G^{\lambda}_{\mu\nu})_{;\lambda} - 3G^{\varepsilon}_{\mu\nu}\Phi_{\varepsilon} - \frac{y}{\theta\sqrt{c}}[(b'_{\mu\alpha}k^{\alpha})_{,\nu} - (b'_{\nu\alpha}k^{\alpha})_{,\mu}],$$

with

$$\tfrac{1}{2} b_{\mu\alpha}(g^{\mu\nu} + g^{\nu\mu}) = \delta^{\nu}_{\alpha}.$$

3. Particular case

When $\Phi_{\lambda} = 0$ but Γ_{μ}, $k^{\mu} \neq 0$ we have

$$a_{12}a_{34} + a_{23}a_{14} + a_{31}a_{24} = 0.$$

Let us make the transformation and observe that

$$\Gamma'_{\mu} = 3\theta\Gamma_{\mu} ;$$

$P^{\lambda\,\cdot}_{\mu\,\nu}$ and $G^{\lambda}_{\mu\,\nu}$ do not change and we obtain

$$E'_{\mu\nu} + t\Gamma'_{\mu}\Gamma'_{\nu} - G^{\lambda}_{\mu\varepsilon} G^{\varepsilon}_{\lambda\nu} = 0;$$

$$(G^{\lambda}_{\mu\nu})_{;\lambda} + t(\Gamma_{\mu,\nu} - \Gamma_{\nu,\mu}) = 0,$$

the only modification is a reduction of the number of arbitrary constants.

The equation of connection is (since $\Phi_{\lambda} = 0$)

$$g^{\mu\lambda}_{\lambda} + g^{\mu\alpha}\Gamma^{\nu}_{\lambda\alpha} + g^{\alpha\nu}\,\Gamma^{\mu}_{\alpha\lambda} = 0.$$

but with $\Gamma_{\mu} \neq 0$.

Thus it is easily shown that

$$(A^{\mu\nu}\sqrt{\Delta})_{,\nu} = \tfrac{1}{2}(g^{\mu\alpha} + g^{\alpha\mu})\Gamma_{\alpha}.$$

4. Approximation

In order to find an approximation we put the equation of connection in the form

$$\gamma^{\mu\nu}_{;\lambda} + \gamma^{\mu a} P^{\nu}_{\lambda a} + \gamma^{a\nu} P^{\mu}_{a\lambda} - \gamma^{\mu\nu} P^{a}_{\lambda a} = -[A^{\mu a} G^{\nu}_{\lambda a} + A^{a\nu} G^{\mu}_{a\lambda}],$$
$$A^{\mu\nu}_{,\lambda} = A^{\mu a} P^{\nu}_{\lambda a} + A^{a\nu} P^{\mu}_{a\lambda} - A^{\mu\nu} P^{a}_{\lambda a} - k^{\mu}\delta^{\nu}_{\lambda} + k^{\nu}\delta^{\mu}_{\lambda} = -[\gamma^{\mu a} G^{\nu}_{\lambda a} + \gamma^{a\nu} G^{\mu}_{a\lambda}].$$

Let us suppose that the antisymmetric coefficients $A^{\mu\nu}$ and $G^{\lambda}_{\mu\nu}$ are small, close to zero.

Thus $P^{\nu}_{\lambda\mu}$ can also be neglected :

$$\gamma^{\mu\nu} = \delta^{\mu}_{\nu},$$

the covariant derivative is replaced by an ordinary derivative, and we finally obtain in the first approximation

$$A^{\mu\nu}_{,\lambda} - k^{\mu}\delta^{\nu}_{\lambda} + k^{\nu}\delta^{\mu}_{\lambda} = -[G^{\nu}_{\lambda\mu} + G^{\mu}_{\nu\lambda}],$$
$$A^{\mu\nu}_{\nu} = 3k^{\mu},$$

from which we can easily deduce

$$(G^{\lambda}_{\mu\nu}) = A^{\mu\nu}_{,\lambda} - k^{\mu}\delta^{\nu}_{\lambda} + k^{\nu}\delta^{\mu}_{\lambda} - \tfrac{1}{2}[A^{\mu\nu}_{,\lambda} + A^{\lambda\mu}_{,\nu} + A^{\nu\lambda}_{\mu}],$$

and finally with

$$(G^{\lambda}_{\mu\nu})_{\lambda} = \frac{2}{3} A^{\mu\nu}_{,\lambda\lambda} - \frac{1}{6}[A^{\mu\nu}_{,\lambda} + A^{\nu\lambda}_{,\mu} + A^{\lambda\mu}_{,\nu}]_{\lambda}$$

$$\Gamma_{\mu} = \frac{k^{\mu}}{\theta},$$

we obtain the following system

$$A^{\mu\nu}_{\nu} = 3k^{\mu} \;;$$

$$A^{\mu\nu}_{\lambda\lambda} - \frac{1}{4}[A^{\mu\nu}_{\lambda} + A^{\nu\lambda}_{\mu} + A^{\lambda\mu}_{\nu}]_{\lambda} = \frac{3y}{2\theta}[k^{\mu}_{,\nu} - k^{\nu}_{,\mu}].$$

By eliminating k^{μ} we obtain

$$A^{\mu\nu}_{\lambda\lambda}\left(1 - \frac{y}{2\theta}\right) - \frac{1}{2}\left(\frac{1}{2} - \frac{y}{\theta}\right)(A^{\mu\nu}_{\lambda} + A^{\nu\lambda}_{\mu} + A^{\lambda\mu}_{\nu})_{\lambda} = 0.$$

If $1 = \dfrac{y}{2\theta}$ we can put

$$A^{\mu\nu}_{\lambda} + A^{\nu\lambda}_{\mu} + A^{\lambda\mu}_{\nu} = 0$$

and the equation becomes identical to Maxwell's equations of electromagnetic fields.

But if $\dfrac{1}{2} - \dfrac{y}{\theta} = 0$, the equations become

with

$$(A^{\mu\nu}_{\nu}) = \rho^{\mu},$$
$$(A^{\mu\nu}_{\lambda} + \ldots)_{\lambda} = -(\rho^{\mu}_{,\nu} - \rho^{\nu}_{,\mu}),$$
$$A^{\mu\nu}_{\lambda\lambda} = 0,$$

(293)

the system being evidently different from that of Maxwell, but which can have solution without singularities at any point.

It is perhaps interesting to note that certain researches on the quantization of tl electromagnetic field have put the subsidiary condition $A^{\mu\nu}_{\lambda\lambda} = 0$, which would imply modification of field equations.

If $\| a_{\lambda\mu} \| = 0$, we obtain an approximation presenting the same characteristics.

Manuscript received 18 July 1953.

(294)

Certain Consequences of the Existence of the Tensor g in the Affine Relativistic Field

By S. N. BOSE,
University of Calcutta.

Summary — Relations are established where only $\Gamma^{\lambda}_{\mu\nu}$ and their derivatives enter from conditions of integrability of the equations which the tensor *g* of the relativistic theory satisfies.

The number of independent relations between the elements of the field is much larger in the unitary theory than in the theory of gravitation.

1. Introduction

The abstract formalism of the theory of relativity developed from the concept of the field of coefficients of affine connection of a four dimensional variety.

If δx^{μ} are the components of a parallel and infinitely small displacement to which a vector A of the chosen variety is subjected, the variations of the components A^{μ} which result are obtained by the rule

$$\delta A^{\mu} = - A^{\alpha} \Gamma^{\mu}_{\alpha\beta} \, \delta x^{\beta} \tag{1.1}$$

In the theory of gravitation the terms $\Gamma^{\mu}_{\alpha\beta}$ are symmetric, or $\Gamma^{\mu}_{\alpha\beta} = \Gamma^{\mu}_{\beta\alpha}$; thus the rule of transport there is unique. On the contrary, as the coefficients are not symmetric in the unitary theory, one has here a second rule of transport :

$$\bar{\delta} A^{\mu} = - \delta x^{\alpha} \, \Gamma^{\mu}_{\alpha\beta} A^{\beta}. \tag{1.2}$$

Based on relation (1.1) we can easily calculate the tensor of curvature R, the components of which

$$\left. \begin{aligned} R^{\lambda}_{\alpha\mu\nu} &= \Gamma^{\lambda}_{\alpha\mu,\nu} - \Gamma^{\lambda}_{\alpha\nu,\mu} + \Gamma^{\varepsilon}_{\alpha\mu}\Gamma^{\lambda}_{\varepsilon\nu} - \Gamma^{\varepsilon}_{\alpha\nu}\Gamma^{\lambda}_{\varepsilon\mu} \\ &\left(\text{ou } \Gamma^{\lambda}_{\alpha u,\nu} = \frac{d}{dx^{\nu}} \Gamma^{\lambda}_{\alpha\mu}, \ldots \right) \end{aligned} \right\} \tag{1.3}$$

are antisymmetric in the indices μ and ν ; and then by contraction we obtain the Einstein tensor E with the components

$$E_{\alpha\mu} = \Gamma^{\lambda}_{\alpha\mu,\lambda} - \Gamma^{\lambda}_{\alpha\lambda,\mu} + \Gamma^{\varepsilon}_{\alpha\mu}\Gamma^{\lambda}_{\varepsilon\lambda} - \Gamma^{\varepsilon}_{\alpha\lambda}\Gamma^{\lambda}_{\varepsilon\mu} \tag{1.4}$$

But the unitary theory also admits (1.2) as a rule of transport. Thus the same reasoning mentioned above will give us from (1.2) another transport S (replacing R) with its components

$$S^{\lambda}_{\mu\alpha\atop\nu} = \Gamma^{\lambda}_{\mu\alpha,\nu} - \Gamma^{\lambda}_{\nu\alpha,\mu} + \Gamma^{\varepsilon}_{\mu\alpha}\Gamma^{\lambda}_{\nu\varepsilon} - \Gamma^{\varepsilon}_{\nu\alpha}\Gamma^{\lambda}_{\mu\varepsilon} \tag{1.5}$$

and finally also the tensor H (replacing E of Einstein) with the components

$$H_{\mu\alpha} = \Gamma^{\lambda}_{\mu\alpha,\lambda} - \Gamma^{\lambda}_{\lambda\alpha,\mu} \mp \Gamma^{\varepsilon}_{\mu\alpha}\Gamma^{\lambda}_{\lambda\varepsilon} - \Gamma^{\varepsilon}_{\lambda\alpha}\Gamma^{\lambda}_{\mu\varepsilon}. \tag{1.6}$$

The equations $E_{\alpha\mu=0}$ are those of the field in the theory of gravitation.

The ambiguity of the law of transport in the unitary theory seems to make it difficult to choose $E_{\alpha\mu} = 0$ as fundamental equations. But the difficulty has been solved by Einstein in the following manner : it imposes four restrictive conditions on the affine field, namely

$$\Gamma^{\lambda}_{\mu\lambda} - \Gamma^{\lambda}_{\lambda\mu} = 2\Gamma_{\mu} = 0. \tag{1.7}$$

Though arbitrary, this hypothesis easily gives

$$\mathbf{E} = \mathbf{H} ;$$

the two laws thus lead to the same tensor. Therefore the field equations can remain the same in both the theories.

Besides the system

$$E_{\mu\nu} = 0 \tag{A}$$

which depends only on the coefficients $\Gamma^{\lambda}_{\mu\nu}$ and their first derivatives, there is also a second system where the components of the potential $g^{\mu\nu}$ and their derivatives are also included ; these are indeed the equations

$$g^{\mu\nu}_{,k} + g^{\mu\lambda}\Gamma^{\nu}_{k\lambda} + g^{\lambda\nu}\Gamma^{\mu}_{\lambda k} = 0. \tag{B}$$

We have demonstrated that (A) and (B) are obtained from one variational principle. We form the invariant integral

$$I = \int \sqrt{|g|}\, g^{\mu\nu} E_{\mu\nu} dV_4$$

with some components of the potential and the Einstein tensor (supposed to be unique due to $\Gamma_{\mu} = 0$) and we hold

$$\delta I = 0$$

for all arbitrary variations of $g^{\mu\nu}$ and $\Gamma^{\lambda}_{\mu\nu}$; we immediately obtain the required equations.

It is evident that the existence of a potential tensor in the affine field would give it a very special character.

Indeed (B), regarded as a system of differential equations, can determine $g^{\mu\nu}$ only when the conditions of integratibility are satisfied.

Hence the relations, which follow as consequences of this hypothesis of integrability between the coefficients $\Gamma^{\lambda}_{\mu\nu}$ and their derivatives will express those relations of the

characteristic properties of the field which are implicitly postulated by the theory of relativity as well as the theory of the unitary field itself.

To obtain them in an explicit form is precisely the aim of this work.

2. Case I. — When the coefficients $\Gamma^{\lambda}_{\mu\nu}$ are symmetric, the conditions of integrability of the equations (B), namely :

$$g^{\mu\nu}_{,k} + g^{\mu\lambda}\Gamma^{\nu}_{k\lambda} + g^{\lambda\nu}\Gamma^{\mu}_{\lambda k} = 0$$

can more easily be expressed in terms of the curvature tensor **R**.

If the components $R^{\lambda}_{lm\,n}$ are expressed by the multiples $g^{\mu\nu}$ and brackets with four indices as

$$R^{\lambda}_{lm\,n} = g^{\lambda k}\,(klmn), \tag{2.1}$$

the conditions of integrability will be well expressed by the following properties of those brackets which are well known :

$$(klmn) = -(lkmn) = (mnkl) = -(nmkl). \tag{2.2}$$

From these we simultaneously eliminate $g^{\mu\nu}$ and the brackets and obtain relations between the components of **R** only.

(i) As

$$R^{\lambda}_{lm\,n} = g^{\lambda k}\,(klmn),$$

we hold $\lambda = l$ and contract. We now observe that the brackets are antisymmetrical in the indices (k and l) while g^{lk} are symmetrical.

We then automatically have :

$$R^{l}_{lm\,n} = 0 \tag{2.3}$$

(ii) Similarly, since

$$R^{\lambda}_{lm\,n} R^{n}_{l'm'\,\lambda} = g^{\lambda k}\,(klmn)\,g^{nk'}\,(k'l'm'\lambda),$$

and

$$R^{k'}_{ml\,k} R^{k}_{m'l'\,k'} = g^{k'n}\,(nmlk)\,g^{k\lambda}\,(\lambda\, m'l'k'),$$

we have

$$R^{\lambda}_{lm\,n} R^{n}_{l'm'\,\lambda} = R^{k'}_{ml\,k}\, R^{k}_{m'l'\,k'} \tag{2.4}$$

due to the relations (2.2) (with regard to the brackets).

(iii) We can write the 96 components of the tensor $R^{\lambda}_{kl\,m}$ explicitly as functions of Γ^{λ}_{mn} and their derivatives, that is to say (1.3), which easily gives the relations

$$R^{\lambda}_{\underset{n}{lm}} + R^{\lambda}_{\underset{l}{mn}} + R^{\lambda}_{\underset{m}{nl}} = 0. \tag{2.5}$$

(2.3), (2.4), (2.5) are characteristic properties of the symmetrric field, expressed by the components of the curvature tensor **R**.

3. Case II. — When the field is non-symmetric, the conditions of integrability of the equations (B) can also be easily deduced.

With (B) we also consider the system of their first derivatives (B_1). We eliminate all the first derivatives of $g^{\mu\nu}$ between (B) and (B_1) and using the conditions

$$g^{\mu\nu}_{,l,k} - g^{\mu\nu}_{,k,l} = 0,$$

we immediately come to the following conditions of integrability :

$$0 = g^{a\lambda} R^{\mu}_{\underset{n}{am}} + g^{\mu d} S^{\lambda}_{\underset{n}{md}}, \tag{3.1}$$

from which we can immediately deduce these :

(i) By multiplying (3.1) with $g_{\mu\lambda}$ and then summing (since $g^{a\lambda} g_{\mu\lambda} = \delta^a_\mu$, $g^{\mu d} g_{\mu\lambda} = \delta^d_\lambda$ and δ 's are the symbols of Kronecker) we get

$$R^{a}_{\underset{n}{am}} + S^{d}_{\underset{n}{md}} = 0$$

or

$$\frac{d}{dx^m}(\Gamma^a_{na} + \Gamma^a_{an}) = \frac{d}{dx^n}(\Gamma^a_{ma} + \Gamma^a_{am}). \tag{3.2}$$

(ii) As m and μ are the antisymmetric indices, (3.1) can also be written as

$$g^{a\lambda} R^{\mu}_{\underset{n}{am}} = g^{\mu d} S^{\lambda}_{\underset{m}{nd}}$$

by putting $\lambda = m$ and $\mu = n$, and then after the contraction indicated, we obtain the relation

$$g^{\lambda\mu}(E_{\lambda\mu} - H_{\lambda\mu}) = 0. \tag{3.3}$$

In order to obtain more general relations, we begin with a change of notations for the components of the tensors R and S.

As there are only six pairs of (m, n), i.e. antisymmetric indices, each $\binom{m}{n}$ in the component is replaced by one of the numbers r (r=1, ... , 6) which replaces the symbol **R** or **S** of the tensor.

Thus (3.1) take the following form =

$$g^{a\lambda} \underset{r}{R}{}^{\mu}_{a} + g^{\mu d} \underset{r}{S}{}^{\lambda}_{d} = 0. \tag{3.4}$$

Following the numbers r, the 96 equations arrange themselves into six groups each having 16 relations.

But $g^{a\lambda} = g^a_\lambda$, R^μ_a, S^λ_d, etc can be interpretated as the elements of certain square matrices G, R and S (the indices now indicate the rows and columns as per the well known conventions), and consequently the relations (3.4) can be represented as a single equation of the matrices :

$$\underset{r}{R}G + G\underset{r}{\overline{S}} = 0 \tag{3.5}$$

where $\overline{\mathbf{S}}$ represents the transposed matrix of $\mathbf{S}$ with

$$\overline{S}^\lambda_a = S^a_\lambda.$$

By considering the transposed forms, we also have

$$\underset{r}{S}G + \overline{G}\underset{r}{\overline{R}} = 0. \tag{3.6}$$

The equations (3.5) and (3.6) have the properties of hermiticity as it was postulated by Einstein for all the relations of the unitary theory.

In fact, we pass from (3.5) to (3.6) by transposing G and changing the tensor $\mathbf{R}$ and $\mathbf{S}$ and vice versa, which implies that the said relations remain unaltered as a whole by the simultaneous changes of $g^{\mu\nu}$ into $g^{\nu\mu}$ and of $\Gamma^a_{\mu\nu}$ into $\Gamma^a_{\nu\mu}$, exactly as per the principle of hermiticity.

Finally, by multiplying the six equations (3.6) by arbitrary numbers λ_r and summing up, we get after an easy transformation :

$$G^{-1}RG = -\overline{S} \quad (\text{where } R = \lambda_r R_r \cdots). \tag{3.7}$$

The matrices R and $\overline{S}$ are therefore linked by a transformation which leaves their invariants intact, from which

$$I_r(R) - (-1)^r I_r(S) = 0 \quad (r = 1, 2, 3, 4). \tag{3.8}$$

As the λ_r are completely arbitrary, it can be said that each coefficient of the polynomial in λ_r in (3.8), taken separately, is equal to zero.

In all the relations that arise only $\Gamma^\lambda_{\mu\nu}$ and their derivatives enter, each of them expressing the characteristic properties of the unitary field conditioned by the hypothesis of the existence of the potential tensor.

If the products of the chain of the $\underset{r}{R}$ or the $\underset{r}{S}$ are given the following notations :

$$\underset{r}{R}^\lambda_\mu \underset{s}{R}^\mu_\lambda = R(r, s); \quad \underset{r}{S}^\lambda_\mu \underset{s}{S}^\mu_\nu \underset{t}{S}^\nu_\lambda = S(rst), \quad \ldots ; \tag{3.9}$$

$$\underset{r}{R}^\lambda_\lambda = R(r), \qquad \underset{r}{S}^\lambda_\lambda = S(r),$$

the relations obtained above can also be written as

$$\left.\begin{aligned} R(r) &= -S(r),\\ R(r,s) &= \quad S(rs) = S(sr),\\ R(rst) &= -S(str) = -S(rts) = -S(srt),\\ R(rstu) &= S(utsr) = S(ruts) = \ldots \end{aligned}\right\} \tag{3.10}$$

4.a. As consequences of the hypothesis of the existence of the potential tensor in the field of symmetric connections, the following relations are obtained :

16 relations :

(i) $$\underset{n}{R^{\lambda}_{lm}} + \underset{l}{R^{\lambda}_{mn}} + \underset{m}{R^{\lambda}_{nl}} = 0;$$

6 relations :

(ii) $$\underset{n}{R^{l}_{lm}} = 0;$$

45 relations :

(iii) $$\underset{\mu}{R^{\lambda}_{ml}}\,\underset{\lambda}{R^{\mu}_{m'l'}} = \underset{\mu}{R^{\lambda}_{lm}}\,\underset{\lambda}{R^{\mu}_{l'm'}}.$$

Taking into account the 10 field equations, we have a total of 77 homogenous relations between the components $\Gamma^{\lambda}_{\mu\nu}$ and their first derivatives.

b. In the unitary field, we have :

6 relations :

$$R(r) = -S(r);$$

21 relations :

$$R(rs) = S(sr)\ ;$$

56 relations :

$$R(rst) = -S(tsr);$$

126 relations :

$$R(rstu) = S(utsr).$$

Coupled with the field equations the total can go upto 225 relations between the same elements of the field.

Thus the conditions which are imposed on the affine elements in the unitary theory are more restrictive than those that are imposed on the elements of the gravitational field.

Manuscript received 18 July 1953.

THE AFFINE CONNECTION IN EINSTEIN'S NEW UNITARY FIELD THEORY

By S. N. Bose

(Received September 29, 1952)

The non-symmetric $\Gamma^a_{\mu\nu}$'s are defined by the 64 equations

$$\frac{\partial g_{\mu\nu}}{\partial x_\sigma} = g_{\mu\lambda}\Gamma^\lambda_{\sigma\nu} + g_{\lambda\nu}\Gamma^\lambda_{\mu\sigma} \tag{1}$$

in the new theory. A method of obtaining explicit expressions of Γ's in terms of $g_{\mu\nu}$ and its derivatives is sketched in this note. Following the usual convention of the relativity theory, the summation symbols are suppressed throughout, it being understood that summation is always implied whenever dummy indices occur.

I. The generalized Christoffel bracket $\left(\begin{smallmatrix}\mu\nu\\ \sigma\end{smallmatrix}\right)$ is defined by the equation

$$\left(\begin{smallmatrix}\mu\nu\\ \sigma\end{smallmatrix}\right) = \tfrac{1}{2}\left(\frac{\partial g_{\mu\sigma}}{\partial x_\nu} + \frac{\partial g_{\sigma\nu}}{\partial x_\mu} - \frac{\partial g_{\nu\mu}}{\partial x_\sigma}\right); \tag{2}$$

(the subscripts are written in a cyclic order (Schrödinger)). Also,

$$\left(\begin{smallmatrix}\mu\nu\\ \sigma\end{smallmatrix}\right) = s_{\sigma\lambda}\Gamma^\lambda_{\mu\nu} + a_{\mu\lambda}\Gamma^\lambda_{\nu\sigma} + a_{\lambda\nu}\Gamma^\lambda_{\sigma\mu} \tag{3}$$

where s, a, denote symmetric and anti-symmetric components of g. If $\Gamma^a_{\mu\nu}$ is similarly decomposed into

$$\Sigma^a_{\mu\nu} = \tfrac{1}{2}(\Gamma^a_{\mu\nu} + \Gamma^a_{\nu\mu})$$

(4) and

$$V^a_{\mu\nu} = \tfrac{1}{2}(\Gamma^a_{\mu\nu} - \Gamma^a_{\nu\mu}),$$

and the bracket $\left(\begin{smallmatrix}\mu\nu\\ \sigma\end{smallmatrix}\right)$ into

$$\left[\begin{smallmatrix}\mu\nu\\ \sigma\end{smallmatrix}\right] = \tfrac{1}{2}\left[\left(\begin{smallmatrix}\mu\nu\\ \sigma\end{smallmatrix}\right) + \left(\begin{smallmatrix}\nu\mu\\ \sigma\end{smallmatrix}\right)\right]$$

(5) and

$$\left\{\begin{smallmatrix}\mu\nu\\ \sigma\end{smallmatrix}\right\} = \tfrac{1}{2}\cdot\left[\left(\begin{smallmatrix}\mu\nu\\ \sigma\end{smallmatrix}\right) - \left(\begin{smallmatrix}\nu\mu\\ \sigma\end{smallmatrix}\right)\right]$$

the following relations follow immediately

$$\text{(A)} \qquad \begin{cases} s_{\sigma\lambda}\Sigma^{\lambda}_{\mu\nu} + a_{\mu\lambda}V^{\lambda}_{\nu\sigma} + a_{\lambda\nu}V^{\lambda}_{\sigma\mu} = [^{\mu\nu}_{\sigma}] \\ s_{\sigma\lambda}V^{\lambda}_{\mu\nu} + a_{\mu\lambda}\Sigma^{\lambda}_{\nu\sigma} + a_{\lambda\nu}\Sigma^{\lambda}_{\sigma\mu} = \{^{\mu\nu}_{\sigma}\}. \end{cases}$$

Again if

$$\text{(6)} \qquad \begin{aligned} s_{\sigma\lambda}\Sigma^{\lambda}_{\mu\nu} &= P_{\sigma\mu\nu}, \text{ or } \Sigma^{\lambda}_{\mu\nu} = s^{\lambda k}P_{k\mu\nu} \\ s_{\sigma\lambda}V^{\lambda}_{\mu\nu} &= T_{\sigma\mu\nu}, \text{ or } V^{\lambda}_{\mu\nu} = s^{\lambda k}T_{k\mu\nu}, \end{aligned}$$

we can express Σ and V in terms of P and T with the help of $s^{\mu\nu}$, the contravariant component of $s_{\mu\nu}$ obtained in the usual manner; also taking

$$C^{\sigma}_{\mu} = s^{\sigma\lambda}a_{\lambda\mu}$$

as the components of the matrix product of s and a, we can rewrite the equations in the form

$$\text{(A}'\text{)} \qquad \begin{aligned} [^{\mu\nu}_{\sigma}] &= P_{\sigma\mu\nu} - C^{k}_{\mu}T_{k\nu\sigma} + C^{k}_{\nu}T_{k\sigma\mu} \\ \{^{\mu\nu}_{\sigma}\} &= T_{\sigma\mu\nu} - C^{k}_{\mu}P_{k\nu\sigma} + C^{k}_{\nu}P_{k\sigma\mu}. \end{aligned}$$

P's can now be easily eliminated, and a tensor equation, involving only T's is obtained in the form

$$\text{(B)} \qquad \begin{aligned} &\{^{\mu\nu}_{\sigma}\} + C^{k}_{\mu}[^{\nu\sigma}_{k}] - C^{k}_{\nu}[^{\sigma\mu}_{k}] \\ &\qquad = T_{\sigma\mu\nu} + C^{\lambda}_{\sigma}(T_{\lambda k\nu}C^{k}_{\mu} + T_{\lambda\mu k}C^{k}_{\nu}) - C^{\lambda}_{\mu}C^{k}_{\nu}(T_{\lambda k\sigma} + T_{k\sigma\lambda}). \end{aligned}$$

This tensor equation may be regarded as characterizing the continuum of the Unitary Theory.

When (B) is solved, yielding T in terms of known functions of g and its derivatives, and also C^{λ}_{k}'s, the substitution in (A) of T immediately gives the P's and hence also the Γ's.

To solve this central equation we first observe that a cyclic addition easily yields

$$\text{(7)} \qquad \begin{aligned} T_{\sigma\mu\nu} + T_{\mu\nu\sigma} + T_{\nu\sigma\mu} &= T(\mu\nu\sigma) = \{^{\mu\nu}_{\sigma}\} + \{^{\nu\sigma}_{\mu}\} + \{^{\sigma\mu}_{\nu}\} = \{\sigma\mu\nu\} \\ \{\sigma\mu\nu\} &= \tfrac{1}{2}\left(\frac{\partial a_{\mu\sigma}}{\partial x_{\nu}} + \frac{\partial a_{\sigma\nu}}{\partial x_{\mu}} + \frac{\partial a_{\nu\mu}}{\partial x_{\sigma}}\right) \end{aligned}$$

which is easily recognized as a tensor.

Also if Christoffel co-efficients $\gamma^{a}_{\mu\nu} = s^{ai}[^{\mu\nu}_{i}]$ (the usual symmetric γ's) are used, the left-hand side of (B) is changed to

$$\{\mu\nu\sigma\} + a_{\mu\nu\sigma}$$

where

(8) $$a_{\mu\nu\sigma} = \frac{\partial a_{\mu\nu}}{\partial x_\sigma} - a_{\mu\lambda}\gamma^{\lambda}_{\sigma\nu} - a_{\lambda\nu}\gamma^{\lambda}_{\mu\sigma}$$

is the familiar covariant derivative with symmetric $\gamma^{a}_{\mu\nu}$'s as in Riemann-Geometry; also observing that

(9) $$T_{\lambda k\sigma} + T_{k\sigma\lambda} = T(\sigma\lambda k) - T_{\sigma\lambda k} = \{\sigma\lambda k\} - T_{\sigma\lambda k}\,,$$

the equation (B) can be transformed into the form

(C) $$\begin{cases} \{\sigma\mu\nu\} + \{\sigma k\lambda\}C^k_\mu C^\lambda_\nu + a_{\mu\nu\sigma} \equiv U_{\sigma\mu\nu} \\ U_{\sigma\mu\nu} \equiv T_{\sigma\mu\nu} + T_{\sigma k\lambda}C^k_\mu C^\lambda_\nu + C^\lambda_\sigma[T_{\lambda k\nu}C^k_\mu + T_{\lambda\mu k}C^k_\nu] \end{cases}$$

where the anti-symmetric suffixes are put in the end on both sides in the tensors with three-index symbols.

II. To solve the equation (C) we remark that the co-efficients C^k_μ that occur in the equation are the components of the matrix

$$C = sa = \| C^k_\mu \| \text{ or } C^k_\mu = s^{kt} a_{t\mu}\,.$$

The general equation for the eigen vector M of the matrix C can be written in the form

(10) $$| C - pE \mid M = 0 \text{ or } C^k_\mu M^\mu_p = p M^k_p\,;$$

in the vector $M = M^\sigma_p$ the superscript σ indicates the component, and the subscript p the associated eigen-value.

The characteristic eigen values are the roots of the quartic

(11) $$x^4 + I_2 x^2 + I_4 = 0$$

as the first and third invariants vanish (the matrix being the product of a symmetric matrix).

A few well-known properties of matrix equations are here re- capitulated for ready reference.

1. A matrix C, and its transposition $\overline{C}$ have the same eigen values, but different eigen vectors ; and the eigen vectors, written with two suffixes, (representing the component and the eigen value) satisfy the following relations :

$$\overline{M}^\sigma_k M^\sigma_{k'} = \delta^k_{k'}\,,\ \overline{M}^\sigma_k M^{\sigma'}_k = \delta^\sigma_{\sigma'}\,.$$

1 (a). The theorem is true even in the case of repeated roots of the characteristic equation of C, if the eigen vectors are suitably constructed.

2. Taking the elements of the matrix C, as C^k_μ, where k denotes the rows and μ the columns of the matrix, a new compound matrix $C \times C$, can be constructed in the following manner :

(12) $$(C \times C)\binom{kl}{\mu\nu} = C^k_\mu C^l_\nu - C^k_\nu C^l_\mu$$

obtained by the usual cross-multiplication rule. The elements can be arranged in a six × six matrix in any proper order ; e.g.,

(23) → 1, (31) → 2, (12) → 3, (14) → 4, (24) → 5, (34) → 6.

3. The eigen vectors of the matrixes $C \times C$ (arranged in any assumed manner) may be obtained by taking the vector product of the eigen vectors M^σ_p of C. $M_a \times M_b$ is thus the eigen six-vector of the six matrix, $C \times C$ corresponding to the eigen value $p_a p_b$.

4. The eigen vector of $\overline{C} \times \overline{C}$ is similarly obtained as $\overline{M}_a \times \overline{M}_b$ corresponding to the same eigen value $p_a p_b$.

5. The eigen vectors of the compound matrices have corresponding orthogonal properties.

The components of the eigen vectors can be written by the notation used before as $M\binom{\sigma\sigma'}{ab}$ and $\overline{M}\binom{\epsilon\epsilon'}{cd}$.

The orthogonality relation is written as

(13) $$\begin{cases} \overline{M}\binom{\sigma\sigma'}{ab} M\binom{\sigma\sigma'}{cd} = \delta^a_c \delta^b_d \\ \overline{M}\binom{\sigma\sigma'}{ab} M\binom{\epsilon\epsilon'}{ab} = \delta^\sigma_\epsilon \delta^{\sigma'}_{\epsilon'} . \end{cases}$$

III. Keeping the above well-known properties of matrices in mind we can tackle the problem of solving the general equation. We multiply the equation (C') by $M^\sigma_a M^\mu_b M^\nu_c$ formed from the eigen vectors of C, corresponding to three eigen values a, b, and c, where $b \neq c$; then remembering the equations

$$C^k_\mu M^\mu_b = p_b M^k_l , \qquad C^k_\nu M^\nu_c = p_c M^k_c , \text{ etc.,} \qquad (b, c \text{ not summed})$$

we get easily, by suitably changing the dummies

(14) $$\begin{cases} M^\sigma_a M^\mu_b M^\nu_c U_{\sigma\mu\nu} = \{hkl\} M^h_a M^k_b M^l_c (1 + p_b p_c) + a_{klh} M^h_a M^k_b M^l_c ; \\ \text{also} \\ M^\sigma_a M^\mu_b M^\nu_c U_{\sigma\mu\nu} = T_{hkl} M^h_a M^k_b M^l_c (1 + p_a p_b + p_b p_c + p_c p_a). \end{cases}$$

Or remembering the anti-symmetric nature of the suffixes k, l we write the same equation with k, l in the reversed order and add, using the symbol

$$M\binom{kl}{bc} = M^k_a M^l_c - M^k_c M^l_b$$

easily getting the result in the form

$$(15)\qquad (1 + p_a p_b + p_b p_c + p_c p_a)^{-1}[(1 + p_b p_c)\{hkl\}M^h_a M\binom{kl}{bc} + a_{kch}M^h_a M\binom{kl}{bc}] = T_{hkl}M^h_a M\binom{kl}{bc}.$$

Remembering the orthogonal properties of the eigen vectors, namely equations (12) and (13), and multiplying by $\bar{M}^\sigma_a \bar{M}\binom{\mu\nu}{bc}$ on both sides followed by summation on a, b, c, we have

$$(16)\qquad T_{\sigma\mu\nu} = (1 + p_a p_b + p_b p_c + p_c p_a)^{-1}[(1 + p_b p_c)\{hkl\}\bar{M}^\sigma_a \bar{M}\binom{\mu\nu}{bc}M^h_a M\binom{kl}{bc} + a_{klh}\bar{M}^\sigma_a \bar{M}\binom{\mu\nu}{bc}M^h_a M\binom{kl}{bc}].$$

An explicit form of $T_{\sigma\mu\nu}$ is thus obtained in terms of the eigen vectors of C and $\bar{C}$ and the covariant derivatives of a_{kl} calculated with the symmetric Christoffel co-efficients, γ's obtained from the symmetric parts of $g_{\mu\nu}$, namely $s_{\mu\nu}$. The discriminant-determinant of the equation (C) is obviously

$$(17)\qquad D_{24} = \prod (1 + p_a p_b + p_b p_c + p_c p_a)(\text{product with roots of (11) with } b \neq c).$$

The value of the determinant can be easily calculated in terms of the invariants of C; in fact,

$$(18)\qquad D_{24} \equiv (1 + I_2 + I_4)^8[(1 + 5I_4 - I_2)^2 - 4I_4(2 - I_4)^2]^2.$$

The determinant D always vanish when $1 + I_2 + I_4 = 0$; but this happens when the determinant of $g = \| g_{\mu\nu} \| = 0$.

This is proved as follows:

Any transformation $s(\quad)s^{-1}$, where $s^{-1} = \bar{s}$, preserves the symmetric or/and the anti-symmetric nature of a matrix; $s_{\mu\nu}$ part of $g_{\mu\nu}$ may be diagonalized, and at the same time the anti-symmetric tensor $a_{\mu\nu}$ preserves its anti-symmetric character. Hence g can be reduced to the form

$$\begin{vmatrix} s_{11} & a_{12} & \cdots \\ a_{41} & \cdots & s_{44} \end{vmatrix} \quad \text{and } a_{\mu\nu} + a_{\nu\mu} = 0$$

by such a transformation.

The values of all invariants remain obviously unaltered. In this special co-ordinates

$$(19)\qquad \Delta = \| g_{\mu\nu} \| = s_{11}s_{22}s_{33}s_{44} + \Sigma s_{11}s_{22}a_{34}^2 + (a_{12}a_{34} + \cdots)^2.$$

The invariants of the matrix C can be easily calculated which gives

$$I_2 = (s_{11}s_{22}s_{33}s_{44})^{-1}\Sigma s_{11}s_{22}a_{34}^2$$

$$I_4 = (s_{11}s_{22}s_{33}s_{44})^{-1}(a_{12}a_{34} + \cdots)^2$$

hence

$$(20)\qquad \Delta = s_1s_2s_3s_4(1 + I_2 + I_4).$$

Hence if $(1 + I_2 + I_4) = 0$

$$\| g_{\mu\nu} \| = 0 \text{ also.}$$

That when T is determined Γ's are also determined follow simply as a consequence of equation (A).

IV. In later developments of his theory Einstein has introduced the additional condition

$$\Gamma_l \equiv \Gamma^\lambda_{l\lambda} - \Gamma^\lambda_{\lambda l} = 0;$$

however, from the contravariant form of equation (1), viz.:

$$(21)\qquad -\frac{\partial g^{\mu\nu}}{\partial x_\sigma} = -g^{\mu\nu}_\sigma = g^{\mu\lambda}\Gamma^\nu_{\sigma\lambda} + g^{\lambda\nu}\Gamma^\mu_{\lambda\sigma}$$

the following relation

$$(22)\qquad [| g |^{\frac{1}{2}}(g^{\mu\lambda} - g^{\nu\mu})]_\mu = \tfrac{1}{2} | g |^{\frac{1}{2}}(g^{\nu\lambda} + g^{\lambda\nu})(\Gamma^\mu_{\lambda\mu} - \Gamma^\mu_{\mu\nu})$$

follows easily. The condition

$$\Gamma^\lambda_{\lambda l} - \Gamma^\lambda_{\lambda l} = 0$$

means four relations involving the differential co-efficients of $a_{\mu\nu}$ and $g_{\mu\nu}$. In the equation (C) for the determination of T, all first order differential co-efficients occur in the left side of the in-homogeneous equation, whereas the homogeneous part involves only C_k^λ, composed of $s^{\mu\nu}$ and $a_{\mu\nu}$ and no differential co-efficients occur here. Hence the additional conditions do not affect the general nature of the solution.

UNIVERSITY OF CALCUTTA

(313)

Solution of a Tensor Equation Occurring in the Unitary Field Theory

By S. N. Bose,
University of Calcutta.

Summary — The solution to the tensor equation on which the determination of $\Gamma^{\lambda}_{\mu\nu}$ depends, is explicitly expressed in terms of the C and their invariants.
Note that the iteration gives a rigorous solution when

det C = 0.

1. In a preceding article ([1]) it has been shown that it is possible to reduce the calculation of the affine coefficients $\Gamma^{\lambda}_{\mu\nu}$ in terms of $g^{\mu\nu}$ and their derivatives to the solution of the tensor equation

(1.1)
$$U_{a\mu\nu} = T_{a\mu\nu} + T_{akl} C^k_\mu C^l_\nu + C^t_a (T_{tk\nu} C^k_\mu + T_{t\mu l} C^l_\nu).$$

The tensors U and T are antisymmetric with regard to the indices μ and ν, C^k_μ are the elements of a matrix C formed from the symmetric and antisymmetric parts of the tensor g :

(1.2)
$$\begin{cases} C = Sa, \quad a_{\mu\nu} = \frac{1}{2}(g_{\mu\nu} - g_{\nu\mu}), \quad s_{tk} = \frac{1}{2}(g_{tk} + g_{kt}), \\ S^{\mu t} S_{tk} = \delta^\mu_k. \end{cases}$$

The eigenvalues x of the matrix C satisfy the equation

(1.3)
$$x^4 + I_2 x^2 + I_4 = 0,$$

I_2 and I_4 are even invariants of C ; from the equation (1.2) it immediately follows that the odd invariants are zero.

The eigenvector M^k_a of C corresponding to the eigenvalue a satisfies the equation

(1.4)
$$C^\mu_k M^k_a = a M^\mu_a.$$

The transposed matrix $\overline{C}$ ($\overline{C}^\mu_k = C^k_\mu$) has the same eigenvalues as C but its eigenvectors $\overline{M}^k_a$ are different.

If the characteristic equation has all its roots distinct, then there exists a complete series of eigenvectors M and $\overline{M}$ of C and $\overline{C}$ and we can express the elements of C in terms of the eigenvectors : that is

(1.5)
$$C^k_\mu = \Sigma\, \lambda \overline{M}^\mu_\lambda M^k_\lambda.$$

The solution is easily calculated from (1.1) :

(1.6)
$$\begin{cases} T_{\sigma\mu\nu} = \sum_{abc} (1+ab+bc+ca)^{-1} \overline{M}^{\sigma}_{a} M^{x}_{a} \overline{M}\binom{\mu\nu}{bc} M\binom{yz}{bc} U_{xyz} \quad (b \neq c), \\ M\binom{yz}{bc} = M^{y}_{b} M^{z}_{c} - M^{y}_{c} M^{z}_{b} \, ; \end{cases}$$

and the summation is over the 24 possible combinations of (abc).
The solution (1.6) can be written in the form of a tensor equation

the components of the tensor B being

(1.7)
$$B^{xyz}_{\sigma\mu\nu} = \Sigma \frac{\overline{M}^{\sigma}_{a} \overline{M}\binom{\mu\nu}{bc} M^{x}_{a} M\binom{yz}{bc}}{1+bc+ca+ab}$$

(1.7) is identical to (1.5).
I intend to carry out the summation in (1.7) and obtain the components of B in terms of the elements of C.

2. By means of C and of the unitary matrix E we can form an algebra $\{C_4\}$ with addition and matrix multiplication.
The general formula of a matrix A in $\{C_4\}$ is

(2.1)
$$\begin{cases} A = \lambda_0 + \mu_1 C_1 + \mu_2 C_2 + \mu_3 C_3; \\ C_1 = C, \quad C_2 = CC, \quad C_3 = CCC. \end{cases}$$

All the matrices A commute and therefore have the same eigenvector M as C.
The matrices $\overline{A}$ of the transposed algebra also have the same eigenvector $\overline{M}$ as $\overline{C}$.
If a , $\bar{a}$, b , $\bar{b}$ are the roots of the characteristic equation (1.3), one can easily see that the basic matrices develop in the following way :

(2.2)
$$\begin{cases} E = f(a)+f(\bar{a})+f(b)+f(\bar{b}); & C_2 = a^2[f(a)+f(\bar{a})]+b^2[f(b)+f(\bar{b})]; \\ C_1 = a[f(a)-f(\bar{a})+b[f(b)+f(\bar{b})]; & C_3 = a^3[(fa)-f(\bar{a})]+b^3[f(b)-f(\bar{b})]; \end{cases}$$

where $f(a) = \overline{M}_a \; M_a$,. . and by adding the lower and upper indices to the above formulae, one gets

$$C^{\lambda}_{\mu} = \left(\overline{M}^{\mu}_{a} M^{\lambda}_{a} - \overline{M}^{\mu}_{\bar{a}} M^{\lambda}_{\bar{a}} \right) a + b \left(\overline{M}^{\mu}_{b} M^{\lambda}_{b} - \overline{M}^{\mu}_{\bar{b}} M^{\lambda}_{\bar{b}} \right)$$

Let us now deduce a new algebra $\{V_6\}$, where the matrix V is formed from any two elements A, B of $\{C_4\}$:

$$V = (AB),$$

with

(2.3)
$$V^{\mu\nu}_{rs} = \frac{1}{2} \left(A^{\mu}_{r} B^{\nu}_{s} + A^{\nu}_{s} B^{\mu}_{r} - A^{\mu}_{s} B^{\nu}_{r} - A^{\nu}_{r} B^{\mu}_{s} \right) \quad (\mu \neq \nu \text{ et } r \neq s).$$

The rule of composition (2.3) shows that all the matrices V of $\{V_6\}$ commute among themselves.

All the matrices V have the same eigenvectors with six components as (CC) ; it is easily seen that the six eigenvalues of (CC) are obtained by multiplying the eigenvalues of C two by two. These are

(2.4) $$a\bar{a},\quad b\bar{b},\quad ab,\quad \bar{a}\bar{b},\quad a\bar{b},\quad \bar{a}b;$$

and the eigenvectors with six components are deduced from the eigenvectors of C through a similar rule to (2.3) ; for example :

(2.5) $$M^{\mu\nu}_{ab} = M^{\mu}_{a}\, M^{\nu}_{b} - M^{\mu}_{b} M^{\nu}_{a},$$

are the components of Mab, the vector which corresponds to the eigenvalue ab of (CC).

The matrices $\bar{V}$ of the transposed algebra $\{\bar{V}_6\}$, also have the same eigenvectors with six components as $(\bar{C}\,\bar{C})$; it is easily noticed that the vectors $\bar{M}\,(ab)$ are deduced from the eigenvectors of $\bar{C}$ by the following rule :

$$\bar{M}^{\mu\nu}_{ab} = \bar{M}^{\mu}_{a}\, \bar{M}^{\nu}_{b} - \bar{M}^{\mu}_{b}\, \bar{M}^{\nu}_{a}, \quad \ldots$$

As the bases of the algebra $\{V_6\}$, we can take any six linearly independent matrices of V_6.

We choose the following six matrices whose expansion we give in terms of the eigenvectors :

(2.6) $$\begin{cases} (EE) = V(a\bar{a} + V(b\bar{b}) + V(ab) + V(\bar{a}\bar{b}) + V(a\bar{b}) + V(\bar{a}b), \\ (CC) = -a^2 V(a\bar{a}) - b^2 V(b\bar{b}) + ab[V(ab) + V(\bar{a}\bar{b})] - ab[V(a\bar{b}) + V(\bar{a}b)] \\ (C_2C_2) = a^4 V(a\bar{a}) + b^4 V(b\bar{b}) + a^2b^2[V(ab) + V(\bar{a}\bar{b})] + a^2b^2[V(a\bar{b}) + V(\bar{a}b)], \\ 2(EC_2) = -2a^2 V(a\bar{a}) - 2b^2 V(b\bar{b}) + (a^2+b^2)[V(ab + V(\bar{a}\bar{b}) + V(a\bar{b}) + V(\bar{a}b)], \\ 2(EC) = -(a+b)[V(ab) - V(\bar{a}\bar{b})] + (a-b)[V(a\bar{b}) - V(\bar{a}b)], \\ 2(CC_2) = ab(a+b)[V(ab) - V(\bar{a}\bar{b})] - ab(a-b)[V(a\bar{b}) - V(\bar{a}b)], \\ \qquad V(ab) = \bar{M}(ab)\, M(ab), \quad \ldots \end{cases}$$

The formulae (2.6) are obtained in a manner completely similar to the manner used to obtain the formulas (2.2).

The four other combinations which can be formed from the basic elements of $\{C_4\}$ depend on the six above mentioned equations by virtue of the four identities mentioned below :

(2.7) $$\begin{cases} (EC_3) + (C\,C_2) + I_2(EC) = 0, \\ 2(CC_3) + (C_2C_2) + I_2(CC) - I_1(EE) = 0, \\ (C_2C_3) - I_4(EC) = 0, \\ (C_3C_3) - I_2(C_2C_2) - I_4[(CC) + 2(EC_2)] = 0. \end{cases}$$

It is also easily shown that by multiplying V_1 and V_2 in $\{V_6\}$ with we have

$$V_1 = (AB), \quad V_2 = (CD),$$

(2.8) $$V_1 V_2 = \tfrac{1}{2}[(AC \cdot BD)+(AD \cdot BC)].$$

With (2.8), the identities in (2.7) and the general identity of Cayley - Hamilton :

(2.9) $$C_4+I_2C_2+I_4E = 0$$

each matrix of the form (C_r C_s) can be reduced to a linear function of the six elements of the basis adopted in (2.6).

We can choose the bases and demonstrate the analogous identities for all algebras $V_{\frac{n(n-1)}{2}}$ composed from $\{C_n\}$ [by a process similar to (2.3)].

With the minors m_3 of any matrix C_4, a conjugate matrix D can be formed.

In the case of C studied here, where C = Sa is the product of the symmetric component S by the antisymmetric a, it is easily seen that D is given by

$$D = A's',$$

where A′ and s' are formed similarly from the minors m_3 of a and S.

We take for granted that always

$$\| S \| = \Delta \quad \text{et} \quad S \neq 0.$$

One can nevertheless have

$$\| C \| = 0 \quad \text{si} \quad \| a \| = 0.$$

The case $\| a \| = 0$ is interesting for the unitary field theory. Therefore we will discuss the particular properties of V_6 when $\| C \| = 0$ as a consequence of $\| a \| = 0$.

a is an antisymmetric matrix of order 4. Therefore Δ which is the determinant of a is a perfect sequare and $\sqrt{\Delta}$ is a factor of all the minors m_3 of a.

If $\| a \| = 0$, A and D, conjugate minors of C, are identically zero. For the conjugate matrix, we have an identity analogous to (2.9) ;

(2.10) $$C_3+I_2C_1+D = 0 \quad (C_3 = CCC).$$

When

$$\| a \| = 0, \quad I_4 = 0 \quad \text{et} \quad D = 0,$$

we have

(2.11) $$C_3+I_2C_1 = 0$$

and from the identities (2.7) we immediately obtain

(2.12) $$(CC_2) = 0, \quad (C_2C_2) = I_2(C_1C_1), \quad (C_3C_3) = I_2(C_2C_2).$$

We will use these relations to deduce the solution of (1.1) in the particular case $I_4 = 0$.

3. Now we carry out the summation indicated in (1.6) and (1.7). Let us note some simple properties of the denominator D (abc) :

$$
(3.1) \qquad \begin{cases} D(a\bar{a}c) = D(ca\bar{a}) = 1-a^2, \\ D(abc) = D(\bar{a}\bar{b}\bar{c}), \\ D(aab) = \lambda_+ - b^2, \qquad \lambda_+ = 1+(a+b)^2, \\ D(aa\bar{b}) = \lambda_- - b^2, \qquad \lambda_- = 1+(a-b)^2, \end{cases}
$$

With the help of these simple relations, we can easily regroup the terms to be summed up :

We use the following notations to write the final results :

$$
(3.2) \qquad \begin{cases} V(ab)+V(\bar{a}\bar{b}) = X, \qquad V(a\bar{b})+V(\bar{a}b) = Y\ ; \\ V(ab)-V(\bar{a}\bar{b}) = U, \qquad V(a\bar{b})-V(\bar{a}b) = V\ ; \\ \bar{M}_a M_a = f(a), \qquad \bar{M}(bc)M(bc), = V(bc), \quad \ldots\ldots \end{cases}
$$

The summation (1.7) is written as

$$
(3.3) \qquad B = \Sigma \frac{f(a)V(bc)}{D(abc)}\,.
$$

After regrouping we get

$$
(3.4) \qquad B = \Sigma f(\lambda)\Big[\frac{V(a\bar{a})}{1-a^2}+\frac{V(b\bar{b})}{1-b^2}\Big]+\frac{1}{2}(X+Y)\Big[\frac{f(a)+f(\bar{a})}{1-a^2}+\frac{f(b)+f(b)}{1-b^2}\Big]
$$

$$
+\frac{1}{2}X\Big[\frac{f(a)+f(\bar{a})}{\lambda_+-b^2}+\frac{f(b)+f(\bar{b})}{\lambda_+-a^2}\Big]+\frac{1}{2}Y\Big[\frac{f(a)+f(\bar{a})}{\lambda_--b^2}+\frac{f(b)+f(\bar{b})}{\lambda_--a^2}\Big]
$$

$$
-\frac{1}{2}U\Big[\frac{f(a)-f(\bar{a})}{1-a^2}+\frac{f(b)-f(\bar{b})}{1-b^2}\Big]-\frac{1}{2}V\Big[\frac{f(a)-f(\bar{a})}{1-a^2}-\frac{f(b)-f(\bar{b})}{1-b^2}\Big]
$$

$$
-\frac{1}{2}U\Big[\frac{f(a)-f(\bar{a})}{\lambda_+-b^2}+\frac{f(b)-f(\bar{b})}{\lambda_+-a^2}\Big]+\frac{1}{2}V\Big[\frac{f(a)-f(\bar{a})}{\lambda_--b^2}-\frac{f(b)-f(\bar{b})}{\lambda_--a^2}\Big].
$$

Let us make use of (2.2) and (2.6) to express X, Y, U, V and $f(a) \pm f(\bar{a}), \ldots$; X, Y, U,V are functions of matrices of {V6} and $\{C_4\}$, and we obtainn after an easy calculation the result in its final form

$$B = E[d_0(EE)+d_1(CC)+d_3(C_2C_2)+2\mu(EC_2)]$$
$$+C_2[g_0(EE)+g_1(CC)+g_2(C_2C_2)+2\nu(EC_2)]$$
$$+C_1[2f_1(EC)+2h_1(CC_2)]+C_3[2f_3(EC)+2h_3(CC_2)], \tag{3.5}$$

where the coefficient d, g, μ, ν, f, h are functions only of the invariants of C. If

$$t_0 \equiv 1+I_2+I_4, \quad t_+ \equiv \lambda_+^2+I_2\lambda_++I_4, \quad t_- = \lambda_-^2+I_2\lambda_-+I_4;$$

$$\frac{t_+-t_-}{2} \equiv x = 1+5I_4-I_2, \quad \frac{t_+-t_-}{4ab} \equiv y = 2-I_2$$

and

$$t_+t_- = (1-I_2+5I_4)^2-4I_4(2-I_2)^2$$

are functions of the invariants

The coefficients of (3.5) are

$$(3.6)\quad \begin{cases} d_0 = \dfrac{1+I_2}{t_0} - \dfrac{I_4 y}{t_+t_-} - \dfrac{I_4(x-2y)}{t_0t_+t_-}, \\[2ex] d_1 = \dfrac{x-y(y-1)}{t_+t_-}, \ d_2 = -\dfrac{y}{t_+t_-} + \dfrac{2y-x}{t_0t_+t_-}\ ; \\[2ex] g_0 = \dfrac{1_4(x-2y)}{t_0t_+t_-}, \ g_1 = \dfrac{y}{t_+t_-}, \ g_2 = \dfrac{x-2y}{t_0t_+t_-}, \ 2f_3 = \dfrac{2[2x+y(2I_2-t_0)]}{t_0t_+t_-}; \\[2ex] 2h_3 = -\dfrac{4(x+yI_2)}{t_0t_+t_-}, \ 2\mu = \dfrac{I}{t_0} + \dfrac{x}{t_+t_-} - \dfrac{2(x-2yI_4)}{t_0t_+t_-}, \ 2\nu = \dfrac{2(x-2I_4y)}{t_0t_+t_-}; \\[2ex] 2f_1 = \dfrac{2[f_3I_2+41_4y-x(1+I_4)]}{t_0t_+t_-} \ \text{et} \ 2h_1 = -\dfrac{4[I_2h_3+x-y(1+I_4)]}{t_0t_+t_-}. \end{cases}$$

4. We have expressed the final result (3.6) in terms of the matrices deduced from C following fixed rules, with coefficients which are functions of the invariants of C. This result is valid in all the cases, even when for particular matrices there is no complete system of eigenvectors and we can no longer develop the resolvent matrix B as per (3.3) but the general result (3.5) always remains true. The expression (3.6) thus gives the general solution of the problem posed in (1.1).

Nevertheless it is interesting to observe that in the particular cases $I_4 = 0$ or $I_2 = 0$, $I_4 = 0$, we can easily put the solution in the form of a total sum of a small number of terms, thanks to a method of iteration.

Let us briefly discuss the case $I_4 = 0$.

We have already deduced the special properties of the matrix C in (2.11) and (2.12) : to wit

$$(CC_2) = 0, \quad (C_2C_2) = I_2(C_1C_1), \quad (C_3C_3) = I_2(C_2C_2),$$
$$C_3+I_2C_1 = 0, \qquad C_4+I_2C_2 = 0.$$

We write (1.1) in the form

$$U = AT = [E\{(EE) + (CC)\}+2C\ EC)]T$$

or

(4.1) $$U = T+[E(CC)+2C(EC)]\ T = T+\delta T;$$

and look for a solution by putting

$$T = U+\Phi_1.$$

The equation having the form

$$U \equiv T+\delta T.$$

we easily deduce for Φ the equation

$$-\delta U = \Phi_1+\delta\Phi_1.$$

By repeating the operation, we successively obtain

(4.2) $$\delta\delta U = \Phi_2+\delta\Phi_1, \quad -\delta\delta\delta U = \Phi_3+\delta\Phi_3.$$

Beacause of special properties of the operator [(2.11) - (2.12)] we easily deduce

(4.3) $$\delta\delta\delta = I_2^2\delta.$$

By putting $\delta = \Delta - E$, where E is the unit operator, the equation (4.2) is written as

$$\Delta\Phi_3 = -I_2^2(\Delta-E)U$$

or

(4.4) $$\Delta(\Phi_3+I_2^2U) = I_2^2U.$$

The original equation is

$$\Delta T = U.$$

With det $\Delta \neq 0$, the solution is unique and one easily sees that the solution of (4.4) is

$$\Phi_3+I_2^2U = I_2^2T.$$

By writing

$$\Phi_3 = T-U+\delta U-\delta\delta U = I_2^2(T-U),.$$

we easily have

(4.5) $$T = U+ \frac{\delta U-\delta\delta U}{I_2^2-1}.$$

By developing this, we have the following result :

$$T_{\sigma\mu\nu} = U_{\sigma\mu\nu} - (1+I_2)^{-1} U_{\sigma kl} C^k_\mu \, C^l_\nu + (I_2^2-1)^{-1}[C^t_\sigma (C^k_\mu U_{tk\nu} + C^l_\nu U_{t\mu l})]$$

(4.6)
$$-(I_2^2-1)^{-1} C^t_{2,\sigma} \, (U_{tk\nu} C^k_{2,\mu} + U_{t\mu l} C^l_{2,\nu}) - 2(I_2^2-1)^{-1} C^t_{2,\sigma} C^k_\mu C^l_\nu U_{tkl} \, .$$

The general solution gives us the same result by putting $I_4 = 0$, and using the particular relations (2.11) and (2.12).

We can similarly treat the other cases $I_2 = 0$, $I_4 = 0$.

Post-script

If the coefficients λ of

$$N \equiv \lambda_0 + \lambda_1 C_1 + \lambda_2 C_2 + \lambda_3 C_3 \quad \text{(any number in } [C]_4)$$

are chosen in the algebra $[V]_6$, these numbers form a new algebra $[cV]_{24}$.

To solve

$$U = [e(EE) + e(cc) + 2c(Ec)] \, T \equiv BT$$

is thus equivalent to finding the inverse of B as an algebraic number. The inverse is also a number in [cV] and we can easily find it in the following way.

First, we verify the following results as per the rule of multiplication and the identities for the algebra $[V]_6$.

If

$$L = (C_2C_2) - I_4(EE), \qquad M = 2(EC_2) + I_2(CC),$$

$$R = 2L + I_2 M, \qquad X = 4(EC)^2 = 2(CC) + 2(EC_2),$$

and $D \equiv I_2^2 - 4I_4$ is the discriminant of the equations

$$x^2 + I_2 x + I_4 = 0,$$

then we have

$$L(EC) = M(EC) = 0; \quad d\text{'où } LX = MX = RX = 0$$

$$RL = DL, \quad RM = DM, \quad RR = DR, \quad M^2 = R$$

and

$$XX + 2I_2 X + D = R.$$

In what follows we use the the following abbreviations : $s = 2 + I_2$, $t_o = 1 + I_2 + I_4$ $x = 1 + 5I_4 - I_2$, and $y = 2 - I_2$.

With B = (EE) + (CC) + 2C(EC), by multiplying it with its conjugate

$$B^* = (EE) + (CC) - 2C(EC),$$

we have

$$BB^* = (EE + CC)^2 - 4C_2(EC)^2$$

or

$$BB^* = \left(1 + \frac{R - sM}{t_0}\right)(t_0 + X - C_2 X)$$

after factorization ; with $\left(1-\frac{R-sM}{t_o}\right)\left(1+\frac{R+sM}{t_o}\right) \equiv t,$

$$\frac{1}{BB^*}=\frac{\left[1+\frac{R+sM}{2t_0}\right][t_0+X(1+I_2)+C_2X]}{[t_0+X-C_2X][t_0+X(1+I_2)+C_2X]}$$

after adding the same factor in the numerator and denominator of the right side of the equation.

By multiplying we obtain for the denominator

$$t_0[x+y(X+I_2)+R].$$

We notice that

$$[x+y(X+I_2)+R][x-y(X+I_2)-\nu R]$$
$$=\Omega=t_+t_-=x^2-4I_4y^2$$

if $\nu=\frac{x-2y}{t_o}$.

We can thus easily express the inverse by the following equation :

$$B^{-1}=\frac{B^*\left[1+\frac{R+sM}{2t_0}\right][t_0+X(1+I_2)+C_2X][x-y(X+I_2)-\nu R]}{t_0t_+t_-}.$$

By eliminating the brackets, we obtain a result equivalent to the one in the main paper.

English translation of *Bull Soc Math France* **83**, pp. 81-88, 1955 (Gauthier - Villars Publishers, Paris).

II. Satyendra Nath Bose

A. Lectures and Addresses

The essays in this section are based on Bose's lectures and addresses that span several decades. They reflect upon his deep and abiding interests in social problems, particularly education and the educational system that had failed to serve the needs of the country. He offers a penetrating analysis of the system that was set up by the colonial masters, the main purpose of which was to prepare a cadre of young and educated to serve their own needs in governing the vast country in an efficient way. It was not construed to meet the social needs of India, to solve the problems of poverty, health, diseases and sanitation. It was not designed to have the scientific and technological training to develop industries and self-employment. Consequently, university education, attained at times on the basis of the slender resources of the family, resulted only in unemployment, frustration and tragedy.

Bose stresses the importance of science education at all levels. He speaks eloquently against those who "pinned their faith on the medieval ideals of the university," or the "other-worldliness" recommended in the ancient seats of learning, in Nalanda and Taxila, and who thought excessive emphasis on science and its pursuit towards materialistic and utilitarian ends was against the old ideals.

Such "other-worldliness," according to Bose, provided a good breeding place for "inefficiency, selfishness and greed," and the neglect of a serious acceptance of life by first-class thinkers "brought second-rate petty people into prominence who gave lip-adoration to philosophy but actually indulged in jealousy, squabble and internal strife." For Bose, there was no line of distinction between the spiritual and the material. Along with its liberating spirit and profoundly humanizing tone of the investigations, science also had an intimate contact with the utilitarian and the mundane. "After all it is not the universe alone," Bose says, "but human life too, that is one whole; we do it harm and injury by dissecting it. What we should always bear in mind is that in so far science is ministering to the economic and social needs of the human race, it is really working out its deep spiritual purpose."

Bose, however, became a controversial figure when he became the champion of mother tongue as the best medium of instruction for education in general, science in particular. The dominance of a foreign language, according to Bose, was the main impediment against grasping and mastering scientific thinking at an early age. The English medium of instruction had become the privilege of "the upper echelons of the educated people," and had left behind vast masses in ignorance.

The other essays in this section contain Bose's historical perspectives on university education, and his thoughts on research opportunities and new directions in building new institutions in Independent India.

Science in Education

Mr President, Ladies and Gentlemen,
I confess at the outset that I had great misgivings in choosing as the subject of this evening's discourse, the place of Science in Education. My difficulty arose on the ground of my ignorance as to what this conference really stood for, what it exactly aimed at; whether it sought to change the present educational system radically or was merely an attempt to revise the arrangements in vogue.

I am aware the Hon'ble Minister of Education took pains to convince us that the present system of education was by no means an ideal one, especially as it was indirectly responsible for the steady increase in the unemployment of our educated youths; and we are grateful to him for his vigorous attempts to devise plans to remedy some of its glaring defects. To be candid, if I had not been assured that you all desire to revise and reconsider the present system, I would not have wasted your time by pointing out to you that science is one of the crying needs of our present educational system. For one can do very well without much science, or any real scientific teaching, if all that is wanted is somehow to impart literacy to the masses. But if we propose to look on education as the means of effecting adjustment between man and his environment, we cannot but recognize the supreme importance of the study of science in schools and universities. I need not remind you how science has hitherto played a very insignificant part in our educational programme. It had no place in schools when we were boys, and though during the last two decades the remodelling of the university courses has offered increased facilities for scientific studies to the college students, the schools are still practically without it. I do not make this point with any controversial bias, and I may be mistaken in my apprehension of facts. But if any plan proposes to go to the roots of our educational needs, it must, you would admit, allot to science and scientific teaching a more important role, not only in the curricula of the universities, but also in those of the schools — I therefore take this opportunity to express my gratitude to the organizers of the week, for extending to me the privilege of pleading before this distinguished gathering of educationists, the case for science in education. I need hardly add that I am not one of those who would dispense with either science or education in life altogether, simply because neither our universities nor our schools have so far succeeded in solving the economic and cultural problems of our lives. As a matter of fact, this failure is writ large on the face of almost every institution and organization in the province. But the remedy for this state of things does not lie in renunciation of all organized social efforts and a return to wilderness and asceticism. We need not put on sack-cloth and ashes and adopt some simplified form of existence hardly distinguishable from primitive patterns of pastoral life. Lack of direction in our educational efforts has, to my mind, brought about this deplorable state of things; and the amelioration lies not in the abandonment of hopes and obligations but in the adoption of a conscious and consistent plan, and a united will to realize it.

The charge that our educated youth is an economic misfit has been often repeated. Assuming that he is so, is it fair to put the whole blame on him alone? What is really responsible for the fact that our young men cannot utilize their learning in any practical way and find for themselves suitable places in the economic arrangement of the country? Is it because we have not got even a decent amount of trade and industry here, which can offer them occupation? How is it then that hundreds and thousands of strangers make their easy pile in this land? Does it not look strange that side by side with this amassing of large fortunes by others, our young men, who spend the best years of their lives and a considerable portion of the slender resources of their families in acquiring academic distinctions, should be practically starving for want of employment? These questions cannot be answered without going into the historical conditions under which the present educational policy was inaugurated. It is an open secret that the policy was neither scientific nor economic in its origin; its object was not to provide education of a type best suited to the social needs, and scientific considerations did not come within the purview of the educational authorities of the bygone days, in determining the interests of the people concerned. But this is not to deny that it sought to implant an ideal in the minds of the young men of those days, and to serve a useful purpose: for youths of the country were necessary to interpret a new civilization and a new culture, and act as cheap but efficient links in the long chain of administration flung far and wide over the land. There was in consequence sufficient employment for them and the question of fit or misfit never arose. The educational experiment, on the contrary, was so effective, that its range expanded by leaps and bounds, as years went by. But changes gradually came over the whole country, and were accentuated during the last two decades. Imperative social and economic needs made the parochial conception of a hyper-literary education obsolete and useless, and the old system, at best suitable for a leisured and affluent few, continued to cater for the toiling millions.

What else should you expect in the circumstances except a misfit? And should you not in all fairness apportion the blame of the sad situation between the authorities and the students, instead of criticizing the students alone? Should we not desist therefore from further complicating the issue by introducing the question of race, congenital inferiority and all such inane erudities? It is, however, futile to discuss who is or who is not to blame for this deplorable state of things. The important point is to devise such changes as will adjust the educational system to altered conditions, and to do this we must approach the question in a dispassionate and scientific manner, which alone will reveal not only the magnitude of our task but also the methods that we have to adopt to surmount our peculiar difficulties. The importance of a critical survey of the actual conditions prevalent in the country cannot in this respect be overemphasized, and in an examination of ways and means, the essential must be separated from the inessential. It is no doubt, for instance, desirable for a budding teacher to know the height of the highest mountain and the length of the longest canal, as well as the intricacies of the lordly game of cricket and the hierarchy that holds together cabbages and kings. But it is more than perilous to miss the interdependence of anopheles and malaria, and forget the connexion between cholera and drinking water. The life of the community depends on a detailed knowledge of its natural resources, and in an agricultural country with an increasing birth-rate the modern methods for improving the yield of the land deserve a much greater attention than they have hitherto received. It is needless to remind you again that in this land superstition is still the prime mover, and no amount of book-learning will overcome it sooner than an intimate contact with the modern miracles of man-made machines. Thus will the supremacy of law and order even in inanimate nature become implanted in our minds and events will cease to be ordained by stars or determined by a malevolent fate. The distinction of science from idle speculation, and its connection with common sense has been too often stressed for me to emphasize anew. Your philosopher is surely justified in dilating on the epistemological difficulties of accepting any recognized cosmology and may even dismiss time out of court for want of what he considers to be logical evidence; but we men in the streets are denied such luxuries and if we fail to provide for the morrow we shall surely starve. The study of ancient history and archaeology, of literature and theology have, no doubt, their ennobling influence and make us conscious of our past heritage. But science with its causal laws and technical mastery over nature is essential, if our aim is to harmonize life with its surroundings and to seek to harness elemental forces to our will and service. The sooner we impress upon the minds of the coming generations that individual destinies are amenable to human control, the nearer we shall be to whatever be the goal.

It is rather late in the day to extol the achievements of the Japanese; and their phenomenal rise in the political and economic sphere is a tale that has lost its point by endless repetition. It is, nevertheless, important to remember that their contact with Western science and culture began not much earlier than ours. Yet Japan today has not only left us behind in the race of progress, but many of the pioneers of European industrialism are unable to keep pace with her. This has been brought about by careful and conscious planning and not by cultivating a fastidious accent in foreign tongues. So we may laugh as much as we like at their mistakes in English and their woeful lack of expression in languages other than their own; but their adaptation of educational methods to national needs should become an object lesson for those who are still opposed to any sort of planning, and adhere to the medieval ideal of a cloistered education. But education and scientific planning can no longer be dissociated with impunity in the modern world; and even in the country which, we heard, won the battle of Waterloo in the playing fields of Eton and Harrow, education tends to become more and more technical and utilitarian and less and less of an expensive luxury. If the paramount importance of planning in an efficient education is now recognized in a conservative country like England, the part it plays in progressive America or Germany need not be stressed here when it is well known that in these countries many a wonderful discovery in science have, in recent years, resulted from well-planned efforts to supplement national resources.

I am aware that planning in a vague sense is not quite absent in our educational system; and many improvements have been or are about to be effected in providing technical education for grown-up college students. But in order that such plans may bear fruit, it is necessary not only to organize the final forms, but also to think out carefully all the steps leading thereto; and unless the ground is assiduously prepared from the initial stage, such efforts are likely to remain sterile. Success in technology is a triumph of the scientific spirit, and a mind nourished on purely literary ideas is less likely to absorb it than one on which classical moulds have not left their indelible mark. Training in the scientific mode of thinking should, therefore, begin as early as possible, preferably from the primary schools. This, however, must not be confused with either technical education or vocational training, but boys should be familiarized with the elements of natural sciences and the principles of hygiene and biology. I have been told that these subjects are touched upon by some of the recommended textbooks and teachers speak on these topics in their classes. But that is not enough. Pictures and fairy tales about science and scientists are worse than useless; and students, if they are to form any idea of the subject should have direct experience of the scientific method which is not, in reality, a dexterity with instruments, nor an initiation into the mysteries of mathematics, but a training of the senses in direct contact with Nature. Experiments are, of course, essential. But they do not require, in the early stages, expensive instruments. Nature herself should be the laboratory; and an ingenious teacher can find a lot to talk about in the everyday world familiar to the student from his birth. The natural setting of the primary school is the village; and for an agricultural country no better background for an educational system could be imagined. For all the fundamental forms of life are assembled there: agriculture and its possibilities, marketing and its various difficulties, public health and its deadly enemies, machines and means of transport in various stages of development, all are there ready for objective study. If therefore the right sort of teacher is forthcoming, the elementary economy of the Bengali village will be no handicap to the youthful seekers of scientific knowledge.

The teacher is all-important in such a system and his example and sympathy will counterbalance the deficiencies in equipment. He must, therefore, be carefully trained, and acquire not merely book-learning in methods of teaching, but a real enthusiasm for his work. For it is useless to cite examples from foreign textbooks when young children are concerned; and in order that knowledge may become vital to them more attention should be paid to local conditions and instruction should be through materials in daily use. The medium of instruction is, therefore, of foremost importance, and it is not enough that children should be taught, as they are now, in their mother tongue; but that the teachers themselves should learn to think in their own language. This, it is needless to point out, requires the adoption of Bengali as the medium of instruction throughout the whole of the educational system. At any rate, so far as the teaching of science is concerned, there is no other alternative, if in ordinary scientific education we should aim at the cultivation of a scientific attitude rather than proficiency in scientific and pseudo-scientific catchwords. I am aware of the immense difficulties involved in this desirable change in the existing order of things. But remembering the solid achievements of a young university like the Osmania, I believe my suggestions are neither unworkable nor incapable of realization in the near future. What is indispensable for success is the goodwill of the whole country, the cooperation of universities and educational institutions, aspirations of the teachers and determination of the educational authorities. . .

I repeat again that the soundness or otherwise of every educational system depends entirely on the teachers, especially in the primary stages, where foundations of all future activities are laid in extremely impressionable soil. They, therefore, should be carefully recruited, furnished with the necessities of life, freed from material anxieties to devote all their energy and time to the realization of educational ideals and made alive to their great responsibility and power for good that they might easily become. For this purpose such meetings as this conference should become a regular feature of our national life. At the same time less complicated methods to foster closer cooperation and promote more frequent exchange of ideas among teachers must be devised. For Bengal is a vast country with insufficient means of communication between distant parts where isolation of workers is more the rule than the exception; and personal meetings, even if not expensive, are impracticable here. An obvious way out of these natural difficulties would be establishment of a periodical through which teachers would state their problems, receive suggestions, keep in touch with modern educational movements, gather information about current achievements in science and to some extent make up for the lack of usual library facilities. So that its appeal may be universal, the journal must perforce be in the vernacular. There should also be state-aided central libraries feeding different groups of schools, judiciously distributed throughout the province, and stocked with books of reference which are ordinarily beyond the means of primary and secondary schools.

But a forum for teachers and co-ordination of primary schools alone are not enough to ensure a sound scientific education. The secondary schools as well as the subsequent institutions have also to be suitably planned. Here the personality and sense of responsibility of the teachers alone will be insufficient, if the teaching of science is to be taken out of its rudimentary state and made suitable for the complexities of modern life. Provision of properly equipped laboratories where students with aptitude can work in natural and biological sciences becomes imperative at this stage. It would perhaps be a Utopian dream to expect such resources in every secondary school, but some selected institutions must be established, one in every district at least, if science is to be of any practical utility in national life. It nevertheless is unnecessary to make teaching of science compulsory in secondary schools as it is suggested to be made for the primary stages. On the other hand, in case of students who choose science as a special subject, some fairly rigorous test of fitness is desirable. For it would never do to repeat the mistake that has resulted in the overcrowding of some of the professions; and as the minimum of science indispensable for the purposes of daily life is already provided for in the early stages, the available resources should not be wasted indiscriminately, but utilized for the really deserving student. The process of selection should not therefore end in mere written examinations, but be supplemented by such practical tests as will determine his talent for application. For science, divorced from pragmatic considerations, is even worse than speculative philosophy.

I am confident that if these suggestions are given effect to sufficiently early, the universities and advanced technological institutions need not be handicapped for want of suitable human material. On the contrary, there is bound to be a sufficient number of capable students who could both be adequately trained and trusted to carry out the programme for higher studies, and this will serve the real need of the country sufficiently if all the other institutions, like the universities, have already chalked out their programme on the basis of a survey of our national needs. For after all, it is not a question of only producing the right sort or number of students in as economical and efficient a way as possible but also of forming a plan on the actual survey of the country's needs. It is not for me to talk about the university methods, the higher studies, in this gathering. But just as it must be emphasized that we cannot be too careful in training our students, it must be equally obvious that the programme of higher training must be directly and intimately in touch with the actual needs. If we really want to solve the problem of unemployment and make our educational institutions sufficiently effective and responsible, we must so build up our universities and schools as if they were expanded departments of our social existence. It follows from this that science cannot be made to face social problems adequately unless and until universities are linked up with the industrial and commercial activities of the country, and co-ordinated with such Government concerns as make practical use of scientific investigations. What we have to bear in mind is that the question of scientific training and that of the industrial and technical activities in the country do not fall apart; on the contrary they are intimately bound up with one another. If, for instance, we do not utilize the myriad centres of industry and agriculture for giving our students an opportunity to work out their more or less theoretical knowledge, we would be leaving their training half finished. Our experts, at best, will be more of the academic type and therefore not sufficiently useful for practical commercial purposes. If, on the other hand, commerce and technical departments have to recruit all their specialized staff and skilled labour from abroad and not draw upon the indigenous sources, they are bound to suffer from the point of view of economy as well as the country as a whole from the point of view of efficiency. What is, therefore, of supreme importance is that we should provide for as direct an interchange of services as possible between our educational institutions and the technical and industrial centres with assurance that they stand in a relation of interdependence and form between them the social order as a whole. I need hardly add that in the nature of things, there must be an unrestricted give and take between the educational institutions and non-academic undertakings. We should never really be ashamed of going down to the mines, factories and railway lines to finish our scientific training, nor should industry fight shy of our universities on the false ground that they produce only theoretical and academic men.

I have talked sufficiently long and it is time for me to wind up. I am not at all sure if I have been able to convey to you what value I attach to science as it is, or its place in education. But I do not want you to carry away the idea that I am one of those who teach science and prepare its principles on purely pragmatic grounds — and hold that science has nothing spiritual about it — that it is altogether utilitarian in its scope and serves no end that is not rooted in what may be called the mundane existence. I have insisted much too often on the place and the value of science in the economy of our social needs. As a matter of fact a good portion of my lecture has gone to explain how from the primary to the university stage science is bound up with its environment, how the village, in the case of the primary, and the town and the city in the secondary and university training form the necessary background for an efficient and all-round scientific education. But in all this it never was for one moment my object to draw any line of distinction between the spiritual and the material, and in so far as I am convinced of the essentially liberating influence of the scientific spirit, and the profoundly humanizing tone of its investigations, I believe also in its intimate contact with the utilitarian and the mundane. After all it is not the universe alone, but human life too, that is one whole; we do it harm and injury by dissecting it. What we should always bear in mind is that in so far as science is ministering to the economic and social needs of the human race, it is really working out its deep spiritual purpose.

Lecture delivered at a conference held in connection with the Bengal Education Week in February 1936 at Calcutta. From the Proceedings of The Bengal Education Week, *1936, vol. I, pp.56-64. Obtained by courtesy of Rathindranath Bose.*

Visva-Bharati

Friends,

It is my proud privilege to welcome you all in our midst on the eve of our Annual Convocation. We are deeply touched and glad beyond measure that our Acharya and our beloved Prime Minister, Sri Jawaharlal Nehru, has consented to conduct the proceedings of this year. We all know how the great tasks of promoting peace and goodwill among nations and of leading our country in the path of progress towards prosperity absorb all his time and attention. His visit is an encouragement and its memory will remain a cherished source of inspiration to all workers of the Visva-Bharati.

Cordial greetings to our Pradhana on her first visit to this University. Our memories go back to the days when Santiniketan was a place of pilgrimage :— the abode of our illustrious founder, the great seer and poet who sacrificed so much for the sake of his experiment to find a place where the people of the East and the West will be welcome as equal partners and labour for the synthesis of all that is good and beautiful in the culture of many lands and races. It had been his dream that humanity will ultimately be able to rise above petty strife and mean selfishness and realize the glorious destiny which the Poet had seen in his beatific vision. Thereafter many had come and drunk at the fountain of inspiration, among them your great mother, the illustrious daughter of Bengal, the great Sarojini, who had walked with the soldiers of freedom working out a bloodless revolution. We hope that this visit will renew an old tie of friendship and that you will continue to extend to our activities at the Visva-Bharati your sympathy, encouragement and benevolent support.

(141)

To my friend and the Chief Guest of this Convocation, Professor Prasantachandra Mahalanobis, our grateful thanks for having consented to speak about the ideals and the aims of Visva-Bharati. You have been a close associate of Gurudeva when the idea of the Visva-Bharati took shape in his hands. Our old workers still remember you as one who, as the first Secretary of the Visva-Bharati, toiled ceaselessly to nurse the infant association and guide its first footsteps. We hope that what you will say to us today will remind us of the noble beginning of a great enterprise and help us to remain steadfast to our ideals.

It is also gratifying to me to an immense degree that in this first Convocation that I attend I am able to present three of our illustrious pioneer workers for the award of the highest honour that Visva-Bharati may offer in recognition of merit.

Pandit Haricharan Bandyopadhyaya has passed his long life in pursuance of the one great task which he had set before himself. His monumental work, the *Bangiya Śabdakosha* will endure as a magnificent contribution to the study of the Bengali language. His simple life at the Guru-Palli is in harmony with the great tradition of our land where Gurus lived simply and in dedication to truth and wisdom.

Our Pronetri Indira Devi Chaudhurani had long guided our artistic departments. Her sure instincts and profound knowledge had been our mainstay in many of our recent artistic endeavours. She had also been our Upacharya during a short but difficult period of the University.

Pandit Vidhusekhar Shastri will be long remembered in Santiniketan for his assiduous scholarship and for his association with the linguistic studies at Visva-Bharati. He initiated here the studies of Buddhism and Tibetan. He has trained a long line of workers, here as well as in Calcutta, who continue to follow his footsteps and reap harvest in the rich fields of linguistic and other studies.

The names of our pioneers remain in Visva-Bharati as shining examples of what can be achieved by pure devotion and valiant efforts. In honouring them we pay homage to true scholarship and single-minded devotion to the high ideals of Visva-Bharati.

I pause here a moment to remember the great-hearted scholar who literally sacrificed his life to the cause of Visva-Bharati — Professor Prabodhchandra Bagchi. He had organized research in Sino-Indian Studies in Visva-Bharati. He had won laurels in his own particular field and made the name of Visva-Bharati an honoured one in the field of Sino-Indian Studies. He consented to be its Upacharya at a critical time and, I repeat, literally sacrificed himself in endeavouring to pull the affairs of the institution out of the confusion into which it had unfortunately fallen.

Only five years have elapsed since the Visva-Bharati has been recognized as an institute of national importance and the Government has agreed to take it under its protection as a centrally administered University. We are as yet in the initial stages of the new development. Much still remains to be done. However, generous aid has begun to flow in.

Out of Government grants we have been able to put up a new building for a

Vidya-Bhavana and a new hostel for students. The block grants-in-aid of 4.5 lakhs, with supplementary aids to cover our deficits, have helped us to meet our recurring expenditure. Additional grants have helped us to stabilize our Tibetan Studies and to procure valuable books in our library. We have now been able to acquire the Uttarayana properties which will be used by the Visva-Bharati to house its Rabindra-Sadana, a Museum entirely devoted to the memory of our illustrious founder and preserve the priceless treasures that the poet has left us : manuscripts, paintings and other objects hallowed by their association with Gurudeva.

During the present year Visva-Bharati has continued to function in the normal manner in spite of great difficulties. Our scholars have continued to contribute to Indological and Sino-Indian Studies. Valuable Bengali manuscripts have been edited and published. The task of village reorganization and social service continue in Sriniketan. The Basic Training and Crafts School has been converted into a Multipurpose Junior Certificate School. We have organized an Adult Education Centre and refresher course for village teachers at the Vinaya-Bhavana. We have extension services for adult-education. We co-operate in the Community Development projects by allowing our experienced workers to take active part in National Extension Service. From December 1956 we have a Higher Rural Institute in the Sriniketan area which promises well to develop into a Rural University.

With the confidence that we have gained from the first five years' work in the new set-up as a centrally administered University it is now necessary to think of our future. It is best perhaps to reiterate our ideal as set forth in the first schedule of the Visva-Bharati Act. We do not want Visva-Bharati to be an exact replica of the other Indian Universities. We have faith in the programme set forth for Visva-Bharati by our illustrious founder and we do believe that one can achieve a high standard of efficient and all-round training by having a close association of the teacher and the taught. We may have to reorganize our studies in the various departments and emphasize the intimate relation between the disciples and the preceptor still more. Visva-Bharati of the former days had faith in this essentially Indian view of education. However, owing to the stress of circumstances it had temporarily to abandon its programme and had endeavoured to fall in step with the other Indian Universities. It had hoped thereby to escape financial embarrassment by catering to the needs of the students in a way familiar to the student community at large.

To accept this as a permanent feature of teaching in Visva-Bharati will be, however, to accept defeat and to declare our original programme of education to be essentially unworkable. We have recently given to this problem our careful consideration in the Academic and Executive Councils of our University and have chalked out a programme of reorganization of studies which has been approved by the Samsad of the year.

To carry out this programme in future we may have to approach the Government for a more generous grant. This will be one of the items on which we shall base our demand for aid during the next five years. However, we seem to have gained the public confidence. Demand for admission to Visva-Bharati is on the increase. Unfortunately, we have not much spare accommodation to offer. Students from distant and foreign lands continue to come but our residential quarters are often not suited to their needs. To maintain our humanistic studies at a proper level we have to organize our existing schools of studies and offer facilities for new studies. This means bigger maintenance grants and larger aids for capital expenditure on more hostels and staff-quarters. A new Library Building will have to be set up and a proper Administrative Block is to be erected to set free the Udayan for a Museum. The problems of sanitation and water-supply will have to be solved satisfactorily.

All these have gone into our demand for development grants for the next five years. It will not be out of place here to append a few comments on the need of introducing scientific studies of a fairly high level in the Visva-Bharati. It cannot be denied that a certain amount of essential scientific knowledge does enter into the mental make-up of a modern man of culture. Gurudeva himself had recognized it, and has advocated its teaching in Visva-Bharati. It would be a pity if science is deliberately left out in the endeavour to bring about a synthesis of the various kinds of culture of different lands and races in an ambitious programme of giving a whole education to the modern man and woman.

I understand that the Grants Commission will consider the demand for the Library and Administrative Block sympathetically. They have already come forward with a handsome grant to help us to solve our problem of water supply. I understand they have in view a Master Plan for Visva-Bharati. I hope that all we have asked for will ultimately go into the final picture. We welcome the Master Plan, but we pray that the Centre does not prescribe a Master-mould for educational experiments here. A demand for conformity to prescribed courses will limit the scope and hinder the spirit of freedom which is so essential to normal growth of Visva-Bharati.

Friends, I have briefly told you of our hopes and fears. I now request our Chancellor to declare the Convocation open.

Convocation address as Upacharya or Vice-Chancellor delivered at the convocation of Visva-Bharati on 15 January 1957, addressed by Jawaharlal Nehru, Prime Minister of India, and Padmaja Naidu, Governor of West Bengal.

Higher Education for Us

Madam Chancellor, Mr Vice-Chancellor, Ladies And Gentlemen,
I am grateful to the authorities here for having asked me to deliver the Convocation Address. This University is my Alma Mater — and I strive to be one of her faithful sons. Here were spent my student days as also the early years of my apprenticeship as a teacher. The invitation of our Vice-Chancellor is thus almost a command and I obey him with very great pleasure. But what to say on an occasion like the present? Words of advice would perhaps be deemed out of place — an infliction — and nowhere would they go far into young minds! I have, therefore, decided to narrate before my young friends bits of past history : of this University as it then was — when as fresh graduates, as they are now, we had ventured on a long and arduous trek in search of knowledge. Few had cared to traverse this unknown road before them. They had many hurdles to cross; they had numerous failures to contend with. But there were also a few successes which encouraged them for still further efforts. Luckily, the old difficulties have now almost disappeared. But the story may be interesting; it will enable you to compare the old days with the present and I confess outright before my friends that they go out now very much better equipped for their lives' work than we were when we set out quite ill-equipped for our ambitious journey. Those were the days before the famous Sadler Commission. The University had not yet opened the PostGraduate Departments, and the old regime continued. More than a hundred years ago Raja Rammohun had written to Lord Amherst asking for the introduction of the study of the Western Sciences in India. The University, a few years after, came into existence in due course. But for the first fifty years in the curricula of the University the principal stress was laid on the

teaching of English. English was adopted as the medium of instruction of all Arts and Sciences in colleges and in high schools. That was also regarded as the best way of advancing the cause of learning in the country. Our foreign masters had wanted intelligent Indians to help them run their administration and their offices economically; and so for many guardians living in the cities, it seemed that the broad way to soft jobs and easy comfortable lives for their wards ran through the portals of the University. This was clearly not a quick way to disseminate knowledge as the statistics of the progress of literacy in a hundred years would reveal. Many years before, the Catholic Missionaries of Serampore had thought of a better alternative. They were the first to erect a Bengali Printing Press and published mainly religious tracts. But they also helped spread literacy among the people. They also published textbooks of Science and Mathematics in Bengali for the beginners. Bengali books on Medical Science were current in schools and were regularly used by the students before the advent of the University education. The task of spreading knowledge was soon taken up by the Indian educationists. Iswarchandra gave us the textbooks and the keys to the traditional lore of Sanskritic learning, Akshaychandra revealed the wonders of creation, and there was soon no dearth of Bengali books in all conceivable subjects. In creative literature, after vain efforts to write in a foreign tongue, Madhusudan and Bankimchandra realized that to gain approval of the people and an easy access to their hearts one has to write in one's own heart's blood — in that native language which springs from the age-old yearnings of the subconscious soul and nourishes the conscious efforts of our people. If things had moved on this rising tide of national effort, if this University had then proclaimed the principle of dispensing learning through the medium of the mother tongue, perhaps our dreams about the advent of a new era in this ancient land would have come true much earlier. But the old Syndics and Senators were not convinced. They left teaching in the charge of colleges and schools, and busied themselves with the maintenance of standards by the rounds of inspection and examinations. But the spirit of times moved ceaselessly on, and soon came the rude shock of the National Movement. In 1905, people wanted economic independence. Lost trades had to be revived and new industries had to be started. But the University studies were not oriented to that end. It could not meet the sudden demand for knowledgeable men in Science and Technology. Immediate reorganization also was not possible. But feelings ran high : the University was condemned as a slave-factory in public meetings. Attempts were made to reorganize education within the province on a broad basis independent of government control. A National Council dispensing the knowledge of Arts and Sciences through the medium of the vernacular came into existence. Funds were also available from patriotic donors but beyond establishing a College of Technology no lasting results could be achieved. Gradually the political struggle moved on to other arenas and the task of reorganization of education independent of government control remained unachieved for many years. The old system of education stood condemned. But there was no satisfactory alternative to replace it immediately. However, the situation was saved and the popular demands were met to a large extent by the genius of Sir Asutosh who was then at the helm of affairs of the University. He understood the urgency of introducing more satisfactory methods of teaching in schools and colleges. The syllabus and the courses of studies were revised, and the teaching of the experimental sciences was given a prominent place within the University. Before 1908, only a few colleges taught sciences. As a result of the reforms, many colleges, both in the city and in the mofussil, undertook to open laboratories and provide the necessary funds for scientific training for which there were such urgent demands. While the schools and colleges tried their best to meet the demands of the undergraduates of the new age, the University had to come forward herself to shoulder the burden of higher education in Arts, Sciences, and Law. From a merely examining body, the University was thus gradually transformed into a vigorous centre of teaching and research. Sir Asutosh was the principal organizer and the leading spirit of the new movement. All went on very well for a time. The government approved his programme, and the people had confidence in his wisdom and foresight. The legal studies were re-organized in the University Law College. Funds were forthcoming for the installation of several University Professorships (e.g. in Mathematics, Economics, History and Philosophy). These few years of constructive efforts restored people's confidence in the University. Large endowments came from two Bengali lawyers, Sir Taraknath Palit, and Sir Rashbehari Ghose — land, building and funds. Sir Asutosh now took a great step forward and laid the foundations of the University Colleges of Science and Technology. There is however a peculiar condition attached to the Palit and Ghose endowments. The Professors have to be Indian scientists. Such suitable persons could not be found immediately for all the new University Chairs at the Science College. C V Raman, elected for the Palit Chair in Physics, was then working in the Finance Department of the Government of India and wished to have some time allowed him to make a clear decision. He was anxious to pursue his own researches in his leisure hours at the Indian Association free from all worries of a teaching post. Sir P C Ray agreed to be the Palit Professor in charge of the Chemical Laboratories only after his retirement in 1915. Professors D M Bose and Agharkar selected for the Ghose Chairs of Physics and Botany respectively wanted to be sent to Germany for their own researches. The University would have thus to wait for some time before the scheme for the higher teaching of science could mature. In the meantime dispute arose between the University and the government regarding the control of schools. Sir Asutosh took a fearless stand against the government in matters affecting the academic freedom of the University. In 1915 he was no longer the Vice-Chancellor but still remained Chairman of the Boards of Trustees of the Palit and the Ghose endowments.

The National Movement had thrown up many idealists imbued with a spirit of adventure. There were a few young graduates who had resolved to devote themselves entirely to Science and soon after 1915 M Sc examinees had approached Sir Asutosh for advice and guidance. The name of one of them had been sent for the post of a Lecturer in a government college in a different province, but was not considered on the ground that his qualifications were too good for the post. He wondered how he could possibly get along with his programme of higher studies. One of them had even the temerity to suggest that the University might now undertake herself the teaching of the various courses in Science which were prescribed but were not taught in the government colleges. It was difficult for the youngsters to guess what the great Sir Asutosh thought about their proposals. However, some got special scholarships sanctioned and direction to study certain special branches of science which had come into prominence. He asked another particularly to prepare a report on the possibility of procuring the necessary equipments locally in view of the World War — he was also to report about the probable expenses necessary in the first year for fittings and other equipments in case the University decided to open classes in Physics. Soon it was evident that the University would be teaching science subjects. Sir Asutosh had won over his colleagues who had doubted the wisdom of this hasty act. They would have preferred to wait till C V Raman joined, or till the interned Professors returned from abroad. But ultimately they were persuaded to support Sir Asutosh. Thus the Postgraduate Departments in Chemistry, Physics and Mathematics were opened even before the Sadler Commission had submitted their report. The members of the Commission came and went on rounds of inspection when regular class work was going on in all the departments, and apparently approved what was going on. It was a risky venture. But the young graduates of the University worked hard to make the programme a success. In Physics the details of the programme were worked out by the Lecturers. C V Raman joined a few months later. He found the routine work going on satisfactorily. He consented to deliver a few lectures but kept himself busy most of the time at the Indian Association where the Palit grant was spent for research purposes. Here some Lecturers worked under his guidance and earned their well-merited Doctorate degrees.

I have told only a part of the story, relying mainly on my personal knowledge. But in all subjects in Arts and in Science, at Ballygunge or at the Darbhanga Building, the beginning of higher teaching took practically the same course. After the Postgraduate Departments had come into existence the young Lecturers carried the principal load of the new task everywhere. The leader had seen the propitious moment arrive. He had confidence in the young scholars, in their capacity to shoulder the heavy responsibility of hewing new pathways to progress. His call had gone forth and the young people responded. By their devotion, and their enthusiasm, they made the scheme a complete success.

The government had not looked on this new venture with much favour. No grants were forthcoming for the maintenance of the Science Departments and the University had to rely on her own fee fund and on the endowments received from the generous donors for this purpose. But the transformation into a teaching University met the urgent needs of a nation growing every day conscious of its own powers. Students came in increasing numbers; they went out trained, and were either absorbed in the growing industries or they started themselves new and successful ventures. Original contributions began to appear. Within five years, young Indians had made their mark in the scientific world. Ghose's papers on the Conductivity of Electrolytes were published in 1919. Saha had his famous theory on the Temperature Ionization of Stars ready by 1922. Numerous contributions on different subjects followed in continuous stream. Sir

Asutosh's hopes were realized, and the name of the Calcutta University soon occupied a respected position in the scientific world. This position has been maintained, the good work has been continued by successive bands of devoted scholars for about half a century. This has not always been an easy task. Public opinion had undergone violent fluctuations about the need and the importance of University education. During the years of struggle for national independence education enjoyed a very low priority value. Whenever needed, and the call came, students ran out of schools and colleges. Teachers were no longer regarded as reliable guides and they preferred to follow the leaders of the national agitation who assured them that once independence was secured all difficulties would disappear and all problems would find quick and easy solutions. A new era of freedom has finally dawned. For Bengal however the chalice of bliss is tinctured with many sorrows and bitter memories. The old land, united by the traditions of centuries of a common culture, now stands divided. This has occasioned movement of people on a tremendous scale. The problem of the rehabilitation of refugees taxes heavily the resources of our administrators. There has been simultaneously a great demand for technical and scientific training. Schools and secondary education are no longer under the care of the University. Large-scale reforms of secondary education are under contemplation. But, for efficient working of the new plan, trained teachers would be necessary and the University has ultimately to supply a large number of well trained teachers to the schools. Besides, a large number of graduates demand higher training at the University. Their training and test also raise problems of great complexity. Many are refused admission for want of accommodation. And almost every year we hear of unseemly disturbances in the examination halls : an unsympathetic examiner has set a particularly stiff paper, or the questions asked are outside the prescribed syllabus. All these problems require a careful review of all University activities. A closer understanding among the teachers and the taught is eminently desirable. And now more than ever has arisen the need of adopting the language of the province as the medium of instruction in all classes within the University. This had appeared perfectly feasible to the leaders of thought in the province more than fifty years ago. It should again be considered carefully by our Senators and our legislators now. There are now a little over thirty universities in India and many more are bound to follow. As higher education is a Central concern, there is a school of thought which urges an approximately uniform standard for all the universities. They demand the working out of a master plan; they desire that all universities should adhere to it as closely as possible. The need of the same medium of instruction follows, according to them, as a necessary corollary. These planners thus urge the continuance indefinitely of English as the medium of instruction in universities as at present.

I have always spoken against unwise adherence to past policies. This foreign language has been a real hindrance to the rapid spread of literacy in the country. In an educational institution, it encourages cram and effectively damps all creative efforts.

It is time now to introduce the regional language as medium of instruction at every

level of the University for all subjects. Fifty years ago, this appeared a viable proposition to our leading thinkers. Now members of the University Senate and members of the State Legislature have to take up seriously the issue of giving to the vernacular its rightful place in education.

Convocation address delivered at the Calcutta University in 1962. Obtained by courtesy of Rathindranath Bose.

Man in the Scientific Age

Friends, first I have to record my gratefulness for having been invited to come to this symposium, and to be among you, all young scientists eager to explore new roads and to tackle all kinds of difficult problems by new methods.

So, I shall begin with an apology, the apology of an old scientist, an elderly person, who, when he began his studies, was educated mostly in the old school. Determinism still had its sway and, as our teacher used to say then, we were not so anxious to define our concepts as we were eager to get results. He used to say jokingly that when he was a student in Cambridge there was a joke like this : 'What is matter ? Never mind. What is mind ? No matter.' You know the *Principia* of Newton. Well, there he tries, not only very profoundly, to define the concepts, but introduces to us valuable methods. And with them, at first, we were getting results and we were happy. We thought that perhaps all the world phenomena would be investigated and analysed in the same way : starting from a set of axioms and following it up by methods of deductive reasoning, basing our views on certain differential equations with certain initial conditions ; thus we could, in principle, successfully explain all physical phenomena. But very soon, when we were still students, the quantum theory cropped up. Physicists began to know that after all there were many events for which the usual description on the basis of the law of cause and effect could not be applied and newer methods had to be devised. Ultimately we gave up Newton's determinism. At least physicists became more humble and gave up the old idea of giving a casual description of phenomena but satisfied themselves by a statistical way of analysing the results. I say this because certain questions have been raised here regarding the difference between the study of material phenomena and the study of human behaviour. I say even in the world of material science we have now given up the older idea of being able to follow more closely the unit processes and very often because of the reason that we do not possess sufficient intimacy with the processes concerned. We are directly concerned with results. And therefore, certain methods have been developed and older ideas have been given up. This new and statistical method not only permeates the whole of physics but also gradually finds its way into all the sciences.

There was formerly always a quarrel between the materialists, or in other words the natural scientists, and the social scientists or the philosophers. As somebody said here, we are very often talking about things we do not know what they really are. Instead, we have got to reconcile ourselves with the idea that even in the material world we are not so much concerned with the essential nature of things, as with how to record the results of experiments and our interpretation and applications of the results that we get. Our standpoint has been more or less utilitarian ; therefore, I feel, while admitting that there are essential differences between the M-concept and the P-concept, that the scientists naturally are concerned only with results. They need not even, in the case of social sciences, think of the correlation, as somebody wanted to find out, between the M-concept and the P-concept, but devote themselves exclusively in the P-plane as Mr Davidson has said, recording, at first, the description of the circumstances and also nothing than the results. So, I mean to say that we are not so very different : the natural and the social scientists after all in the sense that when we try to analyse things we are more or less concerned with what we think about them. The older idea of getting a sort of knowledge by more intimate understanding has now been dropped.

Of course, we, orientalists, are reputed to have a sort of a different way of getting knowledge, a sort of intuitive knowledge and if you want to understand how the tree grows, you try to think yourself at one with the tree, and then once having brought yourself in tune with a particular phenomenon, you will perhaps be better able to understand it than when you simply observe it from outside. I suppose in some biological sciences you still try to explain the behaviour of the cells in a similar way, bringing down to this elemental plane the ideas of human relationship with which we are familiar : attraction, desire, and volition etc. All these things come in. But while it remains always something personal with the scientist, his results are always expressed in terms of measured quantities similar to those of natural scientists.

When the problem of free will or attraction crops up, there are some persons who seem to think that in the elementary attraction between the atoms and the molecules there is just the beginning of a certain tendency which in the higher beings develops into a feeling of attraction or seem to see in the quantum laws a vestige of the free will. But whatever they might be, I am not very competent to talk about them. But what I want to insist on is that, so far as scientific method is concerned, they have taken a similar turn, by which not only the physical but also the biological sciences are now studied in the same way. Even when it is a question of studying human behaviour in the social sciences, we follow more or less the same statistical methods. Well, in statistics very often you have got a large number of elementary acts in which some will succeed, some will go wrong, and there is always a measure that we have to devise in order to record your impression about the large number of individual events that you study from a certain standpoint.

It was interesting for me when the speaker here talked about the judgement of various observers regarding certain events. Here a statistics of the observers would follow, as far as I have understood him, in which by well-thought-of statistical analysis the possibility of eliminating personal equations in some cases had been recorded. However, in such cases in intelligence tests, in the appraisal of student work, there remains always a larger margin of uncertainty. And during the long years of my experience as an examiner we often felt that when there are more than one examiner it is very difficult to correlate the diverse judgements eliminating the personal equations in the process. In our University of Calcutta we had a peculiar rule : if two examiners differ, refer it to the third examiner ; and then, when the third examiner's results are different from both, take the mean. But you see how uncertain is the possibility of eliminating personal equations or personal prejudices in such cases in this way. However, I am glad that attempts are being made, and very soon a way might be found out, suitable specially for your country, because the methods of appraisal of intelligence or evaluation of results differ in different countries. So every one will have to think out, basing his general ideas on the principle of statistics, as to how to proceed in any particular case.

We are thus in a position to say that, in contradiction to what we thought before, in the social sciences we may also study from the same scientific point of view, thereby meaning that once we have devised a method of measuring certain things, we can always subject all our results to the same statistical treatment as in the case of observations of material phenomena.

Well, naturally I was interested to note that the future of the society depends largely here, as elsewhere, upon the education of your graduates, and the problem of the relation of the essentially scientific and technological teaching with the humanistic one is the problem that has been raised here. But I have not been able to gather exactly the reaction to it of all the members of the symposium. In our country, also, the same problems trouble us and some of us felt that early specialization is a hindrance, and, as far as possible, during the earlier years education should be broadly based, neither too technological nor too literary. All the students should participate, as far as possible, in a general course of education in which eyes and hands should be trained. Endeavours should be made that gradually students master a way of expressing what they feel. As such a certain amount of mastery of the language will be necessary, just as they should be required also to have a broad and unprejudiced view of the physical and biological processes that are going on around us. Naturally that raises the question of the role of scientists, as to what the scientists may do in the future society.

I think that the scientist now claims himself to be a person who has got a greater knowledge and more intimate acquaintance with nature. He not only relies upon his personal experience but can fall back upon years or centuries of recorded experience of the human mind. And if we believe in evolution, then we may say that through thousands of years there has been a simultaneous development, a slow process but essentially the same in all the countries, whereby man has gradually woken up to his responsibilities and now, in the twentieth century, feels responsible enough to take his position as one who is able to guide his destiny.

Naturally the social scientists have got a bigger responsibility in such matters. However, the scientists should take up the standpoint that properly belongs to them ; they belong to the class of intelligentsia who have to be consulted whenever grave decisions are taken regarding education, regarding steps that should be taken to reorganize our economic or political life. This is a big claim, and I wonder whether the persons in power will agree to admit the scientists in the council of advisers. If, however, you, young scientists of Japan, contribute to study nature and human behaviour in the same way that we are doing now, I have great hopes that you will be showing to the world a novel way of reorganizing society.

The problem of education is a matter about which I always feel a bit confused, though I have been teaching for well nigh forty years. But the problems of communication of knowledge, the relation between the pupil and the teacher is to me yet something mysterious. How does your pupil learn from you ? He perhaps tries to imitate you. Your example will act as a leaven which produces the necessary ferment, but the pupil must find himself in the sense that as he grows up from boyhood or from childhood through adolescence into manhood, he should be able to utilize all the elements that the society provides for his development.

And here there are different ideas. I personally feel that in such matters, as in all matters, the process of giving as much freedom as possible is perhaps the best thing to do in the long run. You have talked about machines : now, by suitably coupling the influence with the result by means of a feedback mechanism you can make the machine almost simulate a conscious behaviour ; your missile may follow the enemy through his intricate military evolutions or your automation in the factory may regulate the supply of raw materials and at the same time test the quality of the products that are ultimately produced. It is something remarkable. But in all such cases, if one thinks a little, the best results are always obtained when the machines are so conceived as to allow maximum amount of freedom.

In this symposium, you have taken up very big problems. I was at first at a loss to find out what reactions would be. But being among you I have gradually educated myself and I am telling you of my reactions. I think the object of this symposium is not that you should have a sort of lecture *ex cathedra*. But in intimacy and in close collaboration, we will be able to come to a closer approximation to solutions of problems. Maybe your contacts and discussions will bring out new angles of vision of difficult problems and I hope that in your discussion, after these discussions are over, the active scientists, the politician, the social scientist and the legislator, they will all have a sort of intimate understanding of how things look from different viewpoints.

Naturally, society and the place of the scientists in society bring up the question of the ultimate future of the society : what purpose will we be working for ? There, as an optimist, I believe that the scientists of Japan will be able to educate their countrymen to the point of view that ultimately the best thing for the nation is also ultimately the best thing for mankind. You can always note the evidence of history : how shortsighted ways of looking at things, though they have given quick results, may ultimately lead to disaster. It is not only an experience of Japan alone but practically of many countries and we hope that people will become wiser after such events and they will set up a common brotherhood of scientists of all countries who, naturally as they possess a more intimate knowledge of affairs, are able to guess better about the probable results that may follow any course of action. They may thus be able to guide all nations towards a new era, an era when the different nations, though they may be on different levels of material progress, yet ultimately will be united in one common bond of brotherhood and unity. I often think that if we can really compare humankind to a family we should admit that all the members of the family may not be endowed equally well, but ultimately to the mother or to the head of the family they are equally loved and they have got equal right to the affection of the mother, though the one who is backward may not rightfully claim to have such a treatment. The different nations are in that position, and if the same feeling of goodwill can be cultivated among nations as generally is reputed to hold among a brotherhood, we can hope for the best and all our endeavours and our scientific studies may be directed towards a common purpose not only for the good of the motherland but also for the good of the whole human race. Thank you, gentlemen.

Address at a symposium on Science and its Position in Society, in Tokyo, Japan 1962.

The Mother Tongue

I have been involved with education for the greater part of my life. As students we nurtured in our minds the excitement that science, the cultivation of which has enabled the west to progress so rapidly, would soon be introduced in our country and that we shall dedicate our lives to bringing such things here.

Nearly sixty years ago when the Swadeshi Movement began, the great men and patriots who gave leadership to the country hoped to set up an indigenous system of education, at least within the province of Bengal, through national schools. Several years have elapsed and still education remains confined to a limited sphere in our country.

One thing comes to view when visiting other countries. The attempt everywhere is to evolve a system of education rooted in the mother tongue — a language that everybody understands. This policy is followed everywhere. In medieval times, however, a different language was used to impart education; only a few educated men had access to it. The limitation of this system was that the common people did not understand the language. They required interpreters, and whatever the latter understood was conveyed to the people.

But this system resulted in extremely slow dissemination of knowledge. Today people are attempting to make the country march ahead with the help of knowledge and science, to provide the common man with a wide range of material benefits; at the same time they have realized that the pace of progress will be slow if knowledge remains confined to a certain class. Now that we have achieved independence, we have to think out carefully how quickly education can be spread within the country. All those who have studied history know that if efforts are made, ignorance and illiteracy can be eradicated from this country.

I am reminded of my experience during my visit to Japan. I shall tell you briefly about that.

It is about a hundred years ago that Japan, reeling from the blows struck by the West, resolved to acquire all the knowledge and technology which have made the West so powerful. It is less than a century now since then and everyone is already aware of the achievements of Japan. Twenty years ago when Japan was defeated in the Second World War there was no end to her misery, but today, if one visits that country one can hardly imagine that she had to go through such a time.

The question that naturally comes to mind is, how could Japan rise to this present state of extreme affluence within such a short time? Education and progress are inextricably related. I was therefore curious to know the details of their system of education. In Japan children are sent to school for a minimum period of nine years. School education is free and is compulsory for almost all children. At the primary level the cost is borne by the municipality or the department concerned in the district or the state government as in our country. They share the major part of the expenses. The Government runs all institutions that teach the fine arts. In addition at higher levels the universities are mostly funded by the government.

At first I was under the impression that in Japan science and arts were taught in a foreign language. But on arriving there I found that I was wrong. I collected a large number of books, most of them in Japanese; hence I was not able to make out what they were about. Of course I received a few English books too.

The Japanese had convened a conference where educated people — philosophers, scientists and teachers — gathered to discuss what men should do in the modern age of science and what the future of mankind would be. Except for one or two foreigners including myself, the conference was attended mostly by the Japanese and the proceedings were in Japanese. Some of them occasionally spoke in English, but what followed was conducted entirely in Japanese. It will be incorrect to say that they do not know English, because when we spoke in English most of them understood us. However, they were not confident that they could express their thoughts clearly in English. So all the Japanese scientists and philosophers present there expressed themselves in Japanese. It was apparent that it was possible for abstract philosophical theory or advanced ideas of contemporary science to be expressed in Japanese and the people were keen to do so.

Compared to our Indian languages, Japanese has certain problems. Those who are better informed are already aware of it. In our languages writing and printing are done using a handful of letters of the alphabet, but this is not possible in the Japanese system. Apart from their own letters there are three thousand letters of the Chinese alphabet. Those interested in pursuing higher education have to learn them all. In our country a child can pick up the vernacular within a year or two, whereas an average Japanese child takes about six years to learn Japanese. Despite these problems the

Japanese language has reached a stage where every Japanese scientist or philosopher can express his thought in Japanese. This proved to be an advantage when compulsory primary education was introduced in the country, and there was no dearth of teachers —teachers who had already mastered science or philosophy or fine arts in the Japanese language and were competent enough to teach them.

This has contributed to the rapid spread of education in Japan. As a result, Japan could easily acquire all forms of knowledge. If we take the cases of Japan and Germany, the two nations in the world who have made rapid progress from a low level of knowledge to the highest position today — more than ninety per cent of the people in both countries are literate.

If we take a country to stand for the people of the country, not just the educated or the leader class, but the entire mass of the common people of the country, then only if they are educated can we call the nation fully developed. And then they would hold in their hands all the power to defend the country from any danger. Our country has a number of languages. But in every language, specially those recognized by the Constitution as major ones, the total number of educated people will be more or less the same as the population of Japan.

I feel that to start with if we confine ourselves to building our own homes without thinking of the entire country, that is, if we prepare the system of education for people in our respective provinces in such a way that every child gets compulsory education, then the new problems we are facing will easily disappear.

There is an idea in the world today that the nations will unite to build up a greater human community. But nobody thinks that to achieve this each country should adopt a specific language as the medium of education. After all, linguistic barriers do not count when there is a mental affinity.

It has often been said that without the use of the English language there will be no unity in the country, that we became conscious of our national identity only when we lived in subjugation.Hence that sense of unity should be maintained through the English language. But it is not correct to say that we had no sense of national identity before our enslavement.

Those who are at the helm of affairs in our country are perhaps not sufficiently aware of the strong feeling of unity within the country, or how firmly the bond of unity binds the people together. Only when people from various states of this country meet abroad, they note how striking the similarity is in their mental make-up, customs and manners and how different they are from a foreigner's. I have a strong conviction that the unity of the country will never be at stake if education is imparted totally through the regional language of every state.

In our recent history there have been some glorious as well as some shameful episodes. An analysis of the history of these shameful episodes reveals that the persons who were responsible for bringing disgrace were all educated in English.What I am trying to say is that just as there are instances in our recent history of which we can be proud equally there are instances which we should regret and be ashamed of. But if one were to look for the root of such regrettable events, one would find in many instances the activities of people who were arrogant, selfish and educated in English.

There I feel that English has never been a unifying factor. The root of unity resides in mental affinities. If we are responsive to the feelings of others, then we can be friendly with foreigners. On the other hand, lack of brotherly feeling can only go on to build another Pakistan. And the two men who built these two nations came from the same province and both had English education. Attempts were made to awaken the mass through English for many years but they have not succeeded.

Later when the father of our nation came to us, he chose to speak in a language understood by the common people though he could speak and write English better than many. With independence we have now a new responsibility. The fruits of freedom should not be confined to a handful of English-educated people; they must reach one and all.

Those who think that the doors to knowledge and freedom will be shut if English is taught less, are grossly mistaken. They think that India should be surrounded forever by high prison walls and that light will enter from the top to reach only the upper echelons of the educated people and they will transmit that light as they receive it to the ignorant people at the lower strata. It is difficult to achieve progress in this way. It is not advisable to impose the responsibility of the entire population on a small section. People of the country should bear their own burden themselves. We have not made much headway in achieving this in the last fifteen years.

I think that your demand of banishing English from the prime position it has occupied so long should be given serious consideration. Think carefully how this proposal can be implemented in different provinces. My arguments for removing English arise purely from an urge towards self-defence. I strongly feel that this extraordinary emphasis on English would impede the spread of education in the country. Those who choose the vocation of a teacher must make a lifelong effort to see how best one's thoughts can be expressed in one's mother tongue.

We appeal to the educated people. Let them try hard to usher in a new era in the country. Let them figure out what the new medium of instruction should be. Let them introduce such a system in every state as would remove illiteracy and ignorance from our country. A word of caution is needed for the fanatics. Our call for a new era will never materialize simply by substituting one language by another.

Our priority is to educate our countrymen. What I want is that knowledge which is useful to them, which keeps them free from disease, makes them prosperous and provides them their daily bread, should be made easily available to all.

This calls for a sustained effort. The problem will not be solved simply by taking part in discussions. I hope the participants speaking different languages will keep this in mind when they return to their respective homes. The success of this conference will depend on our realization of two points. First, we can achieve our ends quickly only if the mother tongue becomes the medium of instruction; secondly, our regional attachment does not depend on language, it is something rooted in our minds. If you can carry these two observations back with you, then I will consider that your efforts have been worthwhile. Through mutual interactions we hope to become aware of two more facts. First, our unity is not superficial but lies deep. This has not come about by speaking one language. Secondly, there is no problem that cannot be solved through unified and earnest effort.

As I have said before, the integration of our country cannot be brought about through a special language. It is necessary to make it clear that the unity is based on mental affinity. If we wish to build something, the feeling of mental closeness among the workers is essential. Otherwise, what we will try to achieve in different provinces will not hold together as parts of a single structure.

We have tolerated a foreign language and a foreign rule long enough. If this was the way to establish unity, then it could have been attained a long time ago. As it did not occur, voices are now being heard against the acceptance of the constitution and its directives.

Our national heritage is the result of centuries of endeavour and no matter how old it is, we wish to cling to it with our life. Respect for tradition is the foundation of national unity. At least from my own experience I have felt that anybody willing to subdue his self-interests and travel within the country can see that this mental kinship existed though it may not be present in the political sphere. The sages of our country, our pilgrims, our ordinary Hindu boys and girls and even the people of different religions living in the villages have all realized the existence of a common bond that is not to be severed easily. However, sometimes conflict of interests can overwhelm the sense of unity. To keep alive the spirit of integration requires special dedication. Unity cannot be achieved through words alone.

How does one pursue this dedication? It seems to me that people in different provinces have to use their vernaculars to try and understand the nature of our heritage and the common bonds which held them together in the past. The diversity of nature is such that every moment man is faced with a demand to create something new. So the harmony of our country that we wish to build today has to be more complex, profound and powerful than the harmony that was felt in the past. This is my conviction.

We can hear the clarion call : the time has come to know thyself. Let the people from different provinces draw close together and come to know the taste of freedom, what we had lost and what we have gained today.

We have to follow the paths which other nations have adopted to reach the peak of development and move fast. To do this, we need to take the principal message of the modern age to the minds of our people. Prehistoric ideas have to be discarded. The victory of man over nature and the consequences of it that have been felt in many countries have to be realized not only by those whom we regard as educated but conceited but by everybody.

If this is not done, then we cannot keep up with the rapid pace of progress in other countries. And if we are really moved by the term 'Bhāratvarsha' and if our hearts bleed for our countrymen, then we do not need the aid of any foreign language to strengthen the bond between us. As a matter of fact, by the time the English-educated leaders who had once fought to drive a foreign power away from our country themselves gained power, a tendency to drift apart was clearly evident.

It should be easy to create a sense of unity and overcome the separatist mentality through the use of vernaculars. It is not enough to preach national integration from a regal platform. We have to talk this message of unity from every home. Let us not rest content with the fact that we have provided for education — everyone has to realize that it is our common responsibility! This is why we need our mother tongues, we need arduous practice and we need to keep our minds pure. Instead of clashes and conflicts, let us talk about love.

Those of you who realize this, do carry on your work patiently within your own limits. In our childhood we used to say, let us carry on as we have God above us and we have nothing to fear. If you have a clear vision of unity in your minds, then you have nothing to fear. Let the movement that you have started continue to grow. Let our people be conscious of the degree of their responsibilities. Let us hope that compulsory education will be implemented soon so that every child, every young man and woman will have the scope to develop as complete human beings. Let us realize that in the present age into which we have been born, there is great danger in 'little learning'.

I have really nothing more to say, because speaking is not always a good way to reach others. As a writer once said, the aim of language is to conceal our thoughts. There is a belief in our country that if we think of one another's problems intently, we shall find the right answers from within ourselves. I hope that will happen, and your movement will be successful in every sense.

Address delivered in Bengali at the 'Angrezi Hatao Conference' [Banish English Conference] at Hyderabad in October 1962, included in S N Bose, Rachanā Sankalan, *Bangiya Vijnan Parishad, Calcutta 1980. Translated from Bengali.*

Education and Science

Mr Chancellor, Vice-Chancellor and Friends,

I am very glad to be with you all on the Convocation day and to take part in your rejoicings. I congratulate the new graduates of the year and look eagerly forward to know about your young University. It is a wonderful idea to have a residential university at Ranchi. Its beautiful surroundings and its mild climate have always attracted visitors. It has been famed as a haven of the East for mentally ailing and depressed people. A vigorous Agricultural College has now started in its neighbourhood and a research institute of lac has been functioning for some years and you will soon have a factory for heavy engineering tools not many miles from the centre of the town. All these are important developments and will certainly benefit your graduates. However, these important institutions are all of very recent date. Fifty years ago there was only the University of Calcutta to take care of higher education in all the eastern provinces of India. It prescribed courses of study, held examinations and granted degrees to successful candidates. The actual teaching-load, however, was left to the care of college and school teachers. Our recently acquired independence has now increased the tempo of progress. We have now five Universities : within its own jurisdiction, each University superintends and organizes higher learning and research. It selects its own teachers and with the resources placed at its disposal by the State it tries to obtain the best possible results. However, it is not always wise to have mere replicas of older institutions. Every time we have occasion to start afresh a new venture, it is desirable to take account of the results achieved so far and to assess how far the older Universities of India set up after western models have been able to bring about essential improvement and achieve enduring results. I have been all along a University teacher. It had always distressed me to watch the low yield that our earnest endeavours produce in the field of education. If we take an honest and frank stock of our achievement we have to confess that since the introduction of University education in our country, in about a hundred years we have not been able to produce any spectacular results. A large percentage of our people are still illiterate; and in all our attempts of rapid industrialization we have still to lean heavily on foreign support. We have now arrived at a critical moment of our country's history. We have to remember that we can no longer afford to linger peacefully in our nation-building work, but we have to reorganize ourselves quickly if we wish to preserve our self-respect and our recently achieved independence. It is therefore necessary to note the results which have been achieved in a hundred years in other eastern countries and try to analyse the causes which have worked against our quick progress. It will be needless to say that as a nation we are vitally interested in the programme of education. The future of our country depends upon its citizens. The universities are working for the evolution of properly trained men and women. We cannot assert that the advancement of learning and knowledge are the only ideals of our universities. A modern university in India has to think of and to work out a sure way which will enable them to evolve a method to train our future citizens, strong in mind and body, who will be able to resist the influences of disintegration, conserve their forces and intelligently utilize the natural resources of this land for the progress and upliftment of our countrymen. I have mentioned illiteracy which still continues to torment us. Recently there appear other disquieting symptoms which have developed within the universities. Our students and our teachers have often been at loggerheads and the authorities of various universities often pass through days of torment and agony due to a widening gap of misunderstanding and a lack of discipline amongst students.

Students are now sent every year from India to different countries, to universities, factories and industrial establishments in foreign lands. They go abroad for studies in the various branches of mechanical and electrical engineering. But very often the Indian apprentice is found to be lacking the preliminary training which would enable him to absorb rapidly the modern methods of work. An illustrious Indian educationist who has returned recently from America reported that Professors and experts there had remarked to him that our boys are often considerably nervous for their unpreparedness when asked to handle modern engines and machineries. In our own country an analysis of the results of the university examinations would reveal a large percentage of failure in every test. What a tremendous waste of our national effort! I have often thought that all these shortcomings and failures are not really due to anything essentially lacking in the quality of the materials we handle in the student community. On the other hand, given proper guidance and facilities Indian students do shine and show their aptitude and efficiency in all directions and in every field. I feel this loss of time and effort can be avoided if only we analyse and discover the shortcomings in our methods of teaching, and set about rectifying them resolutely in our universities. Some time ago I had occasion to suggest that time has now come when we should revise our method of teaching and utilize the mother tongue as the medium of instruction at all stages in schools, colleges and universities. As an old teacher who had occasion to come in contact with hundreds of students, I am convinced that considerable amount of time could be saved and a better foundation of scientific knowledge prepared in the minds of our students, if at every seat of learning teachers and students cooperate and frankly discuss their problems and work together to solve their difficulties. However, if a foreign language comes in between them there is bound to be considerable handicap. Often an enquiring student is not able to express himself properly nor is the teacher sure that he has been able to make his pupil understand all that he wanted him to do, if he always has to use English as the medium of conversation and teaching. I am convinced that this present method encourages cramming, does not work for a proper understanding of the essential facts of the subjects taught. My opinion expressed with a certain amount of reserve had however raised controversies all over our country and there are many who still think that it would not be to the best interest of our country if we abandon English as the medium of instruction in the universities. A very eminent educationist has put forward the view that if different languages are used in the different universities of India it will hinder 'academic mobility' and thereby work against a free exchange of thought and ideals which alone can eradicate aggressive provincialism. Though our chain of slavery has slipped away, it has left behind the golden thread of the English language, which, as many venture to suggest, has worked for our integration and has helped to re-discover our souls.

These people still pin their faith on the medieval ideals of the University. According to them, the material needs of the hour should not be magnified unnecessarily and oblige us to change the character of our universities or our principal curricula. They think that we should continue to give prominence to the eternal verities, to the philosophy which our forefathers have advocated, and continue to depend upon the study of the classics, on Plato and Aristotle, on the humanities, languages, and jurisprudence as the principal items which should provide the universal background of knowledge to all our students. The importance of scientific studies should not be urged beyond a certain limit. They recognize the usefulness of science. But a modern scientist is a suspect. They believe that science has nothing to offer which would replace the old ideals. Scientists are useful perhaps as builders of efficient tools which would help men to establish welfare states; but they still maintain that culture and purposeful life in a welfare State can best be built on ancient philosophy which, many maintain, meant principally non-violence and *ahimsā*.

In a democratic country anyone is free to have his own views about the lessons of history, and read and interpret them in his own way. Thus it is also permissible to have an alternative interpretation of the decline of Indian civilization.

The 'other-worldliness' which was consistently recommended in our seats of learning, in Nalanda and Taxila, and which formed the backbone of our ancient culture, set a high premium on the sleepless contemplation of the eternal verities — the individual would thus realize the knowledge of self and recognize the world as a very temporary halting place, and this realization would make him yearn for release as soon as possible from the trouble and trails of this life. However much we may extol the philosophic heights of these speculations it cannot be denied that a persistent propaganda that the world is a place of temporary stay for the individual who should think of his own salvation as the principal aim in life has its repercussion in creating a carelessness in all mundane matters. The Indian thereby soon lost grip on the affairs of the world, and was soon overcome and subdued by the barbarians.

The lessons of the centuries should teach the modern Indian not to lay too much emphasis on individual salvation. It is better perhaps to look on life in a different way. The weal or the woes of the individual should not be made the principal object of philosophy, but the men in any country, closely knit by a common heritage, may be regarded as taking part in a relay race, where the advancement of the country and the advancement of the national flag is more important than the fate of the individual runner. Though successive generations toil, and break down in their effort to raise the standard of life, to conquer poverty, disease and death, the important thing that matters is how far such efforts have succeeded for the betterment of the country and the world. It may be urged that this alternative view is also Indian, though it may ultimately deny individual soul at the ultimate reality, and preach the sovereign rule of *Karma*. The 'materialistic' philosopher may also urge that this 'other-worldliness' provides a good breeding place for inefficiency, selfishness and greed. If a man is anxious for his own salvation he will take refuge in a cavern, and go away from the abodes of men. He will thus be saved, and allow the devil to snap up those who linger and toil against earthly ills, and try to procure material prosperity for the community and his country.

While this philosophy flourished in our country, the neglect of a serious acceptance of life by our first-class thinkers brought second-rate petty people into prominence who gave lip-adoration to philosophy but actually indulged in jealousy, squabble and internal strife. As a result foreign invaders came, often by invitation, and established in India centuries of slavedom. However, it is a hopeful sign of the times that we are waking up. There are alternative solutions, and a harmonization of the worldly duties with the other-worldliness may be brought about. Our country has recently celebrated the centenary of the great Swami Vivekananda who bravely put service of suffering humanity even before salvation. May we all remember his message, and gird up our loins to raise up the wheel of the car of humanity from the mire in which it has got stuck and sunk!

I had recently an opportunity to visit Japan, where practically at about the same time as in India efforts were made to introduce the western methods of education, and Japan has now transformed itself into a modern State and its progress causes universal wonder and admiration. I therefore eagerly accepted the opportunity and went to attend a symposium in Tokyo about the place of science in modern life, arranged by the Physics department of the Tokyo University. Here I met many scientists, mathematicians, physicists, biologists and philosophers, mostly Japanese, but also a few foreigners, e.g. a philosopher from America and a mathematician from Yugoslavia. I had thought that in such symposia we would utilize a foreign language but when I arrived

I was told that while most of the scientists of Japan understand English and possibly a few more foreign languages which they often read, the training throughout the length and breadth of the country is based upon the Japanese language and I should make myself ready to hear Japanese principally spoken in the symposium. I should however have the benefit of an interpreter who will interpret the principal points urged by the different speakers in the symposium and when my turn would come the Japanese. Evidently this method worked and I was surprised to see that a most intricate and abstract discussion about the application of science in a modern State was carried out easily in Japanese and when we the foreigners spoke they understood our thoughts perfectly well and put forward their pointed criticism evincing approval or dislike of our theses. I also found Indian students in Japan. I visited the universities where most modern things in Physics were being taught in Japanese side by side with many daring and original experiments rigged up by the Japanese students. In the universities the research scholars discuss science among themselves in their mother tongue. Obviously they use many loan words but they are not ashamed of this habit. I discovered that a book on the effect of the nuclear explosion which was written in English by two Indian scientists has been translated in Japanese and I was told that this book has sold well, about three thousand copies in six months. Ordinary Japanese who perhaps can read only Japanese are anxious to know about the effect of nuclear explosions and would perhaps trust the impartial Indian view more than others. However, in our own country these eminent scientists have continued to write in English and thereby keep about 80 per cent of their own people in ignorance about the danger of nuclear fallouts. I have been also in places where the Japanese students work and try to rebuild what had been destroyed during the last war. The war has been terrible for them. The victorious allied soldiers have broken down their tools of research, and thrown them into the sea. For a time all research institutes were closed where anything with even a faint suggestion of nationalism could be detected. But all these changed after all, and they have now found opportunity to reconstruct during the last seven years what Japan had lost. I was told that most of the research instruments and even the nuclear reactors are manufactured and erected locally by the Japanese themselves. In research institutes students practically build all the apparatus and equipments they need with parts manufactured in their own country. This was a refreshing sight for me who always believed that wonders could be achieved in our own country if only we could utilize our resources in the most efficient way. Japan furnished me with an actual demonstration.

I have returned from Japan firmly convinced about the necessity of adopting the language of the provinces as the medium of instruction in the University. Perhaps in Bihar, the movement for adoption of Hindi as the medium of instruction in all stages in schools and colleges has already begun.

It has often been urged that the absence of suitable technical terms in Indian languages may prove to be a hindrance for the study of the sciences. I am not a purist, and would welcome the continued use of technical and scientific terms in English. If our boys can understand them easily, they will continue as loan words, and I hope they will ultimately be absorbed and enrich our national vocabulary. Many words which were originally of foreign origin are now current in all Indian languages, and are easily understood by our common people.

I hope with the progress of scientific education, many more words could be pressed into service, and ultimately suitably transformed to suit the genius of our people.

Very often the labour of translation of scientific terms would prove to be a vain effort — railways, restaurant, kilogram, centimetre, wheel, lathe, thermometer, pressure gauges, boiler, cutters, electrons, atoms, bacteria, fungus, differential co-efficient integration, are understood by almost everybody; paper, chair, tables, are now in daily use. I need not multiply instances, and I hope the teacher and his pupils will easily come to a mutual understanding when they discuss their subject, and it will be really the duty of the teacher to keep himself abreast with the recent developments in this task of adopting the language for the popularization of the scientific outlook.

A scientist is often forced to join issues with his opponent, and carry on wordy contest — but he is happy when he can set up a suitable experiment whose results would help him to demonstrate his propositions. I feel this very desirable that somewhere in our country, the experiment of running a modern university with a strong bias towards sciences and sociology, where the medium will be an Indian language, be launched. With an eminent scientist as Chairman of the U G C, I have every hope that the scheme will find favour and good support.

In a residential university of the type I have in mind, free intercourse among students and teachers would be easy. If we are all united in our common effort to raise high our national flag in all the spheres of life, the petty misunderstandings and senseless jealousy should easily disappear. The ideal of the modern university should not be only learning and advancement of knowledge. It should preach the essential unity of men, irrespective of caste, creed and religion, of status, wealth or origin.

Man is the ultimate and the supreme truth which we should realize and I would hope that in one of the universities of Bihar, this supreme realistic view preached by the ancient poet of the land should be accepted as a symbol and as an ideal. 'Man is the truth above everything else' (*Sabār uparey mānush satya*) is something to which all western countries are also waking up.

A modern Indian university will I hope be a centre of an age-long sacrifice (*yajna*) where the teacher-priest and the student acolytes would sit together, and try to learn and realize for themselves what everywhere men have realized after centuries of strife, sacrifice and bloodshed. May there be a place found for such penances, and when the final libation to the supreme truth 'man as the ultimate reality' be poured, let there be a reawakening throughout the length and breadth of our land. Let us thus realize one of the fervent prayers of our great poet Rabindranath :

Where the mind is without fear and
the head is held high;
Where knowledge is free;
Where the world has not been broken up into

Fragments by narrow domestic walls;
Where words come out from the depth of truth;
Where tireless striving stretches its arm
towards perfection;
Where the clear stream of reason has not lost
its way into the dreary desert sand of dead habit;
Where the mind is led forward by thee into
ever-widening thought and action—
Into that heaven of freedom, my
Father, let my country awake.

Convocation address delivered at Ranchi University on 11 May 1963.

On his Seventieth Birthday

Respected Chief Minister Prafullachandra, my dear friend Humayun Kabir, Dr Deben Bose who has been like a teacher to me, and all those who have loved me and others who are present here — my greetings to you all.

Seventy years of my life have gone by. One could say every day now is a gift of God. The allotted time span of a man's life is over; the time span over which a man can depend on his own strength to carry on scientific research and social work is also part of the past. But I still have time, perhaps to repent a little for the wasteful activities — of which one or two will find mention here — a little less of which might have allowed for more valuable work to be done. But even the Omniscient cannot tell now if I could have really done something useful. For although psychology deals with how a person's mind works, one cannot always definitely perceive what one's mind is inclined to or what one ought to devote oneself to. In the old days in such cases it was the guru who enlightened one — even that depends now on the mercy of God.

As I was saying yesterday, we started our life at a time when there was a great movement raging through the country. Today, I recall that the few of us who were then the first to feel inclined towards science and were able to achieve one thing or another for the country, all of us went to college at the same time. Many of these friends are no more. Their names have been recorded in our country's history. Not only did they make a name for themselves in science, they also went on to be successful in several constructive activities; Bengalis will never forget them. Particular mention must be made of Dr Saha. On the one hand he changed the character of the Indian Association for the Cultivation of Science founded by Mahendralal Sircar, and on the other hand, added to the University of Calcutta as an integral part a faculty for the teaching and the research of a special subject — the faculty that now bears his name. It was through Dr Meghnad Saha's initiative that the study of nuclear science began in the university, and then in a different way all over India at different centres.

The First World War started soon after we had learnt a few things as students. We heard of many new discoveries after the war ended. We found that both sides had made immense strides in science through competition. We were not aware of them for a long time because in those days not many scientific journals came to India, and our prescribed textbooks also did not carry such information. I remember how eagerly we looked forward to the return of Dr Deben Bose who was interned in Germany for many years and was gathering much information, in the hope that he would bring us news of the latest progress in various areas of science in Germany during the war. I was not yet familiar with the German language, but Dr Saha had studied a little German and acquired his Intermediate Certificate. There was no end to our aspirations. We went to Dr Brühl's house, searched through his library and collected many useful books which Dr Brühl himself might not have read. We were completely enthralled by the information we gleaned from the writings of Maxwell, Boltzmann, Planck and many other scientists and were totally engrossed in them. At this point of time Dr Deben Bose arrived. He brought with him the reports published in Germany to mark the sixtieth birthday of the great Planck. They contained brief accounts of many new aspects of the new science of which we had no idea. We pleaded with him to read them out to us and we tried to translate (particularly me since Dr Saha was not always present) every word of those accounts into our own English. This yielded good results. Along with the new scientific information that I gathered, I also came to be acquainted with the German language at first hand. This gave me a lot of confidence. Some time after the 1919 solar eclipse a startling news appeared in newspapers in our country as well as elsewhere in the world. Einstein's theory had been completely vindicated. We were all amazed. Everybody was eager to know what new things Einstein had to say after Newton. We worked hard to translate those papers — partly Dr Saha and partly me. It took the form of a book. The preface was written by our friend Prasanta Mahalanobis who was then teaching relativity at Presidency College, Calcutta. The book was published by the Calcutta University, a fact which they must have forgotten by now. The book was used in the East for a long time. Then some time before the Second World War German racism assumed such monstrous proportions that all non-Aryan races were driven out of that country, resulting in riots in many cities in which knives were frequently used in secret. My guru, Einstein, who was then not in Germany, protested. As a result, he too had to leave Germany. Our book also went out of circulation thereafter.

When man begins to regard nationalism as the highest treasure of life, his sense of duty and his norms get distorted. The same thing happened in this case as well. I was then in Dhaka, teaching at the University. I had a few German friends there. They said, 'What has your friend done?' I considered it my greatest honour to have Einstein referred to as my friend. For the paper for which I am known might have been triflingly short but Einstein himself had translated it — very few people would be able to boast of such an achievement. In fact, he wrote two papers within three months of reading my paper, using it for all elaborations and extensions possible at the time. Curiously enough, in one of the papers he refers to me as Dr D Bose. I found that quite often the surname Bose meant one family to foreigners. Shortly after this I had the opportunity to go abroad. His words of praise became my passport there. As a result, I could with little effort meet and discuss science with such stalwarts as very few people have had the good fortune to. Among them were Madame Curie, Professor Langevin, even Professor Gehrcke and also others who have become immortal in the world of science. I could not only speak to them about their work but also had access into their well guarded fortress-like laboratories and familiarized myself with secrets that were not usually accessible to foreigners. What is more, I did not have to pay any deposit to borrow books from the State Library of Berlin, the Staat Bibliotheki. On Einstein's letter of recommendation I could borrow three or four books at a time, just like any university professor of that country.

Of course, those were the days before Hitler. My sojourn in Europe lasted for two years. Naturally I remember those days particularly well. We project ourselves as people who honour learning, but I have seen for myself the extent of the respect really learned people in Europe used to command. Things have changed since then. Those who are aware of the facts of history know very well how scientists and professors were harassed by the Hitlerite racists. Some of us had aspired for a level of scientific research that would establish a tradition of science in India, so that she would be recognized and respected everywhere not only as the land of the Vedanta, destined to remain for ever a poor relation of the world-famous Aryans, but as a contributor to modern civilization. The Germans also have great regard for our Upanishads and philosophy. They believe that there is a close relationship between the ancient Aryans of India and the German race of today. At the same time they seem to think that the present day Indians only carry the Aryan name but have no real connection with that great race ; like the present day Greeks, who are descendants of people who lived in the hilly tracts of Macedonia and are not the inheritors of the ancient Greek tradition. At times I had the feeling that similarly the Germans looked down on us with charity. But coming to think of it, do we also really believe in what our forefathers had said ? For instance, we claim that our feeling of equality for all living things is a part of our tradition, yet at the same time defend our caste system as the great solution to the world's problems. There are many who are of this view, including well-known philosophers who will explain the issue very well. While they generally advocate the four *varnas* they are indifferent to the various kinds of spectra that can be produced now which is a modern phenomenon. We are very proud of our Golden Age but we feel reluctant to talk about the backward peoples who lived in this very country at that time and the way we treated them. Still, sometimes these truths get exposed. Take the case of the non-Aryan Shambuk during the reign of Rama. The poor fellow had a tough time in Rāma-rājya. He wanted to acquire through rigorous prayer the same things that the Aryans had made use of to develop themselves. As a result, apparently there was a great drought in the land of the Aryans, and they decided that such a man should not be allowed to live. Therefore he was sentenced to death, even in the reign of the good king Rama. This may sound somewhat irrelevant but I am using it to give you an idea of what we had in mind as students — it is not entirely true to say that we wanted to bring back the same good old days in our country. Actually, most of us wanted to see the country free, and ideas were afloat as to how to get rid of the British.

Like the more recent 'Down with English' movement with which I got involved, we were taken up with the slogan 'Down with the British' in those days and got carried away in various directions by various currents of thought. Some of us were convinced of the need to educate the masses of our country, to create in them a sense of belonging. But this feeling had to grow to be a part of them through thorough education and not merely because somebody in authority like a guru or a zamindar said so. With this in view we had organized some mass education centres in Calcutta — but those night schools were short-lived because the British thought they were training centres for bomb-making. So they had to be closed. That chapter is likely to have been forgotten by now. Then there were those who believed in the power of the bomb, who said — look, don't indulge in such childish things. Let's drive the British out first and then you will see everything will be fine. Now the British have left. Many of my extremist friends have also died. Those who are still around seem quite confused. After all, educating the masses was not as easy as they thought. Sometimes I wonder if some of our old values like our feeling that we had to have the same feeling for all had really persisted, then would we not have been able to nip in the bud many of the newer dangers that are looming in today's troubled times ? One cannot be sure. Even today there are astrologers in our country who believe that whatever we do or whatever happens is pre-destined — our horoscopes tell us not only about this life but about many previous and future lives as well. This is a belief deeply ingrained in many of us. Even a Muslim friend of mine had once gone to Benaras to get his horoscope prepared and found that even the first letter of his wife's name got revealed. He came back very upset. He asked me : How did they do this ? An Indian scientist is often faced with such ticklish problems. For instance, is it possible to reconcile the idea of many births with scientific evolution? The scientist turns to the philosopher who merely smiles. The scientist feels the urge to proclaim that scriptures, philosophy, even religion are not of much consequence. They would kill all initiative in man to work. If you say, no matter what you do, what is going to happen eventually is predestined, nobody would have any initiative. Nobody would even attempt to mitigate the painful tribulations we encounter today. It would be enough to offer prayers at the Kalighat temple or Tarakeswar. The theory of repeated births is a great riddle to scientists. Many of them are breaking their heads over it. I am reminded of an incident. I had once put the same question to a holy man. I happened to have great respect for him. My friend, Bidhubhushan Roy, considered him to be a saint with supernatural powers, for he firmly believed that it was by the holy man's grace that he survived after he lost consciousness in the Science College from some dire disease. Both of us asked him on a suitable occasion, 'Baba, is rebirth a fact?' He said, 'Yes'. But we were not convinced. We kept asking him, 'You say rebirth is true, but have you really found this out for yourself ?' He said that he was told by another yogi. Neither Bidhubhushan nor I was satisfied. We persisted with our question from various angles. Finally he broke into a song, went into a trance and tears rolled down his cheeks. The song went like this — 'Where have you brought me, mother? My entire body is covered with wounds inflicted by the nettles and thorns of this forest. Mother, when will you take me back into your arms?' We came back disappointed that day. Since then I have occasionally felt that the Bengalis today have been wounded in various ways. Perhaps the mother will take them into her arms one day, but before bothering her, it would perhaps be more fair on our part to see what we can do ourselves to alleviate the pain. The thinking and self-confidence required to do this do not necessarily imply that everyone has to become an atheist. One has only to remember that the world has changed a lot over thousands of years. The environment in which happiness once overflowed in our land from prosperity and abundance does not exist any more. People who lived then would not have understood our current problems. Naturally it is futile to look for solutions in the old modes of thinking. Every century throws up new problems to a nation. Finding a solution to these problems, these imbalances, is our responsibility.

The fairy tale prince had one day asked the Sphinx : What is the future of man and what should he do ? Man ought not to have been born, replied the Sphinx, and now that he is born, he should die as soon as possible, that would be the best thing for him. The Sphinx has disappeared but her terrible cynicism springs to our mind from time to time, specially when man is in deep trouble and cannot find a way out. But history indicates that man's attitudes keep changing. It is true that hatred and violence have not abated — man keeps on using firearms and killing — yet efforts are also made in the midst of all that to give relief to the distressed. People are spending money from their own pockets, saving from their own produce to feed the famished. In the utter darkness of despair what holds out hope for a better future for man are perhaps these small incidents.

For various reasons perhaps only a small part of our youthful dreams have turned into reality. Nevertheless, Bengalis cannot afford already to be weighed down by pessimism and think that we can do no more than just turn the dreary pages of our life, or that everything is pre-ordained, there is nothing new for us to strive for. We badly need a change of this attitude. After all we have produced Acharya Jagadischandra, Prafullachandra and Dr Saha who dedicated their entire lives in the service of science. So it is hard to believe that getting a degree or a job will turn out to be the last goal for a Bengali, or that Bengali idealism will be reduced to the demand for the reservation of a certain percentage of government jobs for them. It is not a new thing for Bengalis to be concerned about others. They have often thought nothing of sacrificing their self-interest for the sake of giving concrete shape to a higher ideal. This makes me feel that we can expect great things from them which can help in the solution of national problems. I believe that the Bengalis do have this universal outlook. The history of their heritage will bear me out. My respected friend Prafullachandra is the Chief Minister now. He had done a lot in the cause of society early in life. I think his experience will be of good use to our country and our society. Some cynics might say that the office of a king made even Lord Krishna forget about Vrindavan when he ascended the throne at Mathura. But I am an optimist. I think that a Krishna of the Kaliyuga need not act in the same manner. The humanism and love for all mankind which the Bengalis have cherished is not a quality imported from the West. I have read a story about the great Rammohun Roy, who heard that a certain country in South America had achieved independence and, delighted, he threw a party at once. I would not read it as merely a publicity stunt because the joy that he shared with the invitees ended there — the news did not appear in the next day's papers. Nor are we aware that the Raja received any award for it. Perhaps this feeling of kinship with humankind is shared by everybody — but it is suppressed in practice by petty self-interest.

I have gathered varied experiences from my visits to different countries. I had to move around in Dhaka during communal tensions between the Hindus and Muslims. I found that affection and sympathy between the Hindus and the Muslims were not entirely wanting. Of course, we tend to make a lot of fuss about religion. So I am scared of the religious kind and prefer to give them a wide berth when they are particularly vocal about it. It seems religion does not have such a dominating influence in China. That is why they could achieve so much progress so soon, although it seems the 'great leap forward' has eventually landed them in a ditch.

Bengal has always been distinguished by the absence of religious fanaticism. It was not at all rare to find Hindus and Muslims taking food together — again, Hindu writers jostled with their Muslim colleagues to liven up the literary experience. There was a time when we had hoped to build a harmonious home for the Hindus and Muslims, children of mother Bengal. The soil of mother Bengal has blended the hopes and wishes of God knows how many countless Hindus and Muslims — who can tell now! It was our hope to create a model of communal harmony here that would be our gift to the world. But our hopes have been dashed like castles in the air. Nobody knows what the cruel countenance of time that has gradually revealed itself would lead us to.

Children who have assembled here with a lot of zest for science, I would like to tell you this : do not take up science for the sake of passing examinations. Just as on the one hand, one needs to study literature to appreciate one's own country and its problems, to look for the causes of misery of one's own countrymen and to be acquainted with what men are doing for one another in other countries, so, on the other hand, does one need to study science. It enables one to have an idea of what can be done to get rid of man's miseries. Your interest in science should be rooted in this quest. Today we are constantly being told that man is no longer interested in living on this planet. Soon he is going to fly off to Mars or somewhere far beyond. Maybe we have brought the earth to such a pass that there is no easy way to survive here. Someone from a civilized country has said, 'We are spending billions to launch a rocket'. If they had given us the money without squandering it this way, we could have found some easy means to feed hundreds of thousands of hungry people. But who is going to raise such issues? It is for you to do so. As for me, I shall stop here.

Address given in Bengali in reply to felicitations offered to S N Bose on the occasion of his seventieth birth anniversary, at the Mahajati Sadan auditorium, Calcutta, on 1 January 1964, later published in Jnan O Vijnan, *May 1964, and reprinted in* Rachanā Sankalan. *Translated from Bengali. Prafullachandra Sen, then Chief Minister of the State of West Bengal, was President of the Professor S N Bose Seventieth Birthday Celebration Committee, the body that convened the gathering, and Humayun Kabir (1906–69), then a Union Minister, was one of the Patrons of the committee.*

The Krishnan Memorial Lecture, 1971

Dear Director Verma, I see many familiar faces before me. I won't name them separately but to all of my friends I say, I feel highly honoured for being asked to come among you and to say something on this very memorable day. I had a great sensation of pride being associated with the National Physical Laboratory where, as Dr Verma has said, by dint of merit and hard work, bands of young people are grouped together with the main purpose of solving many problems — either the manufacture of highly sophisticated instruments which may ultimately be utilized for purposes of research, or finding out all about certain commercial processes to enable the manufacturers of our country to produce and thus help our industry, our people and, incidentally, our countrymen also by saving a lot of foreign exchange. We had been discussing these problems together and I remember, years ago when we began our life as students of science, that there were no such facilities offered to the young and meritorious aspirants those days. There had remained, besides the government colleges where alone to some extent facilities were provided for, the solitary Indian Association for the Cultivation of Science, founded by Dr Mahendralal Sircar, a doctor, a patriotic doctor who had thought of the future of the country, of the possibilities that are provided by native talent, and who had collected money and founded the Indian Association where people could be given news about the recent developments of science, and where people may be given facilities for work. Then there was no question of utilizing the resources of the laboratory for research purposes to solve problems which defied the industrialists. It was more or less about a century ago that people were thinking of how to make our countrymen modern in the sense that they would be able to look at science with eyes not half as troubled with superstitions, so that they would able to look at major and natural phenomena and reason as to how things are brought about, how things develop, how things are made and so on.

Dr Sircar had himself a history. He had begun as a doctor trained in western allopathic methods, but very soon he convinced himself that it was possible to effect cures by homeopathic methods. And he was brave enough to adopt this method after he had carefully gone about and tested for himself the efficacies of these drugs. As a result, he ran the risk of being thrown out of the cadre of physicians, because he was a graduate himself of medical science of Calcutta University. The doctors were all up against him for having gone in favour of a method for which they didn't vouchsafe any reasonable efficacy. However, he continued. He was the Dean of Faculty of Medicine. He continued to be in that position and he valiantly collected private subscriptions and erected a modest building and laboratory. In this laboratory, originally, there were many short-course lectures.

Sir Asutosh himself, when he was a young man just out of the university , delivered a course of lectures in mathematics on his research work. When we were students, we found his name in a textbook as a person who had given a very ingenious and brilliant explanation of the equation of mass. He had ambitions. He wanted to do research work but then ultimately, since our country didn't then offer any facilities and no avenue of progress or advancement was available to any scientist, he turned to law. The legal profession was then highly honoured, and somehow or the other, science had curiously got wedded to law in our country, in the sense that not only had Sir Asutosh, a great lawyer as well as a great mathematician, brought about the revolution in the teaching of modern science, there were also two great lawyers, whose statues stand in the foreground of the College of Science, who by their magnificent donations made science possible in those days. Formerly, before our days, the students had mostly to rely on the government facilities being offered, with the solitary exception of two great names which have gone down in the history of science in our country — Sir Jagadischandra Bose and Prafullachandra Ray. They had gone abroad, studied science elsewhere and then returned to our country and continued to work with the native equipment — sharpened by western teaching — and find new directions in their work and in their jurisdiction.

Sir Jagadis's name had remained connected with the applications of science for unravelling the mysteries of plant life. Sir P C Ray's name had remained associated with the first efforts by Indians to manufacture drugs in our country and the first ingredients which go towards basic industry — manufacture of sulphuric acid, etc. But most of the scientists usually had their training and turned to law. However, there was a great change in the history of our country — there came the partition and the Swadeshi Movement. And then people at once thought how futile it was to go about learning things and at the same time remaining thoroughly dependent upon others, even for our own immediate necessities. There was a movement which wanted people to become completely independent so far as the daily necessities were concerned. There were students who had gone abroad to learn the technique of modern industry — to study how things were spun, how mills were erected, how chemical industries were developed, etc. There were also others who had gone abroad to study the sciences. They had hopes and ambitions that they would return and set up for themselves some industry. However, this did not find favour with the rulers of the country then, and there were repressions, persecutions and revolt. There was started the University at Jadavpur where people wanted to give complete education through the mother tongue — where people came, endowed their wherewithals for the furtherance of education in our country. Well, there was suddenly at the same time a sort of revision of the curricula of the Calcutta University, and the training of science was given a more modern look. Students had to work in the laboratories to justify the award of their degrees. Even for the first degree in science you had to do two years' practical work. At that time the Indian Association laboratories offered facilities for students who were anxious to learn science but could not find enough facilities in the modest colleges that then existed in Calcutta. The Indian Association offered facilities for doing intermediate practical courses in physics, chemistry and botany, and students in large numbers thronged to the Association to do work. Well, I tell you all this simply as a sort of introduction, and to remind you that the facilities of the Indian Association were very modest indeed. And we had simply thought of it as a place where beginnings had been made but could not be very much improved because of the failing resources.

Mahendralal Sircar had died, and the laboratories still had very modest resources. When we were students for the Master's degree in Calcutta about 1913, there were new things, new endowments and there were new persons — eminent scientists who had been invited to come to Calcutta. Dr Ganesh Prasad had been appointed Rashbehari Ghose Professor. There were endowments given by the two eminent laywers, Sir Taraknath Palit and Sir Rashbehari Ghose — their names will remain written in letters of gold in our history. The first selections were made of the professors. Of course, curiously enough, when the appointments with the endowments were made and money had been given for the establishment of these professorships, there was a clause which produced a great deal of ire among our rulers, because it said that the person who would be appointed should be an Indian. According to the law, as Indians were defined, it was just barring for ever appointment of anybody other than Indians to these posts. Actually, then most of our science posts had gone to Europeans with the solitary exception of the two illustrious names which I just mentioned — Sir Jagadis Bose and Sir Prafullachandra Ray. All the other professors were mostly Europeans and Scotsmen who did not have very brilliant academic careers, but it was considered sufficient to be educated to teach the Indians a smattering of science. However, some money had already been spent for the teaching of science, and when we were students, we used to look through the almirahs in which were packed things at which we wondered — for example, an Echelon Grating or some other thing — and which were occasionally reverently put on the lecture table and again put back inside the showcase. The professors thought that if they were touched by people, they would be spoiled for ever and so much money wasted. Now, that was science! Well, you had to respect and have reverent superstition and look at these things — and at the same time remember that these apparatus were manufactured not in India but elsewhere — to be shown for our edification, to be treasured and to be always objects of perpetual adulation. At that time, about 1913, when we were students and Sir P C Ray was doing some research on the amine nitrides and Sir J C Bose was doing his magnetic chrescograph, endowments had come, and Sir Asutosh had to seek out certain probable recipients of these endowment chairs.

He had noted a brilliant young man who had come down to Calcutta, not as a scientist but as an officer of the finance department, but who had chosen to pass all the time that he could spare from his official work in the laboratory. I am mentioning Sir C V Raman's early debut in the scientific world as a person, as a volunteer, who came to the Indian Association and utilized all the facilities which the Association had to offer. He chose his dwelling house very near so that he could come and enter the Association by the back door and continue his work in the long hours after sunset.

Sir Asutosh had observed this because he was then the President of the Indian Association. And when Raman was still comparatively unknown among the professors of the various colleges, he had predicted that this young man would go very far. I remember the day when I had gone to see Sir Asutosh and we were wondering about the new appointments, and he said — I know that he will go very far.

So we were all wondering. We occasionally went to the Indian Association, and he was already a name among the aspirants who wanted to do work. He was working like any other student of science — mornings and evenings — making his observations. He was studying what you call asymmetric diffraction and publishing his papers in the *Philosophical Magazine*. And then to the young aspirants who had just passed out of the Calcutta University, he had given problems to solve. He was a very keen observer, Sir C V Raman. For example, I remember there was a very old apparatus in the Indian Association which was set up to show the effects of elasticity. There were two balls which were pulled apart and then allowed to come down and to have a collision and then rebound and jump up to a certain height, depending on how far it could retrace its position. Well, you could see that this one was more elastic than the other, and so on. That was a very old apparatus. Sir C V Raman had taken two of these elastic balls together, knocked one against the other and curiously enough, had observed that it sounded very different if you heard it from one end rather than the broad-side position. It went very far on one side and sounded more metallic than the others, and he was curious. Well, things were all set and balls knocked about in that apparatus, but nobody had ever thought that there was this asymmetry in the production and propagation of sound when two balls collided. But he was curious, and he had observed this thing. And, of course, he was not a master mathematician like Ganesh Prasad, so he asked Ganesh Prasad. They were friends then. Usually before people become great, they remain friendly, and it is only afterwards that they quarrel. So C V Raman was still friendly with Ganesh Prasad and consulted him . He shook his head very wisely and then gave it as a problem to one of the young men. Result — Sudhanshu Banerjee got his doctorate.

Well, he solved the problem. He explained how it had to be explained by regarding these balls as sources of short waves, and he analysed the short waves being reflected one from the other, and then by curious combination getting together, you see, and producing a sort of double source, just like a magnet — you have these magnetic lines proceeding more rapidly, having greater force, in one direction than on the other. He had deduced by means of analysis, of course, as to why these short waves were very much stronger in one direction than the other. I just tell you this to illustrate the keen obvservation power of Sir C V Raman— how out of very modest things he could find certain things which needed explanation, and how he could lead on to very abstruse calculations and thus furnish a sort of enduring result for science. There were such things which I remember as, for example, the Wolfe note he used to show from the violin; how the maintenance of vibration should be very difficult when certain critical vibrations occurred; the colour of thin films, and so on. And for every such little thing which one could observe, he found something out of the argument and gave them as problems to be solved to young men and they were very glad. They were grateful because at last something was given to them to whet their own innate intelligence and to apply the newly acquired scientific methods which they had learnt at the feet of their teachers. But the teachers had not been able to offer them any problems, and now problems were found for them. And that was how people whom we had admired in our young days — like Mitra or Phani Ghosh — were given problems, and they began their life as investigators and got their inspiration from Sir C V Raman. They didn't remain tagged on to him but developed their own capacity for work. Sisir Mitra became one of the pioneers of the study of the ionosphere. And many students who will remember his name with respect and had their first lessons in wireless, in microwaves, were given prizes by the National Physical Laboratory. He had begun the studies of these waves, and he was finally made a Fellow of the Royal Society. Phani Ghosh had developed his powers of organization — he had started the Department of Applied Physics which has grown to be of respectable dimensions now. Every year they are training students who are absorbed by the industries, by government departments, etc. Sir C V Raman continued and observed certain things about which he spoke himself — about the blue colour of the seas which he thought was not reflection of the blue of the heavens — and he went on and ultimately discovered the Raman Effect. And with this discovery was associated a comparatively young man who was a very quiet sort of person. We found out that he was very much younger than us and was an ardent student. When Professor Sommerfeld came and lectured on the new developments of physics, he took down notes which the professor found suitable and valuable for the publication of a brochure containing these lectures. He continued his work in the modest Indian Association and Sir C V Raman continued to work on the problem of the mysterious colour and had guessed that there was not only scattering of light, there was also a change of wavelengths.

He had put people on that problem. I suppose there were many illustrious people who had worked on it, but the problem could not be solved. Ultimately he had found that there was actually a change of wavelength. But somehow or the other, he was not clear as to what was happening. The first paper which he published had talked vaguely about the change and something like the Compton effect — the Compton effect had recently been discovered and Compton had reported about the change of wavelength of X-rays. It was only by dint of very careful work at the Indian Association — with very modest apparatus and with an ordinary spectrograph in which exposures took over 100 hours — that the actual discovery of the Raman Effect could be clinched to a successful issue. Krishnan had worked assiduously and without sparing himself, and as a result he had almost blinded himself by this continuous work and the exposure to the ultraviolet rays. However, that brought fame to India. They were so energetic, and organizationally so able was Sir C V Raman that as soon as he recognized that there was something very new, he put into the problem all the resources that could possibly be utilized and pressed into service and had the problem searched out. And Krishnan had remained associated with him with this work.

Of course, some people thought that it should be called the Raman-Krishnan Effect. But Raman himself would have been very angry if anybody had said it was the Raman-Krishnan Effect. The Germans had thought that Smekal had already thought of the possibility of a similar phenomenon and had talked about it, though, of course, he had done no experiments himself. The first work of Raman was not with atoms but with molecules, and yet the Germans continued to talk about it as the Raman-Smekal effect, making Sir C V Raman very angry. So anybody could talk about it other than as the Raman Effect, though there were the Russians who talked about Landsberg and Mandelstam as the first to notice this effect. They put forward, I suppose, some sort of claim that it was observed in the solids. But, however, so energetic were Raman and his pupils that they convinced the world that here was a novel effect — an effect which neither Stokes nor Rayleigh had thought of — and there was proof positive. Hundreds of experiments were done, all with very modest apparatus at the Indian Association — either the old superannuated apparatus which were treasured there or with some of the new apparatus which were bought for postgraduate work at the College of Science. Sir C V Raman had proved another thing — that it was possible by modest means — if only one really had a sense of the new and the mysterious and were not daunted and went on solving problems — to solve them with even ordinary apparatus. One need not wait for all the apparatus to come from abroad for research purposes; if one was anxious, one could do it. So Raman and Krishnan worked together. Then after some time, when there was the necessity at the newly instituted University at Dhaka to make appointment for a reader of physics, Krishnan was taken in and Raman came and recommended Krishnan as one of the brainiest of his pupils.

Krishnan came. He thought for a few months about continuing the Raman Effect experiments there, but then gave up. And then he took to a new problem where he made his mark, where his pupils have developed methods — the method of making asymmetric magnetic measurements and crystals. One of his pupils talked about it yesterday, and you should remember that though they have now developed and got new things — imported liquefiers etc. from abroad with which they intend to go down to the temperature of liquid helium — the first measurements were made with apparatus with the coarse fibre which were drawn in the laboratory, measurements done with ordinary instruments and so on. That demonstrated once more that if you have problems and if you really want to solve them, there will be methods by means of which you could really make a sort of brilliant beginning. I don't mean to, I don't want to discourage the government from giving grants to science saying that it is not necessary. But what I mean to say is, even before such grants are given, ample facilities already exist in our country to prove the capacity of the various people, and there is no immediate necessity of getting new things imported from abroad. Nowadays the question of getting apparatus from abroad has become rather of second grade importance because of the development of our industries; because of a large number of trained technicians who are now available in our country. We remember when we were students, the professors used to hold on to their mechanics as some of the most precious possessions. For example, there was a mechanic with Sir J C Bose and he did very fine work — indeed to his amazement. Because Sir C V Raman had cast his jealous eyes on this mechanic or another, he grew furious. He was very angry and Sir C V Raman had to content himself with getting somebody who had voluntarily given up service because of the higher emoluments at the Indian Association which he had procured for him .But this Sir J C Bose wouldn't forget and wouldn't forgive.

I tell you these simple, curious little stories and anecdotes to remind you that about seventy/eighty years ago there was a paucity of even technicians in our country, and we used to quarrel about possessing a particularly gifted person or the other to be kept inside the laboratory rather than giving him away to a competitive person.Now there are a lot of technicians in our market and we are now finding it difficult to give suitable employment to these technicians. And if, therefore, there were persons in our country who had problems which they wished to solve — which would require facilities and gifts of a peculiar nature and ability — well, I think among the many thousands or many hundreds of thousands of people who have now been trained during all these years, there would not be any dearth of people who, if suitably used, organized and put to work, will solve problems which the country has to face.

I am telling you all this simply because while reviewing the early history of the introduction of science in our country, I remember great names who began their life very humbly, who had very humble resources at their disposal but who had indomitable patience, who could solve problems for themselves, who would enthuse young men by their example, and thereby place a goal for others and our young men — and place India, as you say, on the map of science. Well, gentlemen, I have talked a great deal. I talked only to remind you of the picture of K S Krishnan as he was then, a modest investigator in the laboratory of Dhaka, who loved his pupils, who lived like an Indian guru surrounded by his students and thereby produced enthusiasm — thereby not only had Krishnan progressed, solved problems, created fresh problems, he became one of the founders of this National Physical Laboratory where we have all assembled. Throughout the country there are now students — his own pupils — who had worked, who had found work, found problems, who had been at it, who had more or less been inspired by his example and thereby had produced for India a rich resource with men and money — men in the sense that laboratories have sprung up all over the country where young men are trained and given suitable opportunities to solve problems. Only somehow or the other, there is something — I suppose it is a sort of evil star in our country — so that though we make very good beginnings, we finish rather a sort of a third or something.

Though we had produced these brilliant young men, at the same time we have not been able to produce or to make as much progress as we could really hope for ultimately in the field of research and industries. Why? I remember that when we were young, people used to point a finger towards Japan, and we had a song that though they were half savages, they could boast of independence — we had no independence, and so on. There was that celebrated song by Hemchandra. However, they had thought of western methods and used them resolutely by virtue of the few very far-seeing statesmen who had chosen upon themselves the task of getting scientific knowledge from abroad. The result — within a hundred years or so, Japan had become a name to be respected in the history of science. And even now, though Japan was humiliated in the last war and devastated and has been under the heels of the aggressor, Japan has been able to maintain its supremacy and its place in the first ranks of the scientists by means of its own efforts. Well, I sometimes wonder why it is that Indians who had very often been able to see very much farther than others — who had seen the sun rise very much earlier than many other people — had not been able to take advantage of the early seers and have not been able to continue this advantage and to maintain their pre-eminent position in the world.

Well, I leave these questions to history and to the new generation of young people who somehow or other feel that they have got nothing in common with the old fogeys. Especially in Bengal they talk about the generation gap. The gap is there. One cannot gainsay it, and it is a challenge. It is a perpetual challenge to the Indian genius as to how, even though the country is endowed with such natural resources and had such a brilliant history, it continues to remain third-rate in spite of so many resources and so much manpower. Well, gentlemen, I stop here with these questions to our young men.

Krishnan Memorial Lecture delivered at the National Physical Laboratory on 12 March 1971.

Education and the New Age

First of all, let me acknowledge with all respect that I feel honoured to be with you today. In this old age I feel deeply touched by your invitation. Times have changed. When we were young, we used to view our motherland in a particular way, and we had also developed certain ideals. We staked all that we had in life and dedicated ourselves completely to the fulfilment of those ideals. Those thoughts and ideas have undergone so much change today, that I cannot help feeling quite overwhelmed.

Respected Madam Vice-Chancellor has spoken of many things. She remarked that initially everything sprang from bliss — all these things. Indeed.

But we have different ways of looking for this bliss. Rabindranath could not perhaps find 'bliss' in these social and domestic surroundings, and so, leaving everything behind, he went to the open fields of Santiniketan. There he tried out a radical reform of the education system in an entirely new way, so that we can develop our future in harmony with the people of the world.

Think of our ancient sages who sacrificed everything for the sake of this bliss. They used to say that unless we could free ourselves from the bonds of attachment, happiness would not find its total fulfilment.

Hence, it is hard to say how much of that happiness can be realized in this city.

Rabindranath had once walked these very grounds. There was a time when he sat in these very rooms in the old world style of his forefathers. Again, through these very rooms came drifting new idioms — the first page wafted into the hands of Maharshi Devendranath — then emerged the great poet with whom we are trying to attune ourselves today.

However, the notes of the new times will not quite ring familiar to old ears. For us the focus of Visva-Bharati or Rabindra Bharati is the culture of mankind. What has sustained us all these years has been the belief that if we could keep faith in people and move forward, then the daily problem of survival would cease to loom so large. It is not possible to say what lies at the root of human culture. The stomach has been placed at the root of modern culture. Yet, even to this old age we have retained the faith that maybe ultimately our inspiration flows from a little above, from where also will flow the spirit of culture — out of which we can hope to build the world anew.

A little above the stomach lies the heart — a lot is being said about it these days. There is a lot of talk about brotherhood of man and of a summons from the heart; yet if we look at the world, what we find is, to quote the poet, a 'delirium of hatred'.

Our students are trying today to discover their Bengali identity — but what constitutes this national identity ? Where is the authentic identity of the Bengali? So many historians, bards and musicians have described the Bengali identity in so many ways.

As children, it made us mad to read about the Bengali character in our English texts! A lot of what was written in history those days made us angry. We used to feel we must build something and prove our worth to these people. This was sheer miscalculation! But this is exactly the way we were led to think those days in our universities where despite all the lack of resources and the grumblings, we thought we would still be able to achieve something significant.

I feel hopeful after listening to the Vice-Chancellor. Perhaps the message of a new era will emerge from the kind of education that is being imparted here at Rabindra Bharati. And we must truly understand and ponder over words that will echo through and stir the minds of the youth as a result. Consequently, I too am rather curious to know how our students feel, what they think, how they are trying to shape themselves in the light of the new dawn.

I live close to a school. I am quite disappointed to see and live in the regular confusion that prevails there. It seems we have not been able to achieve much in the field of education, which is why there is so much unrest.

Anyway, before anybody says anything, I would admit that we have not been able to achieve much after independence, and hence there is so much confusion today. But there should have been confusion long ago. Since this anarchy has surfaced so soon after independence, we must then welcome this independence with the words : Long live Independence!

Even 700 years of subjugation have failed to open our eyes. Today, through independence we have arrived at a new arena, and in its light we can perceive a new direction. Many of the older generation will recall how they dedicated themselves fully to this cause, this independence. Yet, they had of course had a most conventional education. A new ideal has to be set for education, a new system introduced for those who are coming forward to be educated but find the present system irrelevant for a livelihood. The system of education should be so designed today that on completion of studies, they will all be called to serve the nation. Of course, those who have already been called to serve will not want to move away from their special jobs once they settle down. And so there is the apprehension that perhaps a new movement will be needed to remove them. Anyway, one hears a lot of talk today about this education, and this culture and our country's misfortune — it all sounds like some distant music whose meaning is blurred. Today all the universities have been called upon to provide livelihood to the students. Swearing by the old ideals will now sound like a hoax and will no longer work among the students.

Ideals! People are no longer prepared to believe that ideals alone can open up the vistas of a real world beyond. Under the circumstances it is rather embarrassing for you to have invited someone like me to tell you what you ought to do, and consequently, I too am equally embarrassed. Yet, I can at least tell you that the old culture that has survived to this day as well as those you have invited to this special convocation, belong, like me, to the same age.

It is often said that the present generation cannot get along with people from an earlier generation because a generation gap has yawned between them. We can see today all the hopelessness and hunger and we know what to do. Only God knows what this is all about. I don't know what joy there is in it. Who can tell whether the dance-rhythms of Nataraja are audible in the destructive game of Maharudra?

I am an optimist. I hope we will be able to proceed despite all this. My appeal to those who have got their degrees today is — have something novel to say. If there is anything new to say, mere words will not suffice; you must also do something about it, and in so doing, also demonstrate that the effort cannot be merely destructive. So, to those of you who have become inspired today in Rabindra Bharati, those who have become eager to express yourselves, to you my only request is — merely glorifying the revolution will not do, you have to try and train yourselves up too. Hatred and violence alone cannot be the last word of revolution.

The world has seen many revolutions. Everywhere violence and hatred have been dominant. I would not like to believe that this hatred and violence are the last words of our civilization. All the time a nagging thought continues to trouble man — whatever I have done has not been right. Watching the present, people of my generation will say — our ideals allowed us to love each other. That undoubtedly showed us the right direction. Kill the person you cannot love — is that the new order of the day? If this kind of liquidation continues, our minds will become such that there will only be hatred in them and no room for liberality.

Anyway, I will not say more. I am glad I have been able to share my thoughts with you today. I hope what so many workers are trying to build up single-mindedly is successful, that your mutual affection grows, and may God bless you. May all other ventures in Bengal like this one also be successful.

Today my mind goes back specially to those days long ago when many like me had rushed to Dhaka University. The trumpet of destruction that blares there today bears to us also a few tidings of the realization of some of the dreams we had dreamt then. Those of us who had gone then wanted men, irrespective of race or religion, to stand side by side with one another and declare their love for one another as brothers. This wretched person is your brother — you too must be moved by his sorrow — we had tried to accomplish this a lot and for a long time. It all comes back to me today. We have been fighting for one Bengal — the ideal we hear of everywhere today was the same that inspired us then. But those who were groping for it faced a lot of harassment — they had to run here and there.

I have also been to places where today's freedom fighters are locked in combat — of course in the special conditions that prevailed then. But enough, there is no point in talking about times past. All I want to say is that the seeds we sowed a long time ago are today bearing harvest — therein lies the pleasure of sowing.

Therefore it would perhaps be correct to say that one day the seeds of creation we keep sowing will bear fruit in spite of an apparent phase of destruction.

Bearing all these in mind, I appeal to you workers — there will be many obstacles in your way, but cling to your ideals. The students will say many things — they will write much on the walls, but everything is not visible — all is being recorded in the annals of time, and the seeds properly sown will certainly yield harvest in the fitness of time.

With this hope in my mind I await, until the end, the final day.

Address delivered in Bengali at the Convocation of the Rabindra Bharati University, on 10 May 1971, and published in the Rabindra Bharati Patrika, *vol. 9, no.4. Translated from the Bengali. Dr Rama Chaudhuri was then the Vice-Chancellor of the University. At a point in his speach, Bose refers to the two aspects of the god Shiva — as Nataraja or Creator, and Rudra as destroyer.*

At the Calcutta University Convocation 1973

I consider it a great privilege to be present as the Chief Guest at the Convocation of the University. I have reached my eighties, I am physically not very fit; the mind, too, has begun losing its agility. Under the circumstances I feel a little hesitant as to what I could possibly say. Among the learned audience I see lively faces sparkling with intelligence. They have finished their studies in science, history, philosophy, earned fame in research, and have thrown new light on several problems related to the living and non-living. They take my mind back to the past. I, too, received my passport to life at a similar Convocation of this university. Later, along the long course of life, I have had to face various adverse situations; I have gathered various experiences including the humiliation of defeat, the happiness of victory, the company of great men, and especially, the inexpressible satisfaction of being able to wipe off the ignominy of subjugation and sit in the rank of free nations. Perhaps the new travellers will draw some guidance from a veteran predecessor. With this expectation let me begin. . .

We were hardly out of school when the tide of nationalism swept over the country. In our adolescence we used to roam the streets singing songs of the *Rakhi* festival, trying to feel that we were all like brothers, all children of Mother India irrespective of caste and religion. We would have to wipe away the sorrows of our poor Mother India, break the shackles of foreign domination. We would have to save a great nation with an ancient heritage from the ruthless rule and exploitation of foreigners, to instil a feeling of the modern in their old-fashioned descendants, and eradicate illiteracy. That was how the youth of those days conceived the challenge of destiny. We heard great men, eloquent patriots, speak at street meetings, trying to inspire the youth of the country. I feel an urge to tell you those stories.

After passing my Entrance Examination I got admitted to Presidency College. I had the good fortune of studying under eminent and memorable teachers. We heard Acharya Jagadischandra speak about his discovery of radio waves. My first initiation into chemistry was under Acharya Prafullachandra. I had the rare privilege of studying English with Professor Percival. I still have a sense of pride when I recount all this.

The first World War started before we had finished our six years of college. The national movement had brought upon itself all over the country ruthless persecution unleashed by the foreign exploiters, leading to open politics giving way to an era of revolutionary politics all over India, in Bengal, Bihar, Uttar Pradesh, Punjab, Maharashtra. Many noble souls sacrificed their lives, many of their colleagues spread to different countries. Along the long journey of my life I have lost many a friend. Many had staked everything to hold on to their ideals — and sought to achieve the impossible, to liberate the country from the foreign yoke. Now, after independence, it is only natural to remember them.

Sir Asutosh wanted to build a new temple of science within the University where the priests and the masters of the lore would all be Indians. In spite of the antagonism of the foreign rulers there was no dearth of money. The donors have become immortal. The grateful countrymen have erected their marble statues at the portals of the College. A postgraduate section was opened in the newly established College. Chemistry was given the first priority. After retiring from government service Prafullachandra took charge as the Palit Professor. A bachelor totally devoted to his work, he stayed in one of the rooms in the College building even in his leisure hours. Soon a group of young students gathered around him; they were his constant companions, and they looked after him all the time.

Despite the problems posed by the war, Sir Asutosh agreed to begin postgraduate classes in physics with a handful of young enthusiasts like us. Later, of course, C V Raman joined, quitting his government post. In the flush of fresh inspiration, the students plunged into research. Several achievements of the Science College entered the pages of history. From this city of Calcutta Professor Raman won his Nobel Prize. Meanwhile, his colleagues, too, won fame — their researches have been documented in books.

Under the leadership of Sir Asutosh an active centre of research grew up in the Calcutta University in chemistry, biology, history, philosophy and other disciplines. The rule of the country still rested with the foreigners. Yet, in response to the call of the future, young men started competing in international fields and despite adverse conditions could progress steadily, winning honours for the nation. Sir Asutosh had faith in their idealism, self-confidence, intelligence and competence. They also tried to live up to his faith in them. All these may sound irrelevant when the bright picture of the University seems overcast under the burden of several unfortunate circumstances. Yet, as an optimist, I am trying to suggest through these effusive references to the past that even today if we could instil idealism into our students through proper guidance, then they will again be able to stake all and retrieve the lost glory of their Alma Mater.

That was half a century ago. Then destiny drove me to Dhaka. A commission was set up to carry out reforms at the Calcutta University. It consisted of many learned and experienced men from home and abroad. After reviewing many evidences a plan was chalked out for extensive dissemination of higher education in the country. Based on it, new universities came up in many provinces. But Calcutta was not ready to comply with it. A new residential university was set up in Dhaka in accordance with the new recommendations. Some of us from the University of Calcutta received invitations to implement the new system there.

Meanwhile the political atmosphere within the country had undergone considerable change. Gandhiji had appeared on the scene and called for non-cooperation. Some of the talented students left research to take to the spinning wheel and reorganize the country. The wave was felt within the universities too, but it did not lead to any permanent change.

In the meantime seeds of division had been sown by the foreign ruler, dividing the Hindus and Muslims. This had an impact on the political leaders of the country. In the days of the Swadeshi, Hindus, Muslims, Christians were all children of Mother India. The basic principle of the struggle was to strengthen the bond of unity and stand up against the foreign ruler. Later came Khilafat, the dream of a religious state for the Muslims. The rulers cleverly used it as the main instrument of their divisive policy.

The twenty-five years I spent in Dhaka were varied and eventful. That the charisma of a sincere teacher was no match for that of political heroes was proved time and again. Yet I had to keep performing my duty as a teacher in that difficult situation, without paying any heed to caste or creed. But watching the developments in Bangladesh now I feel that perhaps my twenty-five years of teaching have not been in vain. Bangladesh has emerged from the dead East Pakistan. I have heard the hopeful news that they will have a secular administration. But riots and acts of violence are also going on. It is difficult to guess before some time goes by what the future of conflict-ridden Bangladesh is going to be.

I returned to Calcutta after twenty-five years. The Second World War was about to end. The Cripps Mission had gone back. Threats of direct action were in the air. Within a year violence broke out. At last came the independence the Bengalis had dreamt of. In the process their state was also split into two. Millions of families were uprooted in Bengal and Punjab, millions lost all their belongings.

The Bengalis did not hesitate to sacrifice a great deal for freedom. Perhaps some among my audience here, hearing of the losses, will remember their homes in East Bengal with silent grief. Yet I feel it would be good to take stock after twenty-five long years. It might even help in future planning. Some might say that when even nationalism has to be sung on a low key against love for universal man, taking stock of the Bengali's fortunes sounds rather old-fashioned. Perhaps some of my friends would remember a thesis submitted by a scholar who was awarded a doctoral degree and was much appreciated by a university for upholding the view that 'in the old days there was no concept of India as a single nation in the minds of the natives — it grew only with the coming of the British.' At the age of eighty I admit that I am old-fashioned enough to hold the Bengalis so dear and have my mind filled with their joys and sorrows.

Twenty-five years have passed since independence, and Bengalis have lost a lot. They were known for their scholarship abroad. Today there is a feeling of antagonism against them in other provinces. What are we going to do about it? It makes us think again of the greater losses we have suffered compared to other nations. In our time we were drawn towards science in the hope that it would bring prosperity to the country, and increase business. We used to think of such things. We thought P C Ray had done the right thing by establishing Bengal Chemicals. But the industries set up during the Swadeshi era have either closed down or are shifting to other states. The number of truly Bengali organizations of which we can be proud, has dwindled. We are lagging behind other states in the eradication of illiteracy and in industrialization.

How Bengalis can prosper in the changed circumstances should be a matter of concern and responsibility for all the youth of the country. I will now end my talk with an appeal to the youth to try and restore the Bengalis to their old position of eminence.

Convocation address delivered at Calcutta University in Bengali on 16 June 1973. Published in S N Bose, Rachanā Sankalan. *Translated from Bengali.*

Address at Seminar on Scientific Contributions of S N Bose

Professor Kapur, I should be grateful if you allow me to speak seated.

Friends, fifty years ago when we were young, we had the great idea that we would have to prove to the world that Indian intelligence was no less than anybody else's, and therefore we were anxious to show our own intelligence, our originality and to say something which people would accept — that was a great thing in those days, because as you know, this institution was started in the teeth of the objection of our rulers. The first clause that the Professors should be only Indians had set their backs up and they did not really help the Calcutta University to get on as otherwise they would have done. However, we, the Indians, felt that we had our own things to do in this particular matter — we had to work ourselves, we had to create enthusiasm among our people. And fifty years ago we had started the ball rolling which caught people's imagination — I feel really grateful. At the same time our problems have become more complex with the getting of our independence — that is to say, we are not only to show ourselves on paper that we are intelligent people but we are to show in our work, in our hearth and home, in our laboratories and workshops, that we too could rise to the occasion, utilize our own natural resources and build up a country strong, vast and progressive. Of course, a twenty-five years' span of time is really very little compared with hundreds of years of history during which we have left things moving of themselves and allowed ourselves to roll back into ignorance and so on. These things are to be removed. We are very curious people — as somebody has said, we live in this torrid zone and we get easily excited but very easily depressed at the same time. However, I think the cold waves generally coming across the frontier will enable us to remain to be true to ourselves. So, in spite of occasional difficulties, these will help keep our spirit, and at the same time among our young people keep alive the hope that we would be able to remove all our past shackles and difficulties. For example, one thing in this caste-ridden country people quarrel about is how they were born and so on. These things still persist even though Gandhiji had preached the removal of these things. But still we hear of people being persecuted, being condemned, being burnt to death etc. All these things still persist among our people. I think it is not simply the question of solving problems but the question of really changing the way of looking at things that really faces us. And therefore, it is not simply our intelligentsia but also our politicians we have to look forward to. They should be able to take proper guidance from the lessons of history — they should see that it is only by the removal of restrictions that we would be able really to bring out from the Indian spirit the best that can be produced. At one time when there was such a call, people were not wanting who despite difficulties came forward to regain self-respect. When I remember my young friends who had all grown up together — like Saha, Dr Sen and other young people — Mukherjees and Ghoshes — it was really something very surprising and at the same time very very stimulating for us. If such things could be really possible in a small span of time, why could it not be repeated? There is no reason why now that there are greater facilities in our own country, they will not only remember something which was done but do something which people will remember ever afterwards. I hope this occasion when you have started the ball rolling will bring out not only the innermost spirit among our people but also the spirit of cooperation which we feel we are really losing at the present moment. Because, after all in order that we could make the best use of our manpower and our vast country and our large sea line, so to say, we must have to feel like one nation. And everywhere in the country, in every village we should have to feel that we are now going a step forward. Now that there are people who are thinking not only of solving our problems of calculus but also problems of agriculture and other factors, now that the fifth plan is being mooted, I hope in this fifth plan all the difficulties which we had experienced in the last twenty-five years will be remembered and removed so that in another five years we will be looking forward to a country which will make vast progress, which will recognize its own difficulties and which will find strength to overcome them within a short period. Gentlemen, I feel at the age of eighty the weight of the years but I feel very much enthusiastic when I see young faces — when I see Bhaduri, Bose, Dasgupta, Bagchi — people who had been with me when I was really much older than they. We could help ourselves at a time when things were difficult. But now since things are going on getting improved, since now there are various institutions and various universities, there are new factories and new collaborations etc. all over the country, we hope that in the next five years — or perhaps before the end of this Bose year function — you will be able to look forward — to say we have really done something which exceeds very much more the old stage in statistics; that here is something really the country will have got in agriculture, in machineries and in chemistry, in our production. I see here Dr Mukherjee who is now in the Agriculture

Commission — he is now away from the University; I see Mr Basu who is in the Indian Association — the Indian Association has gone a great way further since the time of Sir C V Raman; now the laboratories may be built — and everybody else.

And, of course, Dr Dutta who is still at the helm of things — he is doing what is to be done for the old Calcutta Mathematical Society. I hope that the Calcutta Mathematical Society will still continue not only to be really a sole property of Calcutta mathematicians but really something which the whole of India can regard as their society and the place where they can publish their original contributions. In a similar manner, all our different institutions which are connected with our universities of Bengal should really continue not only to guide our young men but also show a sort of beacon light to all the other universities of India. In India what we really need is really a sort of total abolition of all the racial spirits — the spirits of dissension that still trouble us occasionally to such an extent that we old people really get lost, and we cannot guess as to where our young people are being led on. But I hope, these very few people who are still at the helm, who in spite of all difficulties still really believe in the high ideals in India, still think that India has something to give to the World — the spirit of cooperation and love of humanity — we still believe that in spite of all these dissensions, with all these difficulties, all our fights, perhaps we will able to solve our problems in a sort of bilateral fashion instead of asking people to take to arms — well, to solve our problems ourselves and at the same time to solve the problem of oil which really faces us. At one time we thought that perhaps we could from our splendid remains of coal get something done. But then it looked too expensive as compared with what we could get easily from West Asia, and we left it at that particular stage. But I think now people should really think how to get together, how to utilize, as some people have already done in other countries, the enormous power that is developed by the waves. They tried to see that the sea waves could be utilized for generation of power. I think there should be things done not only to solve our problems for the time to come but also, at any rate, to show that to the World so that science grows. I think it will be possible for us — keeping the right spirit of cooperation among humanity — to keep this humanity living in this civilization progressively in spite of various difficulties that loom forward, in spite of the vanishing reserves in resources. And gentlemen, I think, it is now time that we stop.

I thank you very much from the bottom of my heart for having been all here together so as to give an old man at this age a sort of great satisfaction that life has not really been in vain.

Speech delivered at inaugural session of Seminar on the Scientific Contributions of Professor S N Bose, held at the Calcutta Mathematical Society on 29 December 1973. Published in the Proceedings *of the Seminar, pp. 15-19.*

B. Miscellaneous Pieces

Bose became a champion and a crusader for his mother tongue Bengali. He wrote extensively in Bengali. A collection of his writings in Bengali was published under the title: *Rachana Sankalan*; Satyendranath Bose by *Bangiya Bijnan Parishad* on the occasion of Bose's birth centenary in 1994. Bose was a founding member of the Parishad. The collection includes popular articles on science, articles on education, biographical sketches of eminent scientists, some translations from Einstein, Pauli and Hadamard and from literary figures J. Agnon, Romain Rolland, Tolstoy and others. I have selected a few articles that have been translated into English from this collection as well as from other Bengali publications.

A Scientist's Apologia

A few days ago the Science College celebrated the centenary of Rabindranath Tagore. Chief guest for the occasion was Shri Surajit Lahiri, then Vice-Chancellor of the University. What emerged from his analysis of Tagore's poetic temper and philosophical outlook would amount approximately to this — that the current emphasis in this country and elsewhere on science teaching and scientific research was inordinate, the ultimate outcome of which would not perhaps be beneficial to human society, that the impression we get of Tagore's ideals from his writings has no essential connection with the scientists' vision. Science was perhaps misdirecting us, displacing the desired ideal in favour of the immediately attractive, etc. As a result, the discussions that day were deflected to a different course altogether; and the scientists were called upon to answer the questions raised by the philosopher to the best of their abilities. As Chairman of the session and representative of the scientific community, I had to say a lot of things that day. One cannot remember all of it at this distance of time. Recently my students discovered that a taping machine had been duly installed that day to record the proceedings accurately. So many months later they made an attempt to transcribe the discussions from the magnetic tape in order to satisfy their curiosity. But they were not entirely successful. Regrettably, the Vice-Chancellor's speech could not be deciphered. My own speech too was only partially recovered. Whatever they could make out they recorded and sent to me with a request — if I could spruce up with the help of the incomplete note the plea I had tried to make on behalf of the scientific community. This is what resulted from my attempt.

Our philosopher Vice-Chancellor had raised a very complex question indeed. We of course had not anticipated that such issues could at all rise from the discussion we had envisaged. Nevertheless I feel that scientists ought to say something about this because the question that has occurred in the Vice-Chancellor's mind could very well occur to many of our leaders, especially since many in this country are born philosophers. Their attitude to scientists is somewhat like this : scientists started the railways, so we can travel great distances with ease; they've blackened and soiled themselves to break coal and wrest some energy from Nature; this may have changed the external features of our civilization. But through his absorption in such matters the scientist has lost the right philosophical outlook. He has perhaps never been concerned with the creator's mind that lies behind this creation, about the soul of man, about God, and the many schools of philosophy that have arisen here and abroad from such contemplation do not interest him in the least.

We scientists might admit that we do not understand such things, and therefore evade such questions, or we might think that the creator alone would know the significance of his creation and its essential nature, maybe we find no voice of reassurance from all those philosophical schools that have emerged. Nor do we know the philosophical beliefs of our philosopher-guest. But we cannot help being curious. Does he for instance believe in God alone? Or does the Devil too sneak into his thoughts?

Man has been trying for long to fathom the mysteries of creation; scientists too have been trying. But can he with his limited capacity truly grasp its full power? The scientist cannot conceive of it, maybe the philosopher thinks even that to be possible. But to attain such a level, he must think that this very physical frame itself is Maya and that truly he himself is an inseparable part of that Great Power. Of course in our country this line of thought has been propagated in many forms ; perhaps even now some may do believe in it heart and soul. Once one comes to this realization we say about him, 'Now he is liberated'. The scientist wonders where has he gone then ! For he no longer has any links with our world!

We hear of the lives of many of our sages, who are supposed to have sacrificed themselves totally, even their very mind and soul. Is that how God would have it? To forward this question to Him however, we must enlist the philosopher's help since scientists are blind here, they cannot meet God. Humankind is gradually ascending upwards through a process of evolution. Civilization is on its way up. This the scientist does believe. What he does not believe is that man's vehicle of progress can advance on the mere fancy. Civilization must be founded on the knowledge acquired from Nature's store of power. That is why he is not prepared to cling on to philosophy alone. Otherwise, alternatively we might fall back on philosophy to imagine that we are all human, all alike, and we could live out our lives without envy and animus, but so far this has resulted in a futile mentality. This would be quite apparent if we survey thousands of years of history. The kind of transcendental philosophy our philosopher guest has referred to is not new to this country. It is very much native to India, or so we are proud to believe. The poet D L Roy once wrote, 'You can lambast us if you please, but do try and read the Gita if you can'. But here one is tempted to rejoin that even when this particular philosophical outlook was supposed fully prevalent, even then was there no respite from enmity and strife. Of course those who were pure of heart managed to get around this fact quite charmingly. There is a story I would like to recall here. The Rishi Vashishtha always maintained that the Brahma was Truth, and that his world was an illusion. So the king set an elephant to chase him, and Vashishtha of course promptly took to his heels, his hair lock still unknotted. So his disciple the king asked him, 'How now, O Great Sage, where are you off to? This is nothing after all, just illusion.' However, even if one were to argue here today on similar lines on the essential nature of Truth and Falsehood, we could not dismiss as Maya (or illusion) the widespread poverty and ignorance that prevails so glaringly in the world today. At least as long as man lives on this earth, this ambivalence in his mind must remain. As long as man lives, he will continue to try to wipe these features out of human existence, and seek to build a society in which these accidental dangers do not loom large. For that we need knowledge, we need a large vision.

Today scientists can say that if the world were guided by philosophers alone (as India is supposed to have been guided once) perhaps even then animosities would not have been eliminated. Without scientific research, there may not have been the atom bomb, but the older weapons of death and destruction would still be in use, with effects no less cruel and lethal.

I don't think that through the ages men have showed any special concern for their fellow men as human beings. It is only in this twentieth century that we have started working towards that ideal. But of course what the Vice-Chancellor had to say is very true indeed. There is no future for us unless this concern for one's fellow men is felt by everyone. Scientists believe this too; but unfortunately, the men at the helm in every country are not men of science. Many may be religious, many nationalists. Many a leader believes that his own race was born into this world to fulfil a certain destiny; and his own responsibility is to pulverize the earth and ride it over with his particular notion of civilization. I know (but I am not sure if any of you do) that the man we all wished to put in the dock as the prime accused responsible for the recently concluded World War II (of course his suicide prevented us), that same dictator, Hitler, was in fact a deeply religious man. If we go into his life, we will find that he was a strict vegetarian, and so on and so forth. From all those outward signs that we take into account it would have seemed that whatever we received from this individual would truly benefit mankind. When the German people were in acute peril and sinking to the direst depths, he came forward as their saviour. And those who listened to him on the radio as I did during the war, could realize his wariness. 'If we lost this war, it will be a thousand years before the German nation can rise again,' he warned. Well, so much else fell to Germany's lot. And today one would merely like to say this much, that if this nation which had sunk so low has risen once again it has been in the strength of science. It is the scientists who have pulled her up again. What the German nation have to give us is not philosophy alone; the knowledge the people had acquired through every moment of daily work, was pressed into service for human good so that within a short time they could rebuild their war-shattered economy. The scientist firmly believes that religious practice alone can achieve very little. What man is according to religious doctrine, what his relationship with his anima is, such philosophical questions and their study should be taken up in private. From such discourse man may draw inspiration for work, but once he begins to work, he must work with his whole mind in action ; he has to know the rope for a rope, and never mistake it for a snake. If we from a variety of notions choose to negate everything, there will still remain hundreds and thousands of people suffering from disease, depressed with sorrow, terrified forever of so many dangers, who do not know how to make the land yield four handfuls of rice instead of two. A little knowledge may be of help to them — knowledge which may not give them a complete notion of the Universal Creator, but this little knowledge has helped avert much trauma. This fact our Vice-Chancellor will possibly admit.

We did not wish to launch on this theme today. What we had intended was to show what place Rabindranath himself had given to science — in his writing, his life — not restricted to poetry alone — he had given many lectures at many places, written many books, where his attitude to science is made very clear; that is what this gathering was to discuss. We were not prepared for such deep philosophical questions. But let me say this much at least, that we cannot stop at the eternal disposition of the eastern mind — because he too must be aware that at least three or four hundred years ago St Francis of Assisi would say such things as that the water he considered sister, the wind his brother. That is how, throughout his life, he tried to offer mankind all that came into being within him, and thus draw them into kinship with himself. Just as this sense of universal brotherhood existed in this country, so in fact everywhere in the east as well as in the west. But nowhere do we see men prepared to embrace this philosophy as the truth and hold the whole nation in kinship. But one only hopes that the truth of this will dawn on people when the opposing camps both possess equally powerful weapons.

At times it seems that people are so interested in India only because of the availability of an oil line. But if today we were to discover ways and means to construct lethal weapons from plain water, then this brutal competition for forcible occupation of territories might diminish considerably. When the hydrogen bomb was constructed it had an explosive power two hundred thousand times greater than that of the atomic bomb. When a single atom bomb devastated Hiroshima, a scientist friend of mine said to me: 'God, at least now one could no longer say that people were trying to acquire some rare material from a particular place.' If we are clever enough we can develop the ultimate lethal weapons from any common material. Of course this is not exactly how the scientist's mind works. It would be the job of those who reign over people, all those leaders of men who are engaged through daily propaganda in turning truth to lies, and lies to truth, to build up to this idea, with hand on heart. For then no one nation would have a monopoly over lethal weaponry. If war breaks out both sides would have the power to break havoc on earth using such weaponry. Once an eminent astronomer in Calcutta had said that if you turned your eyes to the sky you would notice how a new bright star emerged all of a sudden, and disappeared again after a few days. He thought that this phenomenon ought to make men cautious. If we are not cautious, if we do not change our attitude, this earth may cease to exist. We may ourselves become such a passing star. The creator alone knows if this star magic will happen. Who knows which way mankind is thinking these days?

But the only cause for hope lies in the fact that many small countries which had made nationalism their sole concern have now learnt to think in terms of all humanity. They are getting together time and again to discuss what duties they owe towards the future of mankind. If man is to survive and develop further we must all change our attitudes accordingly, the way that it has now become possible for America on the one side, and the Soviet Union on the other to think along the same lines. South Africa has not yet come round to this. The southern peoples have usually preferred to stand apart, yet many other countries would like to think together. In a way, none of this would have been possible without science. Which is why I still believe that when the scientist still proceeds with his work through all manner of peril and danger, he shows a certain strength of mind — he is truly a humanist. He will not think that his only duty is to his own group, he loves mankind. He places it at the centre of his daily concern.

Every true scientist who engages in research not merely for his self-satisfaction or egotism dreams all the time that he might discover some fundamental principle that will help us to build a monument to human progress. And again, the scientist who, test tube in hand, pursues the trail of a little known disease, he too tries to discover some means by which he may eradicate a deadly epidemic. That is the mark of a true scientist. Probably all scientists will agree with my evaluation.

Had Rabindranath been concerned solely with his own poetry, he would not have built Sriniketan, nor would he have bothered to go out to the countryside to encourage development of agriculture and cottage industries. He aspired to dedicate all of human knowledge first to the Creator of Life with all reverence and then to put this knowledge into the hands of mankind for them to put to use, with good sense.

Bose's own reworked text of his talk at the Tagore Centenary Celebrations at the Calcutta University Science College in 1961. Published first in Bengali in Shraddhānjali / Homage, Professor Satyendranath Bose Seventieth Birthday volume, *Professor S N Bose Seventieth Birthday Celebration Committee, Calcutta, 1 January 1964. Translated from Bengali by Madhuchhanda Karlekar.*

The Progress of Science in Ancient India

There is now a rush all over India to go to the West to study science. Factories are sprouting everywhere and our boys are going abroad so that they can return equipped to contribute to the country's progress. The government too is keen on hiring scholars from abroad to reform our education system so that we can acquire in every field an approach adequate to the present time. Under the circumstances it is only natural to feel curious about the progress of science in our country in the past.

Soon after the Battle of Plassey, the British gained supremacy in Bengal. The Portuguese, British, French, Dutch and Danish merchants had started coming to India by sea a century earlier. They used to buy fine muslin, silk and other things here at a cheap rate to sell them at a high price in Europe. Indian handicrafts were highly valued abroad. Obviously various technological applications were known in many parts of the country.

Unfortunately all that died out gradually after the coming of the British. At the time of the Sepoy Mutiny in 1857, the British had established their hold all over India. The princes of the Native States who had not yet been ousted accepted the overlordship of the British. They were mere puppets in the hands of the British envoys. However, in spite of this utter humiliation, there were streaks of a silver lining. Contact with the Western civilization shook the nation awake from centuries of deep slumber. The country's intellectuals gathered information from all over the world, and came to know about the remarkable gains in science, wealth, technology and business made by Europe. Anyone with a progressive mind turned towards the West for knowledge. There still persisted in this country a system of education rooted in Sanskrit and confined to the upper caste Hindus. The Muslims went to the madrasas. And, since Persian was the state language during the Muslim period, it was taught and used by Hindus and Muslims alike.

People began to think that with all the time and energy put into it, this system failed to provide knowledge more suited to the times. Instead of grammar, religion, statistics, *nyaya* and astrology, they sought the dissemination of modern science — higher mathematics, chemistry, physics, linguistics and architecture. There was no choice in those days but to study these subjects in English. However, the predominant attitude remained conservative. So, after a lot of debate the suggestions of Raja Rammohun Roy and the forward-looking sections were accepted. The new universities that came up provided for Western education. Most people of the country had by then lost faith in the old traditions. The convention of recording history chronologically was unknown in this country. Besides, many old texts and manuscripts had been lost during periods of political unrest. The common people did not understand Sanskrit. The Brahmans alone used Sanskrit. The priests were the society's lawmakers. All their discussion centred on religion. Most people were illiterate — whatever moral lessons or general knowledge they acquired was through the oral tradition — the songs that constituted part of the religious rituals and ceremonies, the retelling of myths and folk theatre.

At this difficult hour, when we had descended to the very nadir of our culture, fortunately for us there came with the foreigners scholars who wanted to delve into the past of the human race. They learnt Sanskrit, Arabic and Persian from our scholars and rediscovered many new facts, even as they endeavoured to dig out India's lost history. And it was through their efforts that the Asiatic Society was established in Calcutta. They took information about the Sanskrit language and its literature to Europe. They unearthed old coins, inscriptions and copper plaques and other historical evidence in many places. A new vista of knowledge was opened up to Europe. Linguists discovered a close connection between Sanskrit and ancient Greek, Latin and other European languages. They discovered links among the early civilizations of Greece, Rome, Egypt, the Arab world and India. They set out to write a new history of ancient times on the basis of their findings. The information came to circulate in this country. Through western education the information reached India that our old literature, *puranas* and scriptures contain a wealth of facts, enough to excite the curiosity, respect and reverence of western scholars. The Indian mind turned once again to the old heritage. Some of the educated Indians followed the foreign perceptions in their entirety. They went through translations done by foreigners to rediscover their own culture and history. By the time our universities were set up, there had emerged nationalist consciousness among our thinkers. Those who had acquired higher education felt it their duty to write the history of their own civilization. The country's learned had to take this responsibility upon themselves. For foreigners are not always able to grasp the reality of our civilization. They have a different language, a different religion. It is hard for everyone to absorb the culture of an alien community. Hence there was considerable fear that the foreigner may not do justice to Indian culture.

There is deep reverence for the Greco-Roman civilization in the European mind. When they came to study the culture of this country, the Europeans thought that India's advances in several fields were a direct result of coming in contact with the Greeks. They felt that the seeds planted by Alexander, and the relations that he had initiated may have profoundly influenced India's ancient culture.

Nationalism in the modern sense has entrenched itself quite securely in our minds from Bankim's time. In his *Krishna Charitra* and other essays Bankim tried to counter the foreigners' theories. We find Romesh Chandra Dutt coming up with an assessment of our literature. His history of civilization written in English continues to amaze us. His Bengali translation of the Rig Veda stands out as evidence of his profound regard for our ancient heritage. We are also reminded of Radhakanta Deb's *Shabdakalpadrum* and Shourindramohan Tagore's history of Indian music. Balgangadhar Tilak, writing about the first habitation of the ancient Hindus tried to establish on the basis of his studies of the Vedas and Jyotisha that the Aryans were originally settled in the polar regions of the North; some of their experiences from those times could be traced even now in Vedic literature.

All this however concerns literature, religion and history. But a search for our scientific heritage goes further back. Those days astrology and Hindu Mathematics used to be studied in Sanskrit. Ayurvedic medicine was kept alive by the Vaidya community. Doctors following the Charak-Sushruta line served to compensate for a major lack in the country by their diagnosis of and prescription for diseases. The achievements of the ancient Hindus in the field of mathematics were first revealed by Colebrooke. In 1817 he published an English version of Brahmagupta and Bhaskar's texts on algebra and arithmetic with an erudite introduction. Colebrooke sought to assess the level of excellence reached by the Hindus in the ancient period in mathematics and astrology. He had studied the original texts and annotations of Bhaskaracharya's *Līlāvatī* and *Vīja-gaṇita* and Brahmagupta's *Gaṇitādhyay* and *Kuṭṭakādhyay*. He proved that Bhaskar had written *Līlāvati* and *Siddhānta Shiromaṇi* around AD 1150 and it was from these texts that he gathered information about the earlier mathematicians.

Before Colebrooke, Dr William Hunter had spent some time with astrologers in Ujjain and concluded that Brahmagupta lived around the year AD 628. From the elaborate annotations that grew around the ancient texts he came to know of a famous mathematician before Brahmagupta by the name of Aryabhata. None of Aryabhata's original texts have been recovered, but later mathematicians have quoted him extensively in their works, in agreement or in disagreement; providing an impression of the theories of this great mathematician. Aryabhata seems to have been teaching that the earth goes round its axis every day. He had discovered the correct explanations of the lunar and solar eclipses and said that neither the moon nor the planets have any light of their own, actually they are lit up by the sun. Aryabhata calculated the diameter of the earth to be 1050 yojanas. On that basis, holding the ratio of circumference and diameter to be 22/7, he calculated the circumference of the earth to be 3300 yojanas. This comes pretty close to the actual figure, because if one yojana is measured as four crosh, then one yojana will be 1.9 mile. Accordingly, the earth's circumference will stand at 25080 miles.

Colebrooke has given a lot of information about Hindu algebra and geometry. He has shown that the Hindus knew the properties of surds. They even used negative numbers in their calculations. They had derived formulae of quadratic equation, sometimes they solved even more complex equations. They had worked out some extremely complex solutions of problems, which were rediscovered by the Europeans in the middle of the eighteenth century. Later Indian scientists have studied these topics in detail. More than a hundred years after Colebrooke, Dr Bibhuti Dutta and Ayudhesh Narayan wrote a scholarly history of Indian mathematics, which contains thoroughly researched findings on many of the above-mentioned subjects.

After this it were the Indians who took up this study in the history of science. Acharya Prafullachandra Ray and Acharya Brajendranath Seal must be mentioned as pioneers in the field. Acharya Ray went abroad to acquire expertise in chemistry and returned home to dedicate himself to research and teaching. He worked for the rejuvenation of the chemical industry in the country. At the same time he studied ancient texts and manuscripts for information about the past.

Whatever spare time he got after teaching and research he spent on these ancient texts. He took the help of Sanskrit pundits to retrieve a lot of information from a close study of the early works of the Vedic corpus. He learned that the ancient Hindus knew the method of preparation of acid and alkali. They could purify iron, lead, copper, tin, mercury etc. They employed various purification techniques for the purpose. They were aware of highly advanced methods of preparing metal-ash. They had designed and produced the necessary apparatus for all these processes. The chemical inferences they drew from sublimation, distillation and partial distillation astonish us. Several texts describe the many changes that can occur in mercury as well as the preparation of various chemical compounds from the reaction of different metals with sulphur.

The History of Hindu Chemistry, vol.I, was published in 1902. The second volume was completed in 1908. After many years of labour Acharya Prafullachandra won for himself a place of honour among scholars of the world by providing information on an unknown chapter of Hindu scientific studies and practice. That many mathematical formulae reached the West from India via the Arab world had already been shown by Colebrooke. Acharya Ray established that many principles of chemical processes were discovered in India and carried to different parts of the world by the Muslims. With the translation of Charak and Sushruta, many Indian formulae of chemistry reached far and wide. Recently Professor Priyadaranjan Ray, one of P C Ray's able students, has published a revised and recast version of the *History of Hindu Chemistry*.

From a whole range of findings made in the past forty years, we now have a reliable notion of the extent of knowledge of the ancient Aryans. The age of Mohenjodaro dates back to 4,000 years ago. Excavations have revealed many artifacts belonging to that civilization. We have learnt that coloured glass was first produced in this country. We have also found many bronze and copper alloy articles which were used by other people, in other places. Later India used to produce pure iron and export it to other countries. The secret of making steel orginated here. Indians ought to feel proud that Indian steel was used for making the famous swords of Damascus and Toledo.

Now we come to the age of Swadeshi. I will conclude my article with a reference to Acharya Seal's research. He was a world-famous philosopher, born on 3 September, 1864. He was a versatile genius, who had mastered mathematics, science, several languages, poetry and philosophy while still quite young. He helped Acharya Ray when he was writing his *History of Hindu Chemistry,* the last chapter of which contains an account of how the ancient Hindus produced different objects and their theories about the properties of things. The theory of evolution according to the Sankhya school, the speculations on the chemical and physical qualities of matter according to Vedanta, Mimamsa, Buddhism and Charak have been critically presented by Acharya Seal in this chapter. He depended mainly on *Vyasbhāshya*, *Charak Samhitā*, Yuddotkar's *Bartikā*, the Commentaries of Prashastapada and Varahamihir's *Vrihatsamhitā* in his account of these theories. How the world was created from void as a result of differences in the qualites of *satta*, *tamah* and *rajah*; how creation reached the macro-molecular state from the five elements; how material objects came to hold beauty, taste and smell owing to the fusion and break-up of macro-molecules; how the Hindu and Buddhist philosophers have sought to describe the location of sky-air-water-earth — Acharya Seal has addressed himself to all these things. We might see the first principles of modern atomic theory in his account. From the way Acharya Seal has described the attraction and repulsion of atoms, he has given us an early glimpse of chemical affinity. Later, after further research, Dr Seal published his important work, *The Positive Sciences of the Ancient Hindus,* in 1915. In the preface he said that he had gathered together in his work many facts for the use of historians of the different sciences. The Greeks and Hindus had laid the foundation of the laws of natural sciences now in use. The basic facts they discovered regarding the properties of matter are being applied in modern industrial technology. The scientific outlook of the Hindus and the logic they used for drawing conclusions deeply influenced both the West and the East, as will be understood if we think of China, Japan or the Saracen Empire. So it is important to make a comparative study of the Hindu and Greek systems and Dr Seal has been quite successful in doing this.

Acharya Seal sought to make a comparative study of different philosophies, in terms of all those fields of knowledge reflected in philosophy, viz. linguistics, physiology, geometry and mathematics. Just as we now have an idea of the Greek pursuit of knowledge, so it was Dr Seal's objective to introduce readers to the Hindus' mastery in these fundamental sciences. To present his arguments convincingly, he quotes from the original Sanskrit texts proving to his readers that there was nothing in his account of his own invention. This book will thus prove immensely useful for a comparative study of the Hindu shastras.

The book contains a lot of information — in chemistry, the chemical composition of different elements and compounds like mercury, bronze, tin and mica; and their analyses — and calculations of time and distance; information about the transitory motion of planetary bodies. Dr Seal says that the Hindus had developed some principles of differential calculus, such as $d(\mathrm{Sin}\ \Theta)=\cos \Theta d\Theta$, long before Newton. Spottiswoode was persuaded to accept this theory.

He has also dealt with dye technology in this book, describing how the Hindus had invented the science of permanently dyeing silk and other materials with herbal products and alum and how highly valued Indian goods were abroad.

Then again he tells us how the Hindus had classified the plant and animal worlds.

Finally he draws on the medical texts to describe their conceptions of the human anatomy and the nervous system. From the tantras he draws their notions of the Ida, Pingala and Sushumna. He has also tried to explain their concepts of life and spirit, taking from Charbak, Sankhya and the Vedantas. Acharya Seal, in the course of his analysis of the Hindu conventions of argument and proof, refers to the Buddhists and the Vedantists and their approaches. The methods the Hindus adopted to reach the truth, how scholars with different views argued with each other, Dr Seal has analysed in his lucid style. In his autobiography he has singled out this chapter as especially important.

That is all there is in Acharya Seal's history of science. But he always enjoyed being surrounded by inquisitive listeners with whom he could exchange thoughts. He inspired many with valuable advice and led them along the path of investigation. The late Dr Radhakamal Mukhopadhyay is supposed to have been greatly inspired by Acharya Seal when he was writing his famous *History of Ancient Shipping*. So was Dr Prasantachandra Mahalanobis.

This study of the science practised by the ancient Hindus has now at last come to the present. I have discussed the past, I am thinking about the future now. There is a nationwide movement going on to transform India into a mainly industrialized country on the model of other nations. Does a country, whose economy is based on agriculture, have to follow other countries and go the industrial way? I would like to end this article with a few lines from Acharya Seal's unpublished autobiography:

> Our usual notion of a rivalry between industry and agriculture in the national life is basically wrong. Both are essential and fundamental. The latest slogan in England — i.e. 'Back to the soil' — seems to me a healthy trend. In India farming will remain a basic and chief occupation as in China. The ratio between agriculture and industry should be 3:1 in our country. Agricultural methods should be improved with scientific use of machinery. I consider better utilization of solar power and electricity desirable. Thus rate of production would increase — and that would be a fitting reply to the economic argument that continued farming would reduce production and profit. This is cited as the main objection against agriculture *vis-à-vis* industry. Increased use of machinery in Indian agriculture is a good way to make use of our enormous resources — manpower and land — and to take us to the forefront of nations.

Chamberlain had once said that everyone should have three acres of land and a cow. We must raise the same slogan here. At the same time industrial growth is also necessary. Till today industrial projects constitute only four per cent of our economy, against agriculture. This is unnatural and that is why we are so deeply immersed in poverty. At this hour of food crisis we need to pay heed to Acharya Seal's advice.

Previously unpublished material in Bengali published in the Rachanā Sankalan *(Calcutta 1980) for the first time. Translated.*

(229)

The Early Phase of My Explorations in Science

From my childhood I was keen to go abroad. I passed my M Sc at the age of twenty-one. Professor Mullick asked me to see him and said affectionately, 'You have secured embarrassingly high marks, my dear boy'. Perhaps this meant a government scholarship for going abroad, I thought hopefully. But I was out of luck. They all wanted boys to be trained in chemical engineering. That would lead to the growth of indigenous factories, increase of wealth — and we would become self-reliant. My skills were limited to mathematics. That year chemistry won, though the victorious student eventually entered the grooves of academia, but that would not be relevant here.

Meanwhile Acharya Ray had joined the University College of Science after retiring from Presidency College. Europe was at war. Sir Asutosh had already selected Professors of science with the financial backing of Palit and Ghose. But two of his chosen men — Debendramohan Bose and Agharkar — were interned in Germany where they had gone for higher studies. Raman was still working as an efficient government officer in the Accounts Department. In his leisure he was carrying on research in the Association founded by Mahendralal at Bowbazar. His papers were being published in the foreign journal, *Philosophical Magazine*. Asutosh had great faith in him. 'One day he is going to be a great scientist', he told us once. Besides Acharya Ray, Dr Prafulla Mitter had also joined Science College. He had worked for some time in the Bengal Technical in the early days of Swadeshi. The Swadeshi bosses did not see eye to eye with Palit, so the Palits left the Parsibagan house and moved to Muraripukur. Taraknath had so much faith in Sir Asutosh that he donated all his property to the Calcutta University. Sir Rashbehari had already donated one million rupees as a first instalment.

It was Dr Prafulla Mitter who supervised the construction of the Science College in Calcutta. That is why the three-storey building has drains in every room. It had to be, because all of them were to be used as chemistry laboratories. However, worktables had been put in only in a few of the ground floor rooms in the south; and gas, water and electricity connections were being installed. Both Acharya Ray and Dr Prafulla Mitter had joined and started their research. Ray lived in a room on the first floor of the same part of the building. He had left the Bengal Chemical office. He was a bachelor and led a very simple life. His students looked after him. He had free accommodation — quite spacious too. Dr Mitter would often bring food for him in a tiffin carrier. His admirers also brought sweets and other delicacies. After the day's work Acharya Ray would take a ride in a horse- drawn 'compass' carriage to the monument [the Ochterlony Monument, now known as Saheed Minar] in the maidan [open green area adjoining Chowringhee]. His old friends, students and admirers assembled there — it was an exclusive group; others also went along. Our good friend Devaprasad Ghosh was the best student of our time; he was extremely bright and versatile. He was teaching at Ripon College and also attending the High Court to try his luck in the legal profession. He understood the war strategy of the Germans very well. He could visualize what was coming and explained every move to his admiring listeners.

Our batchmates Jnan, Meghnad and others stayed at 110 College Street. Sureshchandra Bandyopadhyay, Nilratan Dhar and their students were all 'Swadeshi' sympathizers and followers of Dr Ray. They were all members of his evening club. Acharya Ray used to return to his room in the college late in the evening. He never switched on the light. His health was ruined by overwork. Some believed it was the result of handling mercurous nitrite. His disciples studied on their own. At times they had heated political discussions. A lot of impostors were around, including, as some claimed, police informers. Prafullachandra Ray had turned Gandhian — he wore a short lungi, and worked at the spinning wheel to develop 'soul-force'. But every morning he went down to the laboratory and spent a few hours doing his research. The students who assisted him were mostly Palit scholars; some were honorary volunteers. There was a poor student Dr Ray had picked and brought up. He helped him in his research. Dr Ray had great confidence in his analysis. He was concerned then with a lot of weird metal compounds — trying to increase the valency of sulphur, dabbling with compounds of gold and platinum; and nearly all his ideas found corroboration in the student's analysis. He ridiculed the students who had degrees. His boy was the best as far as analysis was concerned. Even though Dr Mitter did not always see eye to eye with him, he was too polite to say so. He was the model of a perfect, well-mannered Bengali gentleman. However too much praise went to the student's head. The alcohol in Dr Ray's store started to evaporate a little too quickly. The gold and platinum compounds also started disappearing. Eventually Dr Ray had to change his opinion, but he was kind and fond of his students. The boy merely lost his job; he was not put to any other difficulty.

These things happened a little later. To come back to the beginning, those of us who had not studied chemistry began to ponder about the next course of action.

In 1912/13 (?) Dr Ganesh Prasad joined as Professor of Applied Mathematics in the Rashbehari Chair. He had a very good reputation. Apparently he had worked with Dr Cline in Germany. He came from Benaras — a man with a simple life style modelled on Dr Ray's. He lived in a bare flat in Samabaya Mansion with hardly any furniture in it, but in the midst of German books and encyclopedias. We had all heard about him. His question papers bristled with out of the world problems, difficult enough to floor the most intelligent students. He had joined the newly built Science College at the bidding of Sir Asutosh. Until then Presidency College had monopolized higher studies in the sciences.

In the contentions in the Senate the group opposed to Sir Asutosh belonged to Presidency College; most of them government officers including a number of Englishmen. The government did not approve of Sir Asutosh's efforts to teach science in the University. It had taken umbrage at the stipulations of the Palit and Ghose chairs, that none but Indians were eligible. What do Indians know about the teaching of science? Till then it had been the age of Cunningham, Harrison, Peake, Vredenburg and the rest. Acharya Ray too had worked in the Provincial Service for a long time. Of course Dr Jagadis Bose was an exception. He had turned to botany from physics. But he continued to teach physics, though most of us were under the care of Dwijen-Babu, Charuchandra Bhattacharya and Barada-Babu. But when radio waves were discussed and experiments shown, we felt elated. Puffed with pride we used to disparage Marconi, saying he hadn't done anything that we couldn't have done. Eventually Dr Bose was to search for a scientific basis for the Upanishads — *Sarvam Khalvidam Brahma* [Brahma lives in everything]. However, his mathematical arguments were not particularly strong; he had to fall back on intuitive knowledge. Later he moved to the Bose Institute and plunged headlong into studying the life sciences.

Dr Ganesh Prasad was then a newcomer to the University. The students flocked to him for training in research. They were the best science students of Calcutta, though several of them had not secured high marks in Ganesh's paper in their M Sc. But the fault lay with the teachers of Presidency College, at least Ganesh Prasad thought so. The young students had to stomach adverse comments about their former teachers, too scared to answer back. After my M Sc I too presented myself to Ganesh Prasad who had been one of my examiners, though I had not fared as badly as the others. Dr Prasad was kind to me at first but I was notorious for plain speaking. I found it difficult to bear his tirade against my teachers. I had dared to counter his adverse criticism. This infuriated him. He said, 'You may have done well in the examination but that does not mean you are cut out for research'. Disappointed, I came back. I decided to work on my own. I was trying to do whatever I could when the Bihar government advertised for young teachers in the newspapers. Since I had not studied law, I decided to stick to mathematics and try my luck. I collected letters of recommendation from Principal James and Dr Mullick [Debendranath Mullick taught at Hooghly, Patna and Presidency Colleges. He was the President of the Physics and Mathematics section of the Indian Science Congress in 1919] and others and applied for the job. But my luck was still against me. I was summoned by Dr Mullick one day. The DPI of Bihar had informed him that I happened to be too good a candidate for their requirement. So I was obliged to look around in Calcutta. One of our classmates, Sailen Ghosh, popularly known as 'Mama' [lit. uncle], had done well in physics. He too was without a job, but he was very resourceful. He went up to Sir Asutosh with a suggestion : why not start teaching subjects in the University that are not taught at Presidency? The logicians had in the meantime thought of various plans for the physics department in the University. But who was going to teach? One day we were called up by Sir Asutosh.

Meghnad, Sailen and I went up the steep stairs to the library, to the special chamber where Sir Asutosh sat. We were naturally meek and submissive and overawed by his august presence. He had heard that the younger generation wanted more modern subjects to be introduced in the University curriculum. He asked, 'What subjects are you competent to teach, boys?'

'Sir, we will try our best to teach whatever you want us to.' He smiled. We had only heard of the many new discoveries in physics, most of them made in Germany — new developments and new discoveries. Planck, Einstein, Bohr — we Bengalis had only heard of them. To know more about them one had to read books in German or research journals in other languages. During the war most of these journals did not come to India.

At long last, as the first step to a new career, we were given a special allowance of Rs 125 per month. Meghnad was assigned to study quantum theory and I had to learn Einstein's relativity theory. We came away committing ourselves to being prepared to teach within a year. But where were we to get the books from? There were some books in English on relativity; we got hold of them. But where could we get hold of the writings of Boltzmann, Kirchhoff, Planck? Suddenly, I thought of an idea. Dr Brühl was the answer.

Dr Brühl was then teaching physics at Sibpur [Bengal Engineering] College. He had set up many delicate instruments which even the Presidency College did not have. All the M Sc students had to do practical work in Brühl's laboratory — a very tough test, a difficult hurdle for most students. But some of our predecessors had acquired a great deal of skill from him. We had heard that Brühl was originally a brilliant student of botany. He fell sick, with a problem of the lungs. Doctors advised him to spend some time in a warm climate. So he left his country on foot with the idea of collecting specimens of plants from various countries. He travelled along the Danube and reached the city of the Turks — Constantinople. Crossing the Sea of Marmara, he entered Asia. From the Turkish empire he crossed Persia and finally reached Karachi. Then he took a boat to Calcutta. Many foreign scientists used to visit the Botanical Garden then.

Brühl however got married when he came here. Since this was not done with prior permission of his emperor, his scholarship was terminated. Dr Brühl stayed on in India. He was an expert in all areas of science. From the post of a mere assistant he rose to a high position. He was fond of reading and had an excellent collection of science books in his library, where we discovered many rare books. We borrowed Planck, Boltzmann, Wien; we could not have asked for more. Meghnad had taken great pains to learn

German, and even passed the Intermediate Examination in the language. I had just started. But I read French. We got hold of a few books on relativity in English. Meanwhile, Sailen had become a great favourite of Sir Asutosh and provided him with a lot of information — if new subjects were to be introduced in the Science College, where to find the fine instruments, where to set up the lecture theatres, the electric points, water taps, what the approximate expenditure would be from the University's own budget, and so on. No import was possible during the War. It was to Sailen's credit that he had gathered a lot of information.

The National Council had once attempted to introduce higher education in the sciences. The apparatus they had bought were lying at Muraripukur. There were also some useful tools at St Xaviers College and at Sibpur — all this information was at Sailen's finger-tips. Emboldened by the idea that nobody would refuse Sir Asutosh, he went as far as Krishnanath College in Baharampur to search for apparatus. But the chemists did not approve of these schemes hatched by the younger generation. They apprehended that the three-storeyed building would soon come to accommodate the other sciences too. In their opinion, Asutosh was being too hasty. Prompted by some immature youngsters he was going in before the right time. He should wait for Deben Bose and Raman to come. These youngsters were hardly capable of carrying the heavy burden.

Asutosh felt that if he could win over Acharya Ray to his side, he would not have any difficulty in carrying through his plans in the Senate. Not that Dr Ray had much faith in us, but fortunately our classmate Jnan [Jnanchandra Ghosh] had already earned the Acharya's appreciation by taking the first step towards his famous theory. The old man had faith in the capability of the young. So Asutosh had little problem in winning him over. The new system was introduced. From 1917 postgraduate courses in applied mathematics, physics, chemistry etc. were to commence in Science College as well. Of course, Presidency College would continue with its old syllabus. Students of all subjects who had done well were approached to take up teaching positions at Science College. Ganesh Prasad was the professor of mathematics; he however preferred to spend most of his time on research. We were asked to teach other subjects. It was indeed a peculiar situation. Apart from Meghnad, Sailen and myself who were chosen to teach physics, others who came were Sushil Acharya and Sisir Mitra, both bright students of Jagadis Bose. From Bangabasi College came Joges Mukhopadhyay and Phanindra Ghosh. Jatin Seth had gone to America after completing his studies at the National Council. He had come back qualified, and he too was asked to join.

There was trouble right at the outset. One night the Police Chief, Basanta Chatterjee, was shot at by unknown assailants. The spies suspected the Seths of Beadon Street; their house was ransacked and searched and all members of the family were taken into custody. Of course nothing incriminating was found. But Jatin Seth was interned. Meanwhile Sailen Ghosh was getting ready to sail abroad for higher studies. He was all set to leave when Mr Tegart summoned him. Sailen was not at home when the messenger came. When he heard of it on his way home, he promptly went into hiding. Eventually after crossing many hurdles he reached America through war-torn Europe.

We had thus lost two colleagues. But we were undeterred. We tried our best to honour our commitment to Asutosh. Meanwhile Jnan Ghosh's articles on the Theory of Electrolytic Conduction were being serialized in the *JCS* [*Journal of the Chemical Society*]. They were being appreciated in the scientific circles. Acharya Ray was much impressed by the achievements of the youngsters. Jnan Ghosh left for London on a Palit travelling fellowship. Mukherjee [Jnanendranath Mukherjee] also got a fellowship at the same time.

The moment I went to Asutosh to plead my own case, he smiled and said, 'But you have got married, you see'. The Palit fellowships were meant for bachelors only. Taraknath Palit believed that a married student would try to save money and send it back to please the newly married wife and not spend the entire money on his studies and science. That is why he had stipulated the condition. Once again I had to come back disappointed.

Meghnad and I taught in both the departments of physics and applied mathematics. However, we did not get any additional allowance for that. Things were different in history and other cultural subjects; such were Asutosh's whims. Dr Deben Bose returned in 1919 and joined Science College. More or less at the same time Raman made up his mind to resign from his accountant's job and come to Science College. He summoned the assistants at once. Until then we had been doing all the teaching. He put more emphasis on research. Poor Meghnad was not very good with his hands; he got into the bad books of the Palit Professor. The other teachers were more inclined to work with Raman to get their doctorates. Raman was involved with the diffraction of light. Phani Ghosh and Sisir Mitra got their doctoral degrees. Phani Ghosh had investigated the reasons behind the appearance of various colours in mica. Mitra studied how light waves were scattered by obstacles and how in suitable conditions the scattered waves combined to form strange lines. As for me and Meghnad, mathematics was our forte. We published a few joint papers. But Meghnad had a profound knowledge of astronomy. His articles on the brightness of stars and the related variations in temperature were highly praised. He got his doctorate from the Calcutta University and set out for Europe on a Guru Prasanna Ghosh fellowship.

In 1921, I was at Calcutta. The Dhaka University had come into being. The first Vice-Chancellor, Mr Hartog, was recruiting new teachers. He called me up one day in Calcutta. Someone in a high position had apparently told him about me. I met him and eventually got the offer of a Readership at Dhaka in 1921.

I had been teaching in Science College since its inception, for four years. Now an invitation came to go to Dhaka.

Published originally in Bengali, in Saptahik Basumati, ***in its special autumn issue in 1374 [Bengali era, corresponding to 1968], and included in*** Rachanā Sankalan. ***Translated.***

Einstein (1935)

The paper in which Einstein introduced the theory of relativity thirty years ago has brought about a new age in physics. The main point of this new theory is that the canvas of space and time which a scientist conceives in order to describe the events of the material world and determine their causal relations, is not independent of the observer, and that there is a close and invariant relationship between the properties of the conceived space-time canvas and the velocity of the observer. Although the theory may not sound very new to a philosopher, that physics remains a science even after one has accepted such a position is not quite self-evident or axiomatic. Hence no scientist before Einstein had thought it possible. The belief that it is possible to measure distances and time intervals in a scientific and objective manner, in total disregard of the observer, and that based on such measurements and with the help of mathematics it is possible to predict the positions and velocities of material objects as well as the results of events, lies at the foundation of science. Hence Newton and the other masters laid the foundation of physics on the assumption that the properties of space-time and their measures were objective and absolute. They had recourse to only Euclid's geometry to measure distances. They believed without a shred of doubt that the measuring rod was the same for all observers. That there could be changes in it brought about by alterations in its motion or position was beyond their ken.

They were equally convinced about the invariance of the time marked by a clock. The objectivity of space and time was treated as an axiom in the mechanics of the Principia and the theory of gravitation. The mathematics and physics that were built on the view that the canvas of space-time was the same for all observers and that its properties were not affected by the state of the observer, were accepted by all for two centuries. With improvements in instruments the field of application of science expanded, and there continued constant checking as to whether calculations and observations matched in every sphere. As a result, it became impossible for the old science to reconcile the minute discrepancies that began to be noticed between predictions and observations.

Towards the end of the nineteenth century it became clear to everyone from the discussions and experiments concerning the effect of the observer's motion on light waves, that there was an inconsistency between the old theory and direct experience. Einstein first proposed the theory of relativity to remove this inconsistency. Earlier it was the ether, the carrier of light waves, and the frequencies of its vibrations that were regarded by scientists as the embodiment of Newton's absolute space-time. Einstein was able to explain very simply and beautifully that this idea was wrong and responsible for all the inconsistencies of the theory of light.

In order to demonstrate that science was possible even if one relinquished the absolute character of space-time, Einstein in his first paper began his examination of the principle of relativity by introducing observers in uniform relative motion. He did not take into account the effects of gravity at this stage. Until then no effect of gravity on light had been discovered. As a result, although the sphere of applicability of the relativity principle was restricted as above, the reconciliation that was required was achieved. Moreover, impartial scientists also became convinced that it was possible to build a science based on such a new principle. However, Einstein did not rest content with his success. After ten more years of relentless effort he was able considerably to extend the applicability of the relativity principle and lay the foundation of a new mathematical theory of gravitation which rivalled Newton's theory. Relativity did not any longer remain confined within the limits mentioned above. Einstein was able to show that it was valid for all kinds of variable motion.

In order to understand the difference between the new theory and that of Newton, we have to recall certain basic concepts of the theory of gravitation. Newton advocated the theory of gravitation in order to explain the variety of motion of material bodies. No matter how great the distance is between any two material objects, according to Newton there is a force of attraction between them. The effect of such a gravitational force varies as the inverse square of the distance. But in his theory there is no room for a limitless material medium to transmit the influence of one body on another. In other words, unlike optics the theory of gravitation is not a wave theory. Many found the idea that material objects could influence one another across distances even while remaining detached from one another in every way, mysterious and difficult to understand. It seems Newton himself regarded his theory to be incomplete in this respect. Newton's successors introduced a similar notion of attraction between electrically charged particles. But Faraday and Maxwell established with the help of experiments and logical arguments that in this case the influence spread in the form of waves from one object to the other. Although many found the idea of action-at-a-distance unsatisfactory, nobody before Einstein was able to propose an alternative theory. His theory was able to determine the characteristics of the orbits of planets and satellites and the variations in their motion in a simple and beautiful manner. The remarkable success of this theory ushered in a new era of astronomy. Copernicus banished Ptolemy. The modern astronomers produced ample evidence that the effect of gravity spread far beyond the solar system and reigned unimpaired in the world of very distant stars.

One must remember that the mechanics consistent with the theory of gravitation was based on the concept of objective space-time. While trying to advocate the principle of relativity instead of this concept, Einstein found that there was no room for the Newtonian concept in his new theory. Consequently, there was a need for a different explanation of the motions of matter, planets and stars. The explanation that he found after ten years of labour is in a way incomprehensible without the help of higher mathematics. But he was able to give every natural phenomenon that had an explanation in terms of the old theory of gravitation a different explanation in the new theory. In order to do this Einstein has abandoned Euclid's geometry and adopted Riemann's perception of space. To express it mathematically, one has to say that the reason behind the diverse motions of material objects is the curvature of the space-time canvas conceived by an observer.

The concept of the universe in this theory is completely new and startling. In the old theory Euclidean geometry was the only perception of space. The universe was infinite in extent and unbounded in such a theory. It was impossible to conceive of its magnitude. In the theory of relativity, although the universe is admitted to be unbounded, it is no longer legitimate to call it infinite or unmeasureable. Once the curvature of space-time is admitted, according to Einstein it is no longer impossible to give a measure of its extent.

Like the motion of material objects, the motion of light is also affected by the curvature of space-time. Hence, the so-called effect of gravity on a ray of light is an indispensable part of Einstein's theory. Nearly all the predictions he made mathematically from this theory have been borne out by experiments. Einstein published the results of his new calculations during the First World War. Immediately after the truce English astronomers tested these conclusions during the solar eclipse of 1919 and announced their veracity. That was the beginning of Einstein's international fame. His ideas began to be discussed even in ordinary newspapers. Even laymen became curious and were eager to discuss the content of his relativity theory. As a result Einstein's name is now well known to scientists and non-scientists alike. There is no limit today to people's curiosity to find out about his life and personality.

Long before he became known to common people, his many-sided genius had amazed lovers of science. The significance of many a complicated problem of science is today easy to grasp as a result of his acute and uncommon insight. Each one of his investigations — on the light-quantum hypothesis, the scientific explanation of the ceaseless agitation of microscopic particles discovered by Brown, etc. — spearheads the victorious march of modern science. Einstein's position in the vanguard of all this is universally acknowledged today.

In order to understand the personality of such extraordinary scientists, it is necessary to take into consideration the post-World War I national and international political situation over the last few years. Those who have faith in the future of human civilization, those who regard human life as invaluable irrespective of race, those who are not averse to acknowledging the axiomatic right of every nation to live on this earth irrespective of colour or religion, were alarmed at the all-engulfing, cruel and frightening countenance of the last World War. After the Treaty of Versailles put out the fire that was about to destroy European civilization, true lovers of humanity in every country became anxious and eager to take steps to prevent its recurrence with a vengeance. This feeling lay at the root of international cooperation. Einstein attended the Geneva Convention with the idea and hope that with the efforts of educated and true lovers of humanity in every country, it should be possible to ensure enduring political peace — that for the welfare of the whole human race man's united efforts, knowledge and experience will ever remain engaged against disease, poverty, blind racial hatred, cruel social oppression and all evils. The history of the last few years has been cruel indeed. Those who dreamt of the end of chauvinism, savage warmongering, jealousy, oppression and imperialism and welcomed the international Geneva Convention as the harbinger of a great age of humane civilization, are having a difficult time.

In a state of frustration after many years of untiring hard work, Einstein refused to attend the Geneva Convention again in 1932. With the rise of Hitler in Germany and the beginning of anti-semitism in 1933, he left his country voluntarily, oblivious of any regard for his personal prestige, comfort or happiness. Like many Jews who held on to their religion, he is today an émigré. Even a scientist of his vision has been caught against his own will in the whirlpool of malice and hatred.

The World As I See It is an English translation of the German book *Mein Weltbild*, a collection of Einstein's lectures on various topics, essays and letters, which was published in 1933. The book has five sections separated subjectwise; the essays concerning science fill nearly half the book. The rest of the articles give us a glimpse of his philosophical position and his opinions on current international affairs and the problems of the Jews in Germany. Most of the essays were written at different times. As a result, the consistency of his opinions expressed in different contexts is not always clear, but there is no scope for misunderstanding the author from the simple and lucid style in which every article has been written. Even if the solutions he offers to overcome the current political and economic crises do not appear logical, compulsive or well-founded enough to all, yet the reverence for mankind that surfaces in his essays will touch and move everyone. The essays on science are the most valuable in the whole book. Among these, many essays on the different problems of modern physics, the history of the theory of relativity, the ideals of a scientist and the scientific method, etc. especially deserve the close attention of serious readers.

I have already given a short introduction to the theory of relativity. I will conclude this essay with a short account of Einstein's views on the nature and method of science. Although a complete understanding of the infinite beauty and pure consciousness that our minds can sense just a little and feebly in the boundless mysteries of creation, is beyond the capacity of the human mind, its relentless pursuit is, according to him, the true sign of a scientific attitude. The scientist tries to comprehend at least partially the entire external world beyond the narrow limits of his personality. Hence he is ever busy building to his own satisfaction a mental representation of the external world reflecting its causal connections. He attempts to grasp the true nature of the external world by reaching beyond the sense impressions coloured by his personality with the help of such a causal representation. Since it is beyond the capacity of human intelligence to analyse every complicated law of nature, a scientist has to remain satisfied with the analyses and syntheses of the comparatively simpler events in that representational domain. It is essential that such analyses and syntheses be logical and correct. Hence a physicist has to use the methods and terminology of mathematics. Consequently, although incomplete, there is no room for errors, vagueness and uncertainties in the scientist's representational domain. Although it is beyond the capacity of the finite human mind to explain the totality of events in the world, Einstein believes that it is possible to find explanations of every material and organic event using the method of logical analysis. There is no logical method, however, of making a discovery within the method of logical analysis that a scientist follows to build his world-picture, starting from a few simple and universal principles. They reveal themselves to the scientist's mind as a result of its intimate acquaintance with the events of the external world. Of course, we have to test the reasonableness and correctness of these principles by checking the predictions that follow from them against our experience.

Like Einstein old-fashioned scientists such as Newton also believed that it was possible to explain the events of the world, starting from a few fundamental axioms and using the analytical method. But many of them thought that it was possible for man to extract these axioms from particular natural experiences, using the method of induction by simple enumeration. They believed that the axioms were logically strung to experience. They also considered it impossible to arrive at an explanation of the universe by choosing the fundamental laws of nature freely.

They used to cite the examples of Euclid's geometry and Newtonian mechanics in support of this view. The theory of relativity proved this view of the scientists to be wrong. Although the axioms of the relativity theory were of a different nature from the concepts of the old mechanics, this theory was also able to provide satisfactory explanations of the same class of events. Therefore, there cannot be any invariable logical connection between the fundamental concepts and definitions on which science is built and man's experiences.

This view has however raised a new problem. If there is no invariable logical connection between the concepts and definitions which are linked together to build up the scientist's world-picture and man's experience, and if the right to choose the concepts freely is admitted, then would it be true to say that there is no relation whatsoever between the world imagined by the scientist and the external world? Is it then a futile attempt on the part of man to understand the real nature of the external world by means of a causal picture? Is utility for man then the final justification for science?

Einstein does not believe that. Einstein firmly believes that although there is no logical connection, the external world uniquely determines the concepts and axioms of our world-picture. He also believes that it is possible for man to discover these unique principles. He maintains that we will be able to grasp the true nature of the external world and the inherent and immutable relationships connecting events with the help of the comparatively simpler concepts of mathematics. But such concepts cannot be found by means of logic alone. We have to depend on our intuition for their discovery. We have to find out whether we have found the right concepts by comparing the predictions that follow from the union of the concepts that arise in our minds and seem probable as a result of our experiences, with the nature of the external world. The science that we create by assuming correct concepts and using mathematics is pure and free of inner contradictions. Though it is free of ambiguities, there is no immutable relation between it and the external world. Thus it is an empty construction. We have to ascribe appropriate correspondences between its concepts and the properties of external objects and see whether we are able to comprehend the events of the external world with the help of the theory.

Einstein has tried to explain with the help of his various attempts in the context of the relativity theory that it is neither futile nor a mere daydream to try and comprehend the nature of the external world following the above method.

It should be clear from the above discussion that like his predecessors, Einstein too believes in the independent existence of an external world and the possibility of a causal scientific description. Many have lost faith in causality as a result of the latest quantum theory in physics. Consequently, they consider the consideration of the probabilities of events rather than their inevitability as the proper aim of science. Although this latest theory is able to solve many complicated problems of the world of atoms and molecules, Einstein firmly believes that causality will be reinstated in the future in science. All his endeavours are entirely devoted to such a reinstatement.

First published in the original Bengali, under the title 'Einstein', in Parichay, *Shravan 1342 (corresponding to July-August 1935), and included in* Rachanā Sankalan. *Translated.*

In the Company of Madame Curie

It was October 1924. That was my first opportunity to visit Europe. The University of Dacca had granted me study leave for two years. I had arrived in Paris. A number of friends like Dr Mukherjee and Dr Bagchi who were already there insisted that I stay on with them. A short time ago the paper regarding photon statistics had been published in German. It had attracted the attention of scientists because Einstein himself had translated and published it with a word of appreciation. This had happened while I was in India. On my arrival in Europe I found myself welcome wherever I went. My first stop was Paris. I hoped to learn something new so that on my return I could be of some use to my students.

Paris had many sights worth seeing. I was keen to see them all. My friends came forward as guides.

Langevin, a student of Pierre Curie, was then the Director of the Municipal School where radium was discovered. He had read my paper and welcomed me warmly. He asked me what I wanted to learn, how long I wished to stay, etc. I went to Madame Curie with a letter of introduction from him. I wanted to join her Institute and learn to work with radioactivity. I was allowed entry to her small chamber. The great elderly lady sat there, dressed in black. I could recognize her from pictures of her that I had seen. I handed her the letter of introduction. She greeted me affectionately and said that there was no way she could disregard a recommendation from such a person. You will certainly get an opportunity to work with me, she said. But not right now, after three or four months. Get to know the language, otherwise you will find it difficult to work in the laboratory. You are in no hurry, I presume.

She spoke at ease in chaste English for about ten minutes. I had no opportunity to tell her that I knew a French of sorts already. I had been at it for the last ten years at home. I came away, resigned to carrying out her instructions. I was in Paris for the next five or six months and had a chance to work at the Radium Institute for some time.

First published in the original Bengali in Jnan O Vijnan, *October-November 1967, and included in* Rachanā Sankalan. *Translated.*

III. Historical Perspectives; Reflections

Articles in this section include a historical account of the discovery of new statistics. Since the discovery of Planck's law of black body radiation, several attempts to unravel the true nature of radiation had failed to provide a satisfactory theoretical foundation. Bose provided that, and he was fortunate to have Einstein's instant recognition. For the sake of completeness, I have included articles well-known in the literature that provide different perspectives, and the article on Bose–Einstein Condensation (Nobel Prize Lecture) for its historical importance. I would like to draw special attention to the article by Partha Ghose. To the best of my knowledge, it is the only source that discusses in depth, Bose's second paper, Einstein's objection to it, and Bose's response, the third paper that was never published and was no where to be found.

Papers by C.V. Raman and S. Bhagavantam on the experimental proof of the spin of a photon are of particular historical importance. Only in private conversations did Bose say that in his original article that was sent to Einstein, he had attributed the factor 2 for agreement with Planck's Law to the spin of the zero mass quantum of radiation (photon). Einstein crossed it out and wrote polarization instead, justifiably perhaps, because the concept of intrinsic spin of an elementary quantum was too radical at the time. Raman and Bhagavantam's papers provide evidence that Bose indeed had the idea of intrinsic spin long before it was postulated for the electron (I refer to Partha Ghose's article S.N. Bose; the Man in this volume for more on this this topic.)

Other articles reflect upon Bose as a teacher, an intellectual and a national figure. The interview with Hermann Mark described in Purnima Sinha's article is particularly interesting as it sheds light on Bose's research activities, his colorful personality and the profound impression he made on them during his stay in Berlin in 1925–26.

A Loss of Identity: the Birth of Quantum Statistics

23a. From Boltzmann to Dirac

This episode begins with a letter dated June 1924 [B1], written by a young Bengali. His name was Satyendra Nath Bose. The five papers he had published by then were of no particular distinction. The subject of his letter was his sixth paper. He had sent it to the *Philosophical Magazine.* A referee had rejected it [B2]. Bose's letter was addressed to Einstein, then forty-five years old and already recognized as a world figure by his colleagues and by the public at large. In this chapter I describe what happened in the scientific lives of these two men during the six months following Einstein's receipt of Bose's letter. For Bose the consequences were momentous. Virtually unknown before, he became a physicist whose name will always be remembered. For Einstein this period was only an interlude.* He was already deeply engrossed in his search for a unified theory. Such is the scope of his oeuvre that his discoveries in those six months do not even rank among his five main contributions, yet they alone would have sufficed for Einstein to be remembered forever.

Bose's sixth paper deals with a new derivation of Planck's law. Along with his letter, he had sent Einstein a copy of his manuscript, written in English, and asked him to arrange for publication in the *Zeitschrift für Physik,* if he thought the work of sufficient merit. Einstein acceded to Bose's request. He personally translated the paper into German and submitted it, adding as a translator's note: 'In my opinion, Bose's derivation of the Planck formula constitutes an important advance. The method used here also yields the quantum theory of the ideal gas, as I shall discuss elsewhere in more detail.'

The purpose of this chapter is not to discuss the history of quantum statistics but rather to describe Einstein's contribution to the subject. Nevertheless, I include a brief outline of Bose's work for numerous reasons. (1) It will give us some insight into what made Einstein deviate temporarily from his main pursuits. (2) It will facilitate the account of Einstein's own research on the molecular gas. That work

*In 1925 Einstein said of his work on quantum statistics, 'That's only by the way' [S1].

is discussed in Section 23b with the exception of one major point, which is reserved for the next chapter: Einstein's last encounter with fluctuation questions. (3) It will be of help in explaining Einstein's ambivalence to Bose's work. In a letter to Ehrenfest, written in July, Einstein did not withdraw, but did qualify his praise of Bose's paper: Bose's 'derivation is elegant but the essence remains obscure' [E1]. (4) It will help to make clear how novel the photon concept still was at that time and will throw an interesting sidelight on the question of photon spin.

Bose recalled many years later that he had not been aware of the extent to which his paper defied classical logic. (Such a lack of awareness is not uncommon in times of transition, but it is not the general rule. Einstein's light-quantum paper of 1905 is a brilliant exception.) 'I had no idea that what I had done was really novel. . . . I was not a statistician to the extent of really knowing that I was doing something which was really different from what Boltzmann would have done, from Boltzmann statistics. Instead of thinking of the light-quantum just as a particle, I talked about these states. Somehow this was the same question which Einstein asked when I met him [in October or November 1925]: how had I arrived at this method of deriving Planck's formula?' [M1].

In order to answer Einstein's question and to understand what gave Bose the idea that he was doing what Boltzmann would have done, I need to make a brief digression.

As was discussed in Section 4b, both logically and historically classical statistics developed via the sequence

fine-grained counting → course-grained counting

This is, of course, the logic of quantum statistics as well, but its historical development went the reverse way, from coarse-grained to fine-grained. For the oldest quantum statistics, the Bose–Einstein (BE) statistics, the historical order of events was as follows.

1924–5. Introduction of a new coarse-grained counting, first by Bose, then by Einstein. These new procedures are the main subject of this chapter.

1925–6. Discovery of nonrelativistic quantum mechanics. It is not at once obvious how the new theory should be supplemented with a fine-grained counting principle that would lead to BE statistics [H1].

1926. This principle is discovered by Paul Adrien Maurice Dirac. Recall first Boltzmann's fine-grained counting formula for his discrete model of a classical ideal gas consisting of N particles with total energy E. Let there be n_i particles with energy ϵ_i (see section 4b, especially Eq. 4.5):

$$N \sum_i n_i \qquad E = \sum_i \epsilon_i n_i \tag{23.1}$$

Then the corresponding number w of microstates is given by

$$w = N! \left(\prod_i n_i! \right)^{-1} \qquad \text{(Boltzmann statistics)} \tag{23.2}$$

We owe to Dirac the observation that in the BE case, Eq. 23.2 must be replaced by

$$w = 1 \qquad \text{(BE statistics)} \tag{23.3}$$

only the single microstate that is symmetric in the N particles is allowed. Dirac went on to show that Eq. 23.3 leads to the blackbody radiation law, Eq. 19.6 [D1]. Thus he brought to an end the search—which had lasted just over a quarter of a century—for the foundations of Planck's law.

Equation 23.3 was of course not known at the time Bose and Einstein completed the first papers ever written on quantum statistics. Theirs was guesswork, but of an inspired kind. Let us turn first to Bose's contribution.

23b. Bose

The paper by Bose [B3] is the fourth and last of the revolutionary papers of the old quantum theory (the other three being by, respectively, Planck [P1], Einstein [E2], and Bohr [B4]). Bose's arguments divest Planck's law of all supererogatory elements of electromagnetic theory and base its derivation on the bare essentials. It is the thermal equilibrium law for particles with the following properties: they are massless, they have two states of polarization, the number of particles is not conserved, and the particles obey a new statistics. In Bose's paper, two new ideas enter physics almost stealthily. One, the concept of a particle with two states of polarization, mildly puzzled Bose. The other is the nonconservation of photons. I do not know whether Bose even noticed this fact. It is not explicitly mentioned in his paper.

Bose's letter to Einstein begins as follows: 'Respected Sir, I have ventured to send you the accompanying article for your perusal. I am anxious to know what you think of it. You will see that I have ventured to deduce the coefficient $8\pi\nu^2/c^3$ in Planck's law independent of the classical electrodynamics . . .' [B1]. Einstein's letter to Ehrenfest contains the phrase, 'the Indian Bose has given a beautiful derivation of Planck's law, including the constant [i.e., $8\pi\nu^2/c^3$]' [E1]. Neither letter mentions the other parts of Planck's formula. Why this emphasis on $8\pi\nu^2/c^3$?

In deriving Planck's law, one needs to know the number of states Z^s in the frequency interval between ν^s and $\nu^s + d\nu^s$. It was customary to compute Z^s by counting the number of standing waves in a cavity with volume V. This yields

$$Z^s = 8\pi(\nu^s)^2 V d\nu^s/c^3 \tag{23.4}$$

Bose was so pleased because he had found a new derivation of this expression for Z^s which enabled him to give a new meaning to this quantity in terms of particle language. His derivation rests on the replacing of the counting of wave frequencies by the counting of cells in one-particle phase space. He proceeded as follows. Integrate the one-particle phase space element $d\vec{x}d\vec{p}$ over V and over all momenta between p^s and $p^s + dp^s$. Supply a further factor 2 to count polarizations. This produces the quantity $8\pi V(p^s)^2 dp^s$, which equals $h^3 Z^s$ by virtue of the relation $p^s = h\nu^s/c$. Hence Z^s is the number of cells of size h^3 contained in the particle phase space region being considered. How innocent it looks, yet how new it was. Recall that the kinematics of the Compton effect had been written down only about a year and a half earlier. Here was a new application of $p = h\nu/c$!

Before I turn to the rest of Bose's derivation, I shall comment briefly on the subject of photon spin. When Bose introduced his polarization factor of 2, he noted that 'it seems required' to do so. This slight hesitation is understandable. Who in 1924 had ever heard of a particle with two states of polarization? For some time, this remained a rather obscure issue. After the discovery of the electron spin, Ehrenfest asked Einstein 'to tell [him] how the analogous hypothesis is to be stated for light-corpuscles, in a relativistically correct way' [E3]. As is well known, this is a delicate problem since there exists, of course, no rest frame definition of spin in this instance. Moreover, gauge invariance renders ambiguous the separation into orbital and intrinsic angular momentum (see, e.g., [J1]). It is not surprising, therefore, that in 1926 the question of photon spin seemed quite confusing to Einstein. In fact, he went so far as to say that he was 'inclined to doubt whether the angular momentum law can be maintained in the quantum theory. At any rate, its significance is much less deep than that of the momentum law' [E4]. I believe that this is an interesting comment on the state of the art some fifty years ago and that otherwise not too much should be made of it.

Let us return to Bose. His new interpretation of Z^s was in terms of 'number of cells,' not 'number of particles.' This must have led him to follow Boltzmann's counting but to replace everywhere 'particles' by 'cells,' a procedure he neither did nor could justify—but which gave the right answer. It may help to understand Bose's remark that he did not know that he was 'doing something which was really different from what Boltzmann would have done, from Boltzmann statistics,' if I recall at this point Boltzmann's coarse-grained counting, which is discussed at more length in Section 4b.

Boltzmann. Partition N particles with total energy E over the one-particle phase space cells $\omega_1, \omega_2, \ldots$ There are N_A particles in ω_A. Their mean energy is E_A. We have

$$N = \sum_A N_A \qquad E = \sum_A E_A N_A \tag{23.5}$$

The relative probability W of this coarse-grained state is

$$W = N! \prod_A \frac{\omega_A^{N_A}}{N_A!} \tag{23.6}$$

The equilibrium entropy S is given by

$$S = k \ln W_{\max} + C \tag{23.7}$$

where C is a constant and W_{max} follows from the extremal conditions

$$\sum \delta N_A(\ln\omega_A - \ln N_A + \lambda - \beta^{-1}E_A) = 0 \tag{23.8}$$

which incorporate the constraints (a) hold N fixed and (b) hold E fixed.

Bose. Partition Z^s into numbers p_r^s, where p_r^s is defined as the number of cells which contain r quanta with frequency ν^s. Let there be N^s photons in all with this frequency and let E be the total energy. Then

$$Z^s = \sum_r p_r^s \tag{23.9}$$

$$N^s = \sum_r r p_r^s \tag{23.10}$$

$$E = \sum_s N^s h\nu^s \tag{23.11}$$

and

$$N = \sum_s N^s \tag{23.12}$$

is the total number of photons. Next Bose introduced his new coarse-grained counting:

$$W = \prod_s \frac{Z^s!}{p_0^s! p_1^s! \ldots} \tag{23.13}$$

He then maximized W as a function of the p_r^s holding Z^s and E fixed so that

$$\sum_{s,r} \delta p_r^s \left(1 + \ln p_r^s + \lambda^s + \frac{1}{\beta} r h\nu^s\right) = 0 \tag{23.14}$$

and then derived Planck's law for $E(\nu, T)$ by standard manipulations—and therewith concluded his paper without further comments.

Bose considered his *Ansatz* (Eq. 23.13) to be 'evident' [B3]. Nothing is further from the truth. I venture to guess that to him the cell counting (Eq. 23.13) was the perfect analog of Boltzmann's particle counting (Eq. 23.6) and that his cell constraint, hold Z^s fixed, was similarly the analog of Boltzmann's particle constraint, hold N fixed. Likewise, the two Lagrange parameters in Eq. 23.14 are his analogs of the parameters in Eq. 23.8. Bose's replacement of fixed N by fixed Z^s already implies that N is not conserved. The final irony is that the constraint of fixed Z^s is irrelevant: if one drops this constraint, then one must drop λ^s in Eq. 23.14. Even so, it is easily checked that one still finds Planck's law! This is in accordance with the now-familiar fact that Planck's law follows from Bose statistics with E held fixed as the *only* constraint. In summary, Bose's derivation introduced three new features:

1. Photon number nonconservation.
2. Bose's cell partition numbers p_r^s are defined by asking how many particles are in a cell. Boltzmann's axiom of distinguishability is gone.
3. The *Ansatz* (Eq. 23.13) implies statistical independence of cells. Statistical independence of particles is gone.

The astounding fact is that Bose was correct on all three counts. (In his paper, he commented on none of them.) I believe there had been no such successful shot in the dark since Planck introduced the quantum in 1900. Planck, too, had counted in strange ways, as was subtly recalled by Einstein in his review, written in 1924, of a new edition of Planck's *Wärmestrahlung:* 'Planck's law [was] derived . . . by postulating statistical laws in the treatment of the interaction between ponderable matter and radiation which appear to be justified on the one hand because of their simplicity, on the other hand because of their analogy to the corresponding relations of the classical theory' [E5].

Einstein continued to be intrigued by Bose's paper. In an address given in Lucerne on October 4, 1924, before the Schweizerische Naturforschende Gesellschaft, he stressed 'the particular significance for our theoretical concepts' of Bose's new derivation of Eq. 23.4 [E6]. By this time, he had already published his own first paper on quantum statistics.

23c. Einstein

As long as Einstein lived, he never ceased to struggle with quantum physics. As far as his constructive contributions to this subject are concerned, they came to an end with a triple of papers, the first published in September 1924, the last two in early 1925. In the true Einsteinian style, their conclusions are once again reached by statistical methods, as was the case for all his important earlier contributions to the quantum theory. The best-known result is his derivation of the Bose–Einstein condensation phenomenon. I shall discuss this topic next and shall leave for the subsequent section another result contained in these papers, a result that is perhaps not as widely remembered even though it is more profound.

First, a postscript to Einstein's light-quantum paper of 1905.

Its logic can be schematically represented in the following way.

$$\text{Einstein 1905:}\qquad \left.\begin{matrix}\text{Wien's law}\\ \\ \text{Gas analogy}\end{matrix}\right\}\rightarrow \text{Light-quanta}$$

An issue raised in Section 19c should be dealt with now. We know that BE is the correct statistics when radiation is treated as a photon gas. Then how could Einstein have correctly conjectured the existence of light-quanta using Boltzmann statistics? Answer: according to BE statistics, the most probable value $\langle n_i \rangle$ of n_i for photons is given by $\langle n_i \rangle = [\exp(h\nu_i/kT) - 1]^{-1}$. This implies that $\langle n_i \rangle \ll 1$ in the Wien regime $h\nu_i \gg kT$. Therefore, up to an irrelevant* factor $N!$, Equations 23.2 and 23.3 coincide in the Wien limit. This asymptotic relation in the Wien region fully justifies, *ex post facto,* Einstein's extraordinary step forward in 1905!

Bose's reasoning in 1924 went as follows:

$$\text{Bose 1924:} \quad \left.\begin{array}{r}\text{Photons}\\ \\ \text{Quantum statistics}\end{array}\right\} \rightarrow \text{Planck's law}$$

and in 1924–5 Einstein came full circle:

$$\text{Einstein 1924–5:} \quad \left.\begin{array}{r}\text{Bose statistics}\\ \\ \text{Photon analogy}\end{array}\right\} \rightarrow \text{The quantum gas}$$

It was inevitable, one might say, that he would do so. 'If it is justified to conceive of radiation as a quantum gas, then the analogy between the quantum gas and a molecular gas must be a complete one' [E7].

In his 1924 paper [E8], Einstein adopted Bose's counting formula (Eq. 23.13), but with two modifications. He needed, of course, the Z^s appropriate for nonrelativistic particles with mass m:

$$\begin{aligned} h^3 Z^s &= 2\pi V(2m)^{3/2} (E^s)^{1/2}\, dE^s \\ 2mE^s &= (p^s)^2 \end{aligned} \tag{23.15}$$

Second (and unlike Bose!), he needed the constraint that N be held fixed. This is done by adding a term

$$-\, r \ln A \tag{23.16}$$

inside the parentheses of Eq. 23.14.** One of the consequences of the thus modified Eq. 23.14 is that the Lagrange multiplier ($-\ln A$) is determined by

$$N = \sum_s N^s = \sum_s \left[\frac{1}{A} \exp\left(\frac{E^s}{kT}\right) - 1\right]^{-1} \tag{23.17}$$

Hence, Einstein noted, the 'degeneracy parameter' A must satisfy

$$A \leqslant 1 \tag{23.18}$$

In his first paper [E8], Einstein discussed the régime in which A does not reach

*The $N!$ is irrelevant since it affects only C in Eq. 23.7. The constant C is interesting nevertheless. For example, its value bears on the possibility of defining S in such a way that it becomes an extensive thermodynamic variable. The interesting history of these normalization questions has been discussed in detail by M. Klein [K1].

**The term A^{-1} is defined as exp $(-\mu/kT)$, where μ is the chemical potential. Einstein, of course, never introduced the superfluous λ^s into the parenthetical term. In Eqs. 23.16–23.22, I deviate from Einstein's notation.

the critical value unity. He proceeded to the continuous limit, in which the sum in Eq. 23.17 is replaced by an integral over phase space, and found

$$\text{if } A < 1: \quad \frac{1}{v} = \frac{\phi_{3/2}(A)}{3} \qquad \frac{p}{kT} = \frac{\phi_{5/2}(A)}{3} \qquad \phi_n(A) = \sum_{m=1}^{\infty} m^{-n} A^m \tag{23.19}$$

with $v = V/N$. He then discussed the region $A \ll 1$, where the equation of state (obtained by eliminating A between Eqs. 23.19) shows perturbative deviations from the classical ideal gas. All this is good physics, though unusually straightforward for a man like Einstein.

In his second paper [E7], the most important one of the three, Einstein began with the $v - T$ relation at $A = 1$:

$$kT_0 = \frac{h^2}{2m\,[v_0\phi_{3/2}(1)]^{3/2}} \tag{23.20}$$

and asked what happens if T drops below T_0 (for given v_0). His answer:

> I maintain that, in this case, a number of molecules steadily growing with increasing density goes over in the first quantum state (which has zero kinetic energy) while the remaining molecules distribute themselves according to the parameter value $A = 1$. . . . A separation is effected; one part condenses, the rest remains a 'saturated ideal gas.' [E7]

He had come upon the first purely statistically derived example of a phase transition, which is now called Bose–Einstein condensation. I defer a few comments on this phenomenon to the next section and turn to other important facets of the three Einstein papers.

1. Einstein on Statistical Dependence. After the papers by Bose [B3] and the first one by Einstein [E8] came out, Ehrenfest and others objected (so we read in Einstein's second paper [E7]) that 'the quanta and molecules, respectively, are not treated as statistically independent, a fact that is not particularly emphasized in our papers' (i.e., [B3] and [E8]). Einstein replied, 'This [objection] is entirely correct' [E7]. He went on to stress that the differences between the Boltzmann and the BE counting 'express indirectly a certain hypothesis on a mutual influence of the molecules which for the time being is of a quite mysterious nature.' With this remark, Einstein came to the very threshold of the quantum mechanics of identical particle systems. The mysterious influence is, of course, the correlation induced by the requirement of totally symmetric wave functions.

2. Einstein on Indistinguishability. In order to illustrate further the differences between the new and the old counting of macrostates, Einstein cast W in a form alternative to Eq. 23.13. He counted the number of ways in which N^s indistinguishable particles in the dE^s interval can be partitioned over the Z^s cells. This yields

$$W = \prod_s \frac{(N^s + Z^s - 1)!}{N^s!\,(Z^s - 1)!} \tag{23.21}$$

Einstein's Eq. 23.21 rather than Bose's Eq. 23.13 is the one now used in all textbooks.

3. Einstein on the Third Law of Thermodynamics. As was noted at the end of Section 20c, in 1914 Nernst introduced the hypothesis that the third law of thermodynamics applies to gases. It was also mentioned that no sensible model of a gas with that property was available at that time. In 1925 Einstein made his last contribution to thermodynamics by pointing out that the BE gas does satisfy the third law. (A Boltzmann gas does not do so, Einstein remarked.) Indeed, since all particles go into the zero energy state as $T \rightarrow 0$, we have in this limit $N^0 = N$, all other $N^s = 0$. Hence $W \rightarrow 1$ and $S \rightarrow 0$ as $T \rightarrow 0$. It was as important to him that a molecular BE gas yield Nernst's law as that a BE photon gas yield Planck's law.

4. Einstein and Nonconservation of Photons. After 1917 Einstein ceased to write scientific articles on questions related to radiation.* The only mention of radiation in the 1924–5 papers is that 'the statistical method of Herr Bose and myself is by no means beyond doubt, but seems only *a posteriori* justified by its success for the case of radiation' [E11].

There can be no doubt that he must have noted the nonconservation of photons. In his language, this is implemented by putting $A = 1$ in Eq. 23.16. Yet I have not found any reference to nonconservation, either in his scientific writings or in the correspondence I have seen. I cannot state with certainty why he chose to be silent on this and all further issues regarding photons. However, I do believe that it is a fair guess that Einstein felt he would have nothing fundamental to say about photons until such time as he could find his own way of dealing with the lack of causality he had noted in 1917. Such a time never came.

Other physicists had followed Einstein's work on quantum statistics with interest. Lorentz invited him to speak on this subject at the 1927 Solvay congress. Einstein's reply, written in June 1927, may serve as a most appropriate preliminary to my subsequent discussion of quantum mechanics.

> I recall having committed myself to you to give a report on quantum statistics at the Solvay congress. After much reflection back and forth, I come to the conviction that I am not competent [to give] such a report in a way that really

*Except for a brief refutation of an objection to his work on needle radiation [E9]. I found a notice by Einstein in 1930 announcing a new paper on radiation fluctuations [E10]. This paper was never published, however.

corresponds to the state of things. The reason is that I have not been able to participate as intensively in the modern development of the quantum theory as would be necessary for this purpose. This is in part because I have on the whole too little receptive talent for fully following the stormy developments, in part also because I do not approve of the purely statistical way of thinking on which the new theories are founded. . . . Up until now, I kept hoping to be able to contribute something of value in Brussels; I have now given up that hope. I beg you not to be angry with me because of that; I did not take this lightly but tried with all my strength. . . . Perhaps Herr Fermi in Bologna . . . or Langevin . . . could do a good job. [E12]

23d. Postscript on Bose–Einstein Condensation

(1) In December 1924, Einstein wrote to Ehrenfest, 'From a certain temperature on, the molecules "condense" without attractive forces, that is, they accumulate at zero velocity. The theory is pretty, but is there also some truth to it?' [E13].

(2) In 1925, Einstein mentioned hydrogen, helium, and the electron gas as the best possible candidates in which to observe his condensation phenomenon [E7]. In 1925, these were, of course, sensible proposals. Recall that the Fermi–Dirac statistics was not discovered* until 1926 [F1, D1], following Pauli's enunciation of the exclusion principle in 1925 [P2]. Even then, it took some time until it was sorted out when BE and FD statistics apply respectively: referring to Dirac's paper [D1], Pauli wrote in December 1926, 'We shall take the point of view also advocated by Dirac, that the Fermi, and not the Einstein–Bose, statistics applies to the material gas' [P3]. These matters were cleared up by 1927.

(3) In his 1925 paper, Einstein did not call the condensation phenomenon a phase transition. According to Uhlenbeck (private communication), nobody realized in 1925 that the existence of a phase transition was a 'deep' problem. In 1926, Uhlenbeck himself raised an objection to Einstein's treatment of the condensation problem [U1]. This critique was to lead to a more precise theoretical formulation of the conditions under which phase transitions can occur. Uhlenbeck noted that the quantity N^0 in Eq. 23.17 $\rightarrow \infty$ as $A \rightarrow 1$ (for fixed T); hence also $N \rightarrow \infty$. Thus, if $A \rightarrow 1$, it is impossible to implement the constraint that N is a fixed finite number. Therefore $A = 1$ can be reached only asymptomatically and there is no two-phase regime.

Uhlenbeck recently described the communications between Ehrenfest and Einstein on this question [U2]. Uhlenbeck and Einstein were both right, however. The point is that a sharp phase transition can occur only in the so-called thermodynamic limit $N \rightarrow \infty$, $V \rightarrow \infty$, v fixed. This view emerged in a morning-long debate that took place during the van der Waals Centenary Conference in November 1937. The issue was, Does the partition function contain the information necessary to describe a sharp phase transition? The transition implies the

*Dirac has given a charming account of the time sequence of these discoveries [D2].

existence of analytically distinct parts of isotherms. It was not clear how this could come about. The debate was inconclusive, and Kramers, the chairman, put the question to a vote. Uhlenbeck recalls that the ayes and nays were about evenly divided. However, Kramers' suggestion to go to the thermodynamic limit was eventually realized to be the correct answer. Shortly afterward, Uhlenbeck withdrew his objections to Einstein's result, in a joint paper with his gifted student, the late Boris Kahn (a Nazi victim) [K2].

(4) Until 1928, the BE condensation had 'the reputation of having only a purely imaginary character' [L1]. Recall that the HeI–HeII phase transition was not discovered until 1928, by Willem Hendrik Keesom [K3]. In 1938, Fritz London proposed interpreting this helium transition as a BE condensation. Experimentally, the transition point lies at 2.19 K. It is most encouraging that Eq. 23.20 gives T = 3.1 K [L2]. It is generally believed but not proved that the difference between these two values is due to the neglecting of intermolecular forces in the theoretical derivations.

References

B1. S. N. Bose, letter to A. Einstein, June 4, 1924.
B2. W. Blanpied, *Am. J. Phys.* **40,** 1212 (1972).
B3. S. N. Bose, *Z. Phys.* **26,** 178 (1924).
B4. N. Bohr, *Phil. Mag.* **26,** 1 (1913).
D1. P. A. M. Dirac, *Proc. Roy. Soc.* **112,** 661 (1926).
D2. ——, *History of Twentieth Century Physics,* Varenna Summer School, pp. 133–4. Academic Press, New York, 1977.
E1. A. Einstein, letter to P. Ehrenfest, July 12, 1924.
E2. ——, *AdP* **17,** 132 (1905).
E3. P. Ehrenfest, letter to A. Einstein, April 7, 1926.
E4. A. Einstein, letter to P. Ehrenfest, April 12, 1926.
E5. ——, *Deutsche Literaturzeitung,* 1924, p. 1154.
E6. ——, *Verh. Schw. Naturf. Ges.* **105,** 85 (1024).
E7. ——, *PAW,* 1925, p. 3.
E8. ——, *PAW,* 1924, p. 261.
E9. ——, *Z. Phys.* **31,** 784 (1925).
E10. ——, *PAW,* 1930, p. 543.
E11. ——, *PAW,* 1925, p. 18.
E12. ——, letter to H. A. Lorentz, June 17, 1927.
E13. ——, letter to P. Ehrenfest, November 29, 1924.
F1. E. Fermi, *Z. Phys.* **36,** 902 (1926).
H1. W. Heisenberg, *Z. Phys.* **38,** 411 (1926).
J1. J. M. Jauch and F. Rohrlich, *The Theory of Photons and Electrons,* p. 40. Addison-Wesley, Reading, Mass., 1955.
K1. M. Klein, Proc. Kon. Ned. Akad. Wetensch. Amsterdam, **62,** 41, 51 (1958).
K2. B. Kahn and G. E. Uhlenbeck, *Physica* **4,** 399 (1938).
K3. W. H. Keesom, *Helium.* Elsevier, New York, 1942.

L1. F. London, *Nature* **141,** 643 (1938).
L2. —, *Phys. Rev.* **54,** 1947 (1938).
M1. J. Mehra, *Biogr. Mem. Fell. Roy. Soc.* **21,** 117 (1975).
P1. M. Planck, *Verh. Deutsch. Phys. Ges.* **2,** 237 (1900).
P2. W. Pauli, *Z. Phys.* **31,** 625 (1925).
P3. —, *Z. Phys.* **41,** 81 (1927).
S1. E. Salaman, *Encounter,* April 1979, p. 19.
U1. G. E. Uhlenbeck, 'Over Statistische Methoden in de Theorie der Quanta,' PhD thesis, Nyhoff, the Hague, 1927.
U2. —, *Proceedings Einstein Centennial Symposium 1979* (H. Woolf, Ed.). Addison-Wesley, Reading, Mass., 1980.

Bose Statistics : a historical perspective

a. Planck's derivation of his law of black-body radiation

In order to appreciate the novelty and importance of Bose's work in its historical perspective, it is necessary first to give a brief account of Planck's original derivation [Planck, 14 December 1900] of the law of black-body radiation,

$$\rho(\nu, T) = \frac{8\pi h\nu^3}{c^3} \frac{1}{e^{h\nu/kT} - 1} \tag{1}$$

which he had discovered empirically a few weeks earlier (7 October 1900). It consisted of *three* steps.

First, he established the relation

$$\rho_\nu = \frac{8\pi\nu^2}{c^3} U_\nu \tag{2}$$

between the energy density ρ_ν of incident radiation at temperature T, whose frequency lies between ν and $\nu + d\nu$, and the average energy U_ν of a resonator of the same frequency ν at temperature T, on the basis of *classical* electromagnetic theory. Comparing (1) and (2), he found U_ν :

$$U_\nu = \frac{h\nu}{e^{h\nu/kT} - 1} . \tag{3}$$

In the second step he determined the entropy of these oscillators by integrating $TdS=dU$ where T is taken from (3) as a function of U (for fixed ν). He obtained

$$S = k\left[\left(1+\frac{U_\nu}{h\nu}\right)\ln\left(1+\frac{U_\nu}{h\nu}\right) - \frac{U_\nu}{h\nu}\ln\frac{U_\nu}{h\nu}\right]. \tag{4}$$

The third step was the revolutionary one. He introduced two ideas at this stage that he himself considered as 'acts of desperation'. He assumed that the total energy $U_N = N\,U_\nu$ of N oscillators (resonators) was made up of finite energy elements ϵ such that $U_N = P\epsilon$ with P a large number. Then he searched in Boltzmann's work for a permutation measure W_N (the total number of complexions or distributions) of P discrete equal energy values ϵ among N oscillators that would correspond to the right hand side of (4) when used in $S_N = k\,ln\,W_N$. He found

$$W_N = \frac{(P+N-1)!}{P!(N-1)!}\,. \tag{5}$$

Using $P/N = U_\nu/\epsilon$, $S_N = N\,S$ and applying Stirling's formula, he obtained

$$S = k\left[\left(1+\frac{U_\nu}{\epsilon}\right) ln \left(1+\frac{U_\nu}{\epsilon}\right) - \frac{U_\nu}{\epsilon}\, ln\, \frac{U_\nu}{\epsilon}\right]. \tag{6}$$

Since S is a function of (U_ν/ν) only, it follows from (4) and (6) that

$$\epsilon = h\nu\,.$$

This is how quantum theory was born. Planck had no more justification for using the formula (5) than that it gave him the result that he was after : '. . . a theoretical interpretation had to be found at any price, however high it might be,' he wrote to Robert William Wood in October 1931. [Mehra and Rechenberg, 1982, **1**, *1*, 50]. Actually, Planck's combinatorial approach differed from Boltzmann's probabilistic method in that Planck associated W_N with S_N, the equilibrium entropy, without maximizing it. For Boltzmann W_N was the number of possible complexions corresponding to the macrostate which can be realized by the largest number of complexions. As pointed out by Ehrenfest, a strict adherence to the accepted principles of statistical mechanics would have led Planck to the classical law of Rayleigh and Jeans! [Ehrenfest, 1905]

b. Einstein's light-quantum hypothesis

In 1905 Einstein argued on the basis of his analysis of energy fluctuations of radiation obeying Wien's law that such radiation had statistical properties similar to material particles and so must consist of discrete light-quanta of magnitude $(R/N_o)\,\beta\nu = h\nu$. [Einstein, 1905] This light-quantum hypothesis was immediately applied by him to give a reasonable explanation of several radiation phenomena, specially the photo-electric effect. Since these conclusions contradicted the classical electromagnetic theory on which Planck attempted to base his theory of heat radiation, Einstein considered

Planck's theory 'in some way a counterpart' to his own theory. He subjected Planck's derivation to a critical analysis and came to the following conclusion :

> We must therefore regard the following law as the basis of Planck's quantum theory of radiation : the energy of an elementary resonator can only assume values which are integral multiples of $(R/N_o)\,\beta\nu$; the energy of a resonator changes in jumps by absorption or emission in integral multiples of (R/N_o) $\beta\nu$. . . . If the energy of a resonator can alter only in jumps, then for the evaluation of the average energy of a resonator in a radiation cavity, the usual [electromagnetic] theory cannot be used, for the latter does not admit any distinctive energy values for a resonator. [Einstein, 1906]

Two years later Lorentz also came to the same conclusion [Lorentz to Wien, 6 June 1908; Mehra and Rechenberg, 1982, **1**, *1*, 98]. Planck attempted to modify his theory to take into account the quantum of action *without violating any aspect of classical electrodynamics*, but it gradually became clear from his studies and those of others that it was impossible to establish the theory of black-body radiation entirely on the foundation of Maxwell's classical electrodynamics and the statistical mechanics of Maxwell and Boltzmann.

c. Debye's derivation of Planck's law

In 1910 Debye gave a new derivation of Planck's radiation law in which he sought to avoid the inconsistencies (as viewed from Maxwell's theory) of Planck's earlier derivations as pointed out by Einstein [Debye, 1910]. Instead of using the relation (2) between the radiation density ρ_ν and the average energy of the oscillator U_ν, Debye calculated 'the probability for a given state of radiation and therefore as is well known, the entropy, using the properties of the state alone without employing resonators.' In agreement with Rayleigh and Jeans, he calculated the number $N_\nu\, d\nu$ of elementary states or vibrational modes (*Hohlraum oscillators*) contained in a volume V and obtained

$$N_\nu\, d\nu = \frac{8\pi \nu^2 V\, d\nu}{c^3} . \tag{8}$$

Assuming that an amount of energy $h\nu$ gets distributed over each of these vibrations according to an arbitrary distribution function f_ν, he obtained

$$\rho_\nu\, d\nu = \frac{8\pi h\nu^3}{c^3} f_\nu\, d\nu . \tag{9}$$

Now, 'analogously to Planck', Debye *assumed* that the probability of distributing $N_\nu f_\nu\, d\nu$ quanta of energy $h\nu$ among $N_\nu\, d\nu$ vibrations was given by

$$\omega_\nu = \frac{(N_\nu\, d\nu + N_\nu f_\nu\, d\nu)\,!}{(N_\nu\, d\nu)\,!\ (N_\nu f_\nu\, d\nu)\,!} . \tag{10}$$

He then calculated the distribution function f_ν in the following way. He calculated the maximum of ω_ν with the constraint that the energy remained constant. From this he calculated the equilibrium entropy S_ν to be $k\ ln\ (\omega_\nu^{max})$. Finally, using the definition of entropy to be $(dS_\nu/d\rho_\nu) = T^{-1}$, he obtained

$$f_\nu = \frac{1}{e^{h\nu/kT} - 1}. \tag{11}$$

Together with (9), this yielded Planck's radiation formula (1).

Two important features of Debye's derivation need to be emphasized as background to Bose's derivation that came fourteen years later. First, it became clear from his derivation that Planck's law follows simply from the assumption that the energy transfer from matter to the heat radiation and vice versa is quantized in units of $h\nu$ and that *no knowledge of the properties of material resonators is needed.* In other words, the quantum hypothesis only determined the transfer of energy from one mode of the radiation to another. This was in keeping with Planck's ideas but not Einstein's. Second, he used Planck's definition of the probabilities ω_ν, without analyzing its significance.

d. Indistinguishability of Planck's quanta

In 1911 Ladislas Natanson [1911] subjected the Planck-Debye combinatorial procedure to a critical analysis and showed that it was equivalent to distributing *P indistinguishable* energy elements ϵ among *N* 'receptacles of energy' distinguished only by the numbers *j* of quanta in them so that N_j receptacles contain *j* quanta each, subject to the constraints $\sum_j N_j = N$ and $\sum_j j N_j = P$. He found

$$\omega_\nu = \frac{(N_\nu\, d\nu)!}{\prod_{j=0}^{p} N_j!}. \tag{12}$$

Contrary to popular belief, therefore, the *indistinguishability of the quanta had already been tacitly assumed by Planck and Debye and this was first noticed by Natanson already in 1911.*

Ehrenfest and Kamerlingh Onnes [1914] also came to a similar conclusion in 1914. They however claimed that Planck's procedure was equivalent to distributing *P indistinguishable* energy elements ϵ among *N distinguishable* resonators and arrived at the expression

$$\omega_\nu = \frac{(N-1+P)!}{P!\,(N-1)!} \tag{13}$$

which is equivalent to the expressions (10) and (12) when N>>1. They then proceeded to clarify the distinction between Planck's energy quanta *which were statistically not independent* (they called them 'non-disjointed quanta') and Einstein's hypothetical light-quanta which were believed to be *statistically independent of one another* ('disjointed quanta'). The concluded : *'Planck's formal device* (distribution of P energy-elements ϵ among N resonators) *cannot be interpreted in the sense of Einstein's light-quanta.'*

e. The introduction of phase space cells by Planck

Planck himself had come to the view on his own that classical statistical mechanics had to be modified in order to yield his radiation formula (1) rather than that of Rayleigh and Jeans. He reported his important conclusion at the Solvay Congress held in Brussels in 1911. According to Gibbs, the probability of finding a single particle in the element $d^3p\,d^3q$ of the six-dimensional phase space is given by

$$\frac{e^{-E/kT}\ d^3p\,d^3q}{\int e^{-E/kT}\ d^3p\,d^3q} \tag{14}$$

where $E = E(p,q)$ is the energy of the system. For a classical one-dimensional oscillator, $E = p^2/2m + \beta q^2/2$ and so the average energy of such an oscillator is kT. However, if E can take only the discrete values $E_n = n\epsilon = nh\nu$ with n=0,1,2 . . ., then

$$U = \frac{\sum\limits_n E_n\ e^{-E_n/kT}}{\sum\limits_n e^{-E_n/kT}} = \frac{\epsilon}{e^{\epsilon/kT} - 1}, \tag{15}$$

in agreement with (3). It was this straightforward derivation that led Planck for the first time to make the explicit statement that the energy of an oscillator (and not only its average energy) was an integral multiple of $h\nu$. It also led him to interpret the constant h *as a finite extension of the elementary area in phase space.* The motions of one-dimensional classical oscillators can be described by a family of concentric ellipses of semi-axes $(2E/\beta)^{1/2}$ and $(2\,m\,E)^{1/2}$. Since these oscillators can have any energy E, the ellipses form a continuum. However, if it is assumed that these ellipses are separated from each other so that the nth ellipse encloses an area

$$\int\!\!\int dp\,dq = nh, \tag{16}$$

then the energies of the oscillators must be integral multiples of $h\nu$. Planck therefore drew the revolutionary conclusion that energy quanta were a consequence of the fundamental condition

$$\int\!\!\int_E^{E+\epsilon} dp\,dq = h. \tag{17}$$

This implied a revision of the basic tenets of classical mechanics. He declared : 'The framework of classical mechanics, even if combined with the Lorentz-Einstein principle of relativity, is obviously too narrow to account for all those physical phenomena which are not directly accessible to our coarse senses . . . One should therefore confine oneself to the principle that the elementary region of probability h has an ascertainable finite value and avoid all further speculation about the physical significance of this remarkable constant.' In other words, all attempts at finding a classical mechanical explanation of h must be abandoned. As we shall see later, to Bose these considerations (with which he was familiar) brought home extremely important lessons that would influence his own seminal contribution to the subject. First, he took serious note of the conclusion that classical electrodynamics was essentially incompatible with Planck's radiation formula. Second, since the quantum states of a particle were not continuously distributed throughout phase space (as in classical dynamics), he concluded that their number could be counted by simply dividing the total volume of phase space by h^3, the volume of an elementary cell. Finally, he took seriously Planck's claim that classical statistical mechanics had to be modified in a special way in order to conform to the requirements of quantum theory.

f. Spontaneous and induced transitions

The next important step was taken by Einstein in 1916. In 1913 Niels Bohr had proposed his model of the atom with stationary states with discrete energy values [Bohr,1913, b,c]. The transitions between any pair of these states were assumed to be caused by the exchange of energy quanta $h\nu$ between the atom and the radiation field described by *classical electrodynamics*. Einstein gave an elegant derivation of Planck's radiation formula by considering Bohr atoms in thermal equilibrium with radiation in a cavity but *without making any use of classical electrodynamics* [Einstein, 1916a]. He used Boltzmann's principle to write the probability W_n for an atom to be in a stationary state with quantum-number n in the form

$$W_n = g_n \, exp\left(-\frac{\epsilon_n}{k\,T}\right), \tag{18}$$

g_n being the statistical weight of the state. He then assumed that a stationary state n may pass to a stationary state m of higher energy ($\epsilon_m > \epsilon_n$) by absorbing a light-quantum of frequency $\nu_{n\,m}$, the rate of transition being $B_n^m \, N_n \, \rho_\nu$ (N_n being the number of atoms in the stationary state n and B_n^m a proportionality constant characterizing the absorption of radiation of frequency $\nu_{n\,m}$ in the presence of external radiation of density ρ_ν). Transitions from a stationary state m of higher energy to a stationary state n of lower energy by the emission of a light-quantum of frequency $\nu_{m\,n} = \nu_{n\,m}$ could, however, take place in *two* ways. Atoms can emit this radiation *independent* of the external field, their number being $A_m^n \, N_m$ (A_m^n being the constant of proportionality). About this Einstein said : 'One can hardly think of it in any other way except as a radioactive reaction.' It came to be known later as 'spontaneous emission.' Atoms can also be *induced* to emit radiation by the external field, the rate being $B_m^n \, N_m \, \rho_\nu$. Einstein called this 'negative radiation'. In thermal equilibrium one must have $(A_m^n N_m + B_m^n N_m \rho_\nu) = B_n^m N_n \rho_\nu$. Using (18) to calculate N_n/N_m, one gets

$$A_m^n g_m + \rho_\nu \left[B_n^m g_n \exp\left(\frac{\epsilon_m - \epsilon_n}{kT}\right) - B_m^n g_m \right]. \tag{19}$$

At this stage Einstein had to take his guidance from classical theory by essentially applying the 'correspondence principle' according to which classical theory should be a limiting case of quantum theory. In the classical limit ρ_ν goes to infinity as the temperature T goes to infinity for fixed ν (Rayleigh-Jeans law, $\rho_\nu \alpha T$). This gave $B_m^n g_m = B_n^m g_n$. With this condition (19) could be written as

$$\rho_\nu = \frac{A_m^n / B_m^n}{\exp\left[(\epsilon_m - \epsilon_n)/kT\right] - 1}. \tag{20}$$

Then Einstein used Wien's displacement law to infer that A_m^n/B_m^n must be proportional to ν^3, and the Rayleigh-Jeans law in the limit of low frequencies to determine the constant of proportionality to be $8\pi h/c^3$. Thus he obtained the relation

$$A_m^n = \frac{8\pi h \nu^3}{c^3} B_m^n \tag{21}$$

and

$$\epsilon_m - \epsilon_n = h\nu \tag{22}$$

which is Bohr's frequency condition. These two relations when substituted into (20) gave Planck's law. This is a remarkable derivation that uses an admixture of the light-quantum hypothesis and Bohr's atomic theory which Bohr regarded as being imcompatible, Wien's radiation law and therefore implicitly the Maxwell-Boltzman distribution for the light-quanta, and the correspondence principle which was a heuristic principle without any fundamental basis. Einstein concluded by saying : 'I admit freely, of course, that the three hypotheses concerning outgoing and incoming radiation, do not at all become substantiated results by the mere fact that they lead to Planck's radiation formula. But the simplicitly of the hypotheses, the generality with which the consideration can be carried through easily, as well as the natural connection of the consideration employed with the limiting case of Planck's linear oscillator (in the sense of classical electrodynamics and mechanics), persuaded me to regard it as very probable that all this constitutes the fundamental outline of the future theoretical derivation.' Although the hypotheses concerning the outgoing and incoming radiation turned out to be correct to a large extent (though not entirely so, as we shall see later), the future derivation (given by Bose) did not make use of them at all. In fact, they became the major bone of contention between him and Bose.

By 1916 the light-quantum hypothesis had received a measure of empirical support from Millikan's verification of Einstein's photo-electric equation. Millikan's paper [1916b] appeared in the March 1916 issue of the *Physical Review*. Einstein quickly realized from his new derivation of Planck's law that atoms seemed to interact with radiation as if single atoms collided with light-quanta of energy $h\nu$ and directed momentum $h\nu/c$. He wrote to his friend, Michael Besso :

> The fundamental thing is that the *statistical* consideration, which leads to Planck's formula, has become *unified* and thereby as general as one can imagine, since one has not assumed anything about the nature of the mediating molecules other than the most general quantum idea. From it follows the result (which was not contained in the paper I sent you) that in each energy transfer from radiation to matter the momentum $h\nu/c$ is also transferred to the molecule. Hence we conclude that every such elementary process is a *completely directed* event. With that the light-quanta must be considered as being substantiated. [Einstein to Besso, 6 September 1916; Mehra and Rechenberg, 1982, **1**, *2*, 515]

However, Einstein was not really satisfied. In a subsequent paper he remarked :

> The weakness of the theory lies, on one hand, in the fact that it does not bring us closer to a connection with the undulatory theory, and, on the other hand, in the fact that it leaves the instant and direction of the elementary processes to chance. [Einstein, 1916b, 1917]

g. Reactions to the light-quantum hypothesis: discovery of the Compton effect

The majority of physicists, in fact, did not take the light-quantum hypothesis seriously, even after Millikan's work. Millikan himself remarked in his 1916 paper on the photo-electric effect : 'Yet the semi-corpuscular theory by which Einstein arrived at this equation seems at present to be wholly untenable.' The main difficulty lay, as Einstein himself realized, in accounting reasonably for the interference and diffraction phenomena observed with all kinds of radiation in terms of light-quanta. Evidence in favour of the wave theory was overwhelming and mounting. In 1912 Max von Laue, Walter Friedrich and Paul Knipping had settled the issue in favour of X-rays being electromagnetic waves of extremely short wavelengths through the phenomenon of interference of X-rays in crystals [von Laue, Friedrich and Knipping, 1912]. In 1913 Planck, Nernst, Rubens and Warburg who wished to bring Einstein to Berlin, wrote to the Prussian Ministry of Education:

> That he may sometimes have missed the target in his speculations, as, for example, in his theory of light-quanta, cannot really be held against him. For in the most exact of natural sciences every innovation entails risk. [Clark, 1972, 215]

Bohr had his own reasons for not favouring the light-quantum hypothesis though he realized from the beginning that the laws of electrodynamics had to be abandoned for a proper description of quantum phenomena. The hypothesis ran counter to his plans to build atomic theory on the basis of the correspondence principle.

The general acceptance of the light-quantum hypothesis eventually came in 1923 through the work of Debye [1923] and Compton [1923]. Compton realized that the shift in the wavelength of X-rays scattered by atoms that he observed could not be understood in terms of the classical wave theory of scattering but only in terms of elementary processes of scattering of light-quanta by the electrons in atoms, in which both energy and momentum were conserved. Debye also arrived at the same conclusion independently. Arnold Sommerfeld who was then visiting the University of Wisconsin as Carl Schurz Professor of Physics came to know about Compton's results. On 21 January 1923 he wrote to Bohr : 'The most interesting scientific news I have learned in America is a work of Arthur Compton in St. Louis. According to it the wave theory of X-rays would finally have to be given up.' [Mehra and Rechenberg, **1**, *2*, 529] Wherever he went, Sommerfeld referred to Compton's work and its importance for the quantum theory. On his return to Germany he continued to advocate Compton's work. Several months later he wrote to Compton : 'Your discovery of the change in wavelength of X-rays also occupies the scientific community in Germany most deeply. I met Einstein and Kossel in August, and we mainly discussed your effect.' [Sommerfeld to Compton, October 1923; Mehra and Rechenberg, 1982, **1**, *2*, 530]

h. Pauli's derivation of Planck's law

Wolfgang Pauli came to know of Compton's discovery in the summer of 1923 when he came to Hamburg on several weeks' leave of absence from Copenhagen. He immediately got interested in the problem of thermal equilibrium between radiation and free electrons. Lorentz had reported in the first Solvay Conference that he could not establish the condition of thermal equilibrium for electrons having a Maxwellian velocity distribution [Lorentz's report in Langevin and de Broglie, 1912]. Pauli wished to re-examine the problem by considering reversible collisions between electrons and directed light-quanta, obeying the laws of conservation of energy and momentum. Pauli [1923] was able to show that thermal equilibrium characterized by Planck's law was possible for electrons having a Maxwellian distribution provided the probability of collisions per unit time was given by

$$A\,\rho_\nu + B\,\rho_\nu\,\rho_{\nu'}\,, \qquad (23)$$

where A and B are the coefficients of Einstein's radiation theory [Einstein, 1917] and ρ_ν and $\rho_{\nu'}$ the densities of radiation before and after the collision respectively. Pauli showed that the first term was dominant for radiation satisfying Wien's law while the second term was more important for radiation of longer wavelengths satisfying the Rayleigh-Jeans law. Pauli interpreted this as the effect of interference fluctuations of classical waves. Einstein and Ehrenfest welcomed Pauli's work and clarified and generalized it to cases in which the elementary processes involved more than two light-quanta [Einstein and Ehrenfest, 1923]. In particular, they discussed the 'paradoxical' second term in (23) 'which implies that the number of the elementary processes of scattering occurring per unit time at an electron (which is at quasi rest) increases faster than being proportional to the radiation density ρ' of frequencies ν', which the radiation quantum possesses after being modified by the elementary process.' They concluded that this term followed from a detailed considertion of all interaction processes between radiation and electrons, including the induced emission of light (Einstein's 'negative radiation').

i. Bose's derivation

While the leading European and American physicists were skeptical, even hostile to Einstein's light-quantum hypothesis, two unknown young physicists in India, M N Saha and S N Bose grasped its importance and started using it with great success. Within two years of Einstein's 1917 paper from which he drew the revolutionary conclusion that a light-quantum carried directed momentum $h\nu/c$, M N Saha, a close friend of Bose, applied it to explain the existence of radiation pressure on objects of molecular size [Saha, 1919]. On the basis of the classical theory Nicholson [1914] and Klotz [1918] had shown that radiation pressure on particles of molecular size should become totally 'evanescent'. Saha applied 'quantum theory in the place of the old continuous theory of light. Instead of assuming that "light" is spread continuously over all points of space,' he supposed that 'they are localized in pulses of energy $h\nu$ . . . Let this pulse encounter a molecule m and be absorbed by it. Then in doing so the molecule will be thrust forward with an impulsive momentum of $h\nu/c$, (c=velocity of light); for we may suppose the pulse to have the mass $h\nu/c^2$ and the momentum $h\nu/c$; the absorption of the pulse by the molecule may be taken as a case of inelastic impact, the whole momentum being communicated to the molecule. The velocity with which the molecule will move forward = $h\nu/mc$.' Saha concluded that '*radiation-pressure may exert an effect on the atoms and molecules which are out of all proportion to their actual sizes.*' This must have been the first application of Einstein's hypothesis that light-quanta carried a directed momentum $h\nu/c$.

In March 1924 Saha visited Dhaka and stayed with Bose. While teaching postgraduate students Bose had keenly felt the need for a derivation of Planck's law free of logical difficulties. Saha drew Bose's attention to the papers of Pauli [1923] and their connection to Einstein and Ehrenfest [1923] and their connection to Einstein's paper of 1917. In an interview with J Mehra [Mehra and Rechenberg, **1**, *2*, 565] Bose recalled : 'What seemed to be happening in Pauli's work was that in order to apply the quantum conditions you had to know exactly what was going to happen afterwards.' Stimulated by his discussions with Saha, Bose began to study carefully the works of Planck [1900,1910], Peter Debye [1910,1923], Einstein [1917], Arthur Compton [1923], Pauli [1923] and Einstein and Ehrenfest [1923]. This resulted in two papers that he wrote in June 1924 [Bose, 1924a, 1924b] and sent to Einstein for his views and for publication in *Zeitschrift für Physik*. Einstein translated both of them into German and had them published in *Zeitschrift für Physik* in 1924. To both of them he added his own comments.

It is clear from these two papers of Bose which he always regarded as integral parts of each other that he clearly recognized the following points :

(1) The fundamental assumptions of the quantum theory are incompatible with the laws of classical electrodynamics [Bose, 1924a].

(2) All previous derivations had made use of relation (2) 'between the density of radiation and the average energy of an oscillator, and they make assumptions

about the number of degrees of freedom of ether, which enters the above equation (the first factor on the right hand side). This factor could however be derived only from the classical theory. This is an unsatisfactory feature in all derivations and it is not surprising that efforts are made again and again to give a derivation free from this logical flaw.' [Bose, 1924a]

(3) Even Einstein's 1916/1917 derivation, although 'remarkably elegant', makes use of 'Wien's displacement law and Bohr's correspondence principle. Wien's law is based on classical theory and the correspondence principle assumes that the quantum theory agrees with the classical theory in certain limiting cases.' [Bose, 1924a].

(4) 'In all cases it appears to me,' wrote Bose [1924a] 'that the derivations are not sufficiently justified from a logical point of view. On the other hand, the light-quantum hypothesis combined with statistical mechanics *(as adapted by Planck to conform to the requirements of quantum theory)* appears sufficient for the deduction of the law independent of classical theory' (our italics). In other words, the incompatibility between Einstein's light-quanta ('disjointed quanta') and Planck's law (implying 'undisjointed quanta') can be removed by adapting the statistical mechanics of light-quanta to conform to the requirements of quantum theory, as proposed by Planck.

(5) 'The problem of thermodynamic equilibrium of radiation in the presence of material particles can however be studied using the methods of statistical mechanics, *independently of any special assumption about the mechanism of the elementary processes on which the energy exchange depends.*' [Bose, 1924b]

Bose therefore proceeded in four steps.

First step: He took Einstein's light-quantum hypothesis seriously and treated black-body radiation as a collection of light-quanta enclosed in a volume V. If there are N_s light-quanta of energy $h\nu_s$ ($s = 0 \quad s = \infty$), the total energy E is

$$E = \sum_s N_s h \nu_s = V \int \rho_\nu \, d\nu \,. \tag{24}$$

ρ_ν can be determined from this simple relation if N_s is known. So, the problem is to determine N_s. 'If we can give the probability for each distribution characterized by arbitrary values of N_s, then the solution is determined by the condition that this probability is a maximum subject to the subsidiary condition' (24) [Bose,1924a]. Bose then proceeded to find this probability.

Second step: Before he could do that, he needed to compute the number of *states* of a light-quantum whose frequency lies between ν_s and $\nu_s + d\nu_s$. In order to be able to do this, he *extended* Planck's idea of dividing up the phase space of a material oscillator into elementary cells of volume h^3 *to the phase space of light-quanta.* In other words, *he extended Planck's 'first quantization' of material oscillators to the radiation field itself.* This was the most novel and crucial input. Emboldened by Compton's discovery, he assumed the momentum of a light-quantum of frequency ν to be hν/c in the direction of its motion. He must have also been aware of Saha's success in explaining radiation pressure on molecular-size objects with the help of this hypothesis [Saha,1919]. The instantaneous *state* of such a light-quantum, he stated, is characterized by a point in six-dimensional phase space which is constrained to lie on the surface of a cylinder defined by

$$p_x^2 + p_y^2 + p_z^2 = \frac{h^2 \nu_s^2}{c^2} . \qquad (25)$$

Then the phase space belonging to the frequency interval $d\nu_s$ is

$$\int dx\, dy\, dz\, dp_x\, dp_y\, dp_z = 4\pi V \left(\frac{h\nu_s}{c}\right)^2 \frac{h\, d\nu_s}{c} = 4\pi \frac{h^3 \nu_s^2}{c^3} V d\nu_s . \qquad (26)$$

Dividing this by h^3, the 'ultimate elementary region' [Bose to Einstein, 4 June 1924], Bose obtained $4\pi V\nu_s^2\, d\nu_s/c^3$. Then the following statement occurs : 'In order to take into account the *fact* of polarization, it seems however appropriate to multiply this number once again by 2' (our italics) to obtain $8\pi V\nu_s^2\, d\nu_s/c^3$ as the number of cells belonging to $d\nu_s$. No explanation is offered as to how this 'fact of polarization', an essentially *classical* concept, can be understood in terms of the light-quantum hypothesis, although Bose claimed to deduce this factor *independent of the classical electrodynamics* [Bose to Einstein, 4 June 1924]. Bose had always maintained privately that he did offer a quantum theoretic explanation but Einstein removed it from his translation and substituted it by the statement about the polarization factor. In his letter to Bose, Einstein simply wrote : 'You are the first person to derive the factor quantum theoretically, even though because of the polarization factor 2, not wholly rigorously' [Einstein to Bose, 2 July 1924]. What was the explanation that Bose claimed to have offered? It was that *light-quanta carried an intrinsic spin that could take only the values* $\pm h/2\pi$. There is only one recorded evidence of this because Bose's original manuscript in English is missing from the Einstein archives. It appears in a paper by C V Raman and S Bhagavantam [1931]. The paper is entitled 'Experimental Proof of the Photon Spin'. They write :

> In his well-known derivation of the Planck radiation formula from quantum statistics, Prof. S.N. Bose obtained an expression for the number of cells in phase space occupied by the radiation, and found himself obliged to multiply it by a numerical factor 2 in order to derive from it the correct number of possible arrangements of the quantum in unit volume. The paper as published did not contain a detailed discussion of the necessity for the introduction of this factor, but we understand from a personal communication by Prof. Bose that he envisaged the possibility of the quantum possessing besides energy $h\nu$ and momentum $h\nu/c$ also an intrinsic spin angular momentum $\pm h/2\pi$ round an axis parallel to the direction of its motion. The weight factor 2 thus arises from the possibility of the spin of the quantum

PHYSICS DEPARTMENT.
Dacca University.

Dated, the 4th June 1924.

Respected Sir, I have ventured to send you the accompanying article for your perusal and opinion. I am anxious to know what you think of it. You will see that (I have tried to deduce the coefficient $\frac{8\pi\nu^2}{c^3}$ in Planck's Law independent of the classical electrodynamics) only assuming ~~the finite~~ that the ultimate elementary regions in the Phase-space has the content h^3. I do not know sufficient German to translate the paper. If you think the paper worth publication, I shall be grateful if you arrange for its publication in Zeitschrift für Physik. Though a complete stranger to you, I do not feel any hesitation in making such a request. Because We are all your pupils though profiting only by your teachings through ~~the~~ your writings. I do not know whether you still remember that somebody from Calcutta asked your permission to translate your papers on Relativity in English. You acceded to the request. The book has since been published. I was the one who translated your paper on Generalised Relativity.

Yours faithfully
S. N. Bose

Bose's letter of 4 June 1924 to Einstein

2. VII. 24

Lieber Herr Kollege!

Ich habe Ihre Arbeit übersetzt und der Zeitschrift für Physik zum Druck übergeben. Sie bedeutet einen wichtigen Fortschritt und hat mir sehr gut gefallen. Ihre Einwände gegen meine Arbeit finde ich zwar nicht richtig. Denn das Wien'sche Verschiebungsgesetz setzt die Undulationstheorie nicht voraus und das Bohr-che Korrespondenzprinzip ist überhaupt nicht verwendet. Doch dies thuts nichts. Sie haben als erster den Faktor quantentheoretisch abgeleitet wenn auch wegen des Polarisations-Faktors 2 nicht ganz streng. Es ist ein schöner Fortschritt.

Mit freundlichem Gruss

Ihr A Einstein.

(Sd)

Einstein's postcard of 2 July 1924 to Bose

(48)

> being right-handed or left-handed, corresponding to the two alternative signs of the angular momentum. There is a fundamental difference between this idea, and the well-known result of classical electrodynamics to which attention was drawn by Poynting and more fully developed by Abraham that a beam of light may in certain circumstances possess angular momentum . . . Thus, according to the classical field theory, the angular momentum associated with a quantum of energy is not uniquely defined, while according to the view we are concerned with in the present paper, the photon has always an angular momentum having a definite numerical value of a Bohr unit with one or other of the two possible alternative signs.

Why did Einstein remove Bose's explanation? Probably he felt it was too radical and unnecessary at that stage. However, instead of removing it altogether, he could have added a suitable comment to the one he did make at the end of the paper. In fact, he did not hesitate to add a dissenting note to Bose's second paper [1924b] as we shall see later.

Be that as it may, let us proceed with Bose's derivation. Having divided the phase space volume belonging to the frequency range ν_s and $\nu_s + d\nu_s$ by h^3, he wrote : *'the total number of cells must be regarded as the number of possible arrangements of a quantum in a given volume'* (our italics). Classically there are infinitely many ways of arranging or distributing a particle within a phase space cell of finite extent corresponding to the infinitely many points contained within it. However, according to Planck, as we have seen, quantum states are not continuously distributed in phase space but are separated from one another by finite amounts determined by h. Thus, *an elementary cell of volume h^3 can contain at most one quantum state*, and the problem of counting the number of quantum states reduces to the counting of these elementary cells. This is a totally new interpretation of the factor $A_s = 8\pi \nu_s^2 V d\nu_s / c^3$. (In classical wave theory it is the number of transverse standing modes in the interval $d\nu_s$ in a volume V.) In his covering letter to Einstein Bose wrote :

> *Respected Sir,*
>
> *I have ventured to send you the accompanying article for your perusal and opinion. I am anxious to know what you think. You will see that I have tried to deduce the coefficient in Planck's law independent of the classical electrodynamics, only assuming that the ultimate elementary regions in Phase-space have the content h^3.*
>
> [Bose to Einstein, 4 June 1924].

On 12 July 1924 Einstein wrote to Ehrenfest : 'The Indian Bose has given a beautiful derivation of Planck's law, including the constant [i.e. $8\pi\nu^2/c^3$].'

Third Step: Having demonstrated that the factor $8\pi\nu^2/c^3$ in Planck's law corresponds to the total number of *quantum states* of radiation, Bose then proceeded 'to calculate *the thermodynamic probability (macroscopically defined) of a state*' in the sense of Planck (our italics). This is equal to the number of different microscopically defined states by means of which the macroscopic state can be realized. If all the cells (states) were considered to be distinct from one another, the answer would be A_s! However, this is not the case. *Since each cell is to be counted as a single quantum state,* if follows that the permutations of the light-quanta within a cell do not produce new states. *Consequently, the cells can be partitioned into distinct classes characterized solely by their occupation numbers.* Let there be a particular distribution of N_s quanta among A_s states such that there are p_1^s states that are empty, p_2^s states with one quantum, p_2^s states with two quanta, and so on. Then, *following Boltzmann* and the usual procedure of statistical mechanics for these *states (not the quanta)*, one can immediately write down the possible number of such distributions as

$$W_s = \frac{A_s\ !}{p_o^s\ !\ p_1^s\ !\ p_2^s\ !\ \ldots} \tag{27}$$

which is precisely the expression (12) given by Natanson. The 'thermodynamic probability' of 'the state defined by all p_r^s' is then clearly $W = \prod_s W_s$ which is equivalent to the Debye-Planck expression (10).

Fourth step : Bose then followed the standard procedure, already followed by Debye [1910], of maximizing $ln\ W$ subject to the constraints $E = \sum_s N_s h\nu_s$ and $N_s = \sum_r rp_r^s$ to obtain Planck's law. Notice that the total number of photons $N = \sum_s N_s$ and *there is no constraint on N*. It turns out that the constraint on N_s is not necessary, but Bose did not notice this.

Notice that the only departure of the expression (27) from the usual Boltzmann expression lies not in the product of factorials in the denominator but in putting the multiplicative factor $(g_i)^r$ corresponding to the possible arrangements of r quanta within each cell i having g_i levels equal to unity, as demanded by the division of phase space into elementary cells of volume h^3. (In his second paper [1924b] written more or less at the same time as the paper under discussion and received by Einstein only five days later, Bose used these factors, as we shall see later, for material particles obeying Maxwell-Boltzmann statistics.) This is precisely the adaptation of statistical mechanics by Planck 'to conform to the requirements of the quantum theory' that Bose referred to in the third paragraph of his paper.

Was Bose aware of the analysis of the Planck-Debye definition of the probability by Natanson and by Ehrenfest and Kamerlingh Onnes? One cannot tell. He certainly did not refer to their papers. It is reasonable, however, to assume that he was not aware of the latter work because if he had been, he would have explained how their conclusion, namely that *'Planck's formal device* (distribution of P energy elements ϵ among N resonators) *cannot be interpreted in the sense of Einstein's light-quanta'* [Ehrenfest and Kamerlingh Onnes, 1914, p.873], could be reconciled with his own starting point, namely the combination of the light-quantum hypothesis with Planck's counting device. The explanation, of course, lies in the fact that Planck's law reduces to Wien's law in the limit $h\nu/kT >> 1$, and that it is only in this limit that the light-quanta have Maxwell-Boltzmann behaviour as inferred by Einstein. It is one of those quirks of history that Einstein should have been able to infer the quantum nature of radiation from its approximate classical statistical behaviour in a certain domain and thus missed the true nature of its statistics. It was left to Bose to discover it, possibly uninhibited by the analysis of Ehrenfest and Kamerlingh Onnes and their conclusion. Hence Einstein's enthusiastic welcome of Bose's work as an important and new support to his favourite light-quantum hypothesis, making it compatible with Planck's law.

j. The birth of a new statistics: Einstein applies Bose's method to ideal gases

Bose's counting of states, of course, implied a *new* statistics for the light-quanta. Neither he nor Einstein seemed to have quite grasped the importance of this initially. In his letter of 2 July 1924 to Bose, Einstein compliments him for being the first to derive the factor A_s quantum theoretically but does not write a word about the new statistics. 'In my opinion,' wrote Einstein in a footnote to Bose's paper, 'Bose's derivation signifies an important advance. The method used here gives the quantum theory of an ideal gas as I will work out elsewhere.' He extended Bose's method to material particles in three communications to the Prussian Academy in Berlin on 10 July 1924, 8 January 1925 and 29 January 1925 [Einstein, 1924, 1925a, 1925b] without informing Bose or asking for his collaboration. Bose came to know of the comment in the footnote shortly before he left for Europe and of the first of the communications to the Prussian Academy on his arrival in Europe in October 1924. In the first paper this is how Einstein described Bose's method :

> [First] : The phase space of an elementary object (here of a monatomic molecule and in Bose's case of a light-quantum), associated with a given (three-dimensional) volume, is divided into "cells" of extension h^3. [Second] : If many elementary objects are present then their (microscopic) distribution, which plays a role in thermodynamics, is determined by the particular manner how the elementary objects are distributed among the cells. [Third] : The "probability" of a macroscopically defined state (in the sense of Planck) is equal to the number of different microscopic states by means of which the macroscopic state can be thought to be realized. [Fourth] : The entropy of the macroscopic state and, therefore, the statistical and thermodynamic behaviour of the system is determined by Boltzmann's theorem relating entropy and probability.' [Einstein, 1924, p.261]

Again, there is no mention of the new statistics! In applying Bose's method to ordinary material atoms, Einstein had to introduce an additional parameter (the chemical potential which vanishes for light-quanta) to take into account the conservation of their number. He drew a very important conclusion from this:

> According to the theory presented here, Nernst's theorem is satisfied in the case of ideal gases. To be sure, our formulae cannot immediately be applied to extremely low temperatures, for we have assumed in their derivation that the p_r^s change only relatively infinitely little if s is altered by 1. Still one recognizes at once that the entropy must vanish at the absolute zero of temperature. The reason is that then *all molecules are in the first cell; and for this state there exists only one distribution of molecules according to our counting method*. Hence our assertion is immediately proved to be correct. [Einstein, 1924, p.265] (our italics)

This is a conclusion of far-reaching consequence, yet there is no mention of any new statistics for the molecules; the emphasis is on Bose's counting method for *states* (one quantum state per cell). It was only six months later on 8 January 1925 that Einstein discussed the difference between the counting methods of Boltzmann and Bose and recognized the fact, already pointed out by Ehrenfest [1924], that *the new statistics for molecules was inconsistent with their statistical independence*. However, while Planck [1925] considered the analogy with radiation a weak point in Einstein's gas theory, the statistics of radiation and that of material particles being completely different in his opinion, Einstein regarded this result differently. He concluded : 'This result represents in itself a support of the view concerning the *deep natural relation between radiation and gas, since the same statistical treatment which leads to Planck's formula establishes — when applied to ideal gases — the agreement with Nernst's theorem.*' [Einstein, 1925a, p.7] (our italics) He was of the opinion that an understanding of the 'mysterious influence' among the molecules would come in the future.

k. The birth of wave mechanics

Based on the new counting method, he showed in his second paper communicated on 8 January 1925 that the mean square energy fluctuation of the gas molecules is given by an expression which is the sum of two terms, one corresponding to the Maxwell-Boltzmann statistics of non-interacting molecules and the other to interference fluctuations associated with wave phenomena. (There exists in the Einstein Archives in the Jewish National and University Library, Jerusalem, a two page calculation done by Bose entitled 'Fluctuation in density' which ends with this fundamental result. The document is undated and exists in the Scientific Correspondence File Folder 'B-Misc.-II'. Bose was in Paris from 18 October 1924 until the first part of October 1925.) From this he immediately drew attention to the importance of Louis de Broglie's Doctoral thesis (which he had heard of from Paul Langevin and asked for a copy which he received and read in December 1924 [Jammer, 1966, 249]) in which de Broglie had attached wave properties to ponderable matter in analogy with the wave-particle duality of radiation. Einstein wrote : 'I shall discuss this interpretation in greater detail because I believe that it involves more than a mere analogy.' He then went on to explore various consequences of de Broglie's hypothesis.

When Erwin Schrödinger whose principal research during his early years in Zurich (1921-27) was on the statistical thermodynamics of ideal gases, read Einstein's first paper [1924a] on the quantum theory of an ideal gas, he also did not understand that the molecules were being treated as being indistinguishable. On 5 February 1925 he wrote to Einstein suggesting a possible error in his probability formula. Einstein replied explaining that the 'quanta or molecules are not treated as independent of one another' and even gave a little diagram to explain the difference between Bose's counting method and that of Boltzmann. He concluded his letter with the emphatic statement: 'There is certainly no error in my calculation.' [Walter Moore, 1989, 183] It was Einstein's remark in his second paper that de Broglie's idea 'involves more than a mere analogy' that really induced Schrödinger to study the 'de Broglie-Einstein undulatory theory' as he called it at that time, and eventually led to his discovery of wave mechanics in 1926 [Schrödinger, 1926a, 1926b and 1926c]. Schrödinger later said that

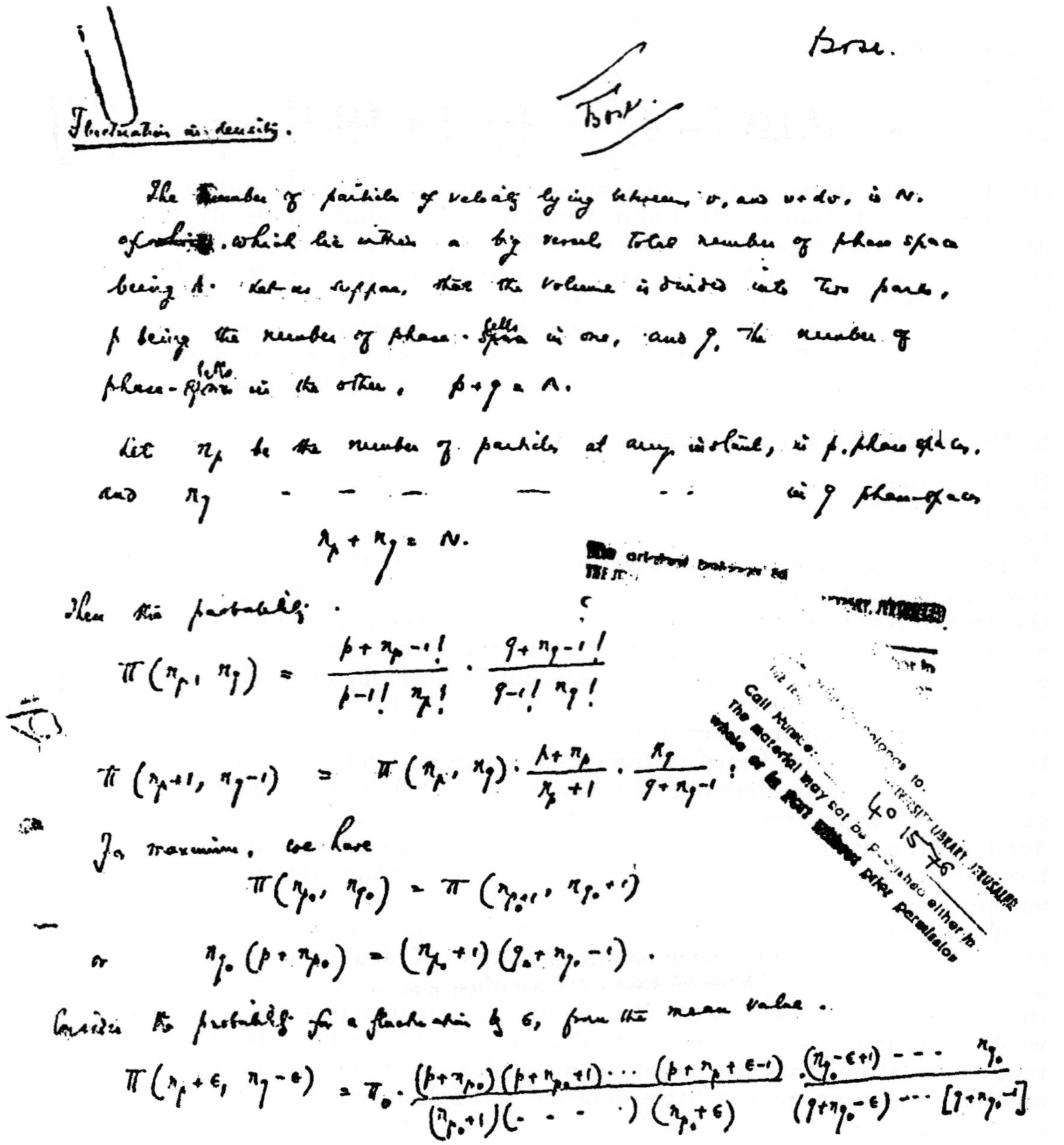

Calculation of 'Fluctuation in density' in Bose's hand (Scientific Correspondence File Folder 'B-Misc.-II'); by courtesy of The Jewish National and University Library, Jerusalem.

$\log \Pi(n_{p_0}+\epsilon, n_{q_0}-\epsilon) - \log \Pi_0$

$$= \frac{(\epsilon-1)(\epsilon)}{2}\left[\frac{1}{p+n_{p_0}} - \frac{1}{n_{p_0}+1}\right] + \frac{\epsilon(\epsilon-1)}{2}\left[\frac{1}{q+n_{q_0}-1} - \frac{1}{n_{q_0}}\right]$$

or $$\Pi(n_{p_0}+\epsilon, n_{q_0}-\epsilon) = \Pi_0 \cdot e^{-\frac{\epsilon^2}{2}\left[\frac{p-1}{(p+n_{p_0})(n_{p_0}+1)} + \frac{q-1}{n_{q_0}(q+n_{q_0}-1)}\right]}$$

$$\therefore \quad \overline{\Delta\epsilon^2} = \frac{1}{\frac{p-1}{(p+n_{p_0})(n_{p_0}+1)} + \frac{q-1}{n_{q_0}(q+n_{q_0}-1)}}.$$

When, the volume containing p phase cells is supposed to be very small compared with the other, $\frac{q-1}{n_{q_0}(q+n_{q_0}-1)}$ can be neglected.

$$\therefore \quad \frac{\overline{\Delta\epsilon^2}}{n_{p_0}^2} = \frac{(p+n_{p_0})\,n_{p_0}}{p\,n_{p_0}^2} = \frac{p+n_{p_0}}{p\,n_{p_0}} = \left(\frac{1}{p} + \frac{1}{n_{p_0}}\right)$$

'wave mechanics was born in statistics.' (our italics) [Walter Moore, 1989, 188] He also wrote to Einstein on 23 April 1926 : 'By the way, the whole thing would not have started at present or at any other time (I mean, as far as I am concerned) had not your second paper on the degenerate gas directed my attention to the importance of de Broglie's ideas.' [Jammer,1966, 257]

l. Further developments

Meanwhile in July 1925 Werner Heisenberg had sent a paper to *Zeitschrift für Physik* in which he gave a preliminary account of matrix mechanics [W Heisenberg,1925]. Schrödinger proved the complete equivalence of his wave mechanics and the Heisenberg–Born–Jordan matrix mechanics [Schrödinger, 1926d]. A whole train of rapid developments occurred. Pauli discovered his Exclusion Principle within six months of Bose's paper [Pauli, 1925]; Fermi published his paper on the statistics of particles obeying Pauli's principle in early 1926 [Fermi, 1926]; Dirac linked the Bose and Fermi statistics of particles to the symmetry properties of their wave functions and named them 'bosons' and 'fermions' [Dirac, 1926]; in 1927 Dirac invented 'second quantization' of the electromagnetic field [Dirac, 1927] which was soon followed by a similar method for Fermi fields developed by Jordan and Wigner [1928]; later Pauli [1940] proved the spin-statistics theorem in relativistic quantum field theory; Pauli [1927] also applied Fermi-Dirac statistics to the paramagnetism of an electron gas and Sommerfeld [1927,1928] applied it to the electrons in a metal; and Fritz London [1938] suggested that the superfluidity of liquid helium (^{4}He) was related to Einstein-Bose condensation. Thus the developments of Schrödinger's wave mechanics and that of quantum statistical mechanics were directly triggered off by Bose's first paper followed by Einstein's far-sighted extension of his method to ideal gases of *material* particles and his championing of de Broglie's ideas which soon received experimental confirmation in the hands of Davisson and Germer [1927] and G P Thomson [1927].

Years later this is how Schrödinger summed up the position :

> The two equivalent ways of looking at (7.19) either as counting the number of quantum states of a particle, or as counting the number of wave-mechanical proper vibrations of the enclosure, interest us for this reason. The second attitude makes us think of the 'n_s particles present in state α_s' as of a proper vibration (or a 'hohlraum' oscillator to use a customary phrase) in its n_sth quantum level. (This attitude really corresponds to so-called second quantization or field quantization.) n_s becomes a quantum number and the stipulation that the system of quantum numbers
>
> $$n_1, n_2, n_3, \ldots, n_s, \ldots,$$
>
> determines only one state of the gas, not a class of
>
> $$\frac{n\,!}{n_1\,!\;n_2\,!\ldots\; n_s\,!\ldots}$$
>
> states, ceases to be a strange new adoption, and comes into line with the ordinary view about quantum states and their statistical weight (viz. equal

for any two of them).

It is the first, the particle attitude, that has suggested the term 'new statistics' which is frequently used. And that is why this idea of new statistics did not, originally, arise in connexion with heat radiation, because here the wave point of view was the historical one, the classical one—nobody thought of any other at the outset. The wave picture was considered to be (and historically was) the classical description. The quantization of the waves therefore duly appeared to be a 'first' quantization and nobody thought of anything like 'second quantization'.

Not until the idea of photons had gained considerable ground did Bose (about 1924) point out that we could, alternatively to the 'hohlraum' oscillator statistics, speak of photon statistics, but then we had to make it 'Bose statistics'. Very soon after, Einstein applied the same to the particles of an ideal gas. And thereupon I pointed out that we could also in this case speak of ordinary statistics, applied to the wave mechanical proper vibrations which correspond to the motion of the particles of the gas [Schrödinger, 1952, 49-50].

We wish only to point out, as explained above, that Bose did not actually 'speak of photon statistics' — it was *implied* by his counting of quantized photon *states*, and this only became gradually clear with Einstein's application of Bose's counting method to ideal gases, although Ehrenfest and Natanson had already pointed this out in 1911 and 1914 respectively in relation to Planck's counting method of December 1900 which Bose showed was applicable to light-quanta.

m. Did Bose make a 'shot in the dark'?

Some recent widely read scientific biographies have given rise to the impression that Bose made 'a shot in the dark' with three assumptions all of which turned out to be correct but none of which he tried to justify 'by arguments of any kind; they seemed to appear intuitively to his mind, perhaps because his thoughts were uncluttered by the ongoing controversies of the European physicists.' [W. Moore, 1989, 181; A. Pais, 1982, 428] The assumptions were that photons were (a) massless particles capable of existing in two states of polarization, (b) indistinguishable and (c) their number was not necessarily conserved. The record must be put right. That photons are massless particles with energy $h\nu$ and directed momentum $h\nu/c$ was proposed by Einstein and only used by Bose. That they exist in two possible 'polarization states' was *inferred* by Bose from phase space considerations which yield for the number of quantum states between ν_s and $\nu_s + d\nu_s$ the factor $4\pi V\nu_s^2\, d\nu_s/c^3$ rather than $8\pi V\nu_s^2\, d\nu_s/c^3$, the factor that occurs in Planck's formula. Actually, as we have already discussed, in order to account for this factor of 2, Bose had proposed that photons carried one unit of intrinsic spin which could only take the values ±1. Whoever had heard of particles with polarization? Moreover, 'polarization' was a wave concept which Bose wanted to avoid. However, Einstein quietly dropped it and the statement regarding the polarization factor was most probably inserted by him. Since Bose's original manuscript in English is missing from the Einstein archives and Bose did not keep a copy, there is no way to verify this.

How about 'indistinguishability'? Again, as we have seen, although Bose did not explicitly use the word 'indistinguishable' which is by itself rather vague, he did use Planck's division of phase space into elementary cells of volume h^3 to justify his counting method for *states* (which implied indistinguishability of the quanta), namely that there cannot be more than one quantum state in an elementary cell. 'In any case,' he wrote, 'the total number of cells must be regarded as the number of possible arrangements of a quantum in the given volume.' What is not usually appreciated in usual expositions of Bose statistics is that Bose did not count the number of ways of distributing individual photons over a set of states. Instead, he calculated the *thermodynamic probability* of a *macroscopic state* by counting the number of ways in which the macroscopic state can be thought to be realized through given distributions of the number of photons over the micro-states (phase space cells).

Finally, what about non-conservation of photons? This is a non-issue. Since, like Debye [1910], Bose considered the statistical thermodynamics of an *open* system of radiation in thermal equilibrium and did not include the molecules of matter with which this radiation constantly exchanged energy (which he did in his second paper which we shall turn to presently), there was no question of conservation of energy quanta even if treated as massless particles. In his 1917 derivation of Planck's law Einstein did not impose the conservation of light-quanta which he certainly regarded as massless particles interacting with molecules. Since Bose was not doing anything new or different in this respect, there was no original or deep comment he could have made on this issue. Nevertheless, it is true that Bose, like most others who have made revolutionary contributions, did not, and could not be expected to, understand the full implications of their discoveries.

n. Bose's second paper (1924)

We now turn to his second paper [1924] on *'Thermal Equilibrium in Radiation Field in the Presence of Matter'*, also translated into German probably by Einstein himself. This paper, completed on 14 June 1924, i.e. more or less at the same time as the first paper, consists of two parts. In the first part Bose derived general conditions for statistical equilibrium of a system consisting of matter and radiation, *independent of any special assumptions about the mechanism of the elementary radiative processes.* In the second part he proposed a new expression for the probability of these elementary radiative processes which differed from that of Einstein. Einstein considered this hypothesis not to be 'applicable' to elementary radiative processes and gave two reasons for it in a footnote to the paper. As a consequence, the paper has been largely ignored. However, Bose remained firmly of the opinion for the rest of his life that Einstein had not done him justice.

Bose started off by giving a quick critical review of the most important derivations of Planck's law until then given by Debye [1910], Einstein [1917], Pauli [1923], and Einstein and Ehrenfest [1923] on the scattering of radiation by electrons. Bose considered Debye's 1910 derivation not to be 'completely independent of classical electrodynamics' since he derived the factor $8\pi\nu_s^2 V d\nu_s/c^3$ from the normal modes of the ether.

He considered the derivations of Einstein [1917], Pauli [1923] and Einstein and Ehrenfest [1923] to be dependent on *ad hoc* special assumptions about the probability of elementary radiative processes. 'The problem of thermodynamic equilibrium of radiation in the presence of material particles,' he wrote, 'can however be studied using the methods of statistical mechanics, independently of any special assumption about the mechanism of the elementary processes on which the energy exchange depends. In this way we obtain a general relation which is valid for all special assumptions about the elementary processes and their probabilities.' This is in keeping with Kirchhoff's law. He then went on to say: 'If it is possible to write down the thermodynamic probability for any particular distribution of quanta of radiation and the probability for any arbitrary energy distribution among the particles, then the thermodynamic probability for the bigger system consisting of matter and radiation is simply the product of the two probabilities. The condition of equilibrium is simply that the resulting probability must be a maximum.' For the thermodynamic probability for radiation, Bose used the expression

$$W = \prod_{\nu} \frac{(A_\nu + N_\nu\, d\nu)\,!}{A_\nu\,!\, N_\nu\, d\nu\,!} \tag{28}$$

where $A_\nu = 8\pi V\nu^2\, d\nu/c^3$ which, he stated, had been derived earlier. As references he quoted his own paper and that of Debye [Debye 1910]. (It is curious that the reference to his previous paper is given as 'The author, to appear in *Phil Mag*'. This is because Bose had submitted his paper first to the *Philosophical Magazine*. Not hearing from them for a while, he sent a copy to Einstein. While translating Bose's second paper, the translator obviously forgot to change the reference of the earlier paper to *Zeitschrift für Physik* **26**, 178-181, 1924). Although expression (28) is indeed the Ansatz used by Debye, it does not appear in Bose's first paper! Instead, the corresponding expression that appears in his first paper is (27), the form given by Natanson. As we have seen earlier, the two are mathematically equivalent, but Bose did not care to explain this, probably because this result must have been fairly well known in the literature in those days.

For the thermodynamic probability for material particles. Bose writes:

> This can be found easily. We want to generalize our assumptions a little so that the case of the Bohr atom with discrete energy levels as well as the case of translational energy of particles can be included. *Let the phase space be divided into cells. For every cell there is a probability g that a particle occupies it. The g's are in general equal except for the case of Bohr's atoms. The thermodynamic probability for any arbitrary distribution* n_1, n_2 etc among the different cells is,

$$\frac{N\,!\; g_1^{n_1}\; g_2^{n_2}\; \ldots}{n_1\,!\; n_2\,!\; \ldots} \tag{29}$$

(our italics). Unlike in the case of radiation (expression (27), Bose now uses the factor $\prod_i g_i^{n_i}$ for material particles and obtains the classical Maxwell-Boltzmann distribution. In other words, he treats material particles as in classical (*distinguishable*) statistical mechanics but light-quanta according to quantum (*indistinguishable*) statistical mechanics.

Finally, he writes down the thermodynamic probability for the total system as the product of (28) and (29) :

$$W = \prod_s \frac{(A_s + N_s)\,!}{A_s\,!\; N_s\,!} \prod_t \frac{g_T^{n_T} N\,!}{n_T\,!} \tag{30}$$

with

$$\Sigma N_s h\nu_s + \Sigma n_T E_T = E \tag{31}$$

and

$$\Sigma N_T = N. \tag{32}$$

He then considers an elementary process in which a particle passes from the rth cell to the sth cell while a light-quantum of frequency ν changes into a light-quantum of frequency ν.′ The stationarity of W subject to the conditions (31) and (32) gives

$$\frac{n_r}{g_r} \prod_\nu \frac{N_\nu}{N_\nu + A_\nu} = \frac{n_s}{g_s} \prod_{\nu'} \frac{N_{\nu'}}{N_{\nu'} + A_{\nu'}} \tag{33}$$

where

$$\Sigma\, h\,\nu' - \Sigma h\nu + E_s - E_r = 0\,. \tag{34}$$

Bose then demonstrated that (33) indeed generalized the results of Pauli [1923] and Einstein and Ehrenfest [1923] without the need for any *ad hoc* assumptions regarding the elementary radiative processes designed to give Planck's formula. In particular, in the case of Bohr's atoms considered by Einstein in 1917, equation (33) reduces to

$$\frac{n_r}{g_r} \frac{N_\nu}{N_\nu + A_\nu} = \frac{n_s}{g_s} \tag{35}$$

which is Einstein's condition derived on the assumption that atoms in higher energy levels make transitions to lower energy levels in two ways:

1. A spontaneous transition as in radioactivity *independent of the state of the external radiation field.*
2. An induced transition whose probability depends on the state of the external radiation field.

Transitions from lower to higher energy levels always take place by induced absorption whose probability depends on the state of the external radiation field. Further, Einstein had to postulate certain relations (namely, the opposite limiting Wien and Rayleigh-Jeans laws) among these transition probabilities to obtain Planck's law. Bose's derivation is independent of all these assumptions. It is for this reason that Einstein took the trouble to have the paper translated into German and communicated it to *Zeitschrift für Physik*. However, his disagreement with Bose started from the *next* part of the paper in which Bose proposed a new expression for the probability of an interaction between a particle and a quantum of radiation.

Bose's starting point was the following observation which he considered to be fundamental :

> . . . *even in a collision no interaction is as probable as the occurrence of any special interaction. . . From the classical theory. . . one would expect that there is some interaction whenever a quantum and a material particle come together. Therefore here it is a question of a departure from classical theory.*
>
> The assumption made above, that even in a collision no interaction need occur, is quite analogous to the assumption of the stability of stationary states which is so fundamental to Bohr's theory of line emission and can be traced back to the same origin—*the probability of existence of the stationary states of the particles*. It is interesting to remark in this connection that in the analogous case of a collision of an electron with an atom, experiments show that the electron goes through the atom without changing either the interatomic or its own motion. (our italics)

Bose was obviously referring to the Ramsauer–Townsend effect. He saw a simple way of realizing the non-classical possibility in Planck's cellularization of phase space. Let p_r be the number of cells with r quanta. 'In order that the radiation and particle may interact, it must be in a cell occupied by a quantum. The particular cell which the molecule under consideration occupies will pass through all possible states if we observe it long enough, that is, sometimees it is empty, sometimes occupied by 1 quantum, sometimes by 2 quanta etc. The duration of these events will be finally proportional to the equilibrium values of p_o, p_1 etc. *When r quanta and a particle are together, then either an exchange of energy takes place or nothing happens.*' (our italics) Therefore $r + 1$ different events are possible, namely the exchange of 1 quantum of energy, 2 quanta etc or no exchange at all. Therefore the total number of possible events is

$$p_o + 2p_1 + 3p_2 + \ldots = A_\nu + N_\nu d\nu_s = \sum_r (r+1)\, p_r\,. \tag{36}$$

The number of cases in which interaction or energy exchange occurs is

$$p_1 + 2p_1 + 3p_3 + \ldots. = N_s d\nu_s = \sum_r rp_r\,. \tag{37}$$

Consequently the probability of an interaction is

$$P = \frac{\Sigma r p_r}{\Sigma (r+1) p_r} = \frac{N_s d\nu_s}{A_s + N_s d\nu_s}. \tag{38}$$

This is Bose's second fundamental result.

Having derived this result, Bose then considered Einstein's 1917 problem, namely Bohr atoms exchanging energy with the radiation field. In order to be consistent with the fundamental equilibrium condition (35), he found it was sufficient to consider (a) transitions from lower to higher states through absorption of quanta of characteristic frequency ν and (b) transitions from higher to lower states to be spontaneous, i.e. independent of the external radiation field. *The additional assumption of induced emission (negative radiation) processes was not necessary.* Let the strength of induced absorption be β. Then the probabillity of induced absorption is

$$\beta \frac{N_\nu d\nu}{A_\nu + N_\nu d\nu}. \tag{39}$$

Let the coefficient of spontaneous emission be α. Then for equilibrium one gets

$$n_r \beta \frac{N_\nu d\nu}{A_\nu + N_\nu d\nu} = \alpha n_s \tag{40}$$

which agrees with (35) provided $g_r \beta = g_s \alpha$. A substitution of the value of n_r/n_s then gives Planck's law. Bose then treated the cases considered by Pauli [1923] and Einstein and Ehrenfest [1923] and showed that the fundamental equilibrium condition (33) followed from the probability law (38) without requiring any further assumptions.

In a note added to the paper Einstein gave two reasons why he thought Bose's hypothesis about the probability of elementary radiative processes was not 'applicable'. In his letter of 3 November 1924 Einstein summarized the two reasons. 'Your principle is not compatible with the following two conditions:

1) The absorption coefficient is independent of the radiation density.
2) The behaviour of a resonator in a radiation field should follow from the statistical laws as a limiting case.' [Einstein to Bose, 3 November 1924]

Let us see to what extent these criticisms were fair. As far as the first point is concerned Einstein was referring to Beer's law. Bose's probability law, however, predicted a dependence of the absorption coefficient on the radiation density , decreasing with it. 'If the behaviour had been such,' wrote Einstein in the note, 'then it would certainly have already been discovered in the case of infra-red radiation from hot light sources.' This is hard to believe. The departure from classical behaviour predicted by Bose's principle should occur only when $N_\nu / A_\nu << 1$, i.e. for very low intensity radiation. It is now well known that the departures from classical behaviour predicted by quantum optics are not seen even when the intensity of light is extremely low, *unless the source happens to be of a special kind* (like a single-photon source). The reason is that light from all classical sources is in a coherent state in which the number fluctuation is extremely large (Poissonian or higher). Therefore only light from 'single-photon states' (Fock states) and 'squeezed states' and other such 'sub-Poissonian situations' show departures from classical behaviour. [Aspect, 1987; Loudon, 1983] Such light sources were certainly unknown in 1924.

The second point concerns the 'correspondence principle' (not necessarily Bohr's) which in this case implies that a radiation field can transfer positive as well as negative energy to a resonator, depending on the phase. 'The probabilities of both these transitions must depend on the density of radiation, that is, on N_ν as opposed to Bose's hypothesis,' wrote Einstein. The discovery of masers and lasers after the Second World War eventually vindicated Einstein's contention. It was nevertheless eventually an *empirical* rather than a sound theoretical refutation. Although the 'correspondence principle' was used as a heuristic tool with great success by Einstein, Planck and later the Copenhagen school prior to the discovery of quantum mechanics around 1925/1926, it is no longer regarded as a fundamental principle. In fact, quantum mechanics itself is known not to satisfy this principle. [Liboff, 1980; Cabrera and Kiwi, 1987; Bohm and Hiley, 1985] For example, the nodes of the eigen-solutions inside a box do not disappear for arbitrarily high quantum numbers so that a classical particle cannot pass through such points. Also the cross-section for hard sphere scattering has no classical limit. Einstein was aware of this and used it against quantum mechanics as well.

o. Bose's third (unpublished) paper (1925)

Although Bose was unhappy, particularly with the first point ragarding Beer's law, he accepted the second point, modified his stance and wrote a third paper which he sent to Einstein from Paris on 27 January 1925. He sent him a letter under separate cover in which he wrote :

> *It seems that the hypothesis of negative Einstrahlung stands, which, as you have yourself expressed, reflects the classical behaviour of a resonator in a fluctuating field. But the additional hypothesis of a spontaneous change, independent of the state of the field, seems to me not necessary. . .*
>
> *I am rather anxious to know your opinion about it. I have shown it to Professor Langevin here and he seems to think it interesting and worth publishing.*

It was never published. Neither did Bose keep a copy of the paper, nor is there one in the Einstein Archives as far as one can ascertain. So, the precise contents and details will remain unknown. Bose arrived in Berlin in October 1925 and discussed his new ideas with Einstein. Einstein was so convinced of the necessity and reality of the two independent processes of emission of light from atoms (spontaneous and induced) that he could not agree with Bose that one of them, the former, was unnecessary to assume in deriving Planck's law. To Bose, the process of emission of light appeared as *a single physical process* and the division into two independent processes appeared artificial. The assumption of one, he claimed, would automatically imply the other in thermal equilibrium, *given the special statistical property of the radiation field.* Initially Bose started with spontaneous emission as input but later changed his position. It was clear to Bose that Planck's law was *a consequence of the special statistical property of the radiation field itself and was independent of the mechanisms of energy transfer with*

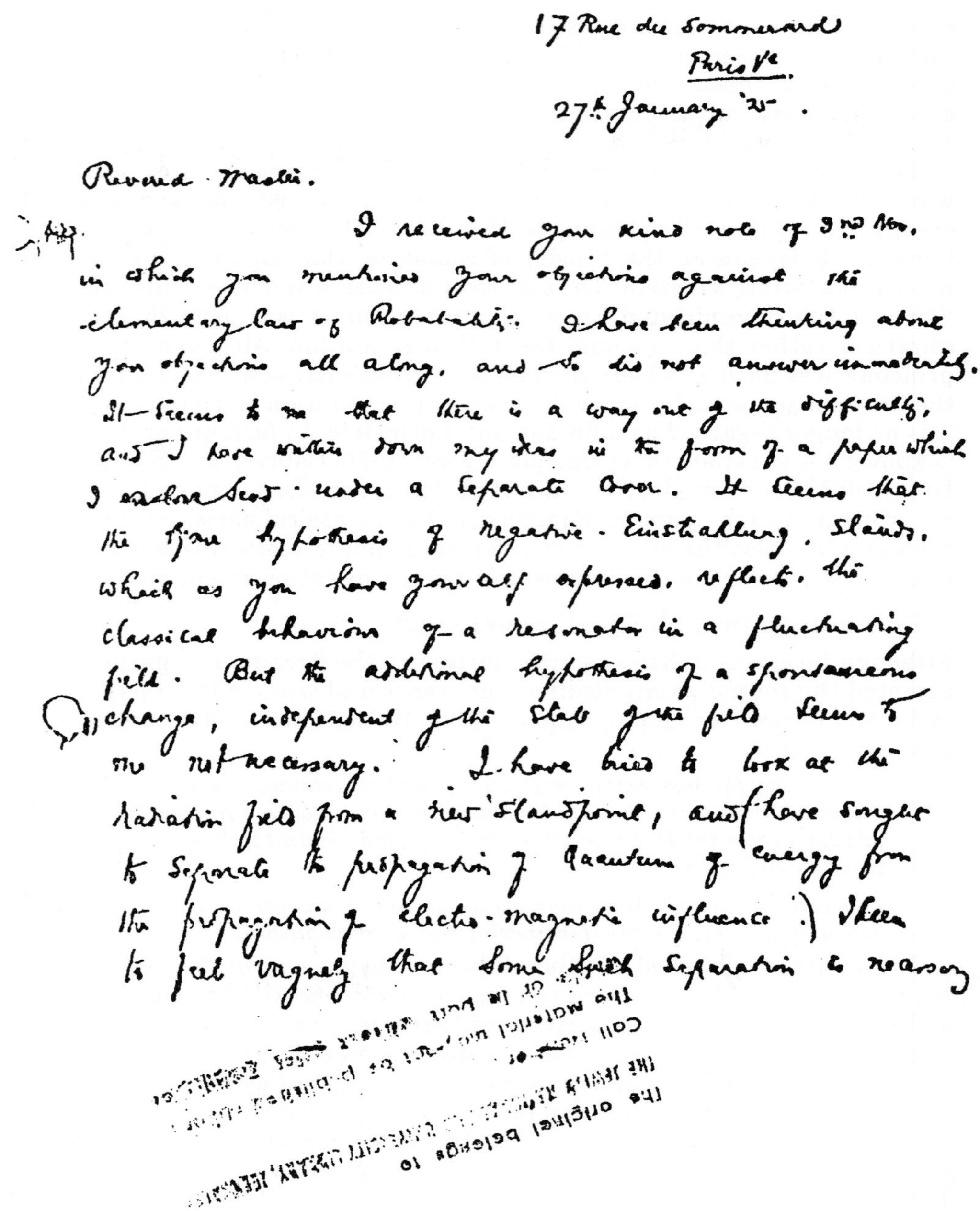

17 Rue du Sommerard
Paris Ve.
27th January '25.

Revered Master.

I received your kind note of 3rd Nov. in which you mentioned your objections against the elementary law of Probability. I have been thinking about your objections all along, and so did not answer immediately. It seems to me that there is a way out of the difficulty, and I have written down my ideas in the form of a paper which I enclose sent under a separate cover. It seems that the first hypothesis of negative-Einstrahlung stands, which as you have yourself expressed, reflects the classical behaviour of a resonator in a fluctuating field. But the additional hypothesis of a spontaneous change, independent of the state of the field seems to me not necessary. I have tried to look at the radiation field from a new standpoint, and I have sought to separate the propagation of quantum of energy from the propagation of electro-magnetic influence.) I seem to feel vaguely that some such separation is necessary

Bose's letter of 27 January 1925 to Einstein

(63)

if Quantum theory is to be brought in line with the Generalised Relativity theory.

The views about the radiation field, which I have ventured to put forward seem to be very much like to what Bohr has recently expressed in Mag. Phil Mag. 1924. But it is only a guess, as I cannot say honestly to have exactly understood all what he means to say, about his virtual fields and virtual oscillators.

I am rather anxious to know your opinion about it. I have shown it to Prof Langevin here and he seems to think it interesting, and worth publishing.

I cannot exactly express how grateful I feel for your encouragement, and the interest you have taken in my papers. Your first p.card came at a critical moment, and it has more than any other made this sojourn to Europe possible for me. I am thinking of going to Berlin at the end of this winter, where I hope to have your inestimable help and guidance.

Yours sincerely
S.N. Bose.

(64)

atoms and molecules. This was indeed his capital contribution to physics and forms the substance of his first paper and the first part of his second paper. To Einstein, on the other hand, it appeared that Planck's law was a consequence of special mechanisms of energy transfer between the radiation field and molecules and that thermal equilibrium was impossible without both induced and spontaneous emission. This was indeed the lesson he derived from his 1916/1917 papers which he considered to contain 'the fundamental outline of the future theoretical derivation.' Einstein was compelled to regard spontaneous emission as being independent of the state of the radiation field and an *inherent* property of atoms and molecules because in 1917 he was unaware of the special quantum statistical property of the radiation field. It is strange that even after championing Bose's method he failed to see Bose's point about spontaneous emission. To Bose it was clear that it depended on the 'environment' in which the atom or molecule was placed. He wrote in his second paper [1924] :

> From classical theory one would expect that no stationary states are possible and that an interaction or coupling (Bohr) between ether and the excited atom and the radiation connected with it is always present. On the other hand, in order to explain spectral emission we must assume it to be possible that no interaction (emission) occurs. Because of the existence of stationary states we are therefore led to assign to every stationary state a probability coefficient or a mean life-time.

Recall the expression (38) for the fundamental probability *P*. It can be written as $\bar{n} / (\bar{n} + 1)$ where $\bar{n} = N_\nu / A_\nu$ is the average number of quanta per cell. For $\bar{n} >> 1$, *P* tends to unity, the classical limit in which stationary states are not possible. It follows from this that the average life-time of a stationary state (and therefore the probability of spontaneous emission) depends on the average radiation density $\bar{n}$. Bose used to recount to his students Einstein's objection to this. 'Suppose there were only one hydrogen atom in the universe in an excited state,' Einstein once remarked to Bose, 'don't you think it would radiate spontaneously and come down to its ground state?' Bose did not disagree, but nevertheless, felt that such a *gedanken* situation did not capture the situation in a Hohlraum or cavity. It was Bose's intuition that eventually turned out to be right in the end, not Einstein's. Firstly, in Dirac's method of 'second quantization' (Dirac, 1927) which is widely used in quantum field theory today, one starts with a classical electromagnetic field in which the equality of the probabilities of induced emission and absorption (determined by the Fourier coefficients in the plane wave expansion) is built in, and obtains the spontaneous emission term as a *consequence* of the fundamental commutation rules. Although this is not exactly what Bose had proposed, it fits in with his basic contention that spontaneous emission is a consequence of the quantal property of the radiation field *itself* and need not be introduced as an independent hypothesis concerning the radiating atoms. Einstein's initial reaction to Dirac's contribution was also decidedly negative. [Pais, 1982, 441]

Secondly, in quantum electrodynamics spontaneous emission is not a property of an isolated atom but of an atom-vacuum system and can be significantly inhibited or enhanced by placing the atoms in a suitable environment. In quantum electrodynamics the vacuum is not empty and devoid of everything—it is full of virtual particles appearing and disappearing, interacting with one another; it provides a non-trivial physical 'environment' to an atom. A whole new branch of quantum optics called 'cavity quantum electrodynamics' has developed since about 1987 utilizing dramatic changes in spontaneous emission rates to construct new kinds of microscopic masers that operate with a single atom and a few photons or with photons emitted in pairs in a two-photon transition. This is what two of the pioneers in the field have to say :

> Ever since Einstein demonstrated that spontaneous emission must occur if matter and radiation are to achieve thermal equilibrium, physicists have generally believed that excited atoms inevitably radiate. Spontaneous emission is so fundamental that it is usually regarded as an inherent property of matter. *This view, however, overlooks the fact that spontaneous emission is not a property of an isolated atom but of an atom-vacuum system.* The most distinctive feature of such emission, irreversibility, comes about because an infinity of vacuum states is available to the radiated photon. If these states are modified — for instance, by placing the excited atom between mirrors or in a cavity– spontaneous emission can be greatly inhibited or enhanced.
>
> Recently developed atomic and optical techniques have made it possible to control and manipulate spontaneous emission. [Haroche and Kleppner, 1989]

In his third unpublished paper Bose had gone further. He wrote in his letter of 27 January 1925 to Einstein :

> *I have tried to look at the radiation field from a new standpoint and have sought to separate the propagation of Quantum of energy from the propagation of electro-magnetic influence. I seem to feel vaguely that some such separation is necessary if Quantum theory is to be brought in line with Generalized Relativity theory.*
>
> *The views about the radiation field, which I have ventured to put forward, seem to be very much like what Bohr has recently expressed in May Phil Mag 1924. But it is only a guess, as I cannot say honestly to have exactly understood all he means to say about virtual fields and virtual oscillators.*

This letter shows that Bose was one of the first persons to have formulated the idea of an 'empty wave' (an electromagnetic wave propagating in space and time but carrying no energy-momentum). He was certainly the first person to have intuitively seen its relationship with the principle of General Relativity, although the connection is not clear. Einstein already had the idea of an empty wave in his light-quantum hypothesis according to which 'a ray of light expands starting from a point, the energy does not distribute on ever increasing volumes, but remains constituted of a finite number of energy quanta localized in space and moving without subdividing, and unable to be absorbed or emitted partially.' [Einstein, 1905]

If the localized quanta carry all the energy and momentum, what happens to the electromagnetic waves which can produce interference? There was no clear-cut answer. The problem was so acute that Einstein referred to these waves as *Gespensterfelder* (ghost waves) guiding the photons. [Bohr, 1945, 206] Louis de Broglie also had a similar idea in his concept of phase or pilot waves 'guiding the propagation of the energy' and enabling a 'synthesis of the waves and the quanta.' [de Broglie, 1923, 549] Slater also arrived at a similar notion. As he recalled : 'A number of scientists — W [illiam] F [rancis] G [ray] Swann, among others — had suggested that the purpose of the electric field was not to carry a continuously distributed density of energy, but to guide the photons in some manner. This was the point of view which appealed to me, and during my visit at the Cavendish Laboratory in the fall of 1923, I elaborated on it. [Slater, 1975, 9] Slater who believed in the existence of light-quanta as well as electromagnetic waves found the following way out of the difficulty : 'I decided you could only have a statistical connection between them because you couldn't set up a vector to represent the photon in electromagnetic theory : that you would have the intensity of the wave governing the probability of finding the photon there. And I wanted to have the wave emitted during the stationary state so as to get it emitted over a long enough period so that it would have a suitable spectral distribution.' [Slater, loc.cit., 30]

When Slater arrived in Copenhagen on 21 December 1923 to collaborate with Hendrik Kramers and Niels Bohr, he was prevailed upon by them to relinquish Einstein's light-quantum hypothesis but to retain his idea of electromagnetic waves emitted by oscillators during their stationary states and carrying no energy. They coined the name *virtual oscillators* for them. The collaboration resulted in the famous Bohr–Kramers–Slater proposal [Bohr, Kramers, Slater, 1924] which Bose referred to in his letter to Einstein. In the BKS proposal the conservation of energy (and momentum) was abandoned for individual transition processes in atoms but was retained as a statistical concept. Most interestingly, they wrote : 'the transitions which in [the Einstein theory of 1917] are designated as spontaneous are, in our view, induced by the virtual field.' Bose must have noted this with satisfaction and this must have been one reason why he regarded the BKS proposal to be 'very much like' his own. The theory was however quickly falsified by experiments carried out by Bothe and Geiger [Bothe and Geiger, 1925] and Compton and Simon [Compton and Simon, 1925a, b] which established the conservation of energy and momentum in individual processes. Nevertheless, their ideas were reformulated and generalized by Born [1926] in his well known statistical interpretation of the wave function in quantum mechanics, an idea now universally accepted. This is what Heisenberg wrote about its historical antecedent :

'The probability wave of Bohr, Kramers and Slater. . . was a quantitative version of the old concept "potentia" in Aristotelian philosophy. It introduced something standing in the middle between the idea of an event and the actual event, a strange kind of physical reality just in the middle between possibility and reality.

Later, when the mathematical framework of quantum theory was fixed, Born took up this idea of the probability wave and gave a clear definition of the mathematical quantity in the formalism.' [Heisenberg, 1958, 29]

On 16 July 1946 , twenty-one years after he received Bose's letter of 27 January 1925 from Paris, Einstein wrote to Schrödinger about the latest status of the unified field theory (of electromagnetism and gravitation based on the General Principle of Relativity):

'It is indeed correct that the energetics occurs quite differently here compared to Maxwell-Poynting. I consider, however, that in this respect the Maxwell theory is really false on account of quantum actualities. In light the energy exists in something like quasi-singularities. *The wave field as such should therefore not be the site of energy. One must simply require this and be happy that in the new theory the transverse wave field is indeed present but as such transports no energy.'* (Our italics) [Moore, 1989, 427]

There is no mention of Bose, his letter or his third paper which Einstein must have rejected outright, although the facsimile of Bose's letter supplied by The Jewish National and University Library, Jerusalem, shows that he had marked the phrase 'have sought to separate the propagation of Quantum of energy from the propagation of electro-magnetic influence' within parentheses.

Louis de Broglie's original interpretation of Schrödinger's equation [de Broglie, 1927] was given a firmer basis by Bohm [1952]. In the de Broglie-Bohm 'casual interpretation' particles are thought to exist independent of observation, embedded in a wave field ψ which satisfies the Schrödinger equation, with a density (over the statistical ensemble) of value $|\psi|^2$. One can show that this approach can explain the interference pattern observed in double-slit experiments without abandoning the notion of well-defined particle trajectories. [Philippidis, Dewdney and Hiley, 1979] Experiments for testing certain consequences of 'empty waves' have been proposed [Selleri, 1990, ch. 4] and some have already been carried out [Zou, Grayson, Wang and Mandel, 1992]. However, such experiments have not yet ruled out all versions of 'empty wave' theories [Vigier and Holland, 1991].

Bose remained ever grateful to Einstein (whom he always addressed as 'Sir' or 'Master') for his encouragement and the interest he had taken in his first paper. Nevertheless, deep within himself, he felt a resentment that rankled all through his life. Three of his most perceptive and original ideas were scotched by Einstein :

1. The idea of the photon spin which is now established beyond doubt. (That Bose proposed the idea in his first paper in 1924 cannot as yet be established beyond every shred of doubt, although circumstantial evidences are strong.)
2. The idea that spontaneous emission is not an inherent property of an isolated atom, totally independent of the radiation field, also confirmed by experiments since 1987.
3. The idea of the propagation of electromagnetic influence carrying no quanta of energy, an area of current theoretical and experimental activity.

Bose could not publish a single paper during his two years in Europe at a time when physics was going through a turmoil and one of its most exciting periods. He returned to Dhaka a disheartened man and did not publish anything in theoretical physics for the next twelve years !

The author is grateful to N Mukunda, V Singh and J C Pati for a critical reading of the manuscript and suggesting improvements.

PARTHA GHOSE
S N Bose National Centre for Basic Sciences, Calcutta

References

Aspect,A.
1987 *Hyperfine Interactions* **37**, 3-18.
Bohm, David
1952 *Phys. Rev.* **85**, 166-179.
Bohm, D. and Hiley, B.J.
1985 *Phys. Rev. Lett.* **55**, 2511.
Bohr, Niels
1913a *Phil Mag.* (6) **26**, 1-25.
1913b *Phil Mag.* (6) **26**, 476-502.
1913c *Phil Mag.* (6) **26**, 857-875.
1949 'Discussion with Einstein', in *Albert Einstein : Philosopher Scientist, ed* P.A Schilpp, The Library of Living Philosophers, vol VIII, Open Court, La Salle, Illinois, London : Cambridge University Press, Third Edition.
Bose, S.N.
1924a *Z. Phys.* **26**, 178-181.
1924b *Z. Phys.* **27**, 384-393.
Broglie, Louis de
1924 *Recherche sur la théorie des quanta*. Thesis presented to the Faculty of Science of the University of Paris for the degree of Doctor of Science. 1927 *J. Phys.* (Paris) **6**, 225.
Cabrera, G.G. and Kiwi, M.
1984 *Phys. Rev.* A**36** 2995.
Clark, Ronald W.
1971 *Einstein : The Life and Times*, New York : World Publishing Company ; paperback edition, New York : Avon Books, 1972.
Davisson, C. J. and Germer, L.H.
1927 *Phys. Rev.* (2) **30**, 705-740.
Debye, P.
1910 *Ann. d. Phys.* **33**, 1427-1434.
1923 *Phys. Zs.* **24**, 161-166.
Dirac, P.A.M.
1926 *Proc. Roy. Soc.* (London) **A112**, 661-677.
1927 *Proc. Roy. Soc.* (London) **A114**, 243-265.
Ehrenfest, Paul
1905 *Sitz. ber. Akad. Wiss* (Wien) **114**, 1301-1314.
Ehrenfest, P. and Kamerlingh Onnes, H.
1914 *Versl. Lon. Akad. Wetensch.* (Amsterdam) **23**. 789-792 ; English translation. *Proc. Lon. Akad. Wetensch.* (Amsterdam) **17**, 870-873 ; reprinted in *Collected Papers*, 353-356.
Einstein, Albert
1905 *Ann.d. Phys.* (4) **17**, 132-148; English translation in D. ter Haar : *The Old Quantum Theory,* Oxford, London, Edinburgh, New York, Toronto, Sydney, Paris, Braumschweig: Pergamon Press, 1967, 91-107.
1906 *Ann. d. Phys.* (4) **20**, 199-207.
1916a *Verh.d. Deutsch. Phys. Ges.* (2) **18**, 318-323.
1916b *Mitteilungen der Physikalischen Gesellschaft Zürich* **16**, 47-62.
1917 *Phys. Zs.* **18**, 121-128.
1924a *Sitz. ber. Preuss. Akad. Wiss.* (Berlin), 261- 267(presented at the meeting of 10 July 1924).

1923 *Z. Phys.* **18**, 272-286.
1925 *Z. Phys.* **31**, 765-783.
1927 *Z. Phys.* **41**, 81-102.
1940 *Phys. Rev.* **58**, 716-722.
Philippidis, C., Dewdney, C. and Hiley, B.J.
1979 *Nuovo Cimento* **52B**, 15.
Planck, Max
1900 *Verh. d. Deutsch. Phys. Ges.* (2) **2**, 237-245.
1910 *'Zur Theorie der Wärmestrahlung., Ann. d. Phys.* (4) 31, 758-768.
1911 'La loi du rayonnement noir et l'hypothèse des quantités élémentaires d'action' in Langevin and de Broglie, 1912, 93-114. 1925 *Sitz. ber. Preuss. Akad. Wiss.* (Berlin), 49-57.
Raman, C. V. and Bhagavantam, S.
1931 *Ind. J. Phys.* **6**, 353-366.
Saha, M.N.
1919 *Astrophys. Jour.* **50**, 220.
Schrödinger, Erwin
1926a *Ann. d. Phys* (4) **79**, 361-376.
1926b *Ann. d. Phys* (4) **79**, 489-527.
1926c *Ann. d. Phys (4)* **80**, 437-490.
1926d *Ann. d. Phys (4)* **79**, 734-756.
1952 *Statistical Mechanics, A course of Seminar Lectures,* London : Cambridge University Press, Second Edition.
Selleri, F.
1990 *Quantum Paradoxes and Physical Reality,* Dordrecht, Boston, London : Kluwer Academic Publishers.
Sommerfeld, A.
1927 *Atti del Congresso Internazionale dei Fisici.* 11-20 Septembre, Bologna : Nicola Zanichelli (1928), **2**, 449-473;
1928 *Z. Phys.* **47**, 43-60; *Ber. dt. chem. Ges.* **61**, 1171-1180.
Thompson, G.P.
1928 *Proc. Roy. Soc.* A **117**, 600.
Vigier, J.P. and Holland, P.
1991 *Phys. Rev. Lett.* **67**, 402.
Zou, X.Y., Grayson, T., Wang, L.J. and Mandel, L.
1992 *Phys. Rev. Lett.* **68**, 3667-3669.

WHO DISCOVERED THE BOSE-EINSTEIN STATISTICS?

Silvio Bergia
Università di Bologna

CONTENTS

1. Foreword 223
2. Bose's first paper of 1924 223
3. The problem of the transmission 227
4. Planck's paper of 1900 228
5. Debye, 1910 232
6. Natanson 233
7. The Wolfke-Krutkow dialogue and the idea of the "light molecules" 235
8. Ehrenfest and Kamerlingh Onnes 236
9. Wolfke, 1921 239
10. De Broglie on the statistics of light atoms 240
11. Radiation and free electrons: Compton, Pauli, Einstein and Ehrenfest 243
12. From Debye to Bose 243

References 246

1. Foreword

The purpose of this paper is to analyse in some detail the work of various authors, who, before the famous studies by Bose and Einstein were published, had already made several steps towards an understanding of the statistical aspects involved in the thermal equilibrium law for photons, i.e. in Planck's law. It is not claimed that the merit for the discovery of the Bose-Einstein statistics should be ascribed to one or the other of these authors rather than to Bose and Einstein. The claim is rather that, when the heritage they had accumulated (or part of it) was taken notice of by Bose and Einstein (which occurred *after* their first papers on the subject had been published), it led to a deeper understanding of the content of the new statistics. Some attention is also devoted to relevant early studies which, apparently, did not receive much attention both when they first appeared and in the Bose-Einstein times.

I should stress from the beginning that I will concentrate my attention on the statistics of photons. This implies that I will not follow in any detail the development of the studies on the ideal gas, from Sackur and Tetrode, to Nernst, Planck and Schrödinger[1], a field which played an important part in focalizing the role of particle identity in statistical mechanics. I should also add that I will restrict my analysis to the period going from the beginning of the century to Bose's papers of 1924: no mention will be made of the subsequent developments and, in particular, of the reassessment of the quantum statistics within the framework of quantum mechanics proper. Even so the number of authors and contributions to be examined was quite high, and I made the choice of basing the analysis almost exclusively on published material and of restricting it to the most relevant contributions. The amount of secondary sources dealing with the subject is also imposing, and some sections of the present paper do not go beyond reviewing earlier pieces of analysis. The points where new information is added, or personal conclusions are drawn, are explicitly indicated.

Finally, I would like to mention that this work is one of the fruits of the activities of the «Unità di Bologna» of the «Gruppo Nazionale di Storia della Fisica del CNR»: I wish to thank A. Desalvo, P. Fantazzini, C. Ferrario, V. Monzoni, G. Morandi and R. Rosa for their very valuable help in preparing this report.

2. Bose's first paper of 1924

A good starting point for our discussion is provided by the synthetic presentation of Bose's famous paper (BOSE 1924a), made by A. Pais in the article he devoted to Einstein's contributions to the quantum physics on the occasion of Einstein's centennial (PAIS 1979; see also PAIS 1982):

> Bose's arguments divest Planck's law of all the supererogatory elements of electromagnetic theory and base its derivation on the bare essentials. It is the thermal equilibrium law for particles with the following properties: they are massless; they have two states of polarization; the number of particles is not conserved, and the particles obey a new statistics.

1 This development forms the object of a parallel investigation by A. Desalvo (DESALVO 1983).

According to Pais, of the four essential ingredients Bose had introduced, he was only aware of the first two. The non-conservation of the number of light quanta was made evident when Einstein introduced the constraint on the number of particles in the first of the papers (EINSTEIN 1924) in which he applied Bose's analysis to a gas of non-relativistic material particles, even though he did not comment explicitly[2] on the meaning of the important difference holding in the case of photons.

Since, by the time the first of Einstein's papers on quantum statistics had appeared, the masslessness of Bose's particles, the necessity of considering two states of polarization and the particle number non-conservation had been recognized, it can be stated that by then every aspect of Bose's paper had been clarified... except for the nature of the new statistics.

Since this assertion may sound unconvincing to many, let me try to summarize the main points of evidence we have for it. Skipping the quantization of the photon phase space, which is the first and perhaps greater innovation introduced by Bose, since it is too well known, let me concentrate on the perhaps less known part of his paper, i.e. the statistical part[3]. Bose was acquainted with Planck's book[4] (the English translation of its second edition (PLANCK 1912) had been available since 1914), where he could find, for the «thermodynamic probability of a macroscopically defined state» (BOSE 1924a) for the case of N gas molecules distributed over the elementary «space» regions 0, 1, 2, ... (N_0, N_1, N_2 molecules in each separate space element), the formula:

$$W = \frac{N!}{N_0!\ N_1!\ N_2! \dots} \tag{1}$$

i.e. Boltzmann's expression. Bose had to deal with a gas of quanta of energies $h\nu_s$ («$s = 0$ to $s = \infty$»), and wanted to define the thermodynamic probability w_s of N_s quanta of frequency ν_s. For this purpose, he took the structure of eq. (1), but with the quantities appearing in it replaced by different entities. Precisely, calling A_s the number of cells at disposal for the quanta at frequency ν_s, $\boldsymbol{P_0^s}$ the numbers of vacant cells, $\boldsymbol{P_1^s}$ the number of cells containing one energy quantum $h\,\nu_s$, etc., he wrote;

2 A very explicit comment about this point appears in a paper on the quantum theory of gases presented by A. Schidlof at the *séance* of October 23, 1924 (SCHIDLOF 1924) of the *Societé de Physique et d'Histoire Naturelle de Genève* (I thank A. Desalvo for pointing out to me the existence of a series of papers by Schidlof on the Bose-Einstein statistics):
«Il y a toutefois deux modifications à introduire dans la théorie de Bose: premièrement, les molécules du gaz sont isotropes, tandis qu'il faut attribuer aux quanta une certaine anisotropie pour tenir compte de la polarisation du rayonnement; en second lieu, le nombre des quanta correspondant à une énergie totale U contenue dans une enceinte de volume V est indéterminé, tandis que le nombre N des molécules du gaz est fixé d'avance». It can be recalled that Schidlof held his communication before the appearance of Einstein's paper. In the series of papers referred above he developed, quite independently of Einstein, although with much less depth, several elements of the quantum statistics of ideal gases.

3 A detailed analysis of the statistical part of Bose's paper is given in TAGLIAFERRI 1982, BERNARDINI, 1981.

4 See MEHRA and RECHENBERG 1982, p. 564.

$$w_s = \frac{A_s\,!}{P_0^s\,!\;\; P_1^s\,!\;\; P_2^s\,!\ldots} \qquad .(2)$$

Then, of course, one must have:

$$A_s = \sum_{r=0}^{\infty} P_r^s \qquad (3)$$

and:

$$N_s = \sum_{r=0}^{\infty} rP_r^s \qquad (4)$$

On the other hand:

$$E = \sum_s N_s \; h\,\nu_s \qquad \text{(total energy)} \qquad (5)$$

and:

$$N = \sum_s N_s \qquad \text{(number of photons)} \qquad (6)$$

Note that eqs. (3) and (4), apart from an algebrically irrelevant constant, are the translation of Boltzmann's constraints

$$N = \sum_k N_k \qquad (7)$$

and

$$E = \sum_k \varepsilon k \; N_k \qquad (8)$$

once Bose's translation codex is adopted.

Now Bose maximizes

$$W = \prod_s w_s \qquad (9)$$

as a function of the P_r^s holding A_s and E fixed, and arrives by standard procedures to Planck's formula. According to Mehra and Rechenberg (MEHRA and RECHENBERG 1982, p. 567), Bose had «changed Planck's definition slightly but consistently with the spirit of probability theory». Let us briefly comment about this point.

Let us first note that in the Boltzmann-Planck procedure, the molecules can be looked at as containers of energy; as such, molecules can be replaced by cells, both for what concerns their total number N and the partial numbers N_k. In this respect Bose's procedure seems legitimate or, in other words, «consistent with the spirit of probability theory».

This observation may give us a clue as to the reason why Bose based himself on formula (2), which he certainly could not have found in the literature, and about which his late testimonies (BLANPIED 1972) are elusive: he did not have to deal with molecules as such; but, inasmuch as in eq. (1) they play the role of receptacles of energy, they may be replaced throughout by cells.

It is however clear that, although in the spirit of probability theory, Bose's procedure is not physically[5] equivalent to Boltzmann's. Bose's formulae, eqs. (2) and (9), should be compared with Boltzmann's formula for course grained counting (PAIS 1979):

$$W = N! \prod_s \frac{(z_s)^{n_s}}{n_s!} , \qquad (10)$$

for the distribution of n_s objects among z_s energy levels in the s-th cell. One gets eq. (10) for the thermodynamic probability if one can answer the question as to *which* particles are in a cell (Boltzmann distinguishability axiom) and if individual molecules have no *a priori* preference for any region in phase space (statistical independence of particles); but, on the one hand, the P_r^s are defined by asking *how many* particles are in a cell, while, on the other hand, Bose's formula implies statistical independence of *cells*. Therefore, both distinguishability axiom and statistical independence of particles have disappeared and we have here indeed a new statistics. The «slight» change introduced in the Boltzmann-Planck definition, when implemented with different constraints, leads to very relevant consequences.

In a late interview recorded by Mehra (reported also by PAIS 1979), Bose gave the following testimony: «I had no idea that what I had done was really novel... I was not a statistician to the extent of really knowing that I was doing something which was really different from what Boltzmann would have done, from Boltzmann statistics.»

On its basis[6] one may agree with Pais' guess that to Bose the cell counting, eq. (2), was «the perfect analog of Boltzmann's particle counting (1) and that his cell constraint (hold A_s fixed) was similarly the analog of Boltzmann's particle constraint (hold N fixed).» (PAIS 1979).

We may however leave aside the question concerning Bose's subjective awareness of what he had objectively introduced and concentrate our attention on a perhaps more stimulating aspect, that is the fact that the procedure Bose had followed was certainly not the most apt to exhibit the peculiar features of the new statistics. No substantial progress was achieved from this point of view with the first of the papers

5 It is not even *formally* equivalent to it: Boltzmann's procedure implies keeping E and N (eqs. (5) and (6)) fixed; once Bose's translation codex is adopted, it is A_s and N_s (eqs. (3) and (4)) which, by analogy, should be kept fixed, but Bose, quite correctly, has constraints on A_s and E. (A. Pais (PAIS 1979) points out, however, that the constraint on A_s is irrelevant).

6 K.C. Wali, however, has warned the present author against the risks of taking a late testimony too literally (both in general and in this particular case).

Einstein devoted to the statistics of the ideal gas, where he basically adopted the statistical scheme of Bose's paper.

The course of events up to this stage would make everyone inclined to ask himself, with M. Delbrück (DELBRÜCK) 1980): «Was Bose-Einstein Statistics Arrived at by Serendipity?»[7].

There is however more to it, and we shall presently begin our journey across these further aspects.

3. The problem of the transmission

Most of the following analysis will deal with the factor:

$$w'_s = \frac{(n_s + z_s - 1)!}{n_s!\ (z_s - 1)!}, \tag{11}$$

where the symbols have the same meaning as in eq. (10). Eq. (11) provides a much more transparent version than eq. (2) of the course grained counting in the Bose-Einstein statistics.

Chronologically, the first universally known paper on the Bose-Einstein statistics in which the factor w'_s is introduced in this form is Einstein's second paper (EINSTEIN 1925) on the ideal gas, in which it is taken to represent the expression of the «number of complexions for the s-th infinitesimal domain» «according to Bose»; the expression is compared with the one which holds «according to the hypothesis of the statistical independence of the molecules». The difference, according to Einstein, confirms what had been pointed out, in particular by Ehrenfest, after the appearance of the first two papers on the new statistics, i.e. that in his gas theory as well as in Bose's radiation theory molecules and quanta could not be considered statistically independent. In commenting this point, Einstein went as far as speaking of a «mutual influence of the molecules which for the time being is of a quite misterious nature». If one wants to fix the moment in which the scientific community took notice of the fact that a profound revolution had been introduced in the basic concepts of the statistics of particles, this may well be the one. Truly, statistical independence is not synonimous of quantum indistinguishability: the two quantum statistics *imply* the loss, in diverging directions, of statistical independence, but they are, first of all, statistics of indistinguishable particles. However, Einstein's second paper started a process which, in a short time, led to the acknowledgment of the essential role of indistinguishability. As stressed by Pais (PAIS 1979), «with this remark Einstein came to the very threshold of the quantum mechanics of identical particle systems. The misterious influence is of course the correlation induced by the requirement of totally symmetric wave functions».

7 «Serendipity: the faculty of making happy and unexpected discoveries by chance»; according to the Oxford Dictionary, the word was coined by Horace Walpole after a fairy tale, «The three Princes of Serendip». I thank R. Rechtman for pointing out to me Delbrück's paper.

There is one little puzzle in what has just been said, i.e. the credit Einstein gives to Bose for the introduction of the factor w_s. The puzzle has, however, a trivial answer: quite simply, Einstein is quoting a second paper by Bose (BOSE 1924b), which is not so well known as his first[8]. It is at this stage that we meet a more substantial question mark; for, in Bose's paper, the factor w_s is not derived nor commented, but simply taken as it stood from a paper published by Debye in 1910 (DEBYE 1910). But this is not the end of the story; since, in turn, Debye quotes the factor from the first edition of Planck's lectures on heat radiation, which sends us, more or less straightly, to Planck's classical paper of 1900 (PLANCK 1900b).

The following sections will be devoted to a more detailed analysis of the process through which the factor (11) was transmitted to Bose and Einstein from his first inventor.

One interesting aspect to be investigated concerns the fact that not all that had been put into focus during the process actually reached Bose and Einstein, nor, for all that matters, captured the attention of the scientific community as a whole when it was first discussed. Therefore, the story we want to tell is also a story of missed opportunities.

This story will allow us to substantiate the hypothesis formulated in the Foreword, which we are now in a position to formulate in the following terms: while it is true that the framework of the new statistics was set by Bose and Einstein, it was the information conveyed by the factor (11), which they inherited from previous work, that produced the awareness that indeed a new statistics had been created and helped in making more transparent the meaning of the whole picture. It should also be stressed that the objective meaning of the factor happened to shift in the process of transmission, so that the way it was considered at the beginning and at the end of its journey changed considerably and the factor played a different role in the initial and final contexts. Since the whole discussion about the factor is more or less directly a discussion about indistinguishability, to isolate this aspect of the historical development seems appropriate for a conference on the theme of symmetries in physics. The mechanism for the transmission and the growth of scientific knowledge which seems to operate in this case could however be of general historiographic relevance.

4. Planck's paper of 1900

The article in which the factor that interests us makes its first appearance in our story, although in a slightly disguised way, is Planck's famous second paper of 1900 (PLANCK 1900b). In order to be able to recognize the objective meaning it had there and to speculate about the way Planck must have looked at it, we need to analyse the paper in some detail. In it, as he had done in the series of papers he had already devoted to the theory of heat radiation, Planck considers oscillators and radiation enclosed in a cavity with perfectly reflecting walls.

8 I wish to thank P. Bernardini for pointing out to me the importance of Bose's second paper.

However, as stressed by Kuhn (KUHN 1978), the paper is not aimed at deriving the equilibrium distribution, but rather at justifying the form Planck had obtained for it in the previous paper (PLANCK 1900a). Very likely (KUHN 1978; MEHRA and RECHENBERG 1982, see also BERGIA 1982), what Planck tried to interpret was the formula he could work back from his distribution for the entropy S_N of N oscillators of frequency

$$S_N = kN \ln \left[\frac{(1 + U/h\nu)^{1 + U/h\nu}}{(U/h\nu)^{U/h\nu}} \right] \tag{12}$$

(U is the mean energy of the oscillator at frequency ν; in the following, $E = NU$).

At any rate, eq. (12) can indeed be immediately given a suggestive form by means of Planck's «most essential» hypothesis: that of considering E as« composed of a very definite number of equal parts» ε, so that denoting with P the number of energy elements which must be divided over the N resonators, one has

$$E = NU = P\varepsilon \tag{13}$$

Indeed, using eq. (13), eq. (12) takes the form

$$S_N = k \ln \left[\frac{(N + P\varepsilon/h\nu)^{N + P\varepsilon/h\nu}}{N^N (P\varepsilon/h\nu)^{P\varepsilon/h\nu}} \right] \tag{14}$$

and, with $\varepsilon = h\nu$,

$$S_N = k \ln \frac{(N + P)^{N+P}}{N^N P^P} \tag{15}$$

Since it is

$$\frac{(N + P - 1)!}{(N - 1)!\ P!} \simeq \frac{(N + P)^{N+P}}{N^N P^P}, \tag{16}$$

(if this was indeed Planck's reasoning) the problem had now become that of justifying the expression in the left hand side of eq. (16).

This question has in Planck's paper an explicit answer we quote almost *verbatim;* «It is clear —says Planck— that the distribution of P energy elements over N resonators can only take place in a finite, well-defined number of ways». Each of these is called by Planck a «complexion» («using an expression introduced by Mr. Boltzmann for a similar quantity»). A possible complexion, for the case $N = 10, P = 100$, is explicitly indicated by Planck with the symbol

1	2	3	4	5	6	7	8	9	10
7	38	11	0	9	2	20	4	4	5

where the oscillators are labelled by the upper index and the number of energy elements pertaining to each of them in the complexion is written underneath. Now, says Planck,

> the number of all possible complexions is clearly equal to the number of all possible sets of numbers which one can obtain for the lower sequence for given N and P. To exclude all misunderstandings, we remark that two complexions must be considered to be different if the corresponding sequences contain the same numbers, but in different order. [Translation as in Ter Haar 1967.]

This is a clear enough statement on the distinguishability of the resonators. With no further comment, Planck states then that, «from combinatorial calculus», one gets for the number of all possible complexions

$$W = \frac{(N + P - 1)!}{(N-1)!\ P!} \tag{17}$$

A more or less familiar image may help us to isolate the important specification which makes Planck's conclusion correct. A complexion, i.e. a given distribution of P energy elements over N resonators, can be likened to the result of a series of P throws (or of the simultaneous throw of P balls) in a roulette with boxes labelled by the sequence of the first N natural numbers. Then the result arises if one stipulates the convention that the outcomes of two series are to be considered as distinct only if they differ for the total number of times each natural number has come out, being irrelevant which particular ball fell in a particular box (balls are not labelled, i.e. indistinguishable). Indeed, eq. (17) expresses the number of ways P indistinguishable objects can be placed in N numbered boxes (with no restrictions on the number per box[9]). It is interesting to recall, that W, eq. (17), can also be directly looked at as the numbers of «combinations with repetitions of N objects taken P at a time». These are the groupings of P elements which can be formed with the given objects, if the same object is allowed to appear up to P times, and two groupings are considered

9 See, for instance, Mayer and Mayer 1940, p. 438.

distinct only if they differ for some object. In our example, the objects are the N integral numbers, and a grouping of P elements with $P>N$ is obtained by allowing a number to come out more than once. In Planck's case, the n-th object is the n-th oscillator inasmuch as it has received one energy element, and in a complexion, a grouping of P elements, this object can appear up to P times (P energy elements attributed to the n-th resonator).

A complexion is the same thing in both ways of looking at expression (17), though with a shade of conceptual difference: in the first way, it arises from counting how many energy elements are attributed to the n-th oscillator, in the second one from counting how many times the n-th of the N objects appears. In one word, the «objects» of the statistical calculations are the *P energy elements* in the first way, the N *oscillators* in the second.

This can make a psychological difference, when coming to the crucial point of the role of indistinguishability, which is clearly exhibited in the first way, but not so much in the second.

To the question as to which was the way Planck looked at the thing, one can give, we believe, only a partial answer[10]. The expression used to indicate the source of the result in the 1900 paper, and quoted above («from combinational calculus»), does not shed much light on the problem. However, in his book[11] the expression «combinations with repetitions of N objects taken P at a time» appears explicitly and the context shows clearly that, at least by the time he wrote it, he definitely looked at the thing from the second point of view. This does not allow to conclude that he had never paid any attention to the aspect of indistinguishability, although the above discussion, implemented by the consideration that Planck had in those days to focus his attention on other even more important features, makes us inclined to think that it was indeed so. At any rate, we cannot resist the conclusion that, had Planck considered the problem, he would have dismissed it as a triviality: after all, the elements are quantities of energy, not physical objects!

Altogether, it seems a fair conclusion about this point, that, to say the least, it was understated in Planck's work, and that this concealed to most physicists the possible existence of a problem there. More specifically, the meaning of the recipe «combinations with repetition» was not transparent to everyone. As late as in 1914, Ehrenfest and Kamerlingh Onnes wrote in a paper we will discuss in more detail in the following (EHRENFEST and KAMERLINGH ONNES 1914):

> Planck proves that the number of distributions [of interest here] must be equal to the number of all «combinations with repetitions of N elements of class P» and for the proof, that this number is given by the expression (A) [i.e. our eq. (17)], he refers to the train of reasoning followed in treatises on combinations for this particular case. In these treatises the expression (A) is arrived at by the aid of the device of «transition from n to $n+1$», and this method taken as a whole does not give an insight into the origin of the final expression.

10 See also HERMANN 1971; KLEIN 1972; MEHRA and RECHENBERG 1982. An interesting essay on the development of the early quantum theory is: G.P. GUIDETTI, *Origini e sviluppi della vecchia teoria dei quanti,* Thesis (Parma 1974) IFPR-SF-04.

11 At least in the second edition.

As we shall see, in the paper just quoted Ehrenfest and Kamerlingh Onnes did provide such an insight. But that happened fourteen years later. There is, of course, a major reason for this delay: as widely discussed, for instance, by Kuhn (KUHN1978), most physicists working on quantum physics in those years concentrated their attention on other aspects of Planck's theory. For several years the message on indistinguishability hidden in Planck's formula, eq. (17), remained in the background and went unobserved. Moreover, as we will see, after a few keen observers were able to spot it, two major reasons produced the effect, that as soon as it was recognized, it rapidly fell again into oblivion.

5. **Debye** 1910

By the time Debye wrote the paper that interests us here, it was clear enough that Planck had indeed quantized his resonator (KUHN 1978). While it was clear to Debye that something had to be «quantized», he did not like Planck's use of resonators, which he considered inessential and artificial. Remember that Planck had derived his fundamental equation, expressing the relation between the spectral density of the black-body radiation and the mean energy of a resonator in thermal equilibrium with it, from the classical electrodynamics of a charged oscillator, with no reference to the counting of the cavity normal modes. But, of course, this equation can be derived immediately in a much simpler way once the number of normal modes for a cavity of volume V in the frequency interval $d\nu$ has been computed as

$$N(\nu)\ d\nu = \frac{8\pi\ \nu^2}{c^3}\ V\ d\nu \qquad (18)$$

(the origins of this approach are to be found in papers by Rayleigh (1900, 1905), Lorentz (1903), Jeans (1905). If we now indicate with U the mean energy of a mode at frequency ν, the total energy $\rho_\nu|V\,d\nu$, [expression (a)], contained in the frequency interval $d\nu$ will be $UN(\nu)\,d\nu$, [expression (b)]. By equating (a) and (b) one then gets Planck's relation. Debye's idea was to quantize directly the cavity normal modes, by assuming that each oscillation receives a share of energy in the form of quanta $h\nu$. This procedure has been analysed elsewhere (KASTLER 1979); we reproduce here for completeness the account given in BERGIA 1982.

Let $f(\nu)$ quanta pertain to the oscillation of frequency ν; then $U = f(\nu)h\nu$ and

$$\rho_\nu\, V\ d\nu = f(\nu)\, h\nu\, N(\nu)\ d\nu \qquad (19)$$

so that

$$\rho_\nu = \frac{8\pi\ \nu^2}{c^3}\ h\nu\cdot f(\nu) \qquad (20)$$

and the problem of determining ρ_ν is reduced to that of determining $f(\nu)$. In the right hand side of eq. (19), $f(\nu)N(\nu)d\nu$ represents the number of energy elements to be distributed over the $N(\nu)d\nu$ oscillations; $N(\nu)d\nu$ and $N(\nu)f(\nu)d\nu$ correspond therefore respectively to Planck's N and P. Debye writes then formally Planck's formula for W:

$$W = \frac{(P+N-1)!}{(N-1)!\ P!} \simeq \frac{(N+P)!}{N!\ P!}, \tag{21}$$

in the form

$$W = \frac{(N\ d\nu + Nf\ d\nu)!}{(N\ d\nu)!\,(Nf\ d\nu)!} \tag{22}$$

which we had found in the first edition of Planck's book.

The he determines the function f which maximizes the entropy subject to the condition of fixed total energy, and finds

$$f(\nu) = \frac{1}{\exp(\alpha h\nu) - 1}, \tag{23}$$

where the constant α is subsequently specified as $1/KT$ by the thermodynamical relation ($dV = 0$) $dS = dU/T$.

To realize that Debye had introduced an essential novelty in the matter, we need only consider the objective support his treatment gave to Einstein's views on the quantization of the energy of radiation (the opportunity that Einstein missed overlooking Debye's paper has been pointed out by Kastler (KASTLER 1979)).

With this novelty, as we shall comment in the following, is linked the basic shift of meaning which Planck's heritage underwent in its transmission to Bose.

On the question of indistinguishability, however, Debye has nothing to say, and does not appear to be conscious of the existence of the problem.

6. Natanson

In a paper which appeared in *Physikalische Zeitschrift* in 1911 (NATANSON 1911), the Polish physicist Natanson set himself the task of computing in how many ways P energy elements can be distributed over N «receptacles of energy» («Energiehaltern»), so that N_j receptacles contained j energy elements, subject to the restrictions $\Sigma_j N_j = N$ and $\Sigma_j\ j.N_j = P$, if it is assumed that each energy element has the same probability «of reaching a given receptable». But then Natanson suggests that one should, *a priori*, distinguish three cases:

1) receptacles and energy elements are both indistinguishable. In this case a microscopic state is defined once the number of receptacles containing a given number of energy elements is specified;

2) receptacles are distinguishable, energy elements indistinguishable. In this case a microscopic state is defined once the number of energy elements for each identifiable receptacle is specified;

3) the receptacles and energy elements are distinguishable (this is Boltzmann's case).

Planck's procedure, continues Natanson, is based on the tacit assumption (tacit, at least, we may add, for what concerns the indistinguishability of the *P* elements) that we are in case (2). This may well be taken, says Natanson, as the «foundation» («Grundlage») of the theory. It so happens that it is confirmed by the experiment. But apparently no one as yet has paid sufficient attention to the circumstance that it is only the experiment which can tell us, *a posteriori*, which is *the correct assumption* in a given physical situation[12].

It must be stressed that Natanson does not suggest a physical interpretation in terms of objects: photons did not exist in 1911[13]. Still, he is certainly to be considered as a forerunner of the Bose statistics.

As we said, his paper appeared in 1911. According to Hund (HUND 1975) «in 1924 Natanson's arguments had been forgotten». We would like to know, however, whether they received any attention in the years immediately following 1911, and why, together with other contributions we will mention, they were forgotten.

First of all, what about the person who should have been the most interested in the matter, i.e. Planck himself? We have not found a reference to Natanson's paper in any of Planck's research papers of those years we have been able to look at. The readers of the English version (1914) of the second edition of Planck's book (1913) can still find, as they could in 1914, Natanson's paper quoted in the appendix containing «a list of the most important papers on the subject treated in this book». The list was however compiled by the translator M. Masius and, although it was given (as one would have expected) «with Professor Planck's permission», does not permit to infer that Planck had anything to say about Natanson's paper. It is on the other hand a hint, although a weak one, that the translator, felt he had to mention Natanson's paper among «the most important papers» on the subject; a not necessarily contrasting hint is contributed by the fact that, while some papers in the list, Debye's paper among them, are briefly introduced and commented, Natanson's paper is just only quoted. Would it be considered a wild speculation to infer that Masius, possibly not unlike some of his contemporaries, suspected that the paper *had to be* important but had not found within himself and the scientific community a sufficient motivation to devote to it the time requested to understand exactly *why it was* important? Although a definite conclusion about the role played by Natanson's paper in the years which immediately followed its publication can hardly be achieved on the basis of such scarce hints, we are here formulating the hypothesis that the

12 The epistemological relevance of this statement —made in 1911!— can hardly be overestimated.

13 There existed, of course, since EINSTEIN's paper of 1905, the light quantum; but most physicists (see, for instance, KLEIN 1970a) did not give much credit to the idea.

point Natanson had caught was not felt of central importance by the majority of the influent physicists who had dedicated some interest to Planck's theory.

There have been, however, early readers of Natanson's paper. Their contributions are interesting in themselves, and now we will briefly examine them.

7. The Wolfke-Krutkow dialogue and the idea of the «light molecules»

As well known, in his famous 1905 paper, Einstein had shown that the thermodynamics of a radiation described by the Wien's law of 1896

$$\rho_\nu(T) = \alpha\nu^3 \ e^{-\beta\nu / T} \tag{24}$$

(in the form used by Einstein) was that of a perfect gas of light quanta of energy $R\beta\nu/N$ (i.e. $h\nu$) (EINSTEIN 1905). He had subsequently shown (EINSTEIN 1909a, b) that, if the radiation is described by Planck's formula, then both a wave and a particle nature of the radiation were needed to account for the energy and momentum fluctuations.

In 1913, another Polish physicist, Mieczysław Wolfke, in a series of papers appeared in *Verh. d. Deutsch. Phys. Ges.* set himself the task of deriving Planck's formula on the basis of the idea of light quanta, which he called light atoms (MEHRA and RECHENBERG 1982, p. 559).

To this, G. Krutkow objected, from Leiden (KRUTKOW 1914a), that Wolfke's assumption of independent light atoms would lead to a probability giving Wien's, and not Planck's law. To show it, he based himself on Natanson's paper and on a paper by Ehrenfest of 1911. The last section of Krutkow's paper contains a very interesting remark.

Let us quote directly: «Our scheme must be radically changed if one wants to obtain Planck's formula... If a resonator has the energy $2h\nu$, this does not mean in fact that it has received one light atom and subsequently another one. And if one wants to preserve the foundations of the 'theory of light atoms', then one must, as Joffé has observed, introduce an association between quanta». One recognizes this easily by giving to Planck's law the form:

$$\frac{8\pi\nu^2}{c^3} \frac{h\nu}{e^{h\nu / KT}-1} = \frac{8\pi\nu^2}{c^3} h\nu \sum_{r=1}^{\infty} e^{-r\,h\nu/KT} \tag{25}$$

(with respect to the formula (25), besides Joffé's paper, published in *Annalen der Physik* in 1911, the author quotes a paper by J. Ishiwara, published in *Zs. Phys.* in 1912). The expansion (25) (in itself, a mathematical triviality) and its possible physical meaning have been rediscovered more than once in the course of our story. We shall have to come back to this point.

Wolfke's reply to Krutkow's criticism is also very interesting (WOLFKE 1914). He recalls that DEBYE 1910, has shown that «for the derivation of Planck's formula it is sufficient to assume that the energy distributes itself over the proper modes in integral quanta $h\nu$». On the other hand, the number of all possible complexions is given by the number of the combinations with repetitions of N objects of class P.

From these two ingredients, says Wolfke, Planck's law ensues. From the formal point of view, he does not add anything to Debye; however, Debye's shares of energy distributed over the normal modes have now become light quanta, or, in Wolfke's terminology light atoms. This is then one step farther along the path marked by the various shifts of meaning undergone by Planck's counting in its transmission to later times. It appears, however, that Wolfke was not fully aware of the meaning of this counting, so that he was puzzled by the fact that indeed, as Krutkow had pointed out to him, the Einstein theory gave a different result. This is witnessed by the following quotation from his paper:

> Mr. Einstein has made me conscious of the difference existing in principle between the Einstein light quantum theory and the above discussion... In my derivation, no restrictive assumption was introduced concerning the distribution of light atoms in space... On the contrary, Einstein's light quantum theory considers the particular case in which the light atoms... are spacially independent of each other...

A decade later (EINSTEIN 1925) Einstein stressed the same point again, in almost the same words: «If one treats the quanta as statistically independent in their localization, one gets Wien's radiation law».

In restating his point of view, Krutkow (KRUTKOW) 1914b), referring to the difference hinted at in Wolfke's last paper, complains about the fact that «Mr. Wolfke... has formulated... such fundamental distinction only in words, and not in explicit formulae».

8. Ehrenfest and Kamerlingh Onnes

The whole matter was taken up again in that same year by Ehrenfest and Kamerlingh Onnes, in the paper mentioned in sect. 4 above (EHRENFEST and KAMERLINGH ONNES 1914). This short, very lucid article has been discussed in detail elsewhere (KLEIN 1970, see also BERGIA 1982); we come back here to its content for completeness. In order to find an explanation of the form $(N\text{-}1)!$ in the denominator of Planck's formula, eq. (17), the authors consider, first, an example ($N=4$, $P=7$). In this case, one of the possible distributions is the one in which the resonator P_1 «contains» the energy 4ε, R_2 the energy $2\,\varepsilon$, R_3 no energy and R_4 the energy ε. It is what is indicated by the symbol.

$$\| \;\varepsilon\;\varepsilon\;\varepsilon\;\varepsilon\;\mathrm{I}\;\varepsilon\;\varepsilon\;\mathrm{I}\;\mathrm{I}\;\varepsilon\; \|$$

Now, continue the authors, giving the first version of the argument that has since become customary, with general values of N and P, the symbol will contain P times the sign ε and $(N\text{–}1)$ times the sign I. The question is, how many different symbols may be formed from the given number of N and P. To answer it, observe that there are altogether $P+(N\text{–}1)$ signs ε and I. If we consider these $(N\text{–}1+P)$ elements as so many distinguishable entities, they may be arranged in $(N-1+P)!$ manners between the ends ||.

However, in the symbol, the elements I, and the elements ε, are not distinguishable among themselves, so that, once a particular arrangement is obtained, a permutation of the P signs ε or of the $(N-1)$ signs I transforms the symbol in itself. The number of different symbols is thus obtained by dividing $(N-1+P)!$ by $(N-1)!P!$, which gives Planck's formula, eq. (17).

Ehrenfest and Kamerlingh Onnes have thus, no doubt, provided us with the clearest possible derivation of Planck's formula. Concerning the physical meaning to be attached to the procedure, their point of view differs from that of Natanson; indeed, in the appendix of their paper, they insist that «the permutation of the elements ε is a purely formal device, just as the permutation of the elements I». This is of course true if one is exchanging *quantities*, numbers, which in no way can be associated with *objects* such as particles. Natanson did not think in terms of an exchange of objects either; as a matter of fact, he does not suggest any physical interpretation. However, he did not exclude one *a priori*, and it seems that only by leaving the possibility of a physical interpretation open one can see that the matter is indeed about a different statistics. This is the point that Ehrenfest and Kamerlingh Onnes seem to miss.

To the «formal device», according to Ehrenfest and Kamerlingh Onnes, «a physical interpretation» has been given «more than once» by «a misunderstanding». «As a matter of fact —continue the authors— Planck's energy elements» were often «almost entirely identified with Einstein's light quanta... The confusion which underlies this view has been more than once pointed out». (The authors quote here Ehrenfest (1911) and Krutkow). But, how can the product of the confusion be measured? This leads us to the second important point contained in this paper. In his 1905 paper, Einstein had considered the thermodynamical properties of a radiation described by Wien's formula eq. (24) and derived, in particular, the expression

$$S-S_0 = \frac{E}{\beta\nu} \ln \frac{V}{V_0} = k \ln \left(\frac{V}{V_0}\right)^{E/h\nu} \tag{26}$$

for the dependence of the entropy of radiation of energy $E = V\rho\, d\nu$ for a change from volume V_0 to volume V. (The second expression in eq. (26) is in modern notation and uses $\beta = h/k$).

Now, take Einstein's formula, eq. (26), and rewrite it in «Planckian» terms. To begin with, call P the number of the light quanta, as the objects which are akin to Planck's energy elements. On the other hand, the number of normal modes at a given frequency in a cavity is proportional to its volume. The normal modes represent the physical states of the system over which the P light quanta are distributed. The number of normal modes at a frequency ν plays therefore the role of the number N of receptacles of energy at disposal, and for what has just been said, it will be

$$V \propto N \tag{27}$$

so that eq. (26) will be rewritten as:

$$S_2 - S_1 = k \ \ln \left(\frac{N_2}{N_1}\right)^P \tag{28}$$

(for a change from volume V_1 to volume V_2).

Read in terms of Boltzmann's relation

$$S = k \ \ln W + \text{const.}, \tag{29}$$

eq. (28) gives then

$$\frac{W_2}{W_1} = \left(\frac{N_2}{N_l}\right)^P \tag{30}$$

On the other hand, in Planck's formula, eq. (17), N represents again the number of receptacles of energy. From it one can then derive, with a similar meaning of the symbols, the formula corresponding to eq. (30) in Planck's scheme:

$$\frac{W_2}{W_1} = \frac{(N_2 + P - 1)!}{(N_2 - 1)! \ \ P!} : \frac{(N_1 + P - 1)!}{(N_1 - 1\,)! \ \ P!} \tag{31}$$

Ehrenfest and Kamerlingh Onnes stress that the profound difference between eqs. (30) and (31) is explained by the fact that «Planck does not deal with really mutually free quanta ε». Ehrenfest and Kamerlingh Onnes do not state explicitly that the distinction is between the number of ways in which P distinguishable and indistinguishable objects may be distributed over N receptacles. Their interest is rather to stress that Einstein's hypothesis leads necessarily to Wien's and not to Planck's law, and that Planck's «formal device» cannot be interpreted in the same sense of Einstein's light quanta.

Is there any sign that the authors' analysis received any attention by the experts of the field in the following years? The hints are here even scantier than in the case of Natanson. None of the treatises which discuss the matter reports of early readers of the paper. It does not appear in Masius' bibliography for the good reason that it appeared in press later, though in the same year. And it is not quoted in any of the classical papers on the Bose-Einstein statistics.

We cannot draw the conclusion that it went completely unobserved. But there is one fundamental reason why it could: the paper was communicated in the meeting of Oct. 31, 1914, of the Amsterdam Academy (in Dutch); its German translation appeared in *Annalen der Physik* in issue 7 of volume 46 of 2nd April 1915 (MEHRA and RECHENBERG 1982). By that time Europe was in fire. In the spring of 1915 there were probably not so many German speaking physicists in the Central Empires capable of spending or willing to spend their time making their best of a scientific paper. The war lasted three and a half more years, and, in the words of Mehra and Rechenberg, «affected the international relations of science more deeply than any other event during the previous hundred years. Scientific researchers active in enemy countries had been cut off from one another almost completely» (MEHRA and RECHENBERG 1982, p. 259). And, we may add, by the time the war was over, problems which had just come into focus when the conflict broke out had very likely been forgotten. When the war was over, physicists either resuming their activity or just starting their careers concentrated mostly on the problems of atomic physics, which looked by far as the most promising field, guided by Sommerfeld's *Atombau und Spektrallinien* (HEILBRON 1980). There was however one further reason why the debate about the meaning of Planck's statistical procedure should be forgotten by most of the physicists: the appearance in 1910 (LORENTZ 1910) of a new derivation by Lorentz of Planck's law, which avoided Planck's unheard-of counting and apparently re-established, apart from the acknowledged discontinuity, a statistical reasoning which looked very close to Boltzmann's and completely masked the role of indistinguishability (KUHN 1978).

Last but not least, an interpretation in line with the Natanson-Wolfke ideas implied the assumption of a corpuscular theory or radiation, and it was more than once (see for instance, KLEIN 1970a) stressed that Einstein was left almost alone in his views on the matter until the discovery and interpretation of the Compton effect (1923).

9. Wolfke 1921

The war introduced a gap in the course of events concerning the radiation problem. One important exception was however Einstein's new derivation of Planck's radiation law (final version: EINSTEIN 1917).

Einstein's paper most likely encouraged Wolfke (who quotes it at the beginning of his paper), at post-war resumption of work, to return to his former studies on the radiation problem and to his corpuscular point of view (WOLFKE 1921). Without quoting either Krutkow or Joffé and Ishiwara, Wolfke introduces the expansion (25), which he writes as

$$\rho_\nu = \frac{8\pi h \nu^3}{c^3} \frac{1}{e^{h\nu/kT}-1} = \sum_{n=1}^{\infty} \rho_\nu^{(n)} \qquad (32)$$

$$\rho_\nu^{(n)} = \frac{8\pi h \nu^3}{c^3} e^{-nh\nu/kT}, \qquad (33)$$

and observes that the «single terms of this series are built in agreement with the Wien's radiation formula». Then Wolfke computes the entropy densities corresponding to the «partial radiations» (33) and shows that their sum gives indeed the entropy densitiy of Planck's black radiation; therefore his partial densities represent indeed radiation energies «which are thermodynamically independent of each other». There follows a physical interpretation of the single partial radiations «according to the Einstein method». That is, Wolfke obtains the entropy change $S-S_0$, undergone by the partial radiation term of energy E_n when condensed from the initial volume V_0 to the smaller volume V, as:

$$S - S_0 = \frac{k\ E_n}{n\ h\nu} \ln \frac{V}{V_0} \tag{34}$$

But this is Einstein's formula, eq. (26), apart from the fact that $nh\nu$ replaces here the quantum $h\nu$. Wolfke therefore concludes that «the cavity radiation, looked at from the point of view of Einstein's light quantum hypothesis, consists of light-molecules $h\nu$, $2h\nu$, $3h\nu$,... which are mutually independent in space.»

The question remained open as to what mechanism caused the light-quanta, or light-atoms, to agglomerate into light molecules.

A possible answer to this question was soon provided by Louis de Broglie. Not only for this reason, but because he fell short of the discovery of the Bose-Einstein statistics as a whole, we shall devote some attention to some of de Broglie's contributions in the years 1922-24.

10. De Broglie on the statistics of light atoms

We do not in the least intend to discuss in any detail the role de Broglie played in the development of quantum physics[14]. The scope of our analysis is limited to what in de Broglie's writings in the years mentioned above directly bears on the matter of the statistical properties of light.

A short paper by the French physicist («On the interferences and the theory of the light quanta», BROGLIE 1922) appeared in *Comptes Rendus* in 1922. In the first sentence of the paper, de Broglie states that «the recent advances in Physics concerning the emission and absorption of radiation direct more and more the attention toward the theory of light quanta...[15]» On the other hand, the explanation of the usual optical phenomena in terms of the light-quantum theory presents obvious difficulties. De Broglie wants to put forward an idea which could «facilitate the construction of a theory of interference in harmony with the existence of the light-quanta». After recalling Einstein's 1909 results on wave-particle dualism (EINSTEIN 1909a,b), as reported at the Solvay Conference of 1911, de Broglie writes once more, without referring to earlier derivations, the expansion of the spectral density into partial densities, eqs. (32), (33). From the technical point of view, he adds to the previous contributions the observation that the formula for the energy fluctuations can be written

$$\overline{\varepsilon^2} = \sum_n n\ h\nu\, E_n\ , \tag{35}$$

14 This role has been recently analysed, for instance, in: RAMAN and FORMAN 1969; MEDICUS 1974; MAC KINNON 1976; HANLE 1977, 1979; BERNARDINI and DONINI 1980; MEHRA and RECHENBERG 1982; DONINI 1982; FER and LOCHAK 1982.

15 The meaning of this sentence is illuminated by the analysis of B. Wheaton (section 11 of these proceedings).

in terms of the partial energies E_n, and that this expression «is just the one that corresponds to a 'gas of light' formed of molecules and atoms». On the other hand, the terms with $n \neq 1$ in eq. (35) sum up to the fluctuation term that Einstein had shown to be of wave origin. De Broglie's conclusion is that «from the point of view of light quanta, the interference phenomena may seem to be connected with the agglomerations of light atoms, whose motions are not independent but coherent» (translation as in MEHRA and RECHENBERG 1982, p. 585).

The question about the continuity between this piece of de Broglie's work and earlier studies like those of Wolfke is in itself of interest for the historian. According to Mehra and Rechenberg, de Broglie read German scientific articles: «For example, in his paper on black-body radiation (also published in 1922) he referred to the article or Robert Emden,' Über Lichtquanten', which was published in *Physikalische Zeitschrift* several months after Wolfke's paper. Emden's paper... contained a derivation of Planck's radiation formula, in which Emden also introduced 'light molecules » (MEHRA and RECHENBERG 1982, p. 586). The hypothesis that, in this particular respect, de Broglie's work could be considered as a development of earlier investigations is supported by his use of terms such as «light atoms» and «light molecules».

Important conclusions[16] along this train of reasoning were reached by de Broglie in the statistical chapter of his thesis (BROGLIE 1924). There, he assumes that the motion of the atoms of a gas is «accompanied» by the propagation of a «phase-wave»; these phasewaves will form, for a gas enclosed in a cavity, a system of stationary waves, whose numbers, in the frequency interval dν, may be computed on the basis of the Rayleigh-Jeans procedure, as adapted by Jeans to the case of a dispersive medium, as

$$n(\nu)\ d\nu = \gamma \frac{4\pi}{UV^2} \nu^2 d\nu , \qquad (36)$$

where U and V are the group and phase velocities, and γ equals 1 or 2 for longitudinal or transversal waves. The next basic idea is to interpret the velocity βc of the atom as the group velocity V and to give the wave the frequency

$$\nu = \frac{1}{h} \frac{m_o c^2}{\sqrt{1-\beta^2}} \qquad (37)$$

(note that the formalism is taken to hold for light quanta as well, since de Broglie attributes to them a very small but finite mass). Then one finds for the number of states (stationary waves *or* atoms *or* light atoms) in the interval dw of kinetic energy

$$u_w\ dw = \gamma \cdot \frac{4\pi}{h^3} m_o c^2 (1 + \alpha) \sqrt{\alpha(\alpha + 2)}\ dw, \qquad (38)$$

16 A former, partially alternative version had been given in a paper published in 1923 (BROGLIE 1923): «Les quanta, la théorie quantique des gaz et le principe de Fermat», which, as suggested by the title, aimed at a unified treatment of the light quantum and the material gases. The development of de Broglie's ideas from this paper to the more mature version of the same topics is analysed in a parallel investigation (DESALVO, MONZONI and ROSA 1983).

where $\alpha = w/(m_o c^2)$. Hereafter, we will disregard the developments concerning the material gas: we have summarized de Broglie's approach up to this point just in order to indicate that, to the extent that he developed, on the basis of eq. (38), a statistics of the light quanta, he could in principle have developed a similar statistics for the material gas[17].

By applying to the set of «atoms» the canonical distribution law, one then obtains, for the number of them contained in the volume $dx\ dy\ dz$ and with kinetic energy between w and $w+dw$:

$$\text{const} \cdot \gamma \cdot \frac{4\pi}{h^3} m_o^2\, c\, (1+\alpha) \sqrt{\alpha(\alpha+2)}\; e^{-w/kT}\, dw\, dx\, dy\, dz \tag{39}$$

For the case of a gas of light atoms, one obtains ($\gamma = 2, \alpha + 2 \sim \alpha + 1 \sim \alpha$):

$$\text{const} \cdot \frac{8\pi}{c^3} \nu^2\, e^{-h\nu/kT}\, d\nu\, dx\, dy\, dz, \tag{40}$$

and, for the spectral density:

$$\rho_\nu\, d\nu = \text{const} \cdot \frac{8\pi h}{c^3} \nu^3\, e^{-h\nu/kT}\, d\nu \tag{41}$$

(de Broglie refers to a previous paper of his for a proof that the constant equals 1).

Unfortunately, observes de Broglie, the law thus obtained is Wien's law, «which is only the first term of the series constituting the (experimentally exact) law of Planck». This result, on the other hand, cannot be a surprise, «because, supposing the motions of the light atoms completely independent, we must necessarily arrive at a law whose exponential factor is identical with that of Maxwell's law». In order to obtain Planck's result, continues de Broglie, one must make here «a new hypothesis», which de Broglie formulates in these terms: «If two of several atoms possess phase waves which superpose exactly, and thus one can say that they are transported by the same wave, then their motions cannot be considered as being completely independent of each other; hence these atoms should not be treated as independent objects in the calculus of probability» (translation as in MEHRA and RECHENBERG 1982, p. 600). «The motions of these atoms 'en onde' (lit.: on wave) will thus present a kind of coherence as a consequence of interactions impossible to specify[18], but probably akin to the mechanism which would make unstable the motions of atoms whose phase waves are not stationary». Up to this stage, it seems to us, de Broglie's ideas have not led him much further than Wolfke, whose «association between quanta» has been substantiated in terms of some rather undefined wave guidance but not derived as the consequence of some fundamental principle. It seems to us that the real conceptual step forward is made by de Broglie at the stage when he states

17 To the question why de Broglie did not realize (what Einstein would soon find) that he had also at hand the theory of degenerate gases (FER and LOCHAK 1982), a tentative answer is given in DESALVO, MONZONI and ROSA 1983.

18 The resemblance between this sentence and that used by Einstein in his 1925 paper (EINSTEIN 1925), concerning a «mutual influence of the molecules which for the time being is of a quite misterious nature» (see our section 3) has been pointed out by Tagliaferri (TAGLIAFERRI 1982, p. 112).

that «we cannot anymore take the single atoms as 'objects' of the general theory, it is the elementary stationary phase waves which must play this role». But then «the canonical distribution law should apply to such waves», so that, since «each elementary wave can carry 0, 1, 2, ... atoms», one finds, for the number of the corresponding atoms

$$N(\nu)\ d\nu = n(\nu)\ d\nu\ \frac{\sum_{1}^{\infty} n\, e^{-nh\nu / KT}}{\sum_{0}^{\infty} e^{-nh\nu / KT}} \qquad (42)$$

($n(\nu)$ dν is the number of waves), which leads directly to Planck's formula for the spectral density.

The use, in de Broglie's derivation, of new statistical concepts justifies, to some extent, the assertion of the editors of *Annales de la Fondation Louis de Broglie* (FER and LOCHAK 1982): «In conclusion, he finds, one year earlier, the result Bose was going to find soon by another argument (that of de Broglie is even better, since it is more physical...)». It is worth recalling, at any rate, that this way of looking at the statistical implications of Planck's law was resumed, after the appearance of the papers by Bose and Einstein, by Landé and Schrödinger in 1925-26 (BARACCA 1980, p. 379). In particular, the latter saw in de Broglie's matter waves a way of freeing the new statistics of its awkward aspects.

The contributions mentioned in our sections 6 to 10 did not have an important part, to our knowledge, in the transmission of information from Planck to Bose. An important role in this respect, on the contrary, was played by other developments which took place about 1923, as we will presently discuss.

11. Radiation and free electrons: Compton, Pauli, Einstein and Ehrenfest

As known, 1923 is the year of the Compton effect. The story of this discovery has been told in detail elsewhere (MEHRA and RECHENBERG 1982; TAGLIAFERRI 1982).

At any rate, it is not its details that interest us here, but only its moral, i.e. the vindication of the light quantum it represented in the eyes of most physicists.

Apart from that, Compton's discovery posed a new problem: in his 1916-17 papers Einstein had found a solution to the problem of thermal equilibrium between atoms and radiation; the scattering of light quanta on free electrons called for a parallel solution of the similar problem which arises when atomic matter is substituted by free electrons. A solution of the problem was first given by Pauli (1923) and then clarified and generalized by Einstein and Ehrenfest (1923). We have mentioned these papers because they have a relevance in Bose's formation and shall briefly come back to them in due time.

12. From Debye to Bose

The most accurate and complete account of Bose's scientific background and early career is given by Mehra and Rechenberg (MEHRA and RECHENBERG 1982; see also

BLANPIED 1972; SUDARSHAN 1974). Bose became fairly early acquainted with Planck's theory; about 1924, however, he was feeling increasingly uneasy for the lack of a self-consistent quantum-theoretical frame-work for the derivation of Planck's formula, and his uneasiness was enhanced in his teaching and tutorial activities. According to Mehra and Rechenberg, a turning point in the course of events which led Bose to his achievement was provided by a visit payed to him in Dacca by his friend and colleague Saha, who «pointed out to Bose the work of Wolfgang Pauli (1923) and of Einstein and Ehrenfest (1923)». Bose was also encouraged by the discussion with Saha «to study carefully» the papers by Peter Debye[19] (1910, 1925) and Compton (1923). No doubt, Compton's analysis of his effect in terms of particles was decisive. Bose's great idea was of course to derive Planck's law as the law of thermal equilibrium for light quanta in a quantized phase space. But the matter is not so clear, in our opinion, as far as the statistical aspects are concerned. There can be little doubt about the circumstance that Bose had become acquainted with the papers of DEBYE (1910), Pauli and Einstein-Ehrenfest by the time he sent his *second* paper (BOSE 1924b) to Einstein, since in it he refers explicitly and significantly to all of these articles. It is however interesting to ask whether he was influenced by them when he wrote his *first* and most famous paper. We would be inclined to conclude that he was not (at least for what concerns Debye) by the substantial fact that his way of attacking the statistical problem is logically independent from Debye's approach, as it was maybe sufficiently documented in our sections 2 and 5.

A corollary to this conclusion is the thesis that Bose's second paper is historically very relevant, first of all, as the one in which Planck's hidden message on indistinguis-hability was transferred, via Debye and Saha, to Bose.

This paper was apparently motivated by Bose's wish to show that the results obtained by Pauli and by Einstein and Ehrenfest could be included in a general scheme, «independent of any special hypothesis concerning the mechanism of the elementary processes in which energy is exchanged» (translation in THEIMER and RAM 1976).

For this purpose, both Bose's original statistical formulation and Debye's can be used. The impression is that, after reading Debye's paper, Bose was quite happy to give up his procedure, which he probably never considered as fundamental. Of course, Bose did not want to follow Debye through: he was up for his central idea of dealing with a gas of light quanta, and there was no place in his approach for a thing like the counting of normal modes. But, as pointed out particularly by Kastler (KASTLER 1979), the two approaches are closely related. The analysis of the nature of this relationship will lead us back to the central point of the essential shift of meaning which took place in this transfer.

Basically, Debye's number of normal modes (in the volume V and in the frequency interval $d\nu$) has now become the number of cells or, as we may say, of photon states, which Bose denotes with A_ν, while he uses $N_\nu d\nu$ to indicate the number of quanta in

19 The first is the paper discussed above in our section 5; the second is Debye's well known paper on the Compton effect.

the same region. He then adapts Debye's formula eq. (22) to his case by assuming, for the thermodynamic probability, the expression

$$W = \prod_{\nu} \frac{(A_\nu + N_\nu \ d\nu)!}{A_\nu! \ (N_\nu d\nu)!} \tag{43}$$

i.e. the same expression, apart for the need of considering the cell structure of the phase space. Since, however, Debye energy elements and normal modes have now become light quanta and light quantum states, Planck's (implicit) formal indistinguishability of the energy elements has now become the (as yet implicit) indistinguishability of particles,which is by no means formal. That this was not immediately evident to the eyes of Bose and of some of his early readers is due, it seems to us, to the fact that the debate on indistinguishability that had flourished about Planck's formula had not attracted the attention of most physicists or had been prematurely forgotten. Planck's message to Bose conveyed however much of the relevant information, and, when it was deciphered, gave new impulse to the course of ideas in physics.

(245)

References

BARACCA, A.
1980 *Manuale critico di meccanica statistica* (Culc, Catania 1980).
BERGIA, S.
1982 «Statistical thermodynamics and the thermal equilibrium law for photons (1894-1924)»; in *Proceedings of the first Escuela Mexicana de Física Estadística* (R. Peralta Fabi Editor, SMF, México 1982).
BERNARDINI, P.
1981 «La statistica di Einstein e la seconda quantizzazione: quale continuità?» *Physis 23,* 337.
BERNARDINI, P. and E. DONINI
1980 «Perché fu Einstein il tramite tra de Broglie e Schrödinger?»; to appear in the Proceedings of the Meeting *La ristrutturazione delle scienze tra le due guerre mondiali* (Firenze-Roma 23 June - 3 July 1980).
BLANPIED, W.A.
1972 «Satiendranath Bose: Co-founder of Quantum Statistics»; *Am. J. Phys. 40,* 1212.
BOSE, S.N.
1924a «Plancks Gesetz und Lichquanten Hypothese»; *Zs. Phys. 26,* 178; English transl. in THEIMER and RAM 1976.
1924b «Wärmegleichgewicht und Strahlungsfeld bei Anwesenheit von Materie»; *Zs. Phys. 27,* 284; English transl. in THEIMER and RAM 1977.
BROGLIE, L. DE
1922 «Sur les interférences et la théorie des quanta de lumière»; *Comptes Rendus* (Paris) *175*, 811; reprinted in *Annales de la Fondation Louis de Broglie* 7, 1982.
1923 «Les quanta, la théorie cinétique des gaz et le principe de Fermat»; *Comptes Rendus* (Paris) *177,* 630; reprinted in *Annales de la Fondation Louis de Broglie* 7, 1982.
1924 *Thesis* (Masson et Cie., Paris 1924).
DEBYE, P.
1910 «Der Warscheinlichkeitsbegriff in der Theorie der Strahlung»; *Ann. d. Phys.* (4) *33,* 1427.
DELBRÜCK, M.
1980 «Was Bose-Einstein Statistics Arrived at by Serendipity?»; *J. Chem. Ed. 57,* 467.
DESALVO, A.
1983 «Il problema del gas perfetto monoatomico e la teoria dei quanti»; communication at the *IV Congresso Nazionale di Storia della Fisica*, Como (to appear in the *Proceedings).*
DESALVO, A., V. MONZONI and R. ROSA
1983 «Onde di fase e meccanica statistica nella tesi di Louis de Broglie»; Communication at the *IV Congresso Nazionale di Storia della Fisica,* Como (to appear in the *Proceedings*).
DONINI, E.
1982 *Il caso dei quanti* (CLUP-CLUED, Milano 1982).
EHRENFEST, P. and H. KAMERLINGH ONNES
1914 *Versl. Kon. Akad. Wetensch.* (Amsterdam) *23,* 789; English transl.: «Simplified deduction of the formula from the theory of combinations which Planck uses as the basis of his radiation theory», reprinted in Ehrenfest's *Collected Scientific Papers*, M.J. Klein ed. (North Holland Pub. Co., Amsterdam).
EINSTEIN, A.
1905 «Über einen die Erzeugung und Verwandlung des Lichtes betreffenden heuristischen Gesichtspunkt»; *Ann. d. Phys.* (4) *17,* 132; English transl.: «On a heuristic point of view about the creation and conversion of light»; in: TER HAAR 1967.
1909a «Zum gegenwärtigen Stand des Strahlungsproblems»; *Phys. Zs. 10,* 185.
1909b «Über die Entwicklung unserer Anschauungen über das Wesen und die Konstitution der Strahlung»; *Phys.Zs. 10,* 817.
1917 «Zur Quantentheorie der Strahlung»; *Phys. Zs. 18,* 121; English transl. in TER HAAR 1967.
1924 «Quantentheorie des einatomigen idealen Gases»; *Sit. ber. Preuss. Akad. Wiss.* (Berlin) 261.
1925 «Quantentheorie des einatomigen idealen Gases. 2. Abhandlung»; *Sit. ber. Press. Akad Wiss.* (Berlin) 13; French transl. in *Ann. Fond. L. de Broglie* 7, 1982.
FER, F. and G. LOCHAK (eds.)
1982 «Sur les statistiques quantiques et la cohérence des ondes»; *Annales de la Fondation Louis de Broglie* 7, 87.

HANLE, P.A.
1977 «The coming of age of Erwin Schrödinger: His Quantum Statistics of Ideal Gases»; *Archive for History of Exact Sciences 17*, 165.
1979 «The Schrödinger-Einstein correspondence and the sources of wave mechanics»; *Am. J. Phys.* 47, 644.
HEILBRON, J.L.
1980 «The Recasting of Physics between the Wars»; to appear in the *Proceedings of the Meeting: La ristrutturazione scientifica tra le due guerre mondiali* (Firenze-Roma 23 June-3 July 1980).
HERMANN, A.
1971 *Frühgeschichte der Quantentheorie (1899-1913);* (Physik Verlag, Mosbach 1969); English transl.: *The Genesis of Quantum Theory (1899-1913)*, (MIT Press, Cambridge).
HUND, F.
1975 *Geschichte der Quantentheorie;* Bibliographisches Institut AG, Zürich. Italian transl.: *Storia della Teoria dei Quanti* (Borighieri, Torino 1980).
KASTLER, A.
1979 «Albert Einstein e il concetto di fotone»; in: *Astrofisica e Cosmologia, Gravitazione, Quanti e Relatività negli sviluppi del pensiero scientifico di Albert Einstein*, M. Pantaleo Ed. (Giunti-Barbera, Firenze 1979) 795-828.
KLEIN, M.J.
1970a «The First Phase of the Bohr-Einstein Dialogue»; *Hist. St. Phys. Sc. 2*, 1.
1970b *Paul Ehrenfest - Volume I: The making of a theoretical physicist* (North Holland Publ. Co., Amsterdam 1970).
1972 «The Beginnings of the Quantum Theory»; in: *Proceedings of the Int. School E. Fermi*, Varenna (Academic Press, New York 1977).
KRUTKOW, G.
1914a «Aus der Annahme unabhängiger Lichtquanten folgt die Wiensche Strahlungsformel»; *Phys. Zs.* 133.
1914b «Bemerkung zu Herrn Wolfkes Note: Welche Strahlungsformel folgt aus der Annahme der Lichtatome?»; *Phys. Zs. 15*, 363.
KUHN, T.
1978 *Black-Body Theory and the Quantum Discontinuity (1894-1912);* (Clarendon Press, Oxford 1978).
LORENTZ, H.A.
1910 «Alte un neue Fragen der Physik»; *Phys. Zs. 11*, 1234.
MAC KINNON, E.
1976 «De Broglie's thesis: A critical retrospective»; *Am. J. Phys. 44*, 1047.
MAYER, E. and M.G. MAYER
1940 *Statistical Mechanics* (Wiley, New York 1940).
MEDICUS, M.A.
1974 «Fifty Years of Matter Waves»; *Physics Today 27*, 38.
MEHRA, J. and H. RECHENBERG
1982 *The Historical Development of Quantum Theory* (Springer Verlag; New York, Heidelberg, Berlin, 1982).
NATANSON, L.
1911 «Uber der statistische Theorie der Strahlung»; *Phys. Zs. 12*, 659.
PAIS, A.
1979 «Einstein and the Quantum theory»; *Rev. Mod. Phys. 51*, 863.
1982 *Subtle is the Lord - The Science and the Life of Albert Einstein* (Oxford Univ. Press/New York, Clarendon Press/Oxford, 1982).
PLANCK, M.
1900a *Verh. Deutsch. Phys. Ges.* (Berlin) *2*, 202; English transl. in TER HAAR 1967.
1900b «Zur Theorie des Gesetzes der Energieverteilung in Normalspektrum»; *Verh. Deutsch. Phys. Ges.* (Berlin) *2*, 237.
1914 *The Theory of Heat Radiation*. English transl. of the II edition of *Wärmestrahlung* (1913);— Dover edition, republished in 1959.
RAMAN, V.V. and P. FORMAN
1969 «Why was it Schrödinger who developed de Broglie's ideas?»; *Hist. St. Phys. Sc. 1*, 291.

SCHIDLOF, A.
1924 «L'entropie des gaz et la théorie des quanta»; *Compte Rendu des Séances de la Société de Physique et d'Histoire Naturelle de Genève, 41,* 127.
SUDARSHAN, E.C.G.
1974 «A world of Bose particles»; *Am. J. Phys. 43,* 69.
TAGLIAFERRI, G.
1982 *Lineamenti di Storia della Fisica Moderna* (CLUED, Milano 1982).
TER HAAR, D.
1967 *The Old Quantum Theory* (Pergamon Press, London 1967).
THEIMER, M. and B. RAM
1976 «The Beginning of Quantum Statistics»; *Am. J. Phys. 44,* 1056.
1977 «Bose's Second Paper: A Conflict with Einstein»; *Am. J. Phys. 45,* 242.
WOLFKE, M.
1914 «Welche Strahlungsformel folgt aus der Annahme der Lichtatom?»; *Phys. Zs. 15,* 308.
1921 «Einsteinsche Lichtquanta und räumliche Struktur der Strahlung»; *Phys. Zs. 22,* 375.

(248)

DISCUSSION

WIGNER:

I would like to make a remark which may be quite interesting, namely that one of the reasons for Einstein's favouring the Bose statistics was the expression for the vapor pressure of a material at a very low temperature. The usual calculation assumed that at very low temperature the solid state was in one single quantum state, and that gave an expression for the vapor pressure which was, of course, in contradiction with experimentally supported real pressure. It pushed the particles out of the solid state, because in this only one state was available for it, but in the gas many states were available for it. So in the formula, which was in contradiction with the formula given I think by Murray, there was a very great discrepancy. I was there and I told Einstein to explain that to several of us. But I want to add, if you forgive me, that Einstein was in error. Because, even in classical theory, in the solid state at very low temperature there is not one state, but $n!$ states, where n is the number of atoms in the solid,—because atoms can be interchanged in the solid, and so the number of states is increased by $n!$. So I must admit as a matter of history, that Einstein's support of the Bose statistics was, to a large extent, based on a common and accepted error. An error which several people probably, and surely I, realized and explained to Einstein; and he was a little bit... —well I don't know what to say! [laughs]. But it is an interesting fact that his support of the Bose statistics was largely based on a commonly accepted and many times published formula for the vapor pressure at very low temperature.

DONINI:

I would like to ask for some more comments about the two issues of statistical independence and identical-particles indistinguishability. You stressed that the year 1925 was a turning point and I agree with you that a deep difference, a deep distinction has to be made in the development of these two issues till the papers of Einstein in 1925. But I would like to know your mind about these points in reference to the problem of accepting or not the light quantum. You quoted indeed a paper of de Broglie, but you had not time, I think, to discuss it. In his papers of 1922[1], having accepted the concept of light quanta, de Broglie could conceive quanta as particles and thus as distinguishable objects, because he had a very Boltzmannian approach to particles. But, at the same time, he realized that quanta had a sort of coherence which implied some statistical dependence; so he began to develop a general point of view about the necessity of connecting waves with any kind of particles. That is why I am interested in understanding better your opinion about the distinction between the two issues.

BERGIA:

The two points are conceptually distinct, of course, and we know that by now. We have two quantum statistics —we will soon hear the history of the second one from Prof. Amaldi. In both these statistics you lose Boltzmann's statistical independence in diverging directions, and this makes the qualitative difference between the two.

Concerning history, I would just like to make this comment. It seems to me that in 1925 attention concentrated initially almost exclusively on the loss of statistical independence, and that the discussion about the second aspect —indistinguishability— began a little later. I am referring in particular to papers by Landé and by Schrödinger which appeared —I do not remember exactly— in 1925 or 26. And I think it is in these papers that the second aspect begins to emerge, even though these authors wanted to reconcile the unusual features of the new statistics with more or less traditional points of view. A final comment: Schrödinger's point of view about the subject was closely related to de Broglie's original ideas. Both thought that there was no statistical mystery hidden in Planck's law provided one considered stationary waves, and not particles, as the object of the statistics. This is the way in which Schrödinger tried to interpret the new statistics along classical lines of thought.

PAIS:

May I perhaps make one brief comment on this. One has to realize that all the discussion we heard about this morning deals with free gases —there are no interactions. Both the Bose gas which Prof. Bergia

1 L. de BROGLIE, «Rayonnement noir et quanta de lumière», *Journal de Physique* 3. 422-428 (1922); and L. de BROGLIE, «Sur les interférences et la théorie des quanta de lumière», *Comptes Rendus de l'Académie des Sciences* 175, 811-813 (1922).

talked about and the Boltzmann gas, with which he compared it, are non-interacting systems. Now the question of distinguishability or not is independent of whether you have interactions or not, whereas the question of statistical independence does depend on whether there is a free gas or an interacting gas. Because physical independence means that there is no preference for any region in phase space. But if you have very strong interactions, and you have something in the corner of free space, then the next wants to get there too. So, I thought that it is perhaps good to say it, as an added illustration of how very different these two concepts are.

STUEWER:

As you may know, Thomas Kuhn has argued at some length that Planck did not in fact introduce energy quantum into physics in 1900. I am wondering, through your own study of Planck's work whether you agree or disagree with that position.

BERGIA:

I tend to agree with that position.

STUEWER:

Could you elaborate slightly on your reasons?

BERGIA:

I prefer to make one comment which is closely related to this point. The comment is the following: it seems to me that one of the reasons why this heritage —the heritage contained in the papers by Natanson, Ehrenfest and so on— was more or less forgotten by the time Bose wrote his first papers, is probably the conclusion reached by Lorentz (as Prof. Kuhn recalls in his book) in a paper of 1910, in which he derives Planck's law by the now standard procedure —I mean the procedure known to every graduate student— which is very close to Boltzmann's procedure. It seems that most physicists were then convinced that there was no further mystery in Planck's law, insofar as it appeared to rely on known procedures of statistical mechanics. Once this conclusion was reached, it was clear to everybody that Planck's theory implied quantization —the quantization as we know it now. But in the preceding years there was a lot of confusion not only about the statistical content of Planck's procedure, but also on the meaning and even the necessity of quantization. Therefore I think that 1910 is a crucial date. Of course Prof. Kuhn has mentioned this point, and I am just stressing his conclusion. I think that, before that time, the confusion involved not only Planck but several other people.

(250)

Addendum

Recently I have come across two accounts of the history of Bose statistics, one by Abraham Pais [1] and the other by S. Begia [2] which bring in two aspects of the history that I was not aware of at the time of writing the Introduction to Bose statistics [5] in 1993. The first concerns the statistical part of de Broglie's Thesis and the second concerns Bose's second paper [6]. There is also a paper by Delbruck with the title 'Was Bose-Einstein Statistics Arrived at by Serendipity?' [4] that has been brought to my notice which, in my opinion, cannot be allowed to pass without comment.

1 de Broglie and Bose

In his book Pais makes the statement:

'One month later de Broglie presented a third short paper which I find particularly interesting because it contains the first derivation of the density of states of electromagnetic radiation by means of a counting procedure in the phase space of photons – one year before Satyendra Nath Bose would independently do the same in his derivation of Planck's radiation law.'

The paper appeared in *C. R. Acad. Sci. Paris*, **177**, 630, 1923. Pais then goes on to quote de Broglie's letters to him of 9 August 1978 and 28 September 1978 in response to his query: 'When did Einstein first know of de Broglie's work and how did he know it?' In the second letter de Broglie states:

'I am therefore certain that Einstein knew of my Thesis *since the spring of 1924*. The necessary formalities for the acceptance and publication of a Thesis did not allow me to defend it until November 1924, but I believe I can affirm that by then Einstein knew the text of my Thesis since at least six months.'

According to Bergia, the statistical part of de Broglie's Thesis contains a more mature version of his earlier ideas contained in the paper that Pais quotes. This statement raises an historically important question that Pais has not dealt with in any detail in his book, namely, did de Broglie's Thesis contain the essential content of Bose's first paper, as would appear to be the case from Pais's statement, and if so, how did Einstein fail to see that precisely at the time when he so enthusiastically accepted and championed Bose's idea?

It is clear from de Broglie's paper and Bergia's account of his Thesis that de Broglie indeed arrived at the correct expression for the density of states factor $8\pi\nu^2 d\nu/c^3$ in Planck's formula from quantized phase space considerations a year before Bose. However, in his reply to Bose on 2nd July, 1924 Einstein wrote: 'You are the first person to derive the factor quantum theoretically, ...'. He also wrote to Ehrenfest on 12th July, 1924 : 'The Indian Bose has given a beautiful derivation of Planck's law, including the constant $[8\pi\nu^2 d\nu/c^3]$'. These show that even though Einstein might have been aware of de Broglie's Thesis since the spring of 1924, he did not give it much importance before receiving Bose's paper. Why?

The historical accounts given in [2] and [5] show that like many before him de Broglie knew that a new statistics of the quanta was required to arrive at Planck's formula, but, like them, *he did not specify what the new statistical distribution should be.* Nevertheless the credit goes to him for foreseeing that it would have to be a consequence of coherence of his phase waves. However, he missed the real significance of his derivation of the density of states factor, which was provided by Bose. Once the division of phase-space into elementary cells of non-zero volume is assumed, *the indistinguishability of the quanta becomes a necessary consequence, not an assumption* because the cells, being 'the ultimate elementary regions of the Phase-space' (vide Bose's covering letter of 4th June 1924 to Einstein) are not accessible to inspection. Bose therefore recognized that ' ... the total number of cells must be regarded as the number of possible arrangements of a quantum in the given volume.' With this interpretation in mind he wrote down an expression for the thermodynamic probability of a quantum 'state' of radiation completely specified, apart from the spatial position, by the occupation numbers of the cells alone, and arrived at Planck's formula using methods of statistical mechanics but *with totally new meanings for the mathematical symbols used earlier* (*cells in phase space or states taking the place of particles*), ushering in a new statistics. This is what Einstein immediately recognized as 'an important advance' and linked it to de Broglie's speculations about matter waves. In other words, Einstein gave credits where they were due, to Bose for the new statistics and to de Broglie for the complementary idea of matter waves which he found to be consequences of Bose statistics applied to material atoms – at last he had found his own reasons for believing in de Broglie's idea.

2 Bose's second paper

Bose's second paper contains the expression

$$W = \Pi_\nu \frac{(A_\nu + N_\nu d\nu)\,!}{A_\nu\,!\, N_\nu d\nu\,!} \tag{1}$$

for the thermodynamic probability for radiation. He says this was derived in an earlier paper and quotes the reference, 'the author, to appear in Phil. Mag. See also Debye, *Ann. d. Phys.* **33**, 1427, (1910) '. The only earlier published paper of Bose containing an expression for the thermodynamic probability of radiation is the first paper. But the expression given in the first paper is different, namely

$$\Pi_s \frac{A^s\,!}{p_0^s\,!p_1^s\,!\ldots} \tag{2}$$

where $A^s = \sum_r p_r^s = 8\pi\nu^{s2} d\nu^s/c^3$, the total number of cells for quanta belonging to the frequency range $d\nu^s$. This is somewhat puzzling. It is difficult to establish precisely which paper he is referring to because no such paper ever appeared in Phil. Mag. The only reasonable assumption is that it was a version of the first paper which was eventually not published. I say a version because it is quite possible that by the time he sent the paper to Einstein he might have incorporated some changes in it. Many historians have recorded that Bose sent his first paper to Phil. Mag. who rejected it, but the reference under discussion clearly contradicts that view. Bose told me personally and quite emphatically that Phil. Mag. had not rejected the paper, but that he was getting anxious at the delay in receiving their response, and so sought Einstein's 'opinion' and help. Probably he was apprehensive it would be rejected by Phil. Mag.. Going by Saha's account of Bose's discovery, one can safely conclude that he must have sent the first version of the paper to Phil. Mag. within days of the discovery, i.e., sometime in March 1924. This explains why Bose became anxious by the beginning of June and wrote to Einstein. Presumably he received Phil. Mag.'s acceptance shortly after despatching a probably revised version of the paper to Einstein on 4th June, 1924 but before despatching the second paper on 15th June, eleven days later. He therefore inserted the phrase 'to appear in Phil. Mag.', still apprehensive whether Einstein had received the first paper and would approve of it. In the covering letter dated 15th June, 1924 [3] he wrote:

> I send herewith another paper of mine for your kind perusal and opinion. I hope my first paper has reached your hands.

I guess Bose withdrew the paper from Phil. Mag. after hearing from Einstein in late July or early August, and Phil. Mag. stopped its publication. However, he was in the dark about the fate of his second paper until early November 1924 when he received Einstein's letter of 3rd November informing him that it had already appeared in print. Hence, the reference remained unchanged.

But why did Einstein retain the statement when he himself had translated both the papers into German and had them published in Zeitschrift für Physik? Carelessness and indifference to such details seem to be the only plausible answer. This is corroborated by the fact that Einstein put only 'Von Bose' as the name of the author of the first paper and omitted the initials altogether, while in the second paper he put 'Von S. N. Bose' as the author's name. Curiously, in his own first paper on the new statistics he refers to the author as 'D. Bose' !

In Bergia's view expression (1) is more transparent regarding the new statistics than expression (2), and Einstein used expression (1) rather than expression (2) in his papers on Bose statistics that followed and that generated so much interest and controversies regarding 'statistical independence' and the 'indistinguishability' of the particles. This is an important historical aspect of Bose's second paper that had not been noticed before.

However, after referring to M. N. Saha's visit to Bose in Dhaka when Saha acquainted him with the papers of Debye, Pauli, Einstein and Ehrenfest, Bergia writes :

> 'There can be little doubt about the circumstance that Bose had got acquainted with the papers of DEBYE (1910) and Pauli and Einstein-Ehrenfest by the time he sent his *second* paper (BOSE 1924b), since in it he refers explicitly and significantly to all these articles. It is however interesting to ask whether he was influenced by them when he wrote his *first* and most famous paper. We would be inclined to conclude that he was not (at least for what concerns Debye) by the substantial fact that his way of attacking the statistical problem is logically independent from Debye's approach, as it was maybe sufficiently documented in our sections 2 and 5.'

The fact is that Saha visited Bose in Dhaka in March 1924, i.e., two months before Bose sent his first paper to Einstein, and it was during this visit that he drew Bose's attention to the papers of Debye and others. This kick-started Bose's search for his now famous solution to the problem. Here is what Saha himself recorded about the incident [7]:

'It was these considerations which led to a collaboration between Einstein and Ehrenfest under the title '*Quanten theorie des Strahlungs-gleichgewichts*' in the Zeitsschrift f. Physik, vol. 19, pp. 301-306. While the author of this note was studying this paper with S. N. Bose in 1924, the latter was led to a very ingenious deduction of the Planck law of radiation. He showed that the number of cells A in phase space can be determined by the expression

$$A = \frac{1}{h^3} \int \int \ldots dp_r \, dq_r \tag{3}$$

the integration extending over the whole phase-space covered by the particles.

'One can then proceed to fill up the phase-space by the number of quanta available, by using combination with repetition (a device already taken by Debye), and thus get the Planck law in a very simple manner. This is in fact what was done by S N Bose.'

Since all other accounts of Saha's visit were orally communicated by Bose, in our opinion Saha's own written version is the most authentic one. It leaves no doubt whatsoever that Bose was aware of the papers of Debye and Einstein and Ehrenfest when he wrote his first paper. In fact, while commenting on the derivation of the density of states factor $8\pi\nu^2 d\nu/c^3$ from classical electro-magnetic theory in all previous papers, he states very clearly in the first paper :

'This is an unsatisfactory feature in *all* derivations, and it is not surprising that efforts are made *again and again* to give a derivation free from this logical flaw.' (italics mine)

Hence there is no reason whatsoever to doubt that Bose was aware of the papers of Debye and others when he wrote his first paper. The statement also shows very clearly that he was, however, not convinced by their logic. The facts that (i) Zeits. f. Physik received Einstein's translation of the second paper within only five days of receiving the first and the fact that (ii) the first paper does not give a specific reference to any previous paper on the subject at all whereas the second paper has them all, rather point to the fact that by the time Bose wrote his second paper he came to view them as a single development and wrote it in two parts with the first part containing no references. This is an additional argument in support of the view that Bose had revised his first paper before sending it to Einstein.

This brings me back to the question of the first and final versions of Bose's first paper. It appears from Saha's account of Bose's discovery and Bose's second paper that he used expression (1) (similar in form to Debye's expression which both Saha and Bose refer to) in the original version of the first paper. By the time he wrote to Einstein on 4th June 1924 he must have revised his first paper and changed the expression to the form (2). Expression (2) is an expression for the number of 'complexions' that correspond to a macrostate of radiation whereas expression (1) is the 'probability of the state'. Both lead to Planck's law on maximization and are statistically equivalent.

3 Delbruck's Paper

After explaining what a *boson* is in terms of a quantum mechanical wave function, Delbruck writes :

> The notion of a boson originated with a paper by S. N. Bose in June 1924 (1), a year before the invention of quantum mechanics, two years before wave mechanics, three years before the Uncertainty Principle (2). How could this be? How could the notion even be formulated without the notion of wave function? It is my contention that it arose from an elementary mistake in statistics that Bose made. Einstein at first copied that mistake without paying much attention to it. He then realized the mistake but saw that it must have a deep meaning since it gives the right answer. A clear justification only came with the working out of quantum mechanics.

Delbruck then goes on to describe how Bose calculated the number of cells in phase-space available to each quantum in the frequency interval $d\nu$, and then writes :

> Clearly, the next step should be to find the number of ways in which a certain number of quanta can be distributed over these cells in accordance with the specifications of the macrostate. The quanta should be treated as statistically independent entities as in the case of the classical statistical mechanics of ideal gases. In other words the microstate should be defined by saying into which cell each photon

> has been put. At this point Bose's mind goes foggy. Instead of carrying out the program as indicated he looks at each *cell* to see how *many quanta* are in it and in this way defines the microstate.

This is slander and misrepresentation of Bose's paper because Delbruck suppresses Bose's explicit statement that he was going to use, not classical statistical mechanics of an ideal gas, but '*statistical mechanics as adapted by Planck to conform to the requirements of quantum theory*' (italics mine). Accordingly, he first shows that once the division of phase-space into cells of non-zero volume h^3 is assumed (the Planck hypothesis), *the 'state' of a quantum in phase-space is completely specified by a cell, not a point.* As we have seen, in his covering letter of 4th June to Einstein he calls these cells 'the ultimate elementary regions in the Phase-space'. In the paper Bose comments: 'Nothing definite can be said about this method of division.' However, once this method is accepted, the cells are no longer accessible to inspection to determine which quantum (Tom, Dick or Harry according to Delbruck) is in which cell. Each cell (and hence each 'state' of the quantum) is completely specified, apart from its spatial location, by the momentum alone, i.e., by the numbers of quanta of specific types that it contains. In bra-ket notation such a state would be written as

$$|r_1^{s_1}, r_2^{s_2}, ..., r_n^{s_n}, ...\rangle \tag{4}$$

where $r_n^{s_n}$ are the numbers of quanta of type s_n or momentum $\hbar k_n$ in the cell/state. This means a collection of $\sum_n r_n^{s_n}$ quanta of different types define a *single* quantum state of radiation. Hence, classical statistical mechanics of an ideal gas is not applicable to light-quanta, and Bose's counting procedure is the correct one. Bose can indeed be said to have been the first to quantize the electromagnetic field in the modern sense of the term.

Delbruck also seemed to be oblivious of the historical fact that the mathematical structure of quantum mechanics had to be specially invented to fit these revolutionary developments, and hence to ask, 'How could the notion even be formulated without the notion of wave function?' is to put the cart before the horse.

References

[1] A. Pais, *Einstein Lived Here*, Clarendon Press, Oxford University Press, New York, 1994, pp. 47-53.

[2] Silvio Bergia, *Who Discovered the Bose-Einstein Statistics?*,

[3] 'S. N. Bose: The Man and His Work, Part II: Life, Lectures and Addresses, Miscellaneous Pieces,", eds C. K. Majumdar *et al*, S. N. Bose National Centre for Basic Sciences, Calcutta 1994.

[4] M. Delbruck, *Jour of Chem. Ed.* vol 57, July 1980, pp. 467-470, first brought to my notice by Asesh Das.

[5] P. Ghose, *Bose Statistics : a historical perspective*, "S. N. Bose: The Man and His Work, Part I: The Collected Scientific Papers,", eds C. K. Majumdar *et al*, S. N. Bose National Centre for Basic Sciences, Calcutta 1994.

[6] S. N. Bose, *Zs. Phys.* **27**,284, 1924.

[7] M. N. Saha, *Albert Einstein (1879-1955)*, Indian Journal of Meteorology and Geophysics, vol. 6, no. 4, October 1955, included in S. Chatterjee (ed), *Collected Works of M N Saha*, vol. 4, Saha Institute of Nuclear Physics and Orient Longman, Calcutta 1993, pp. 620-1.

REVIEWS OF MODERN PHYSICS, VOLUME 74, JULY 2002

Nobel Lecture: Bose-Einstein condensation in a dilute gas, the first 70 years and some recent experiments*

E. A. Cornell and C. E. Wieman

JILA, University of Colorado and National Institute of Standards and Technology, and Department of Physics, University of Colorado, Boulder, Colorado 80309-0440

(Published 19 August 2002)

Bose-Einstein condensation, or BEC, has a long and rich history dating from the early 1920s. In this article we will trace briefly over this history and some of the developments in physics that made possible our successful pursuit of BEC in a gas. We will then discuss what was involved in this quest. In this discussion we will go beyond the usual technical description to try and address certain questions that we now hear frequently, but are not covered in our past research papers. These are questions along the lines of: How did you get the idea and decide to pursue it? Did you know it was going to work? How long did it take you and why? We will review some our favorites from among the experiments we have carried out with BEC. There will then be a brief encore on why we are optimistic that BEC can be created with nearly any species of magnetically trappable atom. Throughout this article we will try to explain what makes BEC in a dilute gas so interesting, unique, and experimentally challenging.[1]

The notion of Bose statistics dates back to a 1924 paper in which Satyendranath Bose used a statistical argument to derive the black-body photon spectrum (Bose, 1924). Unable to publish his work, he sent it to Albert Einstein, who translated it into German and got it published. Einstein then extended the idea of Bose's counting statistics to the case of noninteracting atoms (Einstein, 1924, 1925). The result was Bose-Einstein statistics. Einstein immediately noticed a peculiar feature of the distribution of the atoms over the quantized energy levels predicted by these statistics. At very low but finite temperature a large fraction of the atoms would go into the lowest energy quantum state. In his words, "A separation is effected; one part condenses, the rest remains a saturated ideal gas"[2] (Einstein, 1925). This phenomenon we now know as Bose-Einstein condensation. The condition for this to happen is that the phase-space density must be greater than approximately unity, in natural units. Another way to express this is that the de Broglie wavelength, λ_{dB}, of each atom must be large enough to overlap with its neighbor, or more precisely, $n\lambda_{dB}^3 > 2.61$.

This prediction was not taken terribly seriously, even by Einstein himself, until Fritz London (1938) and Laszlo Tisza (1938) resurrected the idea in the mid 1930s as a possible mechanism underlying superfluidity in liquid helium 4. Their work was the first to bring out the idea of BEC displaying quantum behavior on a macroscopic size scale, the primary reason for much of its current attraction. Although it was a source of debate for decades, it is now recognized that the remarkable properties of superconductivity and superfluidity in both helium 3 and helium 4 are related to BEC, even though these systems are very different from the ideal gas considered by Einstein.

The appeal of the exotic behavior of superconductivity and of superfluidity, along with that of laser light, the third common system in which macroscopic quantum behavior is evident, provided much of our motivation in 1990 when we decided to pursue BEC in a gas. These three systems all have fascinating counterintuitive behavior arising from macroscopic occupation of a single quantum state. Any physicist would consider these phenomena among the most remarkable topics in physics. In 1990 we were confident that the addition of a new member to the family would constitute a major contribution to physics. (Only after we succeeded did we realize that the discovery of each of the original Macroscopic Three had been recognized with a Nobel Prize, and we are grateful that this trend has continued!) Although BEC shares the same underlying mechanism with these other systems, it seemed to us that the properties of BEC in a gas would be quite distinct. It is far more dilute and weakly interacting than liquid-helium superfluids, for example, but far more strongly interacting than the noninteracting light in a laser beam. Perhaps BEC's most distinctive feature (and this was not

*The 2001 Nobel Prize in Physics was shared by E. A. Cornell, Wolfgang Ketterle, and C. E. Wieman.

[1]This article is our "Nobel Lecture" and as such takes a relatively personal approach to the story of the development of experimental Bose-Einstein condensation. For a somewhat more scholarly treatment of the history, the interested reader is referred to E. A. Cornell, J. R. Ensher, and C. E. Wieman, "Experiments in dilute atomic Bose-Einstein condensation in Bose-Einstein Condensation in Atomic Gases," *Proceedings of the International School of Physics "Enrico Fermi" Course CXL*, edited by M. Inguscio, S. Stringari, and C. E. Wieman (Italian Physical Society, 1999), pp. 15–66, which is also available as cond-mat/9903109. For a reasonably complete technical review of the three years of explosive progress that immediately followed the first observation of BEC, we recommend reading the above article in combination with the corresponding review from Ketterle, cond-mat/9904034.

[2]English translation of Einstein's quotes and the historical interpretation are from Pais (1982), *Subtle is the Lord . . . .*

0034-6861/2002/74(3)/875(19)/$35.00

something we sufficiently appreciated, in 1990) is the ease with which its quantum wave function may be directly observed and manipulated. While neither of us was to read C. E. Hecht's prescient 1959 paper (Hecht, 1959) until well after we had observed BEC, we surely would have taken his concluding paragraph as our marching orders:

> The suppositions of this note rest on the possibility of securing, say by atomic beam techniques, substantial quantities of electron-spin-oriented H, T and D atoms. Although the experimental difficulties would be great and the relaxation behavior of such spin-oriented atoms essentially unknown, *the possibility of opening a rich new field for the study of superfluid properties in both liquid and gaseous states would seem to demand the expenditure of maximum experimental effort.*[3]

In any case, by 1990 we were awash in motivation. But this motivation would not have carried us far, had we not been able to take advantage of some key recent advances in science and technology, in particular, the progress in laser cooling and trapping and the extensive achievements of the spin-polarized-hydrogen community.

However, before launching into that story, it is perhaps worthwhile to reflect on just how exotic a system of indistinguishable particles truly is, and why BEC in a gas is such a daunting experimental challenge. It is easy at first to accept that two atoms can be so similar one to the other as to allow no possibility of telling them apart. However, confronting the physical implications of the concept of indistinguishable bosons can be troubling. For example, if there are ten bosonic particles to be arranged in two microstates of a system, the statistical weight of the configuration with ten particles in one state and zero in the other is exactly the same as the weight of the configuration with five particles in one state, five in the other. This 1:1 ratio of statistical weights is very counterintuitive and rather disquieting. The corresponding ratio for distinguishable objects, such as socks in drawers, that we observe every day is 1:252, profoundly different from 1:1. In the second of Einstein's two papers (Einstein, 1925; Pais, 1982) on Bose-Einstein statistics, Einstein comments that "The ... molecules are not treated as statistically independent ..., and the differences between distinguishable and indistinguishable state counting ... express indirectly a certain hypothesis on a mutual influence of the molecules which for the time being is of a quite mysterious nature. This mutual influence is no less mysterious today, even though we can readily observe the variety of exotic behavior it causes such as the well-known enhanced probability for scattering into occupied states and, of course, Bose-Einstein condensation."

Not only does the Bose-Einstein phase transition offend our sensibilities as to how particles ought best to distribute themselves, it also runs counter to an unspo-

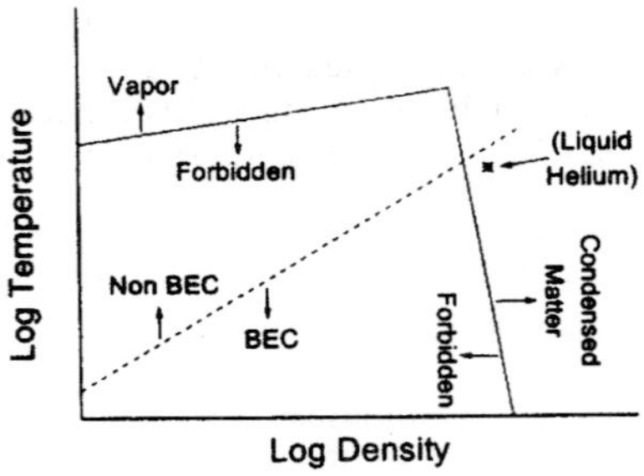

FIG. 1. Generic phase diagram common to all atoms: dotted line, the boundary between non-BEC and BEC; solid line, the boundary between allowed and forbidden regions of the temperature-density space. Note that at low and intermediate densities, BEC exists only in the thermodynamically forbidden regime.

ken assumption that a phase transition somehow involves thermodynamic stability. In fact, the regions immediately above and immediately below the transition in dilute-gas experiments are both deep in the thermodynamically forbidden regime. This point is best made by considering a qualitative phase diagram (Fig. 1), which shows the general features common to any atomic system. At low density and high temperature, there is a vapor phase. At high density there are various condensed phases. But the intermediate densities are thermodynamically forbidden, except at very high temperatures. The Bose-condensed region of the n-T plane is utterly forbidden, except at such high densities that (with one exception) all known atoms or molecules would form a crystalline lattice, which would rule out Bose condensation. The single exception, helium, remains a liquid below the BEC transition. However, reaching BEC under dilute conditions (say, at densities 10 or 100 times lower than conventional liquid helium) is as thermodynamically forbidden to helium as it is to any other atom.

Of course, forbidden is not the same as impossible; indeed, to paraphrase an old Joseph Heller joke, if it were really impossible, they wouldn't have bothered to forbid it. It comes down in the end to differing time scales for different sorts of equilibrium. A gas of atoms can come into kinetic equilibrium via two-body collisions, whereas it requires three-body collisions to achieve chemical equilibrium (i.e., to form molecules and thence solids). At sufficiently low densities, the two-body rate will dominate the three-body rate, and a gas will reach kinetic equilibrium, perhaps in a metastable Bose-Einstein condensate, long before the gas finds its way to the ultimately stable solid-state condition. The need to maintain metastability usually dictates a more stringent upper limit on density than does the desire to create a dilute system. Densities around 10^{20} cm^{-3}, for instance, would be a hundred times more dilute than a condensed-matter helium superfluid. But creating such a gas is quite impractical even at an additional factor-of-1000 lower density, say 10^{17} cm^{-3}, when metastability times would be on the order of a few microseconds;

[3]Emphasis ours.

more realistic are densities on the order of 10^{14} cm^{-3}. The low densities mandated by the need to maintain long-lived metastability in turn make necessary the achievement of still lower temperatures if one is to reach BEC.

Thus the great experimental hurdle that must be overcome to create BEC in a dilute gas is to form and keep a sample that is so deeply forbidden. Since our subsequent discussion will focus only on BEC in dilute gases, we shall refer to this simply as BEC in the sections below and avoid endlessly repeating "in a dilute gas."

Efforts to make a dilute BEC in an atomic gas were sparked by Stwalley and Nosanow (1976). They argued that spin-polarized hydrogen had no bound states and hence would remain a gas down to zero temperature, and so it would be a good candidate for BEC. This stimulated a number of experimental groups (Silvera and Walraven, 1980; Hardy *et al.*, 1982; Hess *et al.*, 1983; Johnson *et al.*, 1984) in the late 1970s and early 1980s to begin pursuing this idea using traditional cryogenics to cool a sample of polarized hydrogen. Spin-polarized hydrogen was first stabilized by Silvera and Walraven in 1980, and by the mid 1980s spin-polarized hydrogen had been brought within a factor of 50 of condensing (Hess *et al.*, 1983). These experiments were performed in a dilution refrigerator, in a cell in which the walls were coated with superfluid liquid helium as a nonstick coating for the hydrogen. The hydrogen gas was compressed using a piston-in-cylinder arrangement (Bell *et al.*, 1986) or inside a helium bubble (Sprik *et al.*, 1985). These attempts failed, however, because when the cell was made very cold the hydrogen stuck to the helium surface and recombined. When one tried to avoid that problem by warming the cell sufficiently to prevent sticking, the density required to reach BEC was correspondingly increased, which led to another problem. The requisite densities could not be reached because the rate of three-body recombination of atoms into hydrogen molecules goes up rapidly with density and the resulting loss of atoms limited the density (Hess, 1986).

Stymied by these problems, Harold Hess (Hess, 1986) from the MIT hydrogen group realized that magnetic trapping of atoms (Migdall *et al.*, 1985; Bagnato *et al.*, 1987) would be an improvement over a cell. Atoms in a magnetic trap have no contact with a physical surface and thus the surface-recombination problem could be circumvented. Moreover, thermally isolated atoms in a magnetic trap would allow cooling by evaporation to far lower temperatures than previously obtained. In a remarkable paper, Hess (1986) laid out most of the important concepts of evaporative cooling of trapped atoms for the attainment of BEC. Let the highest-energy atoms escape from the trap, and the mean energy, and thus the temperature, of the remaining atoms will decrease. For a dilute gas in an inhomogeneous potential, decreasing the temperature will decrease the occupied volume. One can thus actually increase the density of the remaining atoms by removing atoms from the sample. The all important (for BEC) phase-space density is dramatically increased as this happens because density is rising while temperature is decreasing. The Cornell University hydrogen group also considered evaporative cooling (Lovelace *et al.*, 1985). By 1988 the MIT group had demonstrated these virtues of evaporative cooling of magnetically trapped spin-polarized hydrogen. By 1991 they obtained, at a temperature of 100 °K, a density that was only a factor of 5 below BEC (Doyle, 1991a). Further progress was limited by dipolar relaxation, but perhaps more fundamentally by loss of signal-to-noise, and the difficulty of measuring the characteristics of the coldest and smallest clouds (Doyle, 1991b). Evaporative work was also performed by the Amsterdam group (Luiten *et al.*, 1993).

At roughly the same time, but independent from the hydrogen work, an entirely different type of cold-atom physics and technology was being developed. Laser cooling and trapping has been reviewed elsewhere (Arimondo *et al.*, 1991; Chu, 1998; Cohen-Tannoudji, 1998; Phillips, 1998), but here we mention some of the highlights most relevant to our work. The idea that laser light could be used to cool atoms was suggested in early papers by Wineland and Dehmelt (1975), by Hänsch and Schawlow (1975), and by Letokhov's group (Letokhov, 1968). Early optical force experiments were performed by Ashkin (Bjorkholm *et al.*, 1978). Trapped ions were laser-cooled at the University of Washington (Neuhauser *et al.*, 1978) and at the National Bureau of Standards (now NIST) in Boulder (Wineland *et al.*, 1978). Atomic beams were deflected and slowed in the early 1980s (Andreev *et al.*, 1981; Ertmer *et al.*, 1985; Prodan *et al.*, 1985). Optical molasses, where the atoms are cooled to very low temperatures by six perpendicular intersecting laser beams, was first studied at Bell Labs (Chu *et al.*, 1985). Measured temperatures in the early molasses experiments were consistent with the so-called Doppler limit, which amounts to a few hundred microkelvin in most alkalis. Light was first used to hold (trap) atoms using the dipole force exerted by a strongly focused laser beam (Chu *et al.*, 1986). In 1987 and 1988 there were two major advances that became central features of the method of creating BEC. First, a practical spontaneous-force trap, the magneto-optical trap (MOT) was demonstrated (Raab *et al.*, 1987); and second, it was observed that under certain conditions, the temperatures in optical molasses are in fact much colder than the Doppler limit (Lett *et al.*, 1988; Chu *et al.*, 1989; Dalibard *et al.*, 1989). The MOT had the essential elements needed for a widely useful optical trap: it required relatively modest amounts of laser power, it was much deeper than dipole traps, and it could capture and hold relatively large numbers of atoms. These were heady times in the laser-cooling business. With experiment yielding temperatures mysteriously far below what theory would predict, it was clear that we all lived under the authority of a munificent God.

During the mid 1980s one of us (Carl) began investigating how useful the technology of laser trapping and cooling could become for general use in atomic physics. Originally this took the form of just making it cheaper and simpler by replacing the expensive dye lasers with vastly cheaper semiconductor lasers, and then searching for ways to allow atom trapping with these low-cost but also low-power lasers (Pritchard *et al.*, 1986; Watts and Wieman, 1986). With the demonstration of the MOT and sub-Doppler molasses Carl's group began eagerly studying what physics was limiting the coldness and denseness of these trapped atoms, with the hope of extending the limits further. They discovered that several atomic processes were responsible for these limits. Light-assisted collisions were found to be the major loss process from the MOT as the density increased (Sesko *et al.*, 1989). However, even before that became a serious problem, the light pressure from reradiated photons limited the density (Walker *et al.*, 1990; Sesko *et al.*, 1991). At about the same time, the sub-Doppler temperatures of molasses found by Phillips, Chu, and Cohen-Tannoudji were shown to be due to a combination of light-shifts and optical pumping that became known as Sysiphus cooling (Dalibard and Cohen-Tannoudji, 1989). Random momentum fluctuations from the scattered photons limit the ultimate temperature to about a factor of 10 above the recoil limit. In larger samples, the minimum temperature was higher yet, because of the multiple scattering of the photons. While carrying out studies on the density limits of MOT's Carl's group also continued the effort in technology development. This resulted in the creation of a useful MOT in a simple glass vapor cell (Monroe *et al.*, 1990), thereby eliminating the substantial vacuum chamber required for the slowed atomic beam loading that had previously been used.

Seeking to take advantage of the large gains in phase-space density provided by the MOT while avoiding the limitations imposed by the undesirable effects of photons, Carl and his student Chris Monroe decided to try loading the cold MOT atoms into a magnetic trap (Monroe *et al.*, 1990; see Fig. 2). This worked remarkably well. Because further cooling could be carried out as the atoms were transferred between optical and magnetic trap it was possible to get very cold samples, the coldest that had been produced at that time. More importantly, these were not optical molasses samples that were quickly disappearing but rather magnetically trapped samples that could be held and studied for extended periods. These samples were about a hundred times colder than any previous trapped atom samples, with a correspondingly increased phase-space density. This was a satisfying achievement, but as much as the result itself, it was the relative simplicity of the apparatus required that inspired us (including now Eric Cornell, who joined the project as a postdoc in 1990) to see just how far we could push this marriage of laser cooling and trapping and magnetic trapping.

Previous laser traps involved expensive massive laser systems and large vacuum chambers for atomic beam precooling. Previous magnetic traps for atoms were usually (Bagnato *et al.*, 1987; Doyle, 1991) extremely complex and bulky (often with superconducting coils) because of the need to have sufficiently large depths and strong confinement. Laser traps and magnetic traps were both somewhat heroic experiments individually, to be undertaken only by a select handful of well-equipped AMO laboratories. The prospect of trying to get both traps working, and working well, in the same room and on the same day, was daunting. However, in the first JILA magnetic trap experiment our laser sources were simple diode lasers, the vacuum system was a small glass vapor cell, and the magnetic trap was just a few turns of wire wrapped around it. This magnetic field was adequate because of the low temperatures of the laser-cooled and trapped samples. Being able to produce such cold and trapped samples in this manner encouraged one to fantasize wildly about possible things to do with such an atom sample. Inspired by the spin-polarized hydrogen work, our fantasizing quickly turned to the idea of evaporative cooling further to reach BEC. It would require us to increase the phase-space density by 5 orders of magnitude, but since we had just gained about 15 orders of magnitude almost for free with the vapor cell MOT, this did not seem so daunting.

The JILA vapor-cell MOT (Fig. 3), with its superimposed ion pump trap, introduced a number of ideas that are now in common use in the hybrid trapping business (Monroe *et al.*, 1990; Monroe, 1992): (i) Vapor-cell (rather than beam) loading, (ii) fused-glass rather than welded-steel architecture, (iii) extensive use of diode lasers, (iv) magnetic coils located outside the chamber, (v) overall chamber volume measured in cubic centimeters rather than liters, (vi) temperatures measured by imaging an expanded cloud, (vii) magnetic-field curvatures calibrated *in situ* by observing the frequency of dipole and quadrupole (sloshing and pulsing) cloud motion, (viii) the basic approach of a MOT and a magnetic trap which are spatially superimposed (indeed, which often share some magnetic coils) but temporally sequential, and (ix) optional use of additional molasses and optical pumping sequences inserted in time between the MOT and magnetic trapping stages. It is instructive to note how a modern, Ioffe-Pritchard-based BEC device (Fig. 4) resembles its ancestor (Fig. 3).

As we began to think about applying the technique of evaporative cooling with hydrogen to our very cold alkali atoms we looked carefully at the hydrogen work and its lessons. When viewed from our 1990 perspective the previous decade of work on polarized hydrogen provided a number of important insights. It was clear that the unique absence of any bound states for spin-polarized hydrogen was actually not an important issue (other than its being the catalyst for starting the entire field, of course!). Bound states or not, a very cold sample of spin-polarized hydrogen, like every other gas, has a lower-energy state to which it can go, and its survival depends on the preservation of metastability. For hydrogen the lower-energy state is a solid, although from an experimental point of view the rate-limiting process is the formation of diatomic molecules (with appropriately reoriented spins). Given that all atomic gases are only metastable at the BEC transition point, the real experimental issue becomes: How well can one

FIG. 2. Chris Monroe examines an early hybrid MOT-magnetic trap apparatus [Color].

preserve the requisite metastability while still cooling sufficiently far to reach BEC?

The realization that metastability was the key experimental challenge one should focus on was probably at least as important to the attainment of BEC as any of the experimental techniques we subsequently developed to actually achieve it. The work on hydrogen provided an essential guide for evaluating and tackling this challenge. It provided us with a potential cooling technique (evaporative cooling of magnetically trapped atoms) and mapped out many of the processes by which a magnetically trapped atom can be lost from its metastable state.

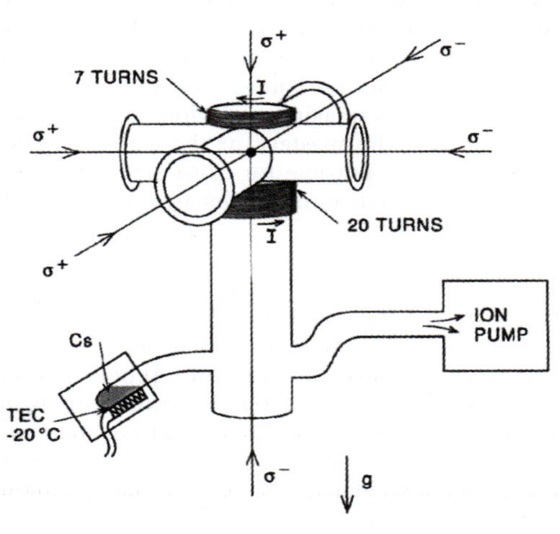

FIG. 3. The glass vapor cell and magnetic coils used in early JILA efforts to hybridize laser cooling and magnetic trapping (see Monroe *et al.*, 1990). The glass tubing is 2.5 cm in diameter. The Ioffe current bars have been omitted for clarity.

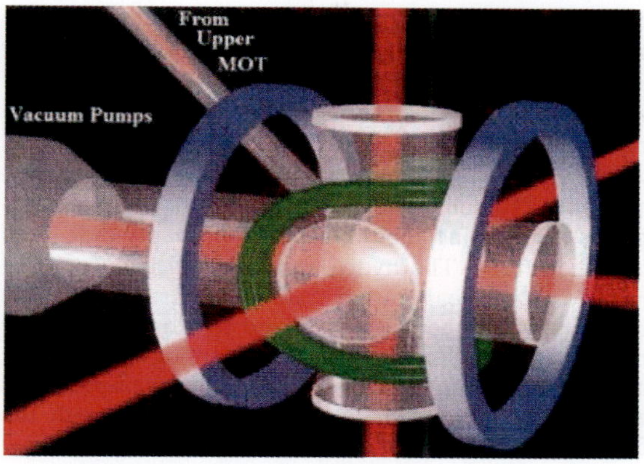

FIG. 4. Modern MOT and magnetic trap apparatus, used by Cornish *et al.*, 2000 [Color].

The hydrogen work made it clear that it was all an issue of good versus bad collisions. The good collisions are elastic collisions that rethermalize the atoms during evaporation. The more collisions there are, the more quickly and efficiently one can cool. The bad collisions are the inelastic collisions that quench the metastability. Hydrogen had already shown that three-body recombination collisions and dipole spin-flip collisions were the major inelastic culprits. The fact that hydrogen researchers were fairly close to reaching BEC was also a strong encouragement. It meant that the goal was not ridiculously distant and that one only had to do a little better in the proportion of good to bad collisions to succeed.

The more we thought about this, the more we began to suspect that our heavy alkali atoms would likely have more favorable collision properties than hydrogen atoms and thus have a good chance of success. Although knowledge of the relevant collision cross sections was totally nonexistent at that time, we were able to come up with arguments for how the cross sections might scale relative to hydrogen. These are discussed in more detail below in the section discussing why collisional concerns make it likely that BEC can be created in a large number of different species. Here we will just give a brief summary consistent with our views circa 1990. The dipole spin-flip collisions that limited hydrogen involve spin-spin interactions and thus could be expected to be similar for the alkalis and for hydrogen because the magnetic moments are all about the same. The good collisions needed for evaporative cooling, however, should be much larger for heavy alkalis with their fat fluffy electron clouds than for hydrogen. The other villain of the hydrogen effort, three-body recombination, was a total mystery, but because it goes as density cubed while the good elastic collisions go as density squared, it seemed as if we should always be able to find a sufficiently low-density and low-temperature regime to avoid it (see Monroe, 1992).

As a minor historical note, we might point out that during these considerations we happily ignored the fact that the temperatures required to achieve BEC in a heavy alkali gas are far colder than those needed for the same density of hydrogen. The critical temperature for ideal-gas BEC is inversely proportional to the mass. It was clear that we would need to cool to well under a microkelvin, and a large three-body recombination rate would have required us to go to possibly far lower temperatures. To someone coming from a traditional cryogenics background this would (and probably did) seem like sheer folly. The hydrogen work had been pushing hard for some years at the state of the art in cryogenic technology, and here we proposed to happily jump far beyond that. Fortunately we were coming to this from an AMO background in a time when temperatures achieved by laser cooling were dropping through the floor. Optimism was in the air. In fact, we later discovered optimism can take one only so far: There were actually considerable experimental difficulties, and further cooling came at some considerable effort and a five-year delay. Nevertheless, it is remarkable that with evaporative cooling a magnetically trapped sample of atoms, surrounded on all sides by a 300-K glass cell, can be cooled to reach temperatures of only a few nanokelvin, and moreover it looks quite feasible to reach even colder temperatures.

General collisional considerations gave us some hope that the evaporative cooling hybrid trap approach with alkali atoms would get us to BEC, or, if not, at least reveal some interesting new physics that would prevent it. Nonetheless, there were powerful arguments against pursuing this. First, our 1990-era arguments in favor of it were based on some very fuzzy intuition; there were no collision data or theories to back it up and there were strong voices in disagreement. Second, the hydrogen experiments seemed to be on the verge of reaching BEC, and in fact we thought it was likely that if BEC could be achieved they would succeed first. However, our belief in the virtues of our technology really carried the day in convincing us to proceed. With convenient lasers in the near-IR, and with the good optical access of a room-temperature glass cell, detection sensitivity could approach single-atom capability. We could take pictures of only a few thousand trapped atoms and immediately know the energy and density distribution. If we wanted to modify our magnetic trap it only required a few hours winding and installing a new coil of wires. This was a dramatic contrast with the hydrogen experiments that, like all state-of-the-art cryogenics experiments, required an apparatus that was the better part of two stories, and the time to modify it was measured in (large) fractions of a year. Also, atomic hydrogen was much more difficult to detect and so the diagnostics were far more limited. This convinced us that although hydrogen would likely succeed first, our hybrid trap approach with easily observed and manipulated alkali samples would be able to carry out important science and so was well worth pursuing in its own right.

From the very beginning in 1990, our work on BEC was heavily involved with cold atomic collisions. This was somewhat ironic since previously both of us had actively avoided the large fraction of AMO work on the subject of atomic collisions. Atomic collisions at very cold temperatures is now a major branch of the discipline of AMO physics, but at the end of the 1980s there were almost no experimental data, and what there was came in fact from the spin-polarized hydrogen experiments (Gillaspy *et al.*, 1989). There was theoretical work on hydrogen from Shlyapnikov and Kagan (Kagan *et al.*, 1981, 1984), and from Silvera and Verhaar (Lagendijk *et al.*, 1986). An early paper by Pritchard (1986) includes estimates on low-temperature collisional properties for alkalis. His estimates were extrapolations from room-temperature results, but in retrospect, several were surprisingly accurate. As we began to work on evaporative cooling, much of our effort was devoted to determining the sizes of all the relevant good and bad collision cross sections. Our efforts were helped by the theoretical efforts of Boudewijn Verhaar, who was among the first to take our efforts seriously and attempt to calculate the rates in question. Chris Greene also provided us with some useful theoretical estimates.

Starting in 1990 we carried out a series of experiments exploring various magnetic traps and measuring the relevant collision cross sections. As this work proceeded we developed a far better understanding of the conditions necessary for evaporative cooling and a much clearer understanding of the relevant collisional issues (Monroe *et al.*, 1993; Newbury *et al.*, 1995). Our experimental concerns evolved accordingly. In the early experiments (Monroe *et al.*, 1990, 1993; Cornell *et al.*, 1991; Monroe, 1992) a number of issues came up that continue to confront all BEC experiments: the importance of aligning the centers of the MOT and the magnetic trap, the density-reducing effects of mode-mismatch, the need to account carefully for the (previously ignored) force of gravity, heating (and not merely loss) from background gas collisions, the usefulness of being able to turn off the magnetic fields rapidly, the need to synchronize many changes in laser status and magnetic fields together with image acquisition, an appreciation for the many issues that can interfere with accurate determinations of density and temperature by optical methods, either florescence or absorption imaging, and careful stabilization of magnetic fields. The mastery of these issues in these early days made it possible for us to proceed relatively quickly to quantitative measurements with the BEC once we had it.

In 1992 we came to realize that dipolar relaxation in alkalis should in principle not be a limiting factor. As explained in the final section of this article, collisional scaling with temperature and magnetic field is such that, except in pathological situations, the problem of good and bad collisions in the evaporative cooling of alkalis is reduced to the ratio of the elastic collision rate to the rate of loss due to imperfect vacuum; dipolar relaxation and three-body recombination can be finessed, particularly since our preliminary data showed they were not enormous. It was reassuring to move ahead on efforts to evaporate with the knowledge that, while we were essentially proceeding in the dark, there were not as many monsters in the dark as we had originally imagined.

It rapidly became clear that the primary concerns would be having sufficient elastic collision rate in the magnetic trap and sufficiently low background pressure to have few background collisions that removed atoms from the trap. To accomplish this it was clear that we needed higher densities in the magnetic trap than we were getting from the MOT. Our first effort to increase the density two years earlier was based on a multiple-loading scheme (Cornell *et al.*, 1991). Multiple MOT-loads of atoms were launched in moving molasses, optically pumped into an untrapped Zeeman level, focused into a magnetic trap, then optically repumped into a trapped level. The repumping represented the necessary dissipation, so that multiple loads of atoms could be inserted in a continuously operating magnetic trap. In practice, each step of the process involved some losses, and the final result was disappointing. Later, however, as discussed below, we resurrected the idea of multiple loading from one MOT to another to good advantage (Gibble *et al.*, 1995; Myatt *et al.*, 1996). This is now a technique currently in widespread practice.

In addition to building up the initial density we realized that the collision rate could be dramatically increased by, after loading into a magnetic trap, compressing the atoms by further increasing the curvature of the confining magnetic fields. In a harmonic trap, the collision rate after adiabatic compression scales as the final confining frequency squared (Monroe, 1992). This method is discussed by Monroe (1992) and was implemented first in early ground-state collisional work (Monroe *et al.*, 1993).

In fall of 1992, Eric's postdoctoral appointment concluded, and, after a tour through the job market, he decided to take the equivalent of an assistant professor position at JILA/NIST. He decided to use his startup money to build a new experimental apparatus that would be designed to put these ideas together to make sure evaporation worked as we expected. Meanwhile, we continued to pursue the possibility of enhanced collision cross sections in cesium using a Feshbach resonance. At that point our Monte Carlo simulations said that a ratio of about 150 elastic collisions per trap lifetime was required to achieve runaway evaporation. This is the condition where the elastic collision rate would continue to increase as the temperature decreased, and hence evaporation would continue to improve as the temperature was reduced. We also had reasonable determinations of the elastic collision cross sections.

So the plan was to build a simple quadrupole trap that would allow very strong squeezing to greatly enhance the collision rate, combined with a good vacuum system in order to make sure evaporative cooling worked as expected. Clearly, there was much to be gained by building a more tightly confining magnetic trap, but the requirement of adequate optical access for the MOT, along with engineering constraints on power dissipation, made the design problem complicated.

When constructing a trap for weak-field-seeking atoms, with the aim of confining the atoms to a spatial size much smaller than the size of the magnets, one would like to use linear gradients. In that case, however, one is confronted with the problem of the minimum in the magnitude of the magnetic fields (and thus of the confining potential) occurring at a local zero in the magnetic field. This zero represents a hole in the trap, a site at which atoms can undergo Majorana transitions (Majorana, 1931) and thus escape from the trap. If one uses the second-order gradients from the magnets to provide the confinement, there is a marked loss of confinement strength. This scaling is discussed by Petrich *et al.* (1995). We knew that once the atoms became cold enough they would leak out the hole in the bottom of the trap, but the plan was to go ahead and get evaporation and worry about the hole later. We also recognized that even with successful evaporative cooling, and presuming we could solve the issue of the hole in the quadrupole trap, there was still the question of the sign of scattering length, which must be positive to ensure the stability of a large condensate.

In setting up the new apparatus Eric chose to use rubidium. Given the modulo arithmetic that goes into determining a scattering length, it seemed fair to treat the scattering lengths of different isotopes as statistically independent events, and rubidium with its two stable isotopes offered two rolls of the dice for the same laser system. Eric then purchased a set of diode lasers for the rubidium wavelength, but of course we kept the original cesium-tuned diode lasers. The wavelengths of cesium and of the two rubidium isotopes are sufficiently similar that in most cases one can use the same optics. Thus we preserved the option of converting from one species to another in a matter of weeks. The chances then of Nature's conspiring to make the scattering length negative, for both hyperfine levels, for all three atoms, seemed very small.

Progress in cold collisions, particularly the experiment and theory of photoassociative collisions, had moved forward so rapidly that by the time we had evaporatively cooled rubidium to close to BEC temperatures a couple of years later there existed, at the 20% level, values for several of the elastic scattering lengths. In particular, we knew that it was positive for the 2,2 state of Rb-87 (Thorsheim *et al.*, 1987; Lett *et al.*, 1993; Miller *et al.*, 1993; Abraham *et al.*, 1995; Gardner *et al.*, 1995; McAlexander *et al.*, 1995).

Our original idea for the quadrupole trap experiment was to pulse a burst of rubidium into our cell, where we would catch a large sample in the MOT and then hold it as the residual rubidium was quickly pumped away, leaving a long trap lifetime. We, particularly Eric's postdoc, Mike Anderson, spent many frustrating months discovering how difficult this seemingly simple idea was to actually implement in practice. The manner in which rubidium interacted with glass and stainless-steel surfaces conspired to make this so difficult we finally gave up. We ended up going with a far-from-optimum situation of working with extremely low rubidium pressure and doing our best at maximizing the number of atoms captured in the MOT from this feeble vapor and enhancing the collision rate for those relatively few atoms as much as possible. We recognized that this was a major compromise, but we had been trying to evaporate for some time, and we were getting impatient! We had no stomach for building another apparatus just to see evaporation. Fortunately we were able to find two key elements to enhance the MOT loading and density. First was the use of a dark-spot MOT in which there is a hole in the center of the MOT beams so the atoms are not excited. This technique had been demonstrated by Ketterle (Ketterle *et al.*, 1993) as a way to greatly enhance the density of atoms in a MOT under conditions of a very high loading rate. The number of atoms we could load in our vapor cell MOT with very low rubidium vapor was determined by the loading rate over the loss rate. In this case the loss rate was the photoassociative collisions we had long before found to be important for losses from MOT's. The dark-spot geometry reduced this two-body photoassociative loss in part because in our conditions it *reduced* the density of atoms in the MOT (Anderson *et al.*, 1994).

Using this approach we were able to obtain 10^8 atoms in the MOT collected out of a very low vapor background (so that magnetic trap lifetime was greater than 100 s). The second key element was the invention of the compressed MOT (CMOT), a technique for substantially enhancing the density of atoms in the MOT on a transient basis. For the CMOT, the MOT was filled and then the field gradient and laser detuning were suddenly changed to greatly suppress the multiple photon scattering. This produced much higher densities and clouds whose shape was a much better match to the desired shape of the cloud in the magnetic trap. This was a very transient effect because the losses from the MOT were much larger under these conditions, but that was not important; the atoms needed only to be held for the milliseconds required before they were transferred to the magnetic trap (Petrich *et al.*, 1994; see Fig. 5). With these improvements and a quadrupole trap that provided substantial squeezing, we were able to finally demonstrate evaporative cooling in rubidium.

Cooling by evaporation is a process found throughout Nature. Whether the material being cooled is an atomic nucleus or the Atlantic Ocean, the rate of natural evaporation and the minimum temperature achievable are limited by the particular fixed value of the work function of the evaporating substance. In magnetically confined atoms, no such limit exists, because the work function is simply the height of the lowest point in the rim of the confining potential. Hess (1986) pointed out that, by perturbing the confining magnetic fields, one could make the work function of a trap arbitrarily low; as long as favorable collisional conditions persist, there is no lower limit to the temperatures attainable in this forced evaporation.

Pritchard (Pritchard *et al.*, 1989) pointed out that evaporation could be performed more conveniently if the rim of the trap were defined by an rf-resonance condition, rather than simply by the topography of the magnetic field; experimentally, his group made first use of position-dependent rf transitions to selectively transfer magnetically trapped sodium atoms between Zeeman levels and thus characterized their temperature (Martin *et al.*, 1988). In our experiment we used Pritchard's technique of an rf field to selectively evaporate.

It was a great relief to see evaporative cooling of laser precooled, magnetically trapped atoms finally work, as we had been anticipating it would for so many years. Unfortunately, it worked exactly as well, but no better, than we had anticipated. The atoms were cooled to about 40 μK and then disappeared, at just the temperature we had estimated they would be lost, through the hole in the bottom of the quadrupole trap. Eric came up with an idea that solved this problem. It was a design for a new type of trap that required relatively little modification to the apparatus and so was quickly implemented. This was the Time Orbiting Potential (TOP) trap in which a small rotating magnetic field was added to the quadrupole field (Petrich *et al.*, 1995). This moved the field zero in an orbit faster than the atoms could follow. It was the perfect solution to our problem.

Mike Anderson, another postdoc, Wolfgang Petrich, and graduate student Jason Ensher quickly implemented this design. Their efforts were spurred on by the realization that there were several other groups who had now demonstrated or were known to be on the verge of demonstrating evaporative cooling in alkalis in the pursuit of BEC. The TOP design worked well, and the samples were cooled far colder, in fact too cold for us to reliably measure. We had been measuring temperature simply by looking at the spatial size of the cloud in the magnetic trap. As the temperature was reduced the size decreased, but we were now reaching temperatures so low that the size had reached the resolution limit of the optical system. We saw dramatic changes in the shapes of the images as the clouds became very small, but we knew that a variety of diffraction and aberration effects could greatly distort images when the sample size became only a few wavelengths in size, so our reaction to these shapes was muted, and we knew we had to have better diagnostics before we could have meaningful results. Here we were helped by our long experience in studying various trapped clouds over the years. We already knew the value of turning the magnetic trap off to let the cloud expand and then imaging the expanded cloud to get a measure of the momentum distribution in the trap. Since the trap was harmonic, the momentum distribution and the original density distribution were nearly interchangeable. Unfortunately, once the magnetic field was off, the atoms not only expanded but also simply fell under the influence of gravity. We found that the atoms tended to fall out of the field of view of our microscope before they had sufficiently expanded. The final addition to the apparatus was a supplementary magnetic coil, which provided sufficient field gradient to cancel the effects of gravity while minimizing any perturbation to the relative ballistic trajectories of the expanding atoms.

Anderson, Ensher, and a new graduate student, Mike Matthews (Fig. 6), worked through a weekend to install the antigravity coil and, after an additional day or two of

FIG. 5. Wolfgang Petrich working on CMOT [Color].

trial and error, got the new field configuration shimmed up. By June 5, 1995 the new technology was working well and we began to look at the now greatly expanded clouds. To our delight, the long-awaited two-component distribution was almost immediately apparent (Fig. 7) when the samples were cooled to the regime where BEC was expected. The excitement was tempered by the concern that after so many years of anticipating two component clouds as a signature of BEC, we might be fooling ourselves.

FIG. 6. From left, Mike Anderson, Debbie Jin, Mike Matthews, and Jason Ensher savor results of early BEC experiment [Color].

Almost from the beginning of the search for BEC, it was recognized (Lovelace and Tommila, 1987) that as the sample started to condense, there would be a spike in the density and momentum distributions corresponding to the macroscopic population of the ground state. This would show up as a second component on top of the much broader normal thermal distribution of uncondensed atoms. This was the signature we had been hoping to see from our first days of contemplating BEC. The size of the BEC component in these first observations also seemed almost too good to be true. In those days it was known that in the much higher density of the condensate, three-body recombination would be a more dominant effect than in the lower-density uncondensed gas. For hydrogen it was calculated that the condensed component could never be more than a few percent of the sample. The three-body rate constants were totally unknown for alkali atoms at that time, but because of the H results it still seemed reasonable to expect the condensate component might only be a modest fraction of the total sample. But in our first samples we saw it could be nearly 100%! In the light of the prevailing myth of unattainability that had grown up around BEC over the years, our observations seemed too good to be true. We were experienced enough to know that when results in experimental physics seem too good to be true, they almost always are! We worried that in our enthusiasm we might confuse the long-desired BEC with some spurious artifact of our imaging system.

However, our worries about the possibility of deluding ourselves were quickly and almost entirely alleviated by the anisotropy of the BEC cloud. This was a very distinctive signature of BEC, the credibility of which was greatly enhanced to us by the fact that it first revealed itself in the experiment, and then we recognized its significance, rather than vice versa. It was a somewhat fortuitous accident that the TOP trap provided a distinctly anisotropic trapping potential, since we did not appreciate its benefits until we saw the BEC data. A normal thermal gas (in the collisionally thin limit) released from an anisotropic potential will spread out isotropically. This is required by the equipartition theorem. However, a Bose-Einstein condensate is a quantum wave and so its expansion is governed by a wave equation. The more tightly confined direction will expand the most rapidly, a manifestation of the uncertainty principle. Seeing the BEC component of our two-component distribution display just this anisotropy, while the broader uncondensed portion of the sample observed at the same time, with the same imaging system, remained perfectly isotropic (as shown in Fig. 8), provided the crucial piece of corroborating evidence that this was the long-awaited BEC.

By coincidence we were scheduled to present progress reports on our efforts to achieve BEC at three international conferences in the few weeks following these observations (Anderson *et al.*, 1996). Nearly all the experts in the field were represented at one or more of these

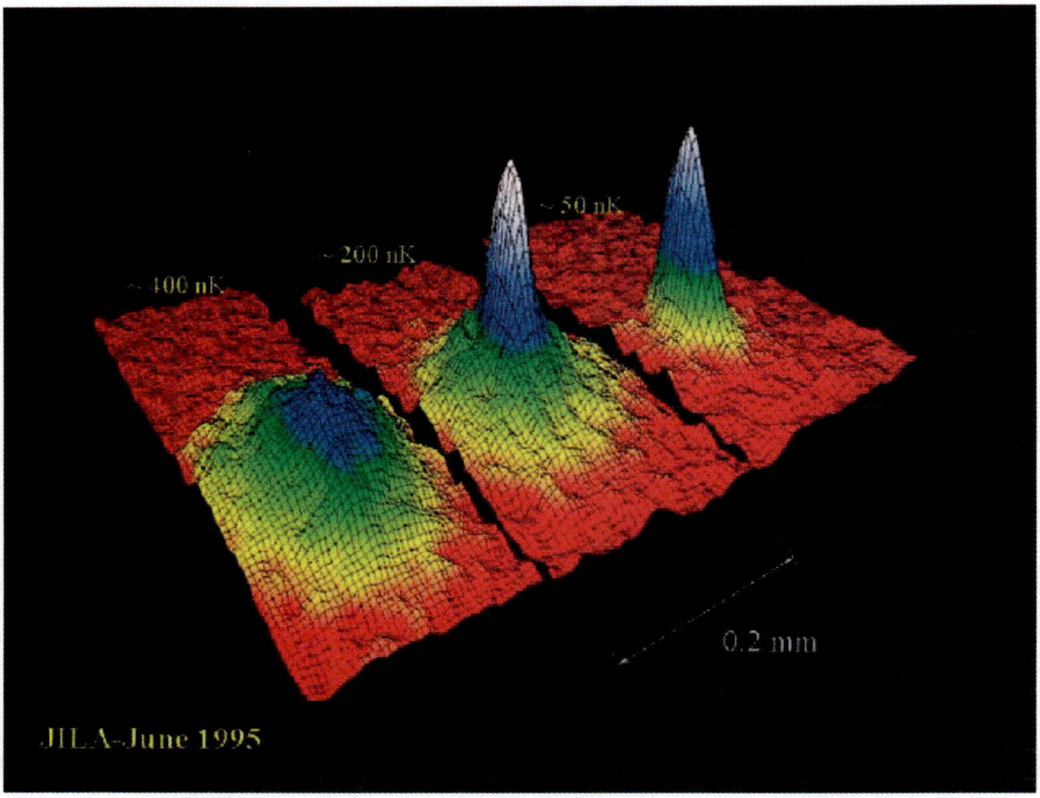

FIG. 7. Three density distributions of the expanded clouds of rubidium atoms at three different temperatures. The appearance of the condensate is apparent as the narrow feature in the middle image. On the far right, nearly all the atoms in the sample are in the condensate. The original experimental data were two-dimensional black and white shadow images, but these images have been converted to three dimensions and given false color density contours [Color].

conferences, and the data were sufficient to convince the most skeptical of them that we had truly observed BEC. This consensus probably facilitated the rapid refereeing and publication of our results.

In the original TOP-trap apparatus we were able to obtain so-called pure condensates of a few thousand atoms. By pure condensates we meant that nearly all the atoms were in the condensed fraction of the sample.

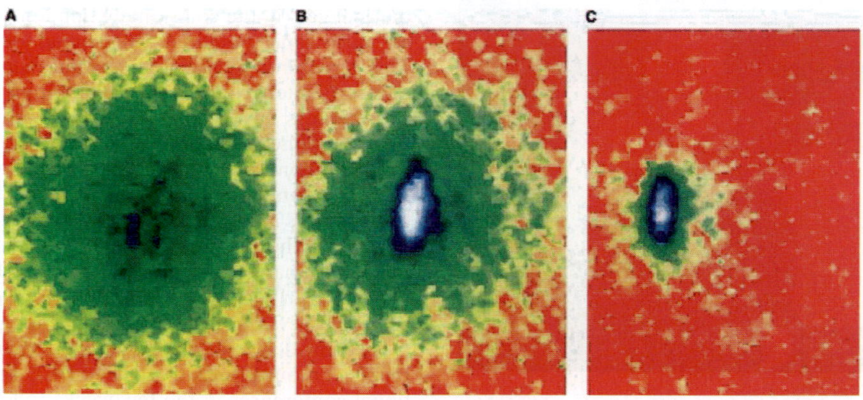

FIG. 8. Looking down on the three images of Figure 7 (Anderson *et al.*, 1995). The condensate in B and C is clearly elliptical in shape [Color].

Samples of this size were easily large enough to image. Over the few months immediately following the original observation, we undertook the process of a technological shoring up of the machine, until the machine reached the level of reliability necessary to crank out condensate after reproducible condensate. This set the stage for the first generation of experiments characterizing the properties of the condensate, most notably the condensate excitation studies discussed below.

Although by 1995 and 1996 we were able to carry out a number of significant BEC experiments with the original TOP-trap machine, even by 1994, well before the original condensates were observed, we had come to realize the limitations of the single-cell design. Our efforts to modulate the vapor pressure were not very successful, which forced us to operate at a steady-state rubidium vapor pressure. Choosing the value of vapor pressure at which to operate represented a compromise between our need to fill the vapor-cell MOT with as many atoms as possible and our need to have the lifetime in the magnetic trap as long as possible. The single-cell design also compelled us to make a second compromise, this time over the size of the glass cell. The laser beams of the MOT enter the cell through the smooth, flat region of the cell; the larger the glass cell, the larger the MOT beams, and the more atoms we could herd into the MOT from the room-temperature background vapor. On the other hand, the smaller the glass cell, the smaller the radii of the magnetic coils wound round the outside of the cell, and the stronger the confinement provided by the magnetic trap. Hans Rohner in the JILA specialty shop had learned how (Rohner, 1994) to create glass cells with the minimum possible wasted area. But even with the dead space between the inner diameter of the magnetic coils and the outer diameter of the clear glass windows made as small as it could be, we were confronted with an unwelcome tradeoff.

Thus, in 1994, in parallel with our efforts to push as hard as we could toward BEC in our original, single-cell TOP trap, we began working on a new technology that would avoid this painful tradeoff. This approach was a modified version of our old multiple loading scheme in which many loads from a MOT were transferred to a magnetic trap in a differentially pumped vacuum chamber. That approach had been defeated by the difficulty in transferring atoms from MOT to magnetic trap without losing phase-space density. There was no dissipation in the magnetic trap to compensate for a slightly too hard or too soft push from one trap to the other. This made us recognize the importance of having dissipation in the second trap, and so we went to a system in which atoms were captured in a large-cell MOT in a region of high rubidium pressure, and then transferred through a small tube into a second, small-cell MOT in a low-pressure region. This eliminated the previous disadvantages while preserving the advantages of multiple loading to get much larger numbers of trapped atoms in a low-vacuum region. The approach worked well, particularly when we found that simple strips of plastic refrigerator magnet material around the outside of the transfer tube between the two traps provided an excellent guide to confine the atoms as they were pushed from one trap to the other (Myatt *et al.*, 1996).

With this scheme we were still able to use inexpensive low-power diode lasers to obtain about one hundred times more atoms in the magnetic trap than in our single MOT-loaded TOP magnetic trap and with a far longer lifetime; we saw trap lifetimes up to 1000 s in the double MOT magnetic trap. This system started working in 1996 and it marked a profound difference in the ease with which we could make BEC (Myatt *et al.*, 1997). In the original BEC experiment everything had to be very well optimized to achieve the conditions necessary for runaway evaporative cooling and thereby BEC. In the double MOT system there were orders of magnitude to spare. Not only did this allow us to routinely obtain million-atom pure condensates, but it also meant that we could dispense with the dark-spot optical configuration with its troublesome alignment. We could be much less precise with many other aspects of the experiment as well.

The first magnetic trap we used with the double-MOT BEC machine was not a TOP trap, but instead was our old baseball-style Ioffe-Pritchard trap. The baseball coil trap is rather complementary to the TOP trap in that each has unique capabilities. For example, the geometry of the TOP trap potential can be changed over a wide range, although the range of dc fields is quite limited. In contrast, the geometry of the baseball coil trap potential can be varied only by small amounts, but the dc bias field can be easily varied over hundreds of gauss. Thus in 1996, when we upgraded the original BEC machine to incorporate the double-MOT technology, we preserved the TOP trap coil design. Each is well suited to certain types of experiments, as will be evident in the discussions below.

With the double-MOT setups we were able to routinely make million-atom condensates in a highly reliable manner in both TOP and baseball-type magnetic traps. These were used to carry out a large number of experiments with condensates over the period from 1996 to the present. Some of our favorite experiments are briefly discussed below.

FAVORITE EXPERIMENTS

Collective excitations

In this section, by excitations we mean coherent fluctuations in the density distribution. Excitation experiments in dilute-gas BEC have been motivated by two main considerations. First, a Bose-Einstein condensate is expected to be a superfluid, and a superfluid is defined by its dynamical behavior. Studying excitations is an obvious initial step toward understanding dynamical behavior. Second, in experimental physics a precision measurement is almost always a frequency measurement, and the easiest way to study an effect with precision is to find an observable frequency that is sensitive to that effect. In the case of dilute-gas BEC, the observed fre-

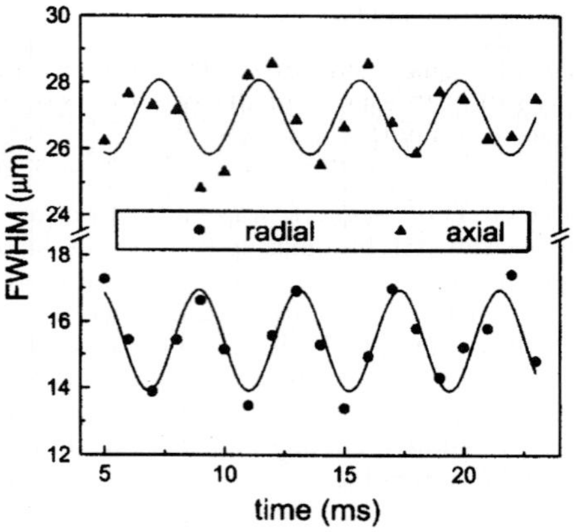

FIG. 9. Zero-temperature excitation data from Jin *et al.* (1996). A weak $m=0$ modulation of the magnetic trapping potential is applied to a 4500-atom condensate in a 132-Hz (radial) trap. Afterward, the freely evolving response of the condensate shows radial oscillations. Also observed is a sympathetic response of the axial width, approximately 180° out of phase. The frequency of the excitation is determined from a sine wave fit to the freely oscillating cloud widths.

quency of standing-wave excitations in a condensate is a precise test of our understanding of the effect of interactions.

BEC excitations were first observed by Jason Ensher, Mike Matthews, and then-postdoc Debbie Jin, using destructive imaging of expanded clouds (Jin *et al.*, 1996). The nearly zero-temperature clouds were coherently excited (see below), then allowed to evolve in the trap for some particular dwell time, and then rapidly expanded and imaged via absorption imaging. By repeating the procedure many times with varying dwell times, the time-evolution of the condensate density profile can be mapped out. From these data, frequencies and damping rates can be extracted. In axially symmetric traps, excitations can be characterized by their projection of angular momentum on the axis. The perturbation on the density distribution caused by the excitation of lowest-lying $m=0$ and $m=2$ modes can be characterized as simple oscillations in the condensate's linear dimensions. Figure 9 shows the widths of an oscillating condensate as a function of dwell time.

A frequency-selective method for driving the excitations is to modulate the trapping potential at the frequency of the excitation to be excited (Jin *et al.*, 1996). Experimentally this is accomplished by summing a small ac component onto the current in the trapping magnets. In a TOP trap, it is convenient enough to independently modulate the three second-order terms in the transverse potential. By controlling the relative phase of these modulations, one can impose $m=0$, $m=2$, or $m=-2$ symmetry on the excitation drive.

There have been a very large number of theory papers published on excitations; much of this work is reviewed by Dalfovo *et al.* (1999). All the zero-temperature, small-amplitude excitation experiments published to date have been very successfully modeled theoretically. Quantitative agreement has been by and large very good; small discrepancies can be accounted for by assuming reasonable experimental imperfections with respect to the $T=0$ and small-amplitude requirements of theory.

The excitation measurements discussed above were then revisited at nonzero temperature (Jin *et al.*, 1997). The frequency of the condensate excitations was clearly observed to depend on the temperature, and the damping rates showed a strong temperature dependence. This work is important because it bears on the little-studied finite-temperature physics of interacting condensates. Connection with theory (Hutchinson *et al.*, 1997; Dodd *et al.*, 1998; Fedichev and Shlyapnikov, 1998) remains somewhat tentative. The damping rates, which are observed to be roughly linear in temperature, have been explained in the context of Landau damping (Liu, 1997; Fedichev *et al.*, 1998). The frequency shifts are difficult to understand, in large part because the data so far have been collected in a theoretically awkward, intermediate regime: the cloud of noncondensate atoms is neither so thin as to have completely negligible effect on the condensate, nor so thick as to be deeply in the hydrodynamic (HD) regime. In this context, hydrodynamic regime means that the classical mean free path in the thermal cloud is much shorter than any of its physical dimensions. In the opposite limit, the collisionless regime, there are conceptual difficulties with describing the observed density fluctuations as collective modes. Recent theoretical work suggests that good agreement with experiment may hinge on correctly including the role of the excitation drive (Stoof, 2000; Jackson and Zaremba, 2002).

Two-component condensates

As mentioned above, the double-MOT system made it possible to produce condensates even if one were quite sloppy with many of the experimental parameters. One such parameter was the spin state in which the atoms are optically pumped before being loaded into the magnetic trap. As our student Chris Myatt was tinkering around setting up the evaporation one day, he noticed, to his surprise, that there seemed to be two different clouds of condensate in the trap. They were roughly at the locations expected for the 2,2 and $1,-1$ spin states to sit, but that seemed impossible to us because these two states could undergo spin-exchange collisions that would cause them to be lost from the trap, and the spin-exchange collision cross sections were thought to be enormous. After extensive further studies to try and identify what strange spurious effect must be responsible for the images of two condensate clouds we came to realize that they had to be those two spin states. By a remarkable coincidence, the triplet and singlet phase shifts are identical and so at ultralow temperatures the spin-exchange collisions are suppressed in ^{87}Rb by three to four orders of magnitude! This suppression meant that the different spin species could coexist and their

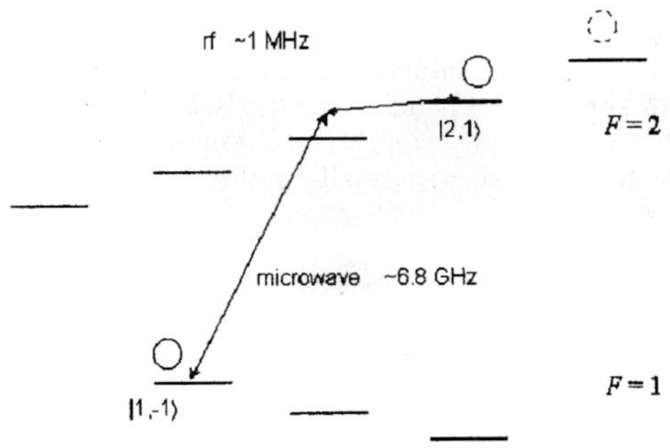

FIG. 10. Energy-level diagram for ground electronic state of ^{87}Rb. The first condensates were created in the 2,2 state. Mixtures containing the 2,2 and 1,−1 state were found to coexist. In later studies we created condensates in the 1,−1 state and then excited it to the 2,1 state using a microwave plus rf two-phonon transition.

mixtures could be studied. In early work we showed that one could carry out sympathetic cooling to make BEC by evaporating only one species and using it as a cooling fluid to chill the second spin state (Myatt *et al.*, 1997). We also were able to see how the two condensates interacted and pushed each other apart, excluding all but a small overlap in spite of the fact that they were highly dilute gases.

These early observations stimulated an extensive program of research on two-component condensates. After Myatt's original measurements (Myatt *et al.*, 1997), our work in this field, led by postdoc David Hall, concentrated on the 1,−1 and 2,+1 states (see Fig. 10) because they could be coherently interconverted using two-photon (microwave plus rf) transitions and they had nearly identical magnetic moments and so saw nearly the same trapping potentials (Matthews *et al.*, 1998). When the two-photon radiation field is turned off, the rate of spontaneous interconversion between the two spin species essentially vanishes, and moreover the optical imaging process readily distinguishes one species from the other, as their difference in energy (6.8 GHz) is very large compared to the excited-state linewidth. In this situation, one may model the condensate dynamics as though there were two distinct quantum fluids in the trap. Small differences in scattering length make the two fluids have a marginal tendency to separate spatially, at least in an inhomogeneous potential, but the interspecies healing length is long so that in the equilibrium configuration there is considerable overlap between the two species (Hall *et al.*, 1998a, 1998b). On the other hand, the presence of a near-resonant two-photon coupling drive effectively brings the two energy levels quite close to one another: on resonance, the corresponding dressed energy levels are separated only by the effective Rabi frequency for the two-photon drive. In this limit, one may in a certain sense think of the condensate as being described by a two-level, spinor field (Cornell *et al.*, 1998; Matthews *et al.*, 1999b).

We got a lot of mileage out of this system and continue to explore its properties today. One of the more dramatic experiments we did in the two-level condensate was the creation, via a sort of wave-function engineering, of a quantized vortex. In this experiment we made use of both aspects of the two-level system—the distinguishable fluids and the spinor gas. Starting with a near-spherical ball of atoms, all in the lower spin state, we applied the two-photon drive for about 100 ms. At the same time, we illuminated the atoms with an off-resonant laser beam whose intensity varied both in time and in space. The laser beam was sufficiently far from resonance that by itself it did not cause the condensate to transition from state to state, but the associated ac Stark shift was large enough to affect the resonant properties of the two-photon drive. The overall scheme is described by Matthews *et al.* (1999a) and Williams and Holland (1999). The net effect was to leave the atoms near the center of the ball of atoms essentially unperturbed, while converting the population in an equatorial belt around the ball into the upper spin state. This conversion process also imposed a winding in the quantum phase, from 0 around to two pi, in such a way that by the time the drive was turned off, the upper-spin-state atoms were in a vortex state, with a single quantum of circulation (see Fig. 11). The central atoms were nonrotating and, like the pimento in a stuffed olive, served only to mark the location of the vortex core. The core atoms could in turn be selectively blasted away, leaving the upper-state atoms in a bare vortex configuration, whose dynamic properties were shown by postdoc Brian Anderson and grad student Paul Haljan to be essentially the same as those of the filled vortex (Anderson *et al.*, 2000).

Coherence and condensate decay

One of our favorite BEC experiments was simply to look at how a condensate goes away (Burt *et al.*, 1997). The attraction of this experiment is its inherent simplicity combined with the far-reaching implications of the results. Although it was well established that condensates lived for a finite period, fractions of a second to many seconds depending on conditions, no one had identified the actual process by which atoms were being lost from the condensate. To do this our co-workers Chris Myatt, Rich Ghrist, and Eric Burt simply made condensates and carefully watched the number of atoms and shape of the condensate as a function of time. From these data we determined that the loss process varied with the cube of the density, and hence must be three-body recombination. This was rather what we had expected, but it was nice to have it confirmed. In the process of this measurement we also determined the three-body rate constant, and this was more interesting. Although three-body rate constants still cannot be accurately calculated, it was predicted long ago (Kagan *et al.*, 1985) that they should depend on the coherence properties of the wave function. In a normal thermal sample there are fluctuations and the three-body recombination predominantly takes place at high-density fluctuations. If there is higher-order coherence, however, as one has in macroscopically occupied quantum states such as a

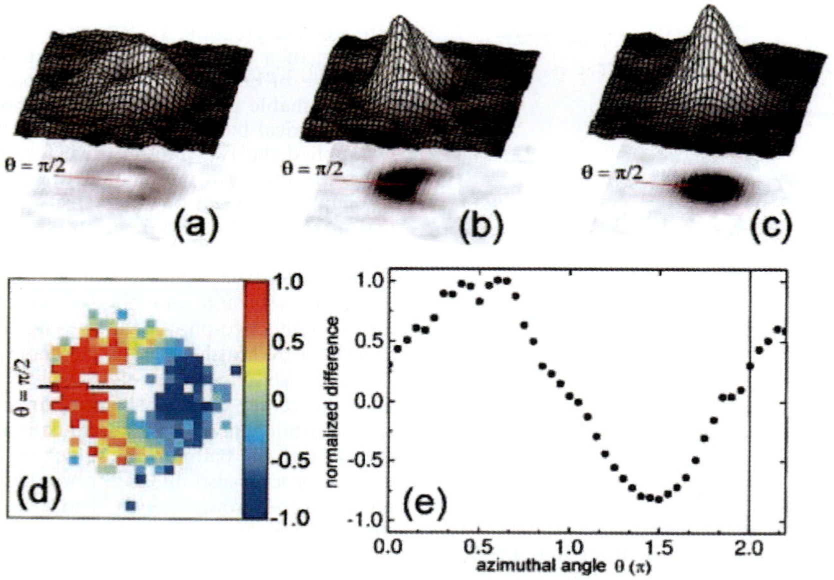

FIG. 11. Condensate images showing the first BEC vortex and the measurement of its phase as a function of azimuthal angle: (a) the density distribution of atoms in the upper hyperfine state after atoms have been put in that state in a way that forms a vortex; (b) the same state after a pi/2 pulse has been applied that mixes upper and lower hyperfine states to give an interferogram reflecting the phase distribution of the upper state; (c) residual condensate in the lower hyperfine state from which the vortex was formed that interferes with a to give the image shown in (b); (d) a color map of the phase difference reflected in (b); (e) radial average at each angle around the ring in (d). The data are repeated after the azimuthal angle 2π to better show the continuity around the ring. This shows that the cloud shown in (a) has the 2π phase winding expected for a quantum vortex with one unit of angular momentum. From Matthews *et al.*, 1999a [Color].

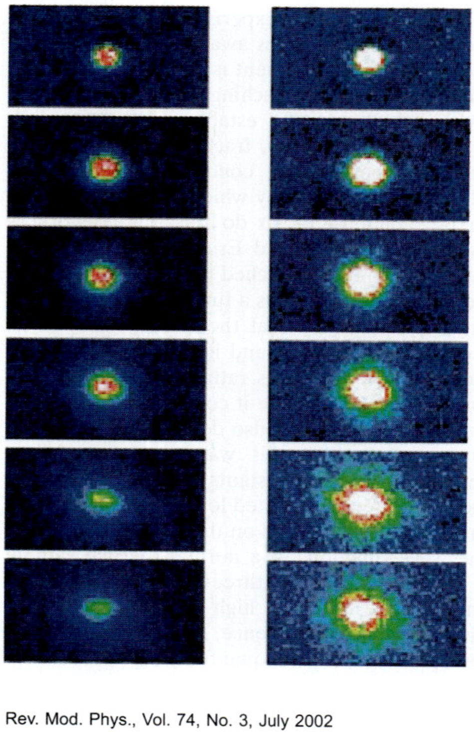

FIG. 12. Bosenova explosion from Roberts *et al.* (2001). From top to bottom these images show the evolution of the cloud from 0.2 to 4.8 ms after the interaction was made negative, triggering a collapse. On the left the explosion products are visible as a blue glow expanding out of the center, leaving a small condensate remnant that is unchanged at subsequent times. On the right is the same image amplified by a factor of 3 to better show the 200 nK explosion products [Color].

single-mode laser, or as was predicted to exist in a dilute gas BEC, there should be no such density fluctuations. On this basis it was predicted that the three-body rate constant in a Bose-Einstein condensate would be 3 factorial or 6 times lower than what it would be for the same atoms in a thermal sample. It is amusing that such a relatively mundane collision process can be used to probe the quantum correlations and coherence in this fashion. After measuring the three-body rate constant in the condensate we then repeated the measurement in a very cold but uncondensed sample. The predicted factor of 6 (actually 7.4 ± 2.6) was observed, thereby confirming the higher-order coherence of BEC (Burt *et al.*, 1997).

Feshbach resonance physics

In 1992 Eric Cornell and Chris Monroe realized that dipole collisions at ultralow temperatures might have interesting dependencies on magnetic field, as discussed in the Appendix. With this in mind we approached Boudwijn Verhaar about calculating the magnetic-field dependencies of collisions between atoms in the lower F spin states. When he did this calculation he discovered (Tiesinga *et al.*, 1993) that there were dramatic resonances in all the cross sections as a function of magnetic field that are now known as Feshbach resonances because of their similarity to scattering resonances described by Herman Feshbach in nuclear collisions. From the beginning Verhaar appreciated that these resonances would allow one to tune the s-wave scattering length of the atoms and thereby change both the elastic collision cross sections and the self-interaction in a condensate, although this was several years before condensates had been created.

In 1992 we hoped that these Feshbach resonances would give us a way to create enormous elastic collision cross sections that would facilitate evaporative cooling. With this in mind we attempted to find Feshbach resonances in the elastic scattering of first cesium and then, with postdoc Nate Newbury, rubidium. These experiments did provide us with elastic scattering cross sections (Monroe *et al.*, 1993; Newbury *et al.*, 1995), but were unable to locate the few-gauss-wide Feshbach resonances in the thousand-gauss range spanned by then theoretical uncertainty.

By 1997 the situation had dramatically changed, however. A large amount of work on cold collisions, BEC properties, and theoretical advances provided accurate values for the interaction potentials, and so we were fairly confident that there was likely to be a reasonably wide Feshbach resonance in rubidium 85 that was within 20 or 30 gauss of 150 G. This was a quite convenient bias field at which to operate our baseball magnetic trap, so we returned to the Feshbach resonance in the hope that we could now use it to make a Bose-Einstein condensate with adjustable interactions.

The time was clearly ripe for Feshbach resonance physics. Within a year Ketterle (Inouye *et al.*, 1998) saw a resonance in sodium through enhanced loss of BEC, Dan Heinzen (Courteille *et al.*, 1998) detected a Feshbach resonance in photoassociation in ^{85}Rb, we (Roberts *et al.*, 1998; notably students Jake Roberts and Neil Claussen) detected the same resonance in the elastic scattering cross section, and Chu (Vuletic *et al.*, 1999) detected Feshbach resonances in cesium. Our expectations that it would be as easy or easier to form BEC in ^{85}Rb as it was in ^{87}Rb and then use this resonance to manipulate the condensate were sadly naive, however. Due to enhancement of bad collisions by the Feshbach resonance, it was far more difficult and could only be accomplished by following a complicated and precarious evaporation path. However, by finding the correct path and cooling to 3 nK we were able to obtain pure ^{85}Rb condensates of 16 000 atoms (Roberts *et al.*, 2001).

The scattering length of these condensates could then be readily adjusted by varying the magnetic field over a few gauss in the vicinity of the Feshbach resonance (Cornish *et al.*, 2000). This has opened up a wide range of possible experiments, from studying the instability of condensates when the self-interaction is sufficiently attractive (negative a) to exploring the development of correlations in the wave function as the interactions are made large and repulsive. This regime provides one with a new way to probe such disparate subjects as molecular Bose-Einstein condensates and the quantum behavior of liquids, where there is a high degree of correlation. This work represents some of the most recent BEC experiments, but almost everything we have explored with this system has shown dramatic and unexpected results. Thus it is clear that we are far from exhausting the full range of interesting experiments that are yet to be carried out with BEC.

In the first of these Feshbach resonance experiments our students Jake Roberts, Neil Claussen, and postdoc Simon Cornish suddenly changed the magnetic field to make a negative. We observed that, as expected, the condensate became unstable and collapsed, losing a large number of atoms (Roberts *et al.*, 2001). The dynamics of the collapse process were quite remarkable. The condensate was observed to shrink slightly and then undergo an explosion in which a substantial fraction of the atoms were blown off (Donley, 2001). A large fraction of the atoms also simply vanished, presumably turning into undetectable molecules or very energetic atoms, and finally a small cold stable remnant was left behind after the completion of the collapse. This process is illustrated in Fig. 12. Because of its resemblance (on a vastly lower energy scale) to a core collapse supernova, we have named this the Bosenova. There is now considerable theoretical effort to model this process and progress is being made. However, as yet there is no clear explanation of the energy and anisotropy of the atoms in the explosion, the fraction of vanished atoms, and the size of the cold remnant. One of the more puzzling aspects is that the cold remnant can be far larger than the condensate stability condition that determines the collapse point would seem to allow (Donley, 2001).

Another very intriguing result of Feshbach resonance studies in ^{85}Rb was observed when our students Neil Claussen and Sarah Thompson and postdoc Elizabeth Donley quickly jumped the magnetic field close to the resonance while keeping the scattering length positive. They found that they could observe the sample oscillate back and forth between being an atomic and a molecular condensate as a function of time after the sudden perturbation (Donley *et al.*, 2002). This curious system of a quantum superposition of two chemically distinct species will no doubt be a subject of considerable future study.

An optimistic appendix

Until a new technology comes along to replace evaporative cooling, the crucial issue in creating BEC with a new atom is collisions. In practice, this means that planning a BEC experiment with a new atom requires learning to cope with ignorance. It is easy to forget that essentially nothing is known about the ultralow-temperature collisional properties of any atomic or molecular species that is not an atom in the first row of the Periodic Table. One cannot expect theorists to relieve one's ignorance. Interatomic potentials derived from room-temperature spectroscopy are generally not adequate to allow theoretical calculations of cold elastic and inelastic collision rates, even at the order-of-magnitude level. Although the cold collisional properties of a new atom can be determined, this is a major endeavor, and in most cases it is easier to discover whether evaporation will work by simply trying it.

Launching into such a major new project without any assurances of success is a daunting prospect, but we believe that, if one works hard enough, the probability that any given species can be evaporatively cooled to the point of BEC is actually quite high. The scaling arguments presented below in support of this assertion are largely the same as those that originally encouraged us to pursue BEC in alkalis, although with a bit more refinement provided by age and experience.

Although there is an extensive literature now on evaporative cooling, the basic requirement is simply that there be on the order of 100 elastic collisions per atom per lifetime of the atoms in the trap. Since the lifetime of the atoms in the trap is usually limited by collisions, the requirement can be restated: the rate of elastic collisions must be about two orders of magnitude higher than the rate of bad collisions. As mentioned above, there are three bad collisional processes, and these each have different dependencies on atomic density in the trap, n: background collisions (independent of n), two-body dipolar relaxation (αn), and three-body recombination (αn^2). The rate for elastic collisions is $n\sigma v$, where n is the mean density, σ is the zero-energy s-wave cross section, and v is the mean relative velocity. The requirement of 100 elastic-to-inelastic collisions must not only be satisfied immediately after the atoms are loaded into the trap, but also as evaporation proceeds toward larger n and smaller v. With respect to evaporating rubidium 87 or the lower hyperfine level of sodium 23, Nature has been kind. One need only arrange for the initial trapped cloud to have sufficiently large n, and design a sufficiently low-pressure vacuum chamber, and evaporation works. The main point of this section, however, is that evaporation is likely to be possible even with less favorable collision properties.

Considering the trap loss processes in order, first examine background loss. Trap lifetimes well in excess of what are needed for ^{87}Rb and Na have been achieved with standard vacuum technology. For example, we now have magnetic trap lifetimes of nearly 1000 s. (This was a requirement to achieve BEC in ^{85}Rb with its less favorable collisions.) If one is willing to accept the added complications of a cryogenic vacuum system, essentially infinite lifetimes are possible. If the background trap loss is low enough to allow evaporative cooling to begin, it will never be a problem at later stages of evaporation because nv increases.

If dipolar relaxation is to be a problem, it will likely be late in the evaporative process when the density is high and velocity low. There is no easy solution to a large dipolar relaxation rate in terms of changing the spring constant of the trap or the pressure of the vacuum chamber. Fortunately, one is not required to accept the value of dipolar collisions that Nature provides. In fact, all one really has to do is operate the trap with a very low magnetic bias field in a magnetic trap, or if one uses an optical trap very far off-resonance (such as CO_2 laser), trap the atoms in the lowest spin state, for which there are no dipole collisions. The bias field dependence comes about because below a field of roughly 5 G, the dipolar rate in the lower hyperfine level drops rapidly to zero. This behavior is simple to understand. At low temperature, the incoming collisional channel must be purely s wave. Dipolar relaxation changes the projection of spin angular momentum, so to conserve angular momentum the outgoing collisional channel must be d wave or higher. The nonzero outgoing angular momentum means that there is an angular momentum barrier in the effective molecular potential, a barrier of a few hundred microkelvin. If the atoms are trapped in the lower hyperfine state ($F=1, m_F=-1$, in rubidium 87) the outgoing energy from a dipolar collision is only the Zeeman energy in the trapping fields, and for B less than about 5 G this energy is insufficient to get the atoms back out over the angular momentum barrier. If relaxation is to occur, it can happen only at interatomic radii larger than the outer turning point of the angular momentum barrier. For smaller and smaller fields, the barrier gets pushed further out, with correspondingly lower transition rates.

It is unlikely that the three-body recombination rate constant could ever be so large that three-body recombination would be a problem when the atoms are first loaded from a MOT into the evaporation trap. As evaporation proceeds, however, just as for the dipolar collisions, it becomes an increasingly serious concern. Because of its density dependence, however, it can always be avoided by manipulating the trapping potential. Adiabatically reducing the trap confinement has no effect on the phase-space density but it reduces both the density and the atom velocity. The ratio of three-body to elastic collisions scales as $1/nv$. Therefore, as long as one can continue to turn down the confining strength of one *s* trap, one can ensure that three-body recombination will not prevent evaporative cooling all the way down to the BEC transition.

To summarize, given (i) a modestly flexible magnetic trap, (ii) an arbitrarily good vacuum, (iii) a true ground state with $F\neq 0$, and (iv) nonpathological collisional properties, almost any magnetically trappable species can be successfully evaporated to BEC. If one is using a very far off-resonance optical trap (such as a CO_2 dipole trap) one can extend these arguments to atoms that cannot be magnetically trapped. In that case, however, current technology makes it more difficult to optimize the evaporation conditions than in magnetic traps, and the requirement to turn the trap down sufficiently to avoid a large three-body recombination rate can be more difficult. Nevertheless, one can plausibly look forward to BEC in a wide variety of atoms and molecules in the future.

ACKNOWLEDGMENTS

We acknowledge support from the National Science Foundation, the Office of Naval Research, and the National Institute of Standards and Technology. We have benefited enormously from the hard work and intellectual stimulation of our many students and postdocs. They include Brian Anderson, Mike Anderson, Steve Bennett, Eric Burt, Neil Claussen, Ian Coddington, Kristan Corwin, Liz Donley, Peter Engels, Jason Ensher, David Hall, Debbie Jin, Tetsuo Kishimoto, Heather Lewandowski, Mike Matthews, Jeff McGuirk, Chris Monroe, Chris Myatt, Nate Newbury, Scott Papp, Cindy Regal, Mike Renn, Jake Roberts, Peter Schwindt, David Sesko, Michelle Stephens, William Swann, Sarah Thompson, Thad Walker, Yingju Wang, Richard Watts, Chris Wood, and Josh Zirbel. We also have had help from many other JILA faculty, including John Bohn, Chris Greene, and Murray Holland.

REFERENCES

Abraham, E. R. I., W. I. McAlexander, C. D. Sackett, and R. G. Hulet, 1995, Phys. Rev. Lett. **74**, 1315.

Anderson, M. H., J. R. Ensher, M. R. Matthews, C. E. Wieman, and E. A. Cornell, 1995, Science **269**, 198.

Anderson, M. H., J. R. Ensher, M. R. Matthews, C. E. Wieman, and E. A. Cornell, 1996, in *Laser Spectroscopy, XII International Conference*, edited by Massimo Inguscio, Maria Allegrini, and Antonio Sasso (World Scientific, Singapore), p. 3.

Anderson, B. P., P. C. Haljan, C. E. Wieman, and E. A. Cornell, 2000, Phys. Rev. Lett. **85**, 2857.

Anderson, M. H., W. Petrich, J. R. Ensher, and E. A. Cornell, 1994, Phys. Rev. A **50**, R3597.

Andreev, S. V., V. I. Balykin, V. S. Letokhov, and V. G. Minogin, 1981, Zh. Eksp. Teor. Fiz. **34**, 463.

Arimondo, E., W. D. Phillips, and F. Strumia, 1991, Eds., *Proceedings of the International School of Physics "Enrico Fermi," Course CXVIII, Laser Manipulation of Atoms and Ions* (North-Holland, Amsterdam).

Bagnato, V. S., G. P. Lafyatis, A. G. Martin, E. L. Raab, R. N. Ahmad-Bitar, and D. E. Pritchard, 1987, Phys. Rev. Lett. **58**, 2194.

Bell, D. A., F. H. Hess, G. P. Kochanski, S. Buchman, L. Pollack, Y. M. Xiao, D. Kleppner, and T. J. Greytak, 1986, Phys. Rev. B **34**, 7670.

Bjorkholm, J. E., R. R. Freeman, A. Ashkin, and D. B. Pearson, 1978, Phys. Rev. Lett. **41**, 1361.

Bose, S., 1924, Z. Phys. **26**, 178.

Burt, E. A., R. W. Ghrist, C. J. Myatt, M. J. Holland, E. A. Cornell, and C. E. Wieman, 1997, Phys. Rev. Lett. **79**, 337.

Chu, S., 1998, Rev. Mod. Phys. **70**, 685.

Chu, S., J. E. Bjorkholm, A. Ashkin, and A. Cable, 1986, Phys. Rev. Lett. **57**, 314.

Chu, S., L. Holberg, J. E. Bjorkholm, A. Cable, and A. Ashkin, 1985, Phys. Rev. Lett. **55**, 48.

Chu, S., D. S. Weiss, Y. Shevy, and P. J. Ungar, 1989, *Proceedings 11th International Conference on Atomic Physics* (World Scientific, Singapore), pp. 636–638.

Cohen-Tannoudji, C. N., 1998, Rev. Mod. Phys. **70**, 707.

Cornell, E. A., J. R. Ensher, and C. E. Wieman, 1999, in *Proceedings of the International School of Physics "Enrico Fermi," Course CXL*, edited by M. Inguscio, S. Stringari, and C. E. Wieman (Italian Physical Society, Bologna), p. 15. Also available as cond-mat/9903109.

Cornell, E. A., D. S. Hall, M. R. Matthews, and C. E. Wieman, 1998, J. Low Temp. Phys. **113**, 151.

Cornell, E. A., C. Monroe, and C. E. Wieman, 1991, Phys. Rev. Lett. **67**, 2439.

Cornish, S. L., N. R. Claussen, J. L. Roberts, E. A. Cornell, and C. E. Wieman, 2000, Phys. Rev. Lett. **85**, 1795.

Courteille, P., R. S. Freeland, D. J. Heinzen, F. A. van-Abeelen, and B. J. Verhaar, 1998, Phys. Rev. Lett. **81**, 69.

Dalfovo, F., S. Giorgini, L. P. Pitaevskii, and S. Stringari, 1999, Rev. Mod. Phys. **71**, 463.

Dalibard, J., and C. Cohen-Tannoudji, 1989, J. Opt. Soc. Am. B **6**, 2023.

Dalibard, J., C. Salomon, A. Aspect, E. Arimondo, R. Kaiser, N. Vansteenkiste, and C. Cohen-Tannoudji, 1989, *Proceedings 11th International Conference on Atomic Physics* (World Scientific, Singapore), pp. 199–214.

Dodd, R. J., M. Edwards, C. W. Clark, and K. Burnett, 1998, Phys. Rev. A **57**, R32.

Donley, E. A., 2001, Nature (London) **412**, 295.

Donley, E. A., N. R. Claussen, S. T. Thompson, and C. E. Wieman, 2002, Nature (London) **417**, 529.

Doyle, J. M., 1991a, Phys. Rev. Lett. **67**, 603.

Doyle, J. M., 1991b, Ph.D. thesis (Massachusetts Institute of Technology).

Einstein, A., 1924, Sitzungsber. K. Preuss. Akad. Wiss., Phys. Math. Kl. **261**.

Einstein, A., 1925, Sitzungsber. K. Preuss. Akad. Wiss., Phys. Math. Kl. **3**.

Ertmer, W., R. Blatt, J. L. Hall, and M. Zhu, 1985, Phys. Rev. Lett. **54**, 996.

Fedichev, P. O. and G. V. Shlyapnikov, 1998, Phys. Rev. A **58**, 3146.

Fedichev, P. O., G. V. Shlyapnikov, and J. T. M. Walraven, 1998, Phys. Rev. Lett. **80**, 2269.

Gardner, J. R., R. A. Cline, J. D. Miller, D. J. Heinzen, H. M. J. M. Boesten, and B. J. Verhaar, 1995, Phys. Rev. Lett. **74**, 3764.

Gibble, K., S. Chang, and R. Legere, 1995, Phys. Rev. Lett. **75**, 2666.

Gillaspy, J. D., I. F. Silvera, and J. S. Brooks, 1989, Phys. Rev. B **40**, 210.

Hall, D. S., M. R. Matthews, J. R. Ensher, C. E. Wieman, and E. A. Cornell, 1998a, Phys. Rev. Lett. **81**, 1539.

Hall, D. S., M. R. Matthews, C. E. Wieman, and E. A. Cornell, 1998b, Phys. Rev. Lett. **81**, 1543.

Hänsch, T. W., and A. L. Schawlow, 1975, Opt. Commun. **13**, 68.

Hardy, W. N., M. Morrow, R. Jochemsen, and A. Berlinsky, 1982, Physica B & C **110**, 1964.

Hecht, C. E., 1959, Physica **25**, 1159.

Hess, H. F., 1986, Phys. Rev. B **34**, 3476.

Hess, H. F., D. A. Bell, G. P. Kochanski, R. W. Cline, D. Kleppner, and T. J. Greytak, 1983, Phys. Rev. Lett. **51**, 483.

Hutchinson, D. A. W., E. Zaremba, and A. Griffin, 1997, Phys. Rev. Lett. **78**, 1842.

Inouye, S., M. R. Andrews, J. Stenger, H. J. Miesner, D. M. Stamper-Kurn, and W. Ketterle, 1998, Nature (London) **392**, 151.

Jackson, B., and E. Zaremba, 2002, Phys. Rev. Lett. **88**, 180402.

Jin, D. S., J. R. Ensher, M. R. Matthews, C. E. Wieman, and E. A. Cornell, 1996, Phys. Rev. Lett. **77**, 420.

Jin, D. S., M. R. Matthews, J. R. Ensher, C. E. Wieman, and E. A. Cornell, 1997, Phys. Rev. Lett. **78**, 764.

Johnson, B. R., J. S. Denker, N. Bigelow, L. P. Levy, J. H. Freed, and D. M. Lee, 1984, Phys. Rev. Lett. **52**, 1508.

Kagan, Y., G. V. Shlyapnikov, and I. A. Vartanyants, 1984, Phys. Lett. A **101**, 27.

Kagan, Y., B. V. Svistunov, and G. V. Shlyapnikov, 1985, JETP Lett. **42**, 209.

Kagan, Y., I. A. Vartanyants, and G. V. Shlyapnikov, 1981, Sov. Phys. JETP **54**, 590.

Ketterle, W., K. B. Davis, M. A. Joffe, A. Martin, and D. E. Pritchard, 1993, Phys. Rev. Lett. **70**, 2253.

Ketterle, W., D. S. Durfee, and D. M. Stamper-Kurn, 1999, in *Proceedings of the International School of Physics "Enrico Fermi," Course CXL,* edited by M. Inguscio, S. Stringari, and C. E. Wieman (Italian Physical Socity, Bologna), p. 176.

Lagendijk, A., I. F. Silvera, and B. J. Verhaar, 1986, Phys. Rev. B **33**, 626.

Letokhov, V., 1968, Zh. Eksp. Teor. Fiz. Pis'ma Red **7**, 348.

Lett, P. D., K. Helmerson, W. D. Phillips, L. P. Ratliff, S. L. Rolston, and M. E. Wagshul, 1993, Phys. Rev. Lett. **71**, 2200.

Lett, P. D., R. N. Watts, C. I. Westbrook, W. D. Phillips, P. L. Gould, and H. J. Metcalf, 1988, Phys. Rev. Lett. **61**, 169.

Liu, W. V., 1997, Phys. Rev. Lett. **79**, 4056.

London, F., 1938, Nature (London) **141**, 643.

Lovelace, R. V. E., C. Mehanian, T. J. Tomilla, and D. M. Lee, 1985, Nature (London) **318**, 30.

Lovelace, R. V. E., and T. J. Tommila, 1987, Phys. Rev. A **35**, 3597.

Luiten, O. J., H. G. C. Werij, I. D. Setija, M. W. Reynolds, T. W. Hijmans, and J. T. M. Walraven, 1993, Phys. Rev. Lett. **70**, 544.

Majorana, E., 1931, Nuovo Cimento **8**, 107.

Martin, A. G., K. Helmerson, V. S. Bagnato, G. P. Lafyatis, and D. E. Pritchard, 1988, Phys. Rev. Lett. **61**, 2431.

Matthews, M. R., B. P. Anderson, P. C. Haljan, D. S. Hall, M. J. Holland, J. E. Williams, C. E. Wieman, and E. A. Cornell, 1999b, Phys. Rev. Lett. **83**, 3358.

Matthews, M. R., B. P. Anderson, P. C. Haljan, D. S. Hall, C. E. Wieman, and E. A. Cornell, 1999a, Phys. Rev. Lett. **83**, 2498.

Matthews, M. R., D. S. Hall, D. S. Jin, J. R. Ensher, C. E. Wieman, E. A. Cornell, F. Dalfovo, C. Minniti, and S. Stringari, 1998, Phys. Rev. Lett. **81**, 243.

McAlexander, W. I., E. R. I. Abraham, N. W. M. Ritchie, C. J. Williams, H. T. C. Stoof, and R. G. Hulet, 1995, Phys. Rev. A **51**, R871.

Migdall, A. L., J. V. Prodan, W. D. Phillips, T. H. Bergeman, and H. J. Metcalf, 1985, Phys. Rev. Lett. **54**, 2596.

Miller, J. D., R. A. Cline, and D. J. Heinzen, 1993, Phys. Rev. Lett. **71**, 2204.

Monroe, C., 1992, Ph.D. thesis (University of Colorado, Boulder).

Monroe, C., W. Swann, H. Robinson, and C. Wieman, 1990, Phys. Rev. Lett. **65**, 1571.

Monroe, C. R., E. A. Cornell, C. A. Sackett, C. J. Myatt, and C. E. Wieman, 1993, Phys. Rev. Lett. **70**, 414.

Myatt, C. J., E. A. Burt, R. W. Ghrist, E. A. Cornell, and C. E. Wieman, 1997, Phys. Rev. Lett. **78**, 586.

Myatt, C. J., N. R. Newberry, R. W. Ghrist, S. Loutzenhiser, and C. E. Wieman, 1996, Opt. Lett. **21**, 290.

Neuhauser, W., M. Hohenstatt, P. Toschek, and H. Dehmelt, 1978, Phys. Rev. Lett. **41**, 233.

Newbury, N. R., C. J. Myatt, and C. E. Wieman, 1995, Phys. Rev. A **51**, R2680.

Pais, A., 1982, *Subtle Is the Lord . . .* (Oxford University Press, Oxford), English translation of Einstein's quotes and historical interpretation.

Petrich, W., M. H. Anderson, J. R. Ensher, and E. A. Cornell, 1994, J. Opt. Soc. Am. B **11**, 1332.

Petrich, W., M. H. Anderson, J. R. Ensher, and E. A. Cornell, 1995, Phys. Rev. Lett. **74**, 3352.

Phillips, W. D., 1998, Rev. Mod. Phys. **70**, 707.

Pritchard, D. E., 1986, in *Electronic and Atomic Collisions*, edited by D. C. Lorents, W. Meyerhof, and J. R. Peterson (North-Holland, Amsterdam), p. 593.

Pritchard, D. E., K. Helmerson, and A. G. Martin, 1989, *Proceedings 11th International Conference on Atomic Physics* (World Scientific, Singapore), pp. 179–197.

Pritchard, D. E., E. L. Raab, V. Bagnato, C. E. Wieman, and R. N. Watts, 1986, Phys. Rev. Lett. **57**, 310.

Prodan, J., A. Migdall, W. D. Phillips, I. So, H. Metcalf, and J. Dalibard, 1985, Phys. Rev. Lett. **54**, 992.

Raab, E. L., M. Prentiss, A. Cable, S. Chu, and D. E. Pritchard, 1987, Phys. Rev. Lett. **59**, 2631.

Roberts, J. L., N. R. Claussen, J. P. Burke, Jr., C. H. Greene, E. A. Cornell, and C. E. Wieman, 1998, Phys. Rev. Lett. **81**, 5109.

Roberts, J. L., N. R. Claussen, S. L. Cornish, E. A. Donley, E. A. Cornell, and C. E. Wieman, 2001, Phys. Rev. Lett. **86**, 4211.

Rohner, H., 1994, *Proceedings of the 39th Symposium on the Art of Glassblowing* (American Scientific Glassblowing Society, Wilmington, Delaware), p. 57.

Sesko, D., T. Walker, C. Monroe, A. Gallagher, and C. Wieman, 1989, Phys. Rev. Lett. **63**, 961.

Sesko, D., T. G. Walker, and C. E. Wieman, 1991, J. Opt. Soc. Am. B **8**, 946.

Silvera, I. F., and J. T. M. Walraven, 1980, Phys. Rev. Lett. **44**, 164.

Sprik, R., J. T. M. Walraven, and I. F. Silvera, 1985, Phys. Rev. B **32**, 5668.
Stoof, H. T. C., 2000, private communication.
Stwalley, W. C., and L. H. Nosanow, 1976, Phys. Rev. Lett. **36**, 910.
Thorsheim, H. R., J. Weiner, and P. S. Julienne, 1987, Phys. Rev. Lett. **58**, 2420.
Tiesinga, E., B. J. Verhaar, and H. T. C. Stoof, 1993, Phys. Rev. A **47**, 4114.
Tisza, L., 1938, Nature (London) **141**, 913.
Vuletic, V., A. J. Kerman, C. Chin, and S. Chu, 1999, Phys. Rev. Lett. **82**, 1406.
Walker, T., D. Sesko, and C. Wieman, 1990, Phys. Rev. Lett. **64**, 408.
Watts, R. N., and C. E. Wieman, 1986, Opt. Lett. **11**, 291.
Williams, J. E., and M. J. Holland, 1999, Nature (London) **401**, 568.
Wineland, D. J., and H. Dehmelt, 1975, Bull. Am. Phys. Soc. **20**, 637.
Wineland, D. J., R. E. Drullinger, and F. L. Walls, 1978, Phys. Rev. Lett. **40**, 1639.

Experimental proof of the spin of the photon*

C V RAMAN and S BHAGAVANTAM

(Received for publication, 24th October 1931)

1. Introduction

The hypothesis that the light-quantum possesses an intrinsic spin in addition to energy and momentum has the merit of enabling the corpuscular concept of radiation to become a complete and intelligible picture. The ratio of the energy to the linear momentum of the quantum determines the velocity of light while the ratio of energy to spin determines its frequency, and the sense of the spin determines the polarisation characters. The energy, momentum and spin of the photon have their counterparts in the Maxwellian field theory and indeed may be regarded as a translation into the language of the quantum theory of ideas derived from the classical electrodynamics. The transference of the classical concepts of radiation energy and radiation momentum into the quantum theory was effected by Einstein, and placed on a firm experimental basis by the work of Millikan and of Compton. The spin of the photon, on the other hand, remains at the present time a somewhat nebulous mathematical abstraction without any convincing experimental support. It is the purpose of the present paper to show how experimental studies on the scattering of light by gases enable us to give precision to the idea of photon spin and firmly to establish its existence as a fact of observation.

2. The theory of photon spin

In his well-known derivation of the Planck radiation formula from quantum statistics, Prof. S N Bose[1] obtained an expression for the number of cells in phase-space occupied by the radiation, and found himself obliged to multiply it by a

*A preliminary note (communicated by cable on 28th September 1931), published in *Nature* of the 3rd October, 1931, on page 576 contained the essential results of this paper.

A further note (communicated to *Nature* by cable on 12th October 1931) contains an account of the results obtained with liquids.

[1] S N Bose, *Z. f. Phys.* (1924) **26**, 178.

numerical factor 2 in order to derive from it the correct number of possible arrangements of the quantum in unit volume. The paper as published did not contain a detailed discussion of the necessity for the introduction of this factor, but we understand from a personal communication by Prof. Bose that he envisaged the possibility of the quantum possessing besides energy $h\nu$ and linear momentum $h\nu/c$ also an intrinsic spin or angular momentum $\pm h/2\pi$ round an axis parallel to the direction of its motion. The weight factor 2 thus arises from the possibility of the spin of the quantum being either right-handed or left-handed, corresponding to the two alternative signs of the angular momentum. There is a fundamental difference between this idea, and the well-known result of classical electrodynamics to which attention was drawn by Poynting[2] and more fully developed by Abraham[3] that a beam of light may in certain circumstances possess angular momentum. The detailed calculation for a spherical wave[4] (assumed to have a total energy $h\nu$), on the basis of the classical field theory shows its angular momentum to be $\pm h/2\pi$ only in the case of circular polarisation, right-handed and left-handed respectively, while for a plane-polarised spherical wave, the angular momentum vanishes, and for elliptically polarised light, it has intermediate values. Thus, according to the classical field theory, the angular momentum associated with a quantum of energy is not uniquely defined, while according to the view we are concerned within the present paper, the photon has always an angular momentum having a definite numerical value of a Bohr unit with one or other of the two possible alternative signs. Following Dirac,[5] we may regard plane polarised light, *not* as a quantum without spin, *but* as a quantum which has an *even* chance of having plus or minus a Bohr unit as its angular momentum. Elliptically polarised light would be similarly regarded as characterised by *unequal* chances of possessing spin with the two alternative signs.

3. Selection rules in light-scattering

It will be recalled[6] that the principle of the conservation of angular momentum was the postulate from which the selection rules for the emission or absorption of radiation from atoms were first derived. The fact that the selection rules were also later established by an application of the correspondence principle of Bohr and follow as mathematical deductions from the wave-mechanics of the atom has tended to diminish the significance of their original derivation in relation to

[2] J H Poynting, *Proc. R. Soc.* (1909) **82**, 560.
[3] M Abraham, *Physikal. Z.* (1914) **15**, 914.
[4] A Sommerfeld, *Atomic Structure and Spectral Lines*, English Edition, Appendix 9 (1923).
[5] P A M Dirac, *Quantum Mechanics*, Cambridge University Press (1930) p. 131.
[6] A Sommerfeld, *Atombau and Spektrallinien*, 4th German Edn. (1924) p. 317.

radiation theory. More recently, however, Oppenheimer[7] has stressed the close connection which exists between the question of photon spin, the selection rules in spectroscopy and the conservation of angular momentum in radiation processes. If we assume that the spin of the photon is a whole Bohr unit (plus or minus), it follows at once in agreement with observation that the *total* angular momentum of the atom may not remain zero during the process of radiation but must increase or decrease by one unit. The simplicity of this derivation is very appealing, and we may therefore extend its principle to the problem of light-scattering. It is an empirically established result that in the scattering of light by various simple diatomic molecules as H_2, O_2, N_2, etc. the rotational quantum number K of the molecule obeys the rule $\Delta K = 0, \pm 2$. The zero value for the change in quantum number gives the QQ branch or the undisplaced Rayleigh line in the spectrum of scattering by the gas, and the value 2 with the alternative signs gives the RR and PP branches of the rotational scattering, which appear as a close group of lines respectively on the low and high frequency sides of the undisplaced line. The selection rule stated was derived by Rasetti, Dieke and others from the conception of the third intermediate level in light-scattering. As pointed out in recent notes in *Nature*,[8] they also follow directly from the principle of conservation of angular momentum. We assume that the spin of the photon has the value $\pm h/2\pi$ and for simplicity confine our attention to the case in which the quantum is scattered without any change in its direction of motion and the molecule has its axis of rotation fixed in the same direction. If the photon has a spin equal to $+h/2\pi$, it might either retain the same value or change to $-h/2\pi$, as the result of the scattering process. If it has initially the value $-h/2\pi$, it might either retain this value or change to $+h/2\pi$. The change in spin of the photon is thus 0 or ± 2 Bohr units. The corresponding change in the rotational quantum number of the molecule would be equal but opposite in sign, and would therefore be 0 or ± 2. The PP and RR branches of the scattering in the forward direction would on this view consist of photons which have experienced a reversal of spin. In other words, if the incident light be circularly polarised, the Rayleigh or undisplaced scattering would be circularly polarised in the same sense, and the rotational scattering would be *reversely* circularly polarised. The fact that this inference from theory is supported by the observations of Bar and of Bhagavantam may be regarded as furnishing strong support for the view that the photon has an intrinsic spin. It must be confessed, however, that as all the results stated can also be derived from the classical radiation theory, either with or without the help of the quantum mechanics, the argument can scarcely be regarded as a conclusive proof of the existence of photon spin.

---

[7] J R Oppenheimer, *Phys. Rev.* (1931) **38**, 726.

[8] C V Raman and S Bhagavantam, *Nature* (1931) **128**, 114; C V Raman *Nature* (1931) **128**, 545.

4. Scattering of spinning photons

A careful considerations shows that a difference between the results of the classical field theory and the idea of spinning photons is to be expected in respect of the spectral distribution of the light scattered by the molecules of a gas. If we regard radiation as a particle possessing an intrinsic spin and impinging on a rotating molecule, the *sense* of the spin which the molecule possesses in relation to that of the photon should obviously enter into the problem in a fundamental way. If the molecule is spinning in a right-handed sense, it is obvious that if it gains energy of spin from the encounter, it must also gain angular momentum from it in a right-handed sense; this obviously it cannot if the photon before the encounter has a left-handed spin, since the latter can take up but not give up right-handed spin. Similarly a molecule with left-handed spin cannot gain both energy and spin from a right-handed photon. We may in fact make out the following scheme of forbidden and allowed transitions:

Molecule spin	Photon spin	PP $K \rightarrow K-2$	QQ $K \rightarrow K$	RR $K \rightarrow K+2$
Right	Left	Allowed	Allowed	Forbidden
Right	Right	Forbidden	Allowed	Allowed
Left	Right	Allowed	Allowed	Forbidden
Left	Left	Forbidden	Allowed	Allowed

It will be seen that while the transitions which give the QQ branch are always allowed, those which could contribute to the intensity of the PP and RR branches are forbidden in half the total number of possible cases and allowed only in the other half. On the other hand, in the treatment of the problem of light-scattering on the Maxwellian field theory, the function of the radiation is assumed to be merely that of furnishing a periodic electric force at the position of the molecule which would polarise it and cause to emit secondary radiation as a rotating Hertzian dipole. The question of the sense in which the molecule rotates does not arise, and has no influence on the final results. In other words, it is implicitly assumed that in the collision between a photon and a molecule, either a gain or a loss of energy by the latter is always allowed. As we have seen, this is not the case if the existence of an intrinsic photon spin and the conservation of angular momentum are taken into account.

To understand what happens in the case of a forbidden transition, we may adopt a hint given by the classical treatment. From the latter, it is clear that the whole of the scattering would appear in the QQ branch if the molecule were optically isotropic, and that the existence of the PP and RR branches in scattering is due entirely to the fact that the molecule is optically anisotropic. It is possible to separate the scattering by an anisotropic molecule into two parts, namely the isotropic part which is completely polarised in a transverse direction, and which appears solely in the QQ branch, and the anisotropic part which consists mainly of unpolarised light, and is divided between the PP, QQ and RR branches. If the anisotropic molecules instead of being allowed to rotate were imagined to be held fixed in randomly distributed orientations, the classical theory tells us that the PP and RR branches would disappear, their contribution being merely transferred to the QQ branch, so that the intensity and state of polarisation of the aggregate scattering would remain unaffected. We may therefore make the assumption that when a $K \rightarrow K - 2$ transition, or a $K \rightarrow K + 2$ transition is forbidden by reason of photon spin, we have simply a $K \rightarrow K$ transition: in other words the photon is scattered without change of energy. Adopting this view *we conclude that the effect of photon spin on the scattering of light by a rotating molecule is to diminish the intensity of the* PP *and* RR *branches to one half the value given by the classical theory*, the intensity of the QQ branch being correspondingly increased, so that the intensity and state of polarisation of the total scattering remain unaffected.

5. Depolarisation of scattering

The foregoing conclusions have a further important consequence. It is clear from what has been said above that the QQ branch of the scattering consists mainly of polarised light, while the PP and RR branches consist mainly of unpolarised light. Hence the defect in polarisation of the QQ branch, when spectroscopically separated from the PP and RR branches would be *smaller* than that of the total unresolved scattering consisting of all the three branches. The extent of the difference would be very considerable if we are to believe the classical theory according to which most of the anisotropic scattering appears in the PP and RR branches. According to the theory of photon spin, on the other hand the intensity of these branches is diminished to one half the value given by the classical theory with a corresponding increase in the intensity of the QQ branch. We should therefore expect the defect of polarisation of the QQ branch to be much nearer that of the total scattering than is indicated by the classical theory. This fact furnishes us with a convenient *experimentum crucis* for the existence of photon spin.

The magnitude of the effects to be expected is readily calculated. As has been explained by Bhagavantam in a recent paper,[9] in the case of simple linear molecules such as N_2, O_2, CO_2, N_2O, CS_2, etc. the anisotropic scattering according to the classical theory, finds a place in the QQ branch only to the extent $\frac{1}{4}$th of the whole. That this result is characteristic of the Maxwellian concepts of

[9]S Bhagavantam, *Indian J. Phys.* (1931) 6, 331.

radiation is shown by the fact that Manneback's theory[10] of scattering by diatomic molecules based on the wave-mechanics and Bhagavantam's treatment based on the classical or Maxwell–Boltzmann statistics of molecular rotation give precisely the same results. For symmetric top molecules such as NH_3, C_6H_6, etc. the ratio should be $\frac{1}{3}$ instead of $\frac{1}{4}$. Following the theory of photon spin, if we transfer one-half of the aggregate intensities of the PP and RR branches to the QQ branch, the anisotropic scattering would be found in the latter to the extent of $\frac{5}{8}$ of the whole in the case of linear molecules and $\frac{2}{3}$ of the whole for symmetric top molecules.

The defect of polarisation in the transversely scattered light is known for the case of various molecules from the earlier investigations in which spectroscopic resolution was not used, and represents the depolarisation of the PP, QQ and RR branches taken together. The scattering with large frequency shifts due to the vibration of molecules is of comparatively negligible intensity. Taking the best accepted values and using the relations stated above, we obtain the following table showing the depolarisation of the QQ branch to be expected according to the classical field theory and the theory of photon spin.

Table I

Molecule	Total depolarisation	Depolarisation of QQ branch (classical)	Depolarisation of QQ branch (spin theory)
N_2	3·6%	3·9%	2·3%
O_2	6·4	1·7	4·1
CO_2	9·8	2·7	6·4
N_2O	12·5	3·5	8·3
CS_2	11·5	3·2	7·6
C_6H_6	4·2	1·4	2·8

6. Experimental method and results

The earliest indication of the failure of the classical field theory was furnished by measurements of the intensity of the lines in the spectrum of the light scattered by hydrogen gas.[11] An attempt to reconcile the data obtained with Manneback's theory and the known depolarisation of the total scattering proved unsuccessful; the observed intensity of the rotational lines was found to be just half that to be expected on the basis of Manneback's theory. It was this discrepancy in fact which suggested the ideas outlined in the preceding paragraphs. Except in the

[10] C Manneback, *Z. f. Phys.* (1930) **62**, 224.
[11] S Bhagavantam, *loc. cit.*

case of hydrogen, it is not easy to determine the intensity of the lines in the PP and RR branches as they are very close to each other and to the QQ branch and are not completely separable from the photographic halation accompanying the latter. The alternative method of the study of the depolarisation of scattering with and without spectroscopic resolution is simpler and also of a more general application, as it can be applied to the case of molecules having high moments of inertia which give unresolved wings instead of discrete lines in the PP and RR branches.

The general plan of the experiments is very simple. The light of the mercury lamp transversely scattered by the gas passes through a square-ended nicol and is focussed on the slit of a high-illumination spectrograph. Two photographs are taken of the spectrum of the scattered light, respectively with the nicol oriented in two perpendicular positions, and the state of polarisation is determined by varying the relative exposures until the horizontal and vertical components are obtained with equal intensity. The depolarisation of the QQ branch is determined by narrowing the slit of the spectrograph so that with the exposures given, only the undisplaced Rayleigh lines are recorded. A second set of measurements made with the slit open as wide as possible gives the depolarisation of the unresolved scattering the comparison of which with the accepted value acts as a check on the perfection of the experimental arrangements.

Numerous experimental difficulties present themselves in the exact determination of such small depolarisations as those observed with gases. The experience gathered from the long series of investigations carried out at Calcutta by I Ramakrishna Rao[12] of the unresolved scattering and the more recent work of Bhagavantam[13] using spectroscopic methods has proved very helpful in dealing with these difficulties. Amongst the major pitfalls which present themselves may be mentioned the following:

(*a*) False light derived from stray illumination of the walls of the container, and specially of the background. This consists mainly of unpolarised light, and if not completely eliminated would result in enormous errors.

(*b*) Errors due to lack of transversality of the illuminating rays to the direction of observation. When using extended sources of light such as a quartz mercury arc, the errors are very serious.

(*c*) Polarisation of the rays occurring within the spectrograph due to oblique refraction at the surfaces of the prisms.

After much preliminary work devoted to elimination of these sources of error, the following arrangements were adopted for the actual measurements. CO_2 and N_2O gases were chosen as the working substances, as they give an intense scattering and large depolarisation: the errors due to false light were therefore far

[12] I R Rao, *Indian J. Phys.* (1927) **2**, 61.
[13] S Bhagavantam, *Indian J. Phys.* (1931) **6**, 319.

smaller than for other gases such as N_2 and O_2. The gases were taken from commercial cylinders under pressure, and were found to be free from dust. They were transferred under pressure into a containing tube of fused quartz blackened outside, armoured in steel and fitted with dark background, diaphragms, etc. as already described and illustrated in a previous paper by Bhagavantam in this Journal. So long as the working pressure was less than 20 atmospheres, the changes in depolarisation due to the deviation from Boyle's law was negligible. The source of illumination was a quartz mercury lamp, the entire luminous area of which was covered up by a metal plate containing an aperture 2 cm in diameter through which the light emerged, and an image of which was focussed by a large glass condenser at the centre of the quartz tube containing the gas. The distance from the centre of the condenser to the illuminated aperture on the one side, and the centre of the gas tube on the other was about 20 cm. The restriction of the illuminating area of the lamp had the effect of greatly diminishing the importance of false light from the back of the observation tube, as was shown by actual trial exposures with a gas evacuated. It also reduced the error due to lack of transversality of the illuminating rays. As was shown by I Ramakrishna Rao,[14] the use of a condenser for focussing a source of light of small angular aperture introduces no error under this head, so long as observations are made of the scattered light at the exact point of focus. The polarisation of light occurring within the spectrograph was carefully determined, and allowed for in making the photographic time exposures.

The experimental results are given in table II.

Table II. Depolarisation in %

Gas	Observed (wide slit)	Observed (narrow slit)	Observed diminution	Calculated diminution (Classical)	Calculated diminution (spin)
CO_2	11·0	7·0	4·0	7·1	3·4
N_2O	12·0	7·0	5·0	9·0	4·2

The measurements are clearly decisive in favour of the spin theory.

7. Observations with liquids

It is well known that the light scattered by liquids at ordinary temperatures exhibits a notable defect in polarisation which is larger and much easier to

[14] I R Rao *loc. cit.*

observe and measure than the corresponding effect in gases. Further when, spectroscopically analysed, the scattered light exhibits very marked wings on either side of the undisplaced line due to the exciting radiation. These wings represent the PP and RR branches of the scattering by molecules of the liquid, and their existence together with their observed depolarisation is *prima facie* evidence that the molecules rotate while scattering radiation. The close packing of the molecules in a liquid (as compared with a gas) has a great influence on the intensity of the coherent part of the scattering which is determined by the compressibility of the fluid and not by the molecular density. This constitutes the 'isotropic' scattering which is fully polarised and appears solely in the QQ branch. The intensity of the anisotropic scattering which is incoherent and therefore in a truer sense molecular in origin, is determined by the density of the liquid, but cannot be assumed to be strictly proportional to it owing to the influence of a complicating factor, namely the local polarisation field acting on the molecules within the liquid. We are however concerned here not with the total intensity of the anisotropic scattering (which may be taken to be that actually observed under the given conditions), but with the manner in which it is shared between the PP, QQ and RR branches. There is reason to suppose that the close packing in a liquid might influence the statistical distribution of the rotational energy of the molecules, making it depart very considerably from that obtaining in the case of a gas. If such were the case, the distribution of intensities *within* the PP and RR branches would be affected. So long, however, as we are free to assume that the material particles which scatter the light (whatever they may be) are free to take up spin from the photons and rotate with appreciable frequencies, there must be a separation of the anisotropic scattering into the PP, QQ and RR branches in a manner consistent with the principles of conservation of energy and angular momentum.

The considerations set out above make it clear that except in the case of highly viscous liquids where large molecular groups may be assumed to exist, we shall probably be justified in regarding the general principles discussed in the earlier part of the paper with reference to gases to be applicable with full force to the case of liquids as well. Especially in the case of non-associated liquids such as CS_2, C_6H_6, etc. we should be justified in determining the part of the total anisotropic scattering which appears in the QQ branch, as a fraction of the whole, assuming the individual chemical molecule to be the scattering particle. The total anisotropic scattering may be determined from the observed depolarisation of the aggregate scattering. In this way the depolarisation of the QQ branch to be expected may be readily calculated. The results along with the observed values[15] for two simple liquids CS_2 and C_6H_6 are given in Table III.

[15]Taken from the recent unpublished work of S Venkateswaran on a large number of liquids.

Table III. Depolarisation in %

Unresolved	Observed (wide slit)	Observed (narrow slit)	QQ branch (spin theory)	QQ branch (classical)
CS_2				
64·0	64·0	56·0	55·5	36·4
C_6H_6				
42·0	41·5	34·0	33·5	20·8

The measurements are clearly decisive in favour of the spin theory.

(494)

Evidence for the spin of the photon from light-scattering

It is well known that in the encounters between molecules and photons, exchanges of energy occur which are rendered evident by the observed changes in frequency of the scattered light. Whenever in such an encounter a molecule gains or loses spin-energy, there is a corresponding change in its angular momentum, which can only be explained on the assumption that there is an equal and opposite change in the spin-moment of the photon. This involves definite consequences for the state of polarisation of the scattered light, and opens up the possibility of demonstrating experimentally the existence of angular momentum in radiation.

The general principle involved is sufficiently illustrated by considering the case of a photon which is circularly polarised and is scattered without change of direction by a molecule the axis of spin of which coincides with the axis of circular polarisation. If the spin of the molecule is unaffected by the encounter, the scattered photon retains its original character. On the other hand, if the molecule gains two additional units of angular momentum, the photon will alter in frequency and at the same time *change over from right-handed to left-handed circular polarisation* or *vice versa*. The considerations are naturally not so simple in the general case in which the molecule is arbitrarily orientated and the photon is scattered in a direction not coinciding with the primary ray. It may be shown that besides the reversal of the sign of circular polarisation in the forward direction, there would also result a depolarisation of the scattered light, which would be specially evident in directions transverse to the primary rays.

It may be remarked that similar phenomena should also be observed when, as the result of an encounter with the photon, a change in spin-energy of the molecule is superimposed on a change in its vibrational state. On the other hand, a change in vibrational state alone would not give rise to the effects indicated.

It may be noted that certain interesting observations recently reported by Hanle and by Bar[1] find a natural explanation in the considerations set out above. In support of the view that the scattered radiations exhibiting strong depolarisation or reversal of circular polarisation involve rotational transitions, it may be

[1] *Die Naturwissenschaften* (1931) **19**, pp. 375 and 463.

mentioned that the lines exhibiting such effects are usually diffuse, while lines which are well polarised are usually sharp and intense.

C V RAMAN
S BHAGAVANTAM

210 Bowbazaar Street
Calcutta (India)
15 June

(482)

S. N. Bose : The Man

Partha Ghose

In the words of the poet Bishnu Dey, Satyendranath Bose was a 'legend in his life time'. This is not only because in times of the British Raj his name got inseparably linked to that of arguably the greatest scientist of all time, Albert Einstein, but also because he was a unique confluence of intellectual brilliance, encyclopedic knowledge, selfless dedication to the motherland and unbounded compassion and love for fellow beings. He was a renaissance figure and a quintessential Bengali. In his *Discovery of India*, this is what Jawaharlal Nehru had to say about Bengal that moulded Satyendranath:

'Bengal had a clear half-century of British rule before it spread over wider areas. During the second half of the eighteenth century and the first half of the nineteenth, Bengal therefore played a dominant role in British Indian life. Not only was Bengal the centre and the heart of the British administration, but it also produced the first groups of English-educated Indians who spread out to other parts of India under the shadow of the British power. A number of very remarkable men rose in Bengal in the nineteenth century, who gave the lead to India in cultural and political matters, and out of whose efforts the new nationalist movement ultimately took shape.'

Satyendranath was born on January 1, 1894 in Kolkata. His father, Surendranath, was a very versatile man and was very much a product of his time. He was so inspired by the national movement that he started a chemical and pharmaceutical company of his own even before the famous *Bengal Chemicals* was founded by Acharya Prafulla Chandra Ray. By the time Satyendranath was eleven, Lord Curzon, the Viceroy and Governor of Bengal, announced the Partition of Bengal in 1905. To divide and rule was a famous British imperial policy. To Bengalis it appeared as a check against the growing nationalism. It sparked widespread protests and kick started the *swadeshi* movement that changed the lives of many young men of that generation.

Surendranath was a dominating father who decreed that Satyendranath, being the only son in a large family, was not to join the *swadeshi* movement nor attend musical soirees. Satyen followed the first decree while remaining a sympathizer, but not the second very strictly – music was in his veins. I had a taste of the father-son relationship when I returned from Imperial College, London in 1963 and joined Satyendranath's research group. He was then the National Professor of Physics. He was of the opinion that the Russians were making remarkable progress in physics and mathematics, and that I had better learn Russian to keep up with the times. He asked his old friend, Nirendranath Roy who had spent considerable time in the Soviet Union and knew Russian very well, to teach me Russian. One late morning shortly after I formally joined Satyendranath's group, Niren Roy was visiting his old friend, and I happened to be there also. Niren Roy was asked to take me up to the first floor of the family house to meet the grand old man, Surendranath. I went up with Niren Roy and entered Surendranath's room sheepishly. He was sitting on his bed wearing dark glasses, probably due to some problem with his eyes. He appeared to be almost blind. When Niren Roy introduced me to him as a young man who had just returned from London and joined his son's research group, he moved his head towards me and said wryly, "you didn't find a single person in the whole world to teach you maths and have come to Bodé!" Bodé was Satyendranath's nickname.

Satyendranath's school life was by and large uneventful, except that his genius was spotted by his school teachers quite early on. According to another boyhood friend, Girijapati Bhattacharya [1],

'Our friendship began in 1908, when we were both in the Hindu School, Calcutta, though he was a year ahead of me. Even at school, Satyen was marked for his extraordinary intellect. Our mathematics teacher predicted that Satyen would one day be a great mathematician like Laplace or Cauchy . . .' In fact, he once gave him 110 out of 100 in mathematics because not only did he get all the sums right, he had done some of them in more than one way!

Satyendranath went through three phases of social and national involvement. The first phase was marked by uncertainty and emotion. In the second phase which covers the period between 1916 and 1945 most of which he spent in Dhaka with stints in Calcutta, Paris and Berlin, he got actively involved in supporting a group of revolutionaries planning armed uprising against the British, with the police following him at times. The two persons he supported most were Abaninath Mukherjee and Sailen Ghosh. Abani Mukherjee had planned to smuggle American arms into India with the help of the Germans during the First World War. The eminent linguist Suniti Kumar Chatterjee met him in Germany in 1922. Chatterjee tried to dissuade him not to return to India because in the ten years that Abani had left India, the nature of politics in India had changed because of the emergence of Gandhi. However, he did not pay heed to this advice and returned to India the same year and established contact with Chatterjee. Not being involved with politics, Chatterjee took him to Satyendranath and through him and their common friend Dilip Kumar Roy, the eminent musician and a close friend of Subhash Chandra Bose, got him to meet members of the *Anushilan Samiti*, a revolutionary nationalist party. The next meeting was in Dhaka where Abani surfaced and was quickly betrayed by his own friends. Satyendranath took a photograph of him urgently needed for his passport, and helped him in other ways. The third meeting was in Paris. Satyendranath carried some money to hand over to him with Suniti Chatterjee playing a vital role and adopting the following strategy [1]:

'I would write to him that this money which I was sending to him was to buy for me some Sanskrit books. He would then understand, and pass the money to Abaninath. His brothers gave me about Rs. 300 and they knew that I was sending it to him through Satyen Bose. This money reached him all right.'

Sailen Ghosh came first in Physics in M. Sc. in the same year that Satyendranath came first in Mixed Mathematics. Sailen was a favourite of the Vice Chancellor Sir Asutosh Mukherjee, and was a very resourceful person. He helped obtain many valuable pieces of apparatus for the newly established Physics Laboratory of Calcutta University from Maharaja Manindra Chandra Nandy of Cossimbazar. Sailen was a member of a revolutionary party, and had to flee to the United States with Satyendranath's assistance when the police caught up with him. He returned to India later with his American wife and Satyendranath helped him get a position in Jagannath College in Dhaka.

The third and final phase was one of disillusionment and intense Bengali nationalism caused by the partition of India in 1947 which he could never accept – the country was freed from the British but mutilated. Following Rabindranath Tagore's lead, he began his crusade for teaching the sciences in the mother tongue. He had to face a lot of criticism. How could one learn all the latest advances in the sciences if one did not become proficient in English? Who would translate the increasing volume of scientific literature in English into the vernaculars? And what about scientific jargon?

I picked up the courage one day to query him about his crusade. His reply was immediate and remarkable. He told me he was not worried about people like me who would take up a career in the sciences. Why, they would have to learn as many languages as necessary, just as I was asked to learn Russian. (Satyendranath and his classmate Meghnad Saha learnt and mastered French and German in their youth in order to study the original works of Einstein, Planck, Poincaré and other great European scientists. They were the first to translate Einstein's and Minkowski's papers on relativity in German into English. The book called *The*

Principle of Relativity was published by the University of Calcutta in 1919 with an Introduction by P. C. Mahalanobis.) He was worried about the common people. A barrage of rhetorical questions followed. Why should they be forced to learn a foreign language to pick up even the elements of science? How could the country acquire a scientific temper if the average citizen had no way of learning about the latest advances in his own language? And who would teach proper English? I have never since doubted the essential validity of the argument, so grossly oversimplified and misunderstood by the opponents.

Satyendranath was often accused of indulging in āddā, often a pejorative word. Santimay and Enakshi Chatterjee, biographers of Satyendranath and his friend Meghnad Saha, have characterized āddā in the following words [1]: 'It is difficult to find an exact English equivalent for āddā which covers a broad range from small talk to serious discourse, but without predetermined purpose or direction. This predominantly Bengali habit is indeed wasteful, but at the same time can be intellectually stimulating depending on the quality of the participants. In its best form the āddā can be an opening out – the mind opens up and reaches out to others with absolutely no regard for gain or material profit, in a leisurely trustful spirit.' I think it sums up āddā admirably. In this sense Satyendranath's life was indeed an extended āddā. Satyendranath was a member of some famous āddās in Calcutta; particularly *Sabujpatra* led by the littérateur Pramatha Chaudhury in his Bright Street residence and *Parichay* started by his friend, the poet Sudhindranath Datta. His role in these āddās has acquired a legendary character. He would listen with rapt attention for the better part of the time and only occasionally make a very cryptic and forceful comment that would stir up a fresh debate, and then reverse into his silence. And, the subject could be anything from art criticism to history to literary criticism.

His versatility knew no bounds. Satyendranath read the draft of Suniti Chatterjee's magnum opus *The Origin and Development of the Bengali Language* and made a few suggestions for improvement which were incorporated in the final version and acknowledged in the Preface. Dhurjatiprasad Mukherjee, the famous historian and musicologist, recounted how Satyendranath had bailed him out before his M. A. examination in History when he suffered from temporary blindness and found it hard to cope with Egyptian history. Satyendranath briefed him on the topic for days and he did so well in the examination that it aroused the examiner's curiosity. 'Dhurjati, I was quite surprised when I went through your script. You never came to my class. You never seemed to have studied the Ancient East. Yet, you wrote so well!' the examiner exclaimed. 'Satyen taught me' explained Dhurjati, upon which the examiner remarked, 'Oh Satyen, that explains it.' [1]

Satyendranath was a great friend of the poet Bishnu Dey who had a remarkable collection of western classical music records. I have heard from the poet himself and his wife, Pranati Dey how Satyendranath would come to their residence for āddā and would listen for hours to western classical music. On one occasion he was listening to Beethoven's Ninth Symphony and became so engrossed that he appeared to fall asleep. The Deys quietly moved to another room lest he be disturbed. Just as it started building up to the grand climax, he slowly opened his eyes again.

While in Dhaka his āddās took on a different form. He would come to the Department around 7 a.m. and be there until late in the evenings, except for a short lunch break. Whenever he was not taking a class or attending a meeting, there would be a continuous stream of visitors to his room, and Satyendranath would get involved with their academic problems for hours. The subject could be any branch of Physics, Chemistry, History, hieroglyphics or indeed any other subject under the sun. He was engaged in creating a culture of intellectual discourse which, unfortunately, came to an abrupt end with the partition of the country and Satyendranath's return to Kolkata in 1945.

Satyendranath was a great connoisseur of music and was quite adept at playing the *esraj*. Dilip Bose who was the Secretary of the Ali Akbar College of Music told me Satyendranath had even composed a new *raga* possibly just mathematically by using permutation and combination of notes. About his love for music this is what his friend Dilip Kumar Roy had to say [1]:

'I recall . . . how we searched together for the houses of the great ustads to hear them sing. Once in Dhaka we sought out together the house of Renuka Sengupta who later became famous all over Bengal for her song 'Gokulachandra braje na elo . . .' Later when I went to her place to give her lessons in music Satyen would often accompany me to hear her angelic voice. He was a real connoisseur of music – he had the perfect discernment to recognize good singing in a flash. He learnt to play the esraj, to follow music better. This was the only field where I could explain things to him with authority, and a little pride.'

He loved French literature. Dr. Adhikari, a former Director of the Indian Statistical Institute recounted to me how once he ran into my *mastermoshai* (a Bengali word for a teacher) sitting alone at a road-side café in Paris, a beret on his silky white mane, a carafe of red wine and a pile of books on the table, reading a French novel. On recognizing Dr. Adhikari, he immediately started to explain how he longed for the latest French novels and how every time he was in France he would buy them and devour them. I recall one occasion when I called on him at his residence shortly after lunch. He was reclining in his bed, bare bodied with a maroon *lungi* tied to his waist. He was chewing *pan* (betel leaf) and reading a book. He was in great mood. He wanted to read out a few pages to me. He made me sit near him and started to read. It was all about the reactions of a cat which was sitting below a banquet table and watching all the delicious food that was being served and consumed. He stopped after a while, smiled, and said to me while putting down the book, delightfully written, isn't it! I concurred and quietly picked up the book to see what its title was and who the author was. I was amazed – it was in French, and yet he had read it out to me in unfaltering English for nearly five minutes! He was in an expansive mood, and even claimed that his French pronunciation was quite good.

His relationship with Einstein was slightly ambivalent. On the one hand, he addressed him in his letters as 'Dear Master', was immensely grateful to him for having appreciated his short paper on photon statistics and for having drawn the attention of the whole scientific world to it, and had the highest respect for his scientific views. On the other hand, he did nurse a hurt feeling too for not having had his Master's full blessings for his other ideas, which he kept secret. I am privy to some of these ideas. On one occasion he told me he was going to confide something in me which I must never disclose. I admit I am therefore breaking his trust in making the story public, only because I think the story must be recorded for the sake of history. Moreover, I found later on that the essential idea was already in print as far back as 1931 in a paper of Raman and Bhagavantam [2]. I now wonder if Professor Bose knew about this paper at all, or had clean forgotten about it. Any way, he got up and closed all the doors and windows so that nobody could hear what he was about to tell me. Then he started to explain to me in a low voice how in his derivation of Planck's formula he got the first factor $4 \pi V \nu^2 / c^3$ instead of $8 \pi V \nu^2 / c^3$ as required, and that he had proposed in his paper that this factor of 2 could come from the photon having a spin of one unit which could be either parallel or anti-parallel to its direction of propagation. But he said to me with a sad smile, the old man crossed it out! Einstein apparently replaced it by the statement that this factor came from the two states of polarization of light. There was no need to talk about the photon spin at that stage, was probably Einstein's stand. And then he went on to remark with a dismissive smile particle polarization what on earth can the polarization of a particle mean? I was shocked. I asked him why he had not pointed this out to Einstein when photon spin was eventually discovered. Einstein would have surely stood by your priority, I remonstrated with him. "How does it matter who discovered it? It's been found, hasn't it?" was his reply with a sense of satisfaction. That was Satyen Bose.

Incidentally, this is what Raman and Bhagavantam wrote to motivate their 1931 paper [2]:

'... we understand from a personal communication from Prof. Bose that he envisaged the possibility of the quantum possessing besides energy $h\upsilon$ and linear momentum $h\upsilon/c$ also angular momentum $\pm h/2\pi$ around an axis parallel to the direction of motion. The weight factor 2 thus arises from the possibility of the spin of the quantum being either right-handed or left-handed corresponding to the two alternative signs of the angular momentum.'

There was an encounter of a different kind between these two giants in Berlin in 1926 that is worth recalling as an indicator of Satyendranath's nationalistic moorings. Those were days of intense revolutionary activities in India and anti-British sentiments ran very high. This is an account of the encounter in Satyendranath's own words [1]:

'One day he confided to me. 'I think Englishmen are better than other Western colonial nations, and I feel that they are far better than the French and Dutch. You should not be surprised at a German like me praising the English (after the World War). Now tell me, do you really want the British should quit your country?' I said, 'Of course – we all want to determine our own destinies ourselves.' He was not quite convinced. He raised a hypothetical question. He said, 'Suppose there was a button near you and all the Englishmen would quit India if you were to press it. So, would you press the button?' I smiled and said, 'If God were to grant me such an opportunity, I will not hesitate even for a moment to press it.' He said, 'Really?' and kept quiet for a little while. I asked him, 'Well, why do you Jews then want to establish a new Israeli State? Even you seem to be fairly inclined in its favour.' He said, 'Of course, I can now understand what you are saying – it is an emotional matter and cannot be understood rationally.'

I will end with a couple of personal anecdotes. One day he asked me to bring my car to take him to the Park Street area of Calcutta for some shopping. When I arrived at his north Calcutta residence in the early afternoon, he quickly got up, put on an orange baggy raw silk coat over his pyjama, fixed a beret on his silky white hair, turned to me and said, "Let's go". His elder son Rathindranath who happened to be lurking around the room stepped in somewhat embarrassed and said, "Dad, should you go out in this dress?" "Why, it's fine". He first went to a shop to buy a wrist watch from the HMT shop on Park Street (the shop no longer exists) and then asked me to take him to the New Market. There he got down from the car, put one of his arms round my shoulder and started walking me towards the row of dry fruit stalls owned by the Afghans some of whom wore *mehndi* (henna) on their beards. There he sat himself down on a shopkeeper's tool by the passage and started a conversation with them, much to the amazement of passersby some of whom recognized him to be the great scientist, stopped for a while in utter disbelief and then carried on without a word in amazement. He was, of course, totally oblivious of all this.

The last story is also about a trip in my car to visit his friends in south Calcutta. He first visited one friend for an hour during which I had my own bit of āddā with my friends, and then he asked me to go to Sadananda Road near Kalighat. When I turned into that road, he said his target was a sweet-meat shop at the corner of a lane to the left. I was to turn left into that lane. In due course we dutifully turned into that lane. However, he could not find the house he was looking for. So he asked me to turn the car around which I did. Before proceeding any further, he asked me, 'Obvious inference Tell me, what's the obvious inference?' I kept quiet, not knowing what to say. So I went back to Sadananda Road, turned left and carried on for a while. There was indeed another a little down Sadananda Roads at the corner of yet another lane to the left. He was overjoyed, "I told you" he said triumphantly. This time he found the house he was looking for, and my day's job was done.

Finally, let me end with a few words about his relationship with the poet Rabindranath Tagore. Unlike most poets, Rabindranath was steeped in science, particularly its philosophical implications, and had a lot of time for Indian scientists like Sir J. C. Bose, Meghnad Saha and Satyendranath. He wrote a whole book on science called *Visva Parichay* which he dedicated to Satyendranath. To give you a glimpse of their relationship, let me display some correspondence between them (by courtesy of Rabindra Bhavana, Visva-Bharati).

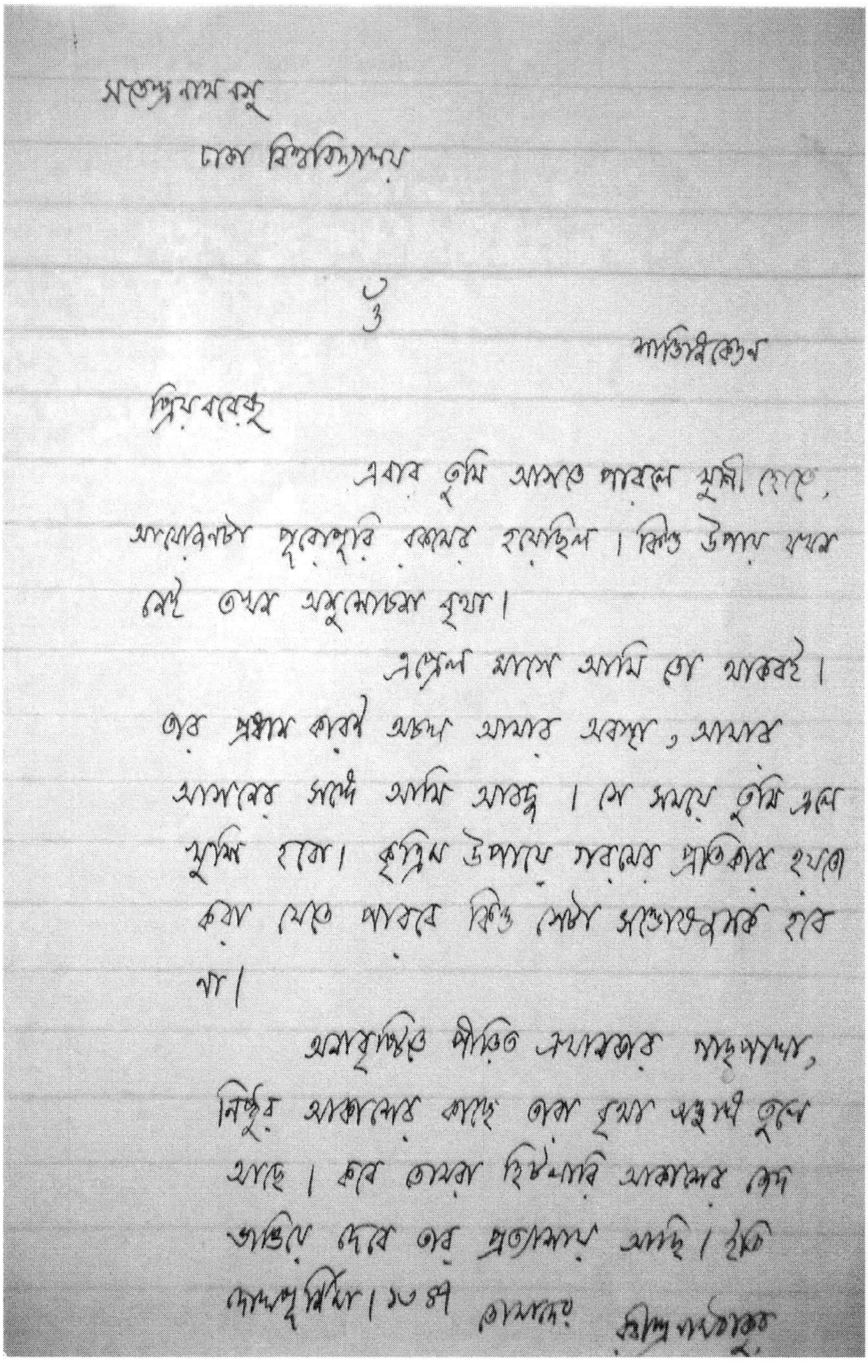

সত্যেন্দ্রনাথ বসু

ঢাকা বিশ্ববিদ্যালয়

ওঁ

শান্তিনিকেতন

প্রিয়বরেষু

এবার তুমি আসতে পারলে খুশী হোতুম, আয়োজনটা পুরোপুরি রকমের হয়েছিল। কিন্তু উপায় যখন নেই তখন অনুশোচনা বৃথা।

এপ্রেল মাসে আমি তো থাকবই। তার প্রধান কারণ এখন আমার অবস্থা, আমার আসবাবের সঙ্গে আমি আবদ্ধ। সে সময়ে তুমি এলে খুশি হব। কৃত্রিম উপায়ে গরমের প্রতিকার হয়তো করা যেতে পারবে কিন্তু সেটা সন্তোষজনক হবে না।

অনাবৃষ্টিতে পীড়িত এখানকার সাঁওতালরা, নিষ্ঠুর আকাশের কাছে তাদের দুহাত ঊর্ধ্বমুখী তুলে আছে। তবে তোমরা বিজ্ঞানীরা আকাশের মেঘ ভাঙিয়ে দেবে তার প্রত্যাশায় আছি। ইতি

দোলপূর্ণিমা। ১৩৪১ তোমাদের রবীন্দ্রনাথ ঠাকুর

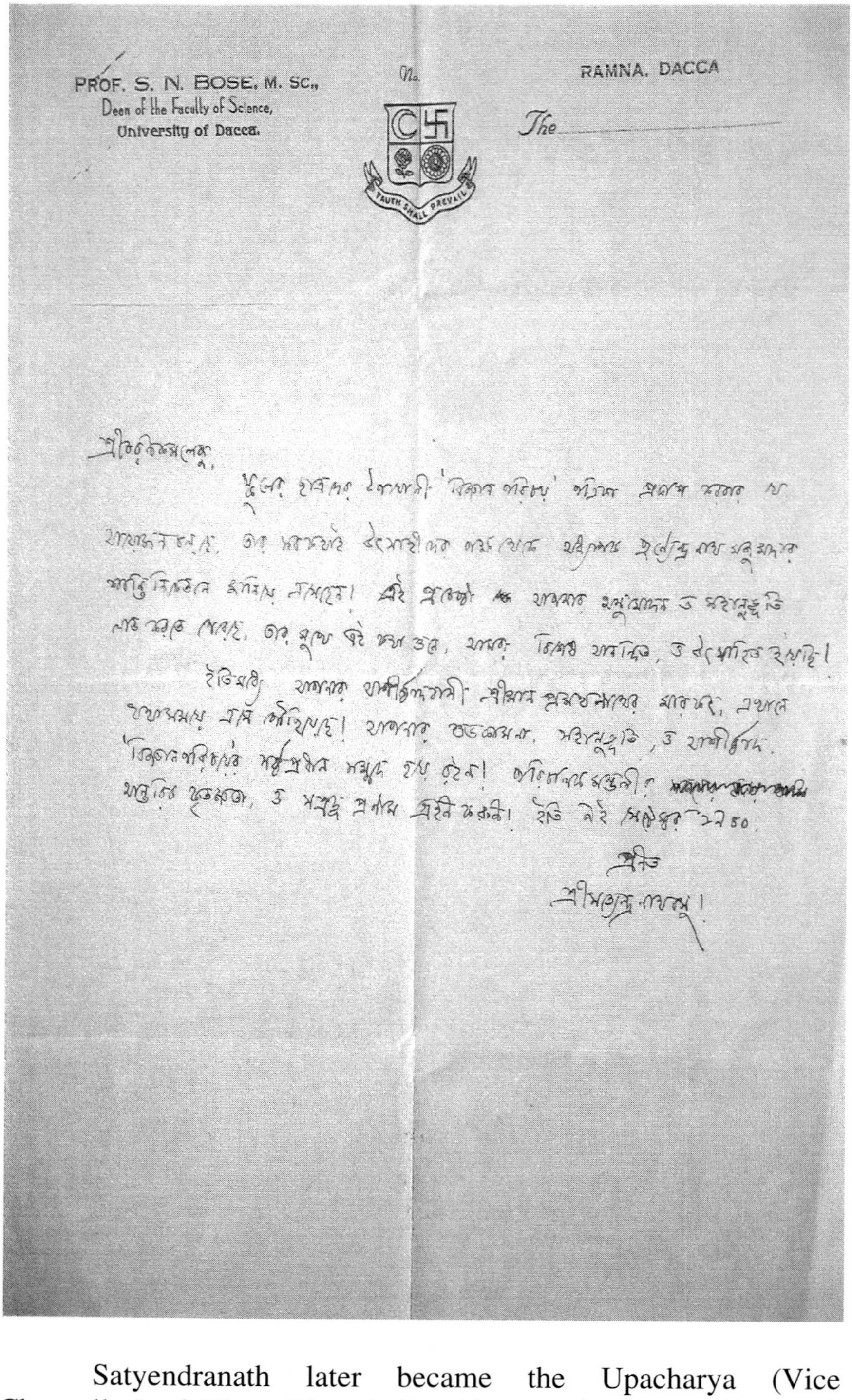
PROF. S. N. BOSE, M. Sc.,
Dean of the Faculty of Science,
University of Dacca.

No.

RAMNA, DACCA

The

TRUTH SHALL PREVAIL

Satyendranath later became the Upacharya (Vice Chancellor) of Visva-Bharati. But his experience there was not very happy, and he resigned well before his term of office was over. He was then offered the Emeritus position of a National Professor of Physics for the rest of his life.

References

1. *S. N. Bose: The Man and his Works*, Part II, eds. C. K. Majumdar, Partha Ghose, Enakshi Chatterjee, Samik Bandyopadhyay, Santimay Chatterjee, (S. N. Bose National Centre for Basic Sciences, 1994).
2. C. V. Raman and S. Bhagavantam, *Experimental Proof of the Spin of the Photon, Ind. J. of Phys.*, vol. 6 (1931), p. 355.

Satyendranath Bose: Co-Founder of Quantum Statistics

WILLIAM A. BLANPIED*
Department of Physics
Case Western Reserve University
Cleveland, Ohio 44106
(Received 26 January 1972; revised 11 April 1972)

Satyendranath Bose's recognition of the fact that the photons in a cavity radiator are indistinguishable led him to a derivation of the Planck law free from any assumptions of classical electrodynamics. His result, published in 1924, was a significant vindication of the photon concept that Einstein had championed since 1905. It also laid the foundations of quantum statistics, as was freely acknowledged by Einstein and Dirac, among others. Various aspects of Bose's career, including his apparent failure to understand the full implications of his work, may be of interest both to physicists and historians of science.

I. INTRODUCTION: BOSE AND THE PLANCK LAW

On 4 June 1924, Satyendranath (S.N.) Bose, then a thirty-year-old reader (associate professor) at Dacca University in East Bengal, sent a short paper in English to Albert Einstein, then professor of physics at Berlin. The paper was accompanied by a handwritten covering letter which began[1]:

> Respected Sir:
>
> I have ventured to send you the accompanying article for your perusal and opinion. I am anxious to know what you think of it. You will see that I have tried to deduce the coefficient $8\pi\nu^2/c^3$ in Planck's Law independent of the classical electrodynamics, only assuming that the ultimate elementary regions in the phase–space has the content h^3. I do not know sufficient German to translate the paper. If you think the paper worth publication I shall be grateful if you arrange for its publication in *Zeitschrift für Physik*. Though a complete stranger to you, I do not hesitate in making such a request. Because we are all your pupils though profiting only from your teachings through your writings. . . .

Einstein was impressed, in part because Bose's result vindicated a view he had championed for almost twenty years. Einstein himself translated, and in early July 1924 submitted the paper to the *Zeitschrift* in Bose's name, where it was published under the title "Plancks Gesetz und Lichtquantenhypothese" ("Planck's Law and the Hypothesis of Light Quanta").[2] He also added a note stating, "In my opinion Bose's derivation of the Planck formula signifies an important advance. The method used also yields the quantum theory of the ideal gas, as I will work out in detail elsewhere."

Undergraduate texts in statistical mechanics frequently derive the Planck black body radiation law by two different methods. The first, introduced by Debye in 1910, quantizes the classical result obtained by Rayleigh in 1900 by applying the Maxwell–Boltzmann statistics to the number of normal modes of electromagnetic radiation in a cavity at an (absolute) equilibrium temperature T. It assumes all the laws of classical electrodynamics except for the proviso that the energy and frequency of the radiation are to be linked by the relation $E=h\nu$, where h is Planck's constant.[3]

The second method is due to Bose. As noted in his 4 June letter to Einstein it is free from all assumptions of classical electrodynamics. Instead it relies on phase space arguments, treating the radiation in the cavity as an ideal photon gas, each photon having energy $h\nu$ and momentum $h\nu/c$. The frequency distribution of the radiation at an absolute temperature T is then deduced by finding the distribution in phase space that maximizes the entropy of the system.

When the same treatment is used to find the equilibrium distribution for a dilute system of noninteracting atoms, one of the results obtained is the well known ideal gas law. However, a photon gas differs from the more usual dilute ideal gas in a number of respects. First, the particles must be treated relativistically; second, photons are massless; third, their number is not conserved. Most important, perhaps, the photons must be treated as *indistinguishable*. Because this is so, all arrangements of a given number of phase points within an elementary phase space cell of volume h^3 are to be regarded as identical arrangements, whereas in classical statistics they would be regarded as distinct. Thus, the numbers of permutations of phase points for systems of distinguishable and nondistinguishable particles differ considerably and result in quite different equilibrium distributions. In the case of a photon gas the resultant distribution leads to the correct form of the Planck law, as Bose's first *Zeitschrift* paper demonstrated.[2]

It was no mere whim which led Bose to submit his paper to Einstein for criticism. He was certainly aware that the result was a logical development of Einstein's work, which had, however, eluded him for nineteen years. In his so-called "photoelectric effect" paper of 1905 Einstein had used thermodynamic arguments to show that electromagnetic radiation has quantum characteristics and thus introduced the photon concept into physics, though the word photon was not coined until 1926.[4] His results had far more conceptual significance than Planck's celebrated paper of 1900 which had assumed that the oscillators which emit and absorb radiation do so in a quantized manner but had not quantized the radiation field itself. During the succeeding years many, if not most, leading physicists (including Planck himself) either disregarded Einstein's paper or failed to take it seriously for, indeed, it did seem to undermine the magnificent structure of Maxwellian electrodynamics.[5] Yet as Einstein pointed out as early as 1906, any attempt to graft an *ad hoc* quantization assumption onto the body classical electrodynamics contradicted the basic continuum assumption on which that system rested. Thus in his view a complete reconstruction of electrodynamics was necessary. Since historically the black body problem had led to the introduction of the quantum of action and to Einstein's quantization of the radiation field itself, then clearly if his view were to prevail it was essential that the Planck law be derivable along the lines suggested by his 1905 paper. Bose's paper provided that derivation. His results, coming only a year after A. H. Compton's x-ray scattering experiments showed that photons could transfer momentum to electrons in collision processes,[6] vindicated Einstein's views and indicated that henceforth electrodynamics would have to develop along the lines he had suggested in 1905.

Clearly, S. N. Bose deserves credit for taking the first step in the development of quantum statistical mechanics, credit that was unstintingly granted him by Albert Einstein, who was soon to generalize his results to particles of finite mass, and by P. A. M. Dirac who first coined the word "boson." His name is one of six which undergraduate physics majors learn to associate with a type of statistics (the others being Maxwell, Boltzmann, Einstein, Fermi, and Dirac). Yet outside India very few know anything about him besides his name. In fact, as he himself puts it, many want to place an umlaut over the "o" and turn him into a German, while most of those who know he is an Indian confuse him with the nineteenth century botanist J. C. (Jagdish Chandra) Bose.

II. BOSE AT CALCUTTA UNIVERSITY

S. N. Bose resents being turned into a German, who, as he says, still talks about the "Einstein–Bose" statistics. It offends both his sense of national identity and his avowed francophile leanings. But he laughs off the merging of his identity with that of J. C. Bose because he regards the botanist as one of the first Indian scientists to win international recognition.[7] Besides, few Westerners could know that Bose is a very common name in Bengal, and S. N. Bose, like so many of his compatriots, regards himself first and foremost as a Bengali.[8]

Bengal, and particularly Calcutta which was the seat of the British Imperial government and residence of the Viceroy until 1921, has long been one of the most volatile, colorful, tragic, and altogether exciting regions of the Indian subcontinent. The Bengalis, with some justice, still regard Calcutta as the artistic and intellectual capital of the subcontinent. Its university, founded in 1857, is the oldest in the country and still has the largest number of students—over 200 000 in 1970. In March 1914 the Vice Chancellor, Sir Asutosh Mookerjee, having obtained considerable independent funding from several wealthy Bengalis, laid the foundation of the University College of Science. These endowments stipulated that all professorships were to be held by Indians rather than by Europeans, as had been the normal practice theretofore. Bose recalls that previous to that time most occupants of university posts in Calcutta had been Scotsmen (and third-rate Scotsmen at best). He also recalls that his physics laboratories used to consist mainly in the reverent display of some piece of apparatus on the lecture table: "The Professors thought that if it would be touched by any student there it would be spoiled forever and so much money wasted! Now that was science."[9]

Sir Asutosh sought to change all that by offering respectable academic appointments to qualified Indians and by establishing first-rate laboratories and libraries. In 1916 S. N. Bose and M. N. (Meghnad) Saha, another Bengali who had been his classmate at Presidency College in Calcutta, became two of the first lecturers appointed in the new University College. The following year, with the establishment of departments within the College, they were both appointed lecturers in physics.

Bose immediately showed his interest in statistical mechanics. In 1918 he and Saha published a joint paper in the *Philosophical Magazine* entitled "On the Influence of the Finite Volume of Molecules on the Equation of State"[10]; two years later they published another joint paper in the same journal entitled "On the Equation of State."[11] In 1919 they published an anthology of Einstein's papers on relativity, one of the first such collections published in English anywhere.[12] Bose closes his 4 June 1924 letter to Einstein with a reference to that collection[1]:

> I don't know whether you still remember that somebody from Calcutta asked your permission to translate your papers on relativity in English. You acceded to the request. The book has since been published. I was the one who translated your paper on Generalized Relativity."

Evidently Sir Asutosh had shown considerable insight in appointing the scholarly Bose and the terse Saha to his faculty. In 1918 he demonstrated his insight once again by securing the appointment of C. V. Raman as Palit Professor of Physics.

Raman, Saha, and Bose were all destined to make considerable contributions both to physics and to their country. Colorful and erratic, Raman (a South Indian by birth) was a civil servant in the Indian Finance Department who up to the time of his appointment had spent most of his free time carrying out research in acoustics at the nearby, privately funded Indian Institute for the Advancement of Science. Soon after securing his professorship he began his researches in optics that ultimately led to the discovery of the Raman effect, fellowship in the Royal Society, knighthood, and the Nobel Prize. Raman's students at Calcutta included K. S. Krishnan, who later joined Bose at Dacca University and ultimately became the first Director of the National Physical Laboratory at New Delhi.

Saha left Calcutta in 1923 to found an outstanding school of physics at Allahabad University where he carried out his renowned theoretical studies on thermal ionization processes, became an intimate of Jawaharlal Nehru, and made serious contributions to Indian archeology. In 1934 he helped to found the monthly journal *Science and Culture*. His unsigned editorials covering a twenty year period from 1935 to his death in 1956, had a considerable influence on public science policy in India. In 1938 he returned to Calcutta University as Palit Professor of Physics and soon established the Institute of Nuclear Studies which now bears his name. In 1951, after breaking with Nehru, he was elected to the Lok Sabha (Parliment) from Calcutta as an independent candidate. An ardent advocate of economic planning, he became a member of the Indian Planning Commission soon after Independence and served until his death in 1956.[13]

Although his name is coupled with that of Einstein, S. N. Bose's career was less spectacular than either Raman's or Saha's. His first quantum theoretical paper, entitled "On the Deduction of Rydberg's Law from the Quantum Theory of Spectral Emission," was published in the *Philosophical Magazine* in 1920.[14] In it he showed that the empirical formulas for the spectra of the alkali atoms could be derived from the Bohr–Sommerfeld quantization rules and the assumption that the effective potential seen by the valence electrons in these atoms consists of the superposition of a point charge and an electric dipole potential. A year later Bose left Calcutta to become reader in physics at the newly organized University of Dacca in East Bengal. Here he lectured, read, thought, "spent many sleepless nights" thinking about the Planck law, and late in 1923 submitted a paper on the subject to the *Philosophical Magazine.* Six months later the editors of that journal informed him that (regretably) the referee's report on his paper was negative. Undeterred, he sent the rejected manuscript to Einstein and thus took the first step in the formulation of quantum statistics.

III. A BOSE–EINSTEIN COLLABORATION?

Einstein's endorsement of his paper also gave Bose his first opportunity to see the world outside India. He had applied to the university for a two-year study leave earlier in 1924, but by June had yet to receive a definite reply. A handwritten postcard from Einstein stating that he regarded the paper as a most important contribution settled the study leave question in Bose's favor—and at once! (Bose also recounts that on the strength of that postcard the German Consulate in Calcutta issued his visa without requiring payment of the customary fee.) Thus in early September 1924 he sailed from Bombay to Europe aboard a steamer of the Lloyd Triestino line.

Upon his arrival in Paris during mid-October, some Bengali friends drew Bose into a circle presided over by a noted French Indianologist who in turn provided him with an introduction to Paul Langevin, then Professor Suppleant at the College de France and "thesis adviser" to no less a person than Louis deBroglie. Yet it appears that Bose was anxious to move on and, perhaps, play the role that history might demand he play. On 26 October 1924 he addressed a second letter to Einstein, which began[1]:

> Dear Master,
>
> My heartfelt gratitude for taking the trouble of translating the paper yourself and publishing it. I just saw it in print before I left India. I have sent you about the middle of June a second paper entitled "Thermal Equilibrium in the Radiation Field in the Presence of Matter."
>
> I am rather anxious to know your opinion about it, as I think it to be rather important. I don't know whether it will be possible also to have this paper published in *Zeit. für Physik.*
>
> I have been granted study leave by my university for 2 years. I have arrived just a week ago in Paris. I don't know whether it will be possible for me to work under you in Germany. I shall be glad however if you grant me the permission to work under you, for it will mean for me the realization of a long cherished hope.
>
> I shall wait for your decision as well as your opinion of my second paper here in Paris. . . .

In fact, Einstein had already translated this second paper and sent it to the *Zeitschrift,* which published it later that year.[15] This time, however, Einstein added a remark stating that he could not agree with this author's conclusions, and went on to give his reasons in some detail. Apparently he also communicated these objections privately, for in a letter from Paris dated 27 January 1925 Bose thanks Einstein for his communication of 3 November and states that he is about to send off another paper which he hopes will satisfy the objections raised with regard to the first. This promised paper was never published, nor does any further mention of it occur in Bose's correspondence.

Since no copy of Einstein's 3 November letter is extant there is no evidence that he encouraged Bose's "long-cherished hope" of working with him in Berlin. But it seems certain that Bose himself had not yet given up the idea, for his 27 January letter closes by stating that he hopes to come to Berlin at winter's end and there obtain Einstein's "inestimable help and guidance." However, in retrospect, it seems clear that the opportunity for a true Bose–Einstein collaboration was already gone by the time Bose wrote his January 1925 letter. In July 1924, about the time Bose was finally settling the study leave question with the Dacca University authorities, Einstein read a paper before the Prussian Academy that applied Bose's statistical method to an ideal gas with particles of nonzero rest mass and suggested some of its nonclassical thermodynamic properties.[16]

This nonclassical gas continued to fascinate him for several months. In September 1924 (while Bose was at sea) Einstein wrote to Ehrenfest stating that at low temperatures the molecules of the gas would condense into the zero energy state even in the absence of attractive forces between them. By early December (while Bose was pondering Einstein's objections to the second *Zeitschrift* paper) Einstein wrote Ehrenfest that "The matter of the quantum gas is getting very interesting."[17] Finally, in January 1925, Einstein published a paper that Klein has called "another of his masterworks."[18] It not only rounded out his theory of the ideal quantum gas by predicting the now well known Bose–Einstein condensation, but in addition, it attempted to rationalize Bose's unique method of counting microstates by appealing to the wave–particle hypothesis which had been suggested by Louis deBroglie in the PhD thesis he had submitted to the Sorbonne in November 1924. Thereafter Einstein turned his attention to other related matters.[5]

IV. BOSE IN EUROPE, AND AFTERWARDS

Although S. N. Bose's dream of working with Einstein never materialized, he did come into contact with several of the luminaries of European physics. He had felt it would be important for him to devote some portion of his two year study leave to experimental work. After all, he had hardly touched a piece of apparatus during his college days, or thereafter. Langevin suggested he discuss the possibility of working at Madame Curie's laboratory with the great lady herself. Bose recalls the interview with amusement. She conducted it in English and would scarcely let him say a word. Instead, she told him of another Indian student who had worked with her and had had considerable difficulty because he had not taken the trouble to learn proper French. Thereupon she suggested to Bose that he concentrate on the language for six months and then come back to her. In fact, he was given no opportunity to inform her that he had been studying French for fifteen years! Thus ended his idea of doing experiments in radioactivity.

Through Langevin, Bose also came into contact with the deBroglie brothers—Maurice, who was then doing excellent x-ray crystallography work, and the younger Louis, who had just submitted his revolutionary thesis to the Sorbonne. Bose was once a guest at the deBroglie ancestral estate —"even some of the bathrooms" had been turned into Maurice's private laboratory with apparatus superior to that existing in most European university laboratories. The superb, four-phase generator used to power the x-ray tubes was housed in a stable![19]

Under Maurice's direction Bose did learn something of x-ray crystallography and began to become interested in the theoretical properties of crystals. He feels that these experiences permitted him to be of considerable service at Dacca where K. S. Krishnan and his students were beginning to carry out some significant experiments on crystal magnetism.

A year after he arrived in Europe Bose left France, took up residence in Berlin, and on 8 October 1925 wrote a one-line message to Einstein stating that he would welcome an appointment.[1] Unfortunately, Einstein was then at Leiden with Ehrenfest and did not return for several weeks. When the meeting Bose had dreamed about finally did take place it must have been somewhat anticlimactic. As a result of that meeting Bose obtained a letter granting him certain privileges enjoyed by students at Berlin, including the right to take books out of the university library! (He claims that he made good use of that right.) A high point of his year in Germany came in the early summer of 1926 when he went to Göttingen and heard Max Born lecture on the promising new matrix mechanics of his protege, Werner Heisenberg.

In the late summer of 1926 S. N. Bose returned to Dacca, and a year later was appointed professor of physics, a post he held until 1945 when he returned to Calcutta University as Khaira Professor of Physics. During his years at Dacca he lectured, guided, and interacted with colleagues and students, and delved into Indian and Western philosophy and literature. He also became increasingly involved in both the growing Indian independence and the Bengali renaissance movements. Bose visited France for the second time in 1951, and for the next several years traveled extensively. But he never had any desire to see Germany again. The memories of 1926 when young German girls confided that their sons would revenge the "betrayal of 1918" remains too strong for him. Why had he never visited the United States? "Your Senator [Joseph] McCarthy objected to the fact that I had seen Russia first."

Although Bose has retired from active teaching he continues to live in Calcutta as one of the last survivors of the group of bright young men that Sir Asutosh brought to his new College of Science almost sixty years ago. With what might be regarded as poetic logic his interests during the years preceeding his retirement again flowed in the same channel as Einstein's, for of the six papers Bose published between 1953 and his retirement, five deal with Einstein's unified field theory. In fact, the final extant communication between the two men is a two page German draft of Einstein's dated 22 October 1953, which comments on a preprint of Bose's[20] and thanks him for the "friendly letter" which had accompanied it.

V. COULD BOSE HAVE EXPLOITED HIS WORK AS EINSTEIN DID?

S. N. Bose is revered by Indian scientists and feels his life has been a full one. Nevertheless, one might have hoped that the initiator of quantum statistics would have become a physicist of international as well as national renown. After all, he arrived in Europe in the fall of 1924 bearing Einstein's endorsement and was in contact with deBroglie and Langevin. One might ask why he failed to become one of the lesser luminaries of the new physics?

In mid-1924, when Bose arrived in Europe, theoretical physics was poised on the threshold of the most critical and exciting five year period of the century. The pace was rapid and accelerating as witnessed, for example, by the fact that only seven months elapsed from Einstein's receipt of Bose's first paper to the time when he had satisfied himself that he understood the quantum gas as well as he cared to. Bose was probably not prepared emotionally for the pace of these developments. By his own account he had pondered the Planck law for a long while before arriving at the derivation that had so impressed Einstein. Furthermore, even more than forty years later one still has the impression that the young Bose was terribly intimidated by most Europeans. For example, although he was in Paris with Langevin while the latter was communicating with Einstein on deBroglie's thesis, there is no evidence that Bose ever tried to impress upon Langevin his dream of working with Einstein. Presumably Madame Curie would have accepted him as a research assistant had he been able to convince her that he knew sufficient French. Yet he was either too polite or too frightened to interrupt her English monolog by replying in French and thus (perhaps) convincing her. Finally —perhaps most tellingly—Bose remarks that Franck was his favorite German physicist because his complexion was dark!

When one recalls the circumstances under which Bose grew up it is not particularly difficult to understand why he felt as he did. Whether it was a conscious policy or not, the nature of British rule in India during the early years of the present century had the effect of making the subject people believe that they really *were* inferior. This was certainly true in academic circles. Virtually all university and college administrative offices and teaching posts were filled by Englishmen and Scotsmen. As Bose was ultimately to learn, many of these teachers were third rate at best. For several years the obviously gifted Raman had been unable to obtain an academic post and had been forced to carry out his early research during the time he could spare from his duties as an employee of the British government. If Sir Asutosh's appointment as vice chancellor of a prestigious university such as Calcutta raised several British eyebrows, the fact that he succeeded in reserving a number of first-rate academic positions for Indians must have been considered revolutionary.

As already noted, S. N. Bose and his classmate M. N. Saha were among the first to benefit from the endowments secured by Sir Asutosh. During the next few years additional appointments (such as those held by Bose at Dacca) were opened to qualified Indians. Yet British dominance continued to remain strong in academic circles, and even certification of his work by the world's greatest scientist might have been insufficient to erase from the mind of a sensitive, retiring man like Bose the impression that he was indeed inferior to any European scientist.

Even if Bose had considered himself the intellectual equal of his European colleagues, it is still not at all clear that he would have proceeded to exploit his first paper in the same manner that Einstein did. Instances in which a now generally accepted idea was at first either ignored or rejected by most practicing physicists are too well known to bear repetition here. Instances in which the originator of a unique idea failed to see its full implications himself are less well documented, but probably almost as common. Bose's idea seems to fall into this latter category. He had pondered the Planck law for years and had ultimately arrived at a unique derivation that made no reference to classical electrodynamics. Having succeeded where many famous men had tried and failed it was quite natural for him to turn his attention to the application of his method to more involved thermal radiation problems. It was only clear in retrospect that Einstein's generalization of his method to ideal gases was to be a more fruitful path. Indeed, the properties of Einstein's quantum gas seemed so peculiar and unanticipated that at one point Ehrenfest warned Einstein to proceed with caution.[5] Bose's initial impressions of Einstein's papers on the ideal quantum gas are unrecorded. However, since Bose's major interests had been related to the statistical interpretation of experimental data it is conceivable that he may have regarded this exotic "Bose–Einstein" gas as amusing, though not terribly important. More than forty years later Bose noted that he had always tried to give everything a statistical form, and perhaps had often become interested in the mathematical arguments at the expense of the physics.

If, as seems reasonable, Einstein understood the physical implications of Bose's work better than Bose himself, then Bose was in good company. As already noted, Einstein's second paper on the ideal quantum gas[18] used deBroglie's two month old wave–particle hypothesis to rationalize Bose's counting method. Yet deBroglie himself, who was close to Bose in Paris, had failed to perceive the vital connection!

VI. A PERSONAL NOTE

As an American physicist interested in the evolution of physics in India, I sought and ultimately obtained a private interview with the legendary S. N. Bose. My opportunity came on 25 March 1971, shortly after Bose had delivered the annual Krishnan Memorial Lecture at the National Physical Laboratory in New Delhi.[9] By prior arrangement my good friend Dr. A. R. Verma, the laboratory director, introduced me to Bose after the lecture. I suggested we meet the following day. But Bose was in the mood to talk. He waved aside the many anxious suggestions that he needed rest, led me to his car, and thence to his New Delhi residence. For two hours he sat yoga-fashion on the charpoy answering my questions, with impatience when he found me a bit too impolite, more often with laughter at my naivety, and frequently with long monologs recalling Europe in the mid-twenties and Dacca in the thirties and early forties.[21]

I had hoped that my interview might unearth some material of real objective value to the history of quantum mechanics. It failed to do so, for Bose saved few papers and confessed that his memory for details was failing. But I did have an opportunity to spend two hours in the presence of a unique human being and as a result was led to ponder the basic question: What is the meaning of a successful career? For it is difficult to judge S. N. Bose by the standard criteria which both Western and Westernized Indian physicists apply to their peers.

In 1964 when Calcutta University celebrated his 70th birthday, Bose was asked for his scientific bibliography. The list, covering the period 1918–1955, contains a mere 25 entries—seven of them

* Present address: Jefferson Physical Laboratory, Harvard University, Cambridge, Massachusetts 02138.

[1] I am grateful to Miss Helen Dukas, custodian of the Einstein papers at the Institute for Advanced Study in Princeton, New Jersey, for providing me with copies of the private letters from S. N. Bose contained in the collection.

[2] S. N. Bose, Z. Physik **26,** 178 (1924).

[3] P. Debye, Ann. Physik **33,** 1427 (1910).

[4] A. Einstein, Ann. Physik **17,** 132 (1905).

[5] Most of these comments on Einstein's work are taken from M. J. Klein, Natural Philosopher **3,** 26 (1963). For a critical review on the development of both the black body radiation problem and the photon concept see M. Jammer, *The Conceptual Development of Quantum Mechanics* (McGraw–Hill, New York, 1966), pp. 10–56.

[6] A. H. Compton, Phys. Rev. **21,** 483 (1923).

[7] All unreferenced comments of S. N. Bose as well as most unreferenced statements concerning his career are based on a private interview held on 25 March 1971 (see Sec. VI). Since Professor Bose has neither seen the present author's notes on the interview nor read any draft of this paper, responsibility for all such comments and statements rests with the author.

[8] Two examples will serve to illustrate Bose's loyalty to Bengal. First, Professor J. Mahanty of the Indian Institute of Technology in Kanpur, who attended Bose's lectures at Calcutta in the early 1950's, has informed the author that those lectures were frequently in Bengali rather than in English or Hindi. Second, in May 1971, during the early weeks of the rebellion which ultimately led to the creation of Bangladesh, Bose stated publically that the people of East Bengal were beginning to act out the ideas which he and so many others had talked about in Dacca during his years as a professor there.

[9] S. N. Bose, *Krishnan Memorial Lecture,* delivered at the National Physical Laboratory, New Delhi, 25 March 1971 (unpublished).

[10] S. N. Bose and M. N. Saha, Phil. Mag. **36,** 199 (1918).

[11] S. N. Bose and M. N. Saha, Phil. Mag. **39,** 456 (1920).

[12] S. N. Bose and M. N. Saha, *The Principle of Relativity* (Calcutta U. P., Calcutta, Ind., 1919).

[13] Most of these notes on Saha are taken from his obituary by D. S. Kothari published in the Biograph. Mem. Fellows Roy. Soc. **5,** 217 (1960).

[14] S. N. Bose, Phil. Mag. **40,** 619 (1920).

[15] S. N. Bose, Z. Physik **27,** 384 (1924).

[16] A. Einstein, Berl. Ber. 261 (1924).

[17] M. J. Klein, in Ref. 5, p. 31.

[18] A. Einstein, Berl. Ber. 3 (1925).

[19] In December 1923 Maurice deBroglie came to the United States for the express purpose of presenting x-ray scattering results at a meeting of the American Physical Society. He did so in order to lend support to A. H. Compton whose results and their interpretation had been bitterly attacked by William Duane of Harvard. (Jammer, Ref. 5, p. 162, and R. Shankland, private communication.)

[20] S. N. Bose, Compt. Rend. **236,** 1333 (1953).

[21] Ironically, our relaxed conversation took place only a few hours before Pakistani troops initiated their attempt to suppress the liberation movement in East Bengal. In so doing they virtually destroyed the University at Dacca where Bose had spent his most productive years.

[22] *Satyendranath Bose 70th Birthday Commemoration Volume* (Calcutta U. P., Calcutta, Ind., 1964).

[23] Santimay Chatterjee (private communication).

[24] Bose's ability to deliver polished lectures from memory was considered exceptional, even among Indians who take great pride in their ability in that art. He trained himself to do so at an early age in part, perhaps, because his poor eyesight made it difficult for him to consult notes. In answer to a direct question he confessed that he had been very proud of his memory and was increasingly distressed by indications that it was failing. For this reason he requested that I not tape record our conversation.

[25] S. N. Bose, Bull. Calcutta Math. Soc. **37,** 51 (1945).

Einstein and Bose

John Stachel

Introduction

Abraham Pais calls Bose's derivation of Planck's law "the fourth and last of the revolutionary papers of the old quantum theory (the other three being by, respectively Planck, Einstein and Bohr)" (Pais 1982, p. 425). This is high praise indeed, when one considers that: 1) The 1900 paper by Planck introduced the concept of the quantum of action and derived the law for the equilibrium spectrum of electromagnetic radiation that bears his name; 2) the 1905 paper by Einstein introduced the concept of the quantized structure of that radiation, and derived from it (among other things) the laws of the photoelectric effect; 3) the 1913 paper by Bohr introduced the concept of the quantized energy-level structure of the atom, and derived from it the atomic spectrum of hydrogen. Yet Pais's praise is surely deserved, for as he says "Bose's arguments divest Planck's law of all supererogatory elements of electromagnetic theory and base its derivation on the bare essentials" (*ibid*). In this paper, I shall review Bose's work and his contacts with Einstein, both epistolary and personal. But I will devote equal time to Einstein's reception of Bose's work. I shall discuss how and why Einstein was primed, so to speak, to immediately recognize the significance of Bose's approach and exploit its implications for a quantum theory of the ideal gas, as well as to recognize the limitations of that approach and the need for a deeper insight into the nature of things than the old quantum theory could provide. Finally I shall briefly indicate how the new quantum mechanics provides such a deeper insight, but at a price—quantum entanglement—that Einstein, for one, was unwilling to pay.

Bose Derives Planck's Formula from the Photon Model

Bose's oral reminiscences, recorded by Jagdish Mehra (Mehra 1975, cited here from Mehra and Rechenberg 1982), together with the documentary record consisting of Bose's published papers and his correspondence with Einstein, allow reconstruction of at least the outlines of the path Bose took to his derivation of Planck's formula.

Satyendra Nath Bose must be seen as part of a generation of Bengali students, for whom patriotism and personal achievement were inseparable. He was one of the

> particularly brilliant lot—the famous 1909 batch of Presidency College [Calcutta, which] in all its hallowed history has not seen the likes of this batch since. These inspired youngsters were all destined to contribute to Indian Science, but perhaps their greatest contribution was a growing sense of national pride" (Chatterjee 1994).[1]

After getting his M.Sc. in 1915, Bose continued his study of mathematics and physics at University College, just founded in 1914 as part of the University of Calcutta, and began his teaching career there in 1917 as a lecturer in physics (Venkataraman 1992, pp. 3–9). While his close friend, fellow student, and then fellow lecturer Megh Nad Saha concentrated on thermodynamics, statistical mechanics and spectroscopy, Bose initially "decided upon electromagnetism and relativity." In the aftermath of the stir that followed the announcement in November 1919 of the results of the Eddington eclipse expeditions, Bose and Saha published the first English translation of a collection of original papers on relativity, Bose translating Einstein's 1916 review of the general theory. After an English edition of *The Principle of Relativity* was published, its publisher made an attempt to suppress the similar Indian edition, but as Bose recalled: "Einstein was very generous. He said that so long as the book remained in circulation inside India, he had no objection."

Though initially occupied with relativity, Bose did not neglect statistical mechanics. He recalled studying such books as Planck's *Vorlesungen über die Theorie der Wärmestrahlung* [*Lectures on the Theory of Thermal Radiation*] and *Vorlesungen über Thermodynamik* [*Lectures on Thermodynamics*], Boltzmann's *Vorlesungen über Gastheorie* [*Lectures on Gas Theory*], and Gibbs's *Elementary Principles in Statistical Mechanics*, as well as numerous papers on the topic. In 1918 he published his first paper, in collaboration with Saha, on a topic in statistical mechanics. He also read extensively in the literature of the then rather new quantum theory, to which we now refer as the *old* quantum theory. In addition to Planck's works, his known readings include papers by Einstein, Bohr, and Sommerfeld, and he soon published on that topic.

In 1921 Bose moved to the newly-established Dacca University, where he began to guide students in experimental work, as well as continuing his study and teaching of theoretical physics. He advised his students: "Never accept an idea as long as you are yourself not satisfied with its consistency and the logical structure upon which the concepts are based. Study the masters. These are the people who have made significant contributions to the subject. Lesser authorities clev-

E and $E + dE$, multiplying the result by two to take into account the photon's two possible independent states of polarization. "The rest of the calculation," far from being "essentially the same" as Debye's derivation, is based on the substitution of energy quanta for quanta of energy. This raises deep problems, to which I shall return in the third section; but first let us turn to Bose's contacts with Einstein.

Bose and Einstein

There is a conflict between two accounts, both based on discussions with Bose, of the sequence of events that led Bose to send the manuscript of his paper on Planck's law to Einstein. They differ by a half-year on when it was written, and on where it was first sent. According to Blanpied's account, based on an interview with Bose in 1971, "late in 1923 [Bose] submitted a paper on the subject to the *Philosophical Magazine.* Six months later the editors of the journal informed him that (regrettably) the referee's report on his paper was negative. Undeterred, he sent the rejected manuscript to Einstein..." (Blanpied 1972, p. 1215). According to Mehra and Rechenberg, it was only after a visit by Saha (who had moved to Allahabad shortly after Bose left for Dacca) in March 1924 that Bose finished his work: "In the beginning of June 1924 he had the manuscript of a paper ready, which he entitled 'Planck's Law and the Light Quantum Hypothesis.' Since he had the impression that his treatment very much followed Einstein's thought—indeed, it completed the proof which Einstein had been searching for many years—Bose decided to send his paper to Einstein..." (Mehra and Rechenberg 1982, p. 565).

I believe Blanpied's story is correct,[2] and that Mehra and Rechenberg's account actually concerns the circumstances in which Bose composed his *second* paper on the topic of thermal radiation, "Wärmegleichgewicht im Strahlungsfeld bei Anwesenheit von Materie" ["Thermal Equilibrium of the Radiation Field in the Presence of Matter"], also translated and sent to the *Zeitschrift für Physik* by Einstein. It refers to the first paper as "to appear in Phil. Mag." (Bose 1924b, p. 386, note 1), which is hard to square with the story Mehra and Rechenberg tell about the first paper, while its contents fit well with the assumption that the story applies to the second paper.

After finishing his first paper late in 1923 and sending it to *The Philosophical Magazine*, Bose continued to work on the radiation problem, turning his attention to the interaction between radiation and (ordinary) matter. In March 1924 Saha visited Bose in Dacca, and directed Bose's attention to recent papers by Pauli (1923) and by Einstein and Ehrenfest (1923), both of which attempted to generalize Einstein's famous 1917 derivation of Planck's law; and to Compton's 1923 explanation of the anomalous scattering of X-rays, which was based on the photon picture. Bose then prepared his second paper, which must have been completed before 4 June 1924, the date of his first letter to Einstein. By then he presumably had received notice of rejection by the *Philosophical Magazine* (if the second paper had not been finished before he received the rejection, it is inexplicable why it should refer to the first as forthcoming in that journal). The letter reads:

Respected Sir,

I have ventured to send you the accompanying article for your perusal and opinion. I am anxious to know what you think of it. You will see that I have tried to deduce the coefficient $8\pi\nu^2/c^3$ in Planck's Law independent of the classical electrodynamics, only assuming [*that finite* (crossed out in letter)] that the ultimate elementary regions in the Phase-space has the content h^3. I do not know sufficient German to translate the paper. If you think the paper worth publication, I shall be grateful if you arrange for its publication in Zeitschrift für Physik. Though a complete stranger to you, I do not feel any hesitation in making such a request. Because we are all your pupils through your writings. I do not know whether you still remember that somebody from Calcutta asked your permission to translate your papers on Relativity in English. You acceded to the request. The book has since been published. I was the one who translated your paper on Generalised Relativity.

Yours faithfully,
SN Bose
(Einstein Archive, Doc. 6-127).

Only eleven days later, on 15 June, Bose sent Einstein his second paper, with the following letter:

Respected Master,

I send herewith another paper of mine for your kind perusal and opinion. I hope my first paper has reached your hand. The result to which I have arrived seems rather important (to me at any rate). You will see that I have dealt with the problem of thermal equilibrium between Radiation and Matter in a different way, and have arrived at a different law for the probability for the elementary processes [presumably, different from the law proposed in Einstein and Ehrenfest 1923], which seems to have simplicity in its favor. I have ventured to send you the typewritten paper in English; it being beyond me to express myself in German (which will be intelligible to you). I shall be glad [*to* (crossed out in letter)] if its publication in Zeitschrift für Physik or any other German journal can be managed. I myself know not how to manage it. In any case, I shall be grateful if you express any opinion on the papers, and send it to me at the above address.

Yours truly
SN Bose
(Einstein Archive, Doc. 6-128).

On 2 July Einstein sent a postcard in response to Bose's first paper:

Dear Colleague:

I have translated your paper and given it to the *Zeitschrift für Physik* for publication. It signifies an important step forward and pleases me very much. However, I do not find your objection to my paper correct. For Wien's displacement law does not presuppose the wave theory and Bohr's correspondence principle is not used. But this is unimportant. You have derived the first factor quantum-theoretically even if not quite rigorously on account of the polarization factor 2. It is a beautiful step forward.

With friendly greetings   your A. Einstein.
(copy in another hand, Dacca University Archives).[3]

Einstein said as much in a note he appended to his translation of the article: "Bose's derivation of Planck's law appears to me to be an important step forward.

The method used here also yields the quantum theory of the ideal gas, as I shall show elsewhere" (Bose 1924a, p. 181).

With the help of this postcard Bose was able to get quick approval of a study leave abroad, for which he had applied some time earlier.

> As soon as [the Senate Council] showed it to Hartog [the Vice-Chancellor], it solved all problems. As a student Hartog had spent some time at the University of Paris and he understood something of what such an experience could do for a young man. That little thing [the postcard from Einstein] gave me a sort of passport to the study leave. They gave me leave for two years and rather generous terms. ... Then I also got a visa from the German consulate just by showing them Einstein's card. They did not require me to pay the fee for the visa! (Mehra and Rechenberg 1982, p. 569).

Bose sailed for Europe in September 1924, going first to Paris, where he stayed for about a year. According to Mehra and Rechenberg, "Bose had not gone to Berlin straight away from Dacca because he felt that his knowledge of German was not good enough" (Mehra and Rechenberg 1982, p. 569). According to a physics student with whom he became friendly, he spoke excellent French by the time he left Paris; while there, he attended Paul Langevin's lectures at the Collège de France, but otherwise did not work too hard at physics (interview with Mme. Jacqueline Eisenmann, Paris, 21 May 1994).[4]

Soon after reaching Paris, he contacted Einstein:

> Dear Master,
>
> My heart-felt gratitude for taking the trouble of translating the paper yourself and publishing it. I just saw it in print before I left India. I have also sent you about the middle of June, a second paper entitled "Thermal Equilibrium in the Radiation field in the presence of matter".
>
> I am rather anxious to know your opinion about it [*before* (crossed out in letter)] as I think it to be rather important. I don't know whether it will be possible also to have this paper published in Zeit. für Physik. I have been granted study leave by my University for 2 years.
>
> I have [*just* (crossed out in letter)] arrived just a week ago in Paris. I don't know whether it will be possible for me to work with you in Germany. [*If* (crossed out in letter)] I shall be glad however if you grant me the permission to work under you, for it will mean for me the realisation of a long-cherished hope.
>
> I shall wait for your decision as well as your opinion of my second paper here in Paris.
>
> If the second paper has not reached you by any chance. Please let me know, I shall send you the copy I have with me.
>
> With respect
> yours sincerely
> SN Bose (Prof.)
> (26 October 1924, Einstein Archive, Doc. 6-129).

Einstein replied on 3 November, assuring Bose that his second paper was already in print:

> Most esteemed colleague!
>
> Friendly thanks for your letter of 26.10. I am happy to have the opportunity soon to meet you personally. Your papers have appeared already some time ago: unfortu-

nately the offprints were sent to me instead of you. You can have them at any time. I am not in agreement with your elementary law of probability for the interaction between radiation and matter, and have given the reasons in a note that appeared together with your article. To wit, your law is not compatible with the following two conditions:

1. The coefficient of absorption is independent of the density of the radiation.

2. The behavior of an oscillator in a radiation field must result from the statistical laws as a limiting case.

We can discuss this in more detail when you come here.

With friendly greetings

(retained carbon copy, Einstein Archive, Doc. 6-130).

On 27 January 1925 Bose replied from Paris:

Revered Master,

I received your kind note of 3rd Nov. in which you mentioned your objections against the elementary law of Probability. I have been thinking about your objections all along, and so did not answer immediately. It seems to me that there is a way out of the difficulty, and I have written my ideas in the form of a paper which I [*enclose* (crossed out in letter)] send under a separate cover.

Bose never published this third paper, and no copy of the manuscript has been found. As we shall see in a moment, Einstein was unmoved by Bose's new paper, and it is possible that when they met in Berlin he convinced Bose, but later remarks by Bose suggest that he still felt the paper contained important new results. Returning to the letter, it concludes:

I cannot exactly express how grateful I feel for your encouragement, and the interest you have taken in my papers. Your first postcard came at a critical moment, and it has more than any other made this sojourn to Europe possible for me. I am thinking of going to Berlin at the end of the winter, where I hope to have your inestimable help and guidance.

Yours sincerely

SN Bose

(27 January 1925, Einstein Archive, Doc. 6-131).

By the following October, Bose was in Berlin:

Dear Sir,

I have just arrived in Berlin, and am very anxious to see you. I shall be glad if you kindly write when it will be convenient for you.

Yours sincerely

S.N. Bose

(8 October, 1925, Einstein Archive, Doc. 6-132).

Here he met Einstein, and during the next year was plunged into one of the most exciting periods in physics. The following May, he wrote a friend:

Everybody (every physicist) seems to be quite excited in Berlin, about the way things have been going on with Physics, first on the 28th [of April] last Heisenberg spoke in the Colloquium about his theory, then, in the last Colloquium, there was a long lecture on the recent hypothesis of the spinning electron, (perhaps you have heard of it.) Everybody is quite bewildered, and then there is going to be very soon a discussion

> of Schrödinger's papers. Einstein seems quite excited about it; the other day coming from the Colloquium, we suddenly found him, jumping in the same compartment, where we were, and forthwith he began to talk excitedly about the things we have just heard. He has to admit that it seems a tremendous thing, considering the lot of things which these new theories correlate and explain, but he is very much troubled by the unreasonableness of it all. We were all silent, but he talked almost all the time; unconscious of the interest [*everybody is taking* (crossed out)] and wonder that he is exciting in the mind of the other passengers (Bose to Jacqueline Zadoc-Kahn, May 1926).

According to Mehra and Rechenberg: "Einstein introduced Bose to a number of people in Berlin, including Fritz Haber, Otto Hahn, Lise Meitner and Walter Gordon. He also proposed two problems to Bose to work on: first, the question of whether the new statistics implied a novel type of interaction between the light-quanta" (Mehra and Rechenberg 1982, p. 571). The second problem was described by Bose himself: "He wanted me to try to see what the statistics of light-quanta and the transition probabilities of radiation would look like in the new theory [i.e., quantum mechanics]" (cited from Venkataraman 1992, p. 120). As we shall see in the next section, these two questions go to the heart of Einstein's concerns about the interpretation of Bose's work.

Apparently Bose made no progress on either problem. In May 1926 he wrote:

> I have made an honest resolution of working hard during these months, but it is so hard to begin, when once you have given up the habit (letter cited above).

But the reasons for his failure probably lie deeper. Like many contemporaries who had mastered the old quantum theory, he seems never to have been comfortable with the new quantum mechanics.

A two-page note in Bose's hand, on "Fluctuations in density," in the Einstein Archive, which seems related to Einstein's second paper (see the next section), may be from Bose's stay in Berlin. But Bose published nothing during his stay in Europe, returning to Dacca in the late summer of 1926. A year later he was appointed Professor of Physics, supported by a letter from Einstein (Mehra and Rechenberg 1982, p. 571).

There seems to have been no further contact between the two until 1945, by which time Einstein had moved to Princeton and Bose to University College of Science in Calcutta, when Bose sent Einstein a paper "by an esteemed friend of mine Principal B. M. Sen" of Presidency College in Calcutta. He ended his letter on a personal note:

> As one who owes much to you and your guidance in life, I have always been anxious to have news of yourself, especially under the new conditions when you are in a new country. Some of my friends who are more lucky than me, had chances to visit the States recently, and had the good fortune of seeing you in your new house. I am glad to know that you are well, and still take a lot of interest in India and Indians (9 December 1945, Einstein Archive, Doc. 6-134).

During the 1950's Bose worked on Einstein's last unified field theory. A copy of a manuscript on this topic and drafts of two replies by Einstein are in the Einstein Archive.

Einstein Ponders Bose's Results

Now let us return to 1924 and look at matters from Einstein's point of view. As noted above, Bose's first letter to Einstein is dated 4 June 1924. Mehra and Rechenberg estimate that mail from Dacca to Berlin probably took two to three weeks at the time, so that Einstein received the letter in late June. By 2 July, when he sent his postcard to Bose, he had submitted his translation of Bose's paper to the *Zeitschrift für Physik*. Einstein's first paper based on Bose's work, "On the Quantum Theory of the Monatomic Ideal Gas" (Einstein 1924a), was presented to the Prussian Academy of Sciences at its session of 10 July. Thus, it took Einstein no more than a week or two to digest and translate Bose's work, and conceive and complete his own extension of it to the quantum ideal gas. Clearly, Bose's work triggered a creative outburst in a mind already well primed to respond to it. Let us try to understand why Einstein was so well prepared.

In his second paper on the subject, "On the Quantum Theory of the Monatomic Ideal Gas. Second Paper" (Einstein 1924b), Einstein notes that his theory "is based on the hypothesis of a far-reaching formal relationship between radiation and gas. According to this theory, the degenerate gas deviates from the gas of [classical-JS] statistical mechanics in a way that is analogous to that in which radiation obeying Planck's law deviates from radiation obeying Wien's law. If Bose's derivation of Planck's radiation formula is taken seriously, then this theory of the ideal gas is inevitable; for if it is justified to regard radiation as a quantum gas, then the analogy between quantum gas and gas of molecules must be complete" (Einstein 1924b, p. 3). In his third paper on the subject, "On the Theory of the Ideal Gas" (Einstein 1925), he reiterates this point: "This theory seems justified if one starts with the conviction that (apart from its property of being polarized) a light quantum differs essentially from a monatomic molecule only in that its rest mass is vanishingly small"—but hastens to add that this analogy is by no means sanctioned by all researchers (he may well have had Bohr—a well known opponent of the photon concept—in mind).

He also notes a second objection: "[T]he statistical method applied by Bose and myself is by no means indubitable, but on the contrary only appears to be justified *a posteriori* by its success in the case of radiation" (Einstein 1925, p. 18). Einstein thus highlighted two issues: 1) The analogy between radiation and matter (calculation of the first factor), and 2) the method of counting equiprobable states employed in deriving the probabilities W for use in Boltzmann's principle, $S = k \ln W$ (calculation of the second factor). By looking at the earlier history of these two issues in Einstein's work, we may hope to understand his ability to accept and apply Bose's method so quickly, as well as his abiding conviction that much still remained obscure about the reasons for its success.

Einstein's speculations on the possible analogy between matter and radiation can be documented as early as 1901, when he wrote: "Recently the idea came to me that in the production of light perhaps a direct transformation of motional energy into light takes place due to the parallelism kinetic energy of molecules—absolute temperature—spectrum (spatial radiant energy in the state of equilibrium" (Einstein to Marić, 30 April 1901, Doc. 101, Einstein 1987, p. 293). Since he was convinced early on of the correctness of the atomic model for matter, the analogy suggests the need for an atomic model of radiation.

In 1905 he developed this idea in his light quantum paper, "On a Heuristic Point of View Concerning the Production and Transformation of Light" (reproduced as Doc. 14, Einstein 1989, pp. 149–169). In it, he shows that:

> Monochromatic radiation of low density (within the range of validity of Wien's radiation formula) [high-frequency, low-temperature limit of Planck's: $\rho(\nu, T) = 8\pi\nu^2(h\nu)/\{c^3\exp(-h\nu/kT)\}$], behaves thermodynamically as if it consisted of energy quanta that are independent of each other and of the magnitude, [$h\nu$, in modern notation] (ibid., p. 161).

He shows this by an ingenious inversion of Boltzmann's principle [which Einstein so named] $S = k\ln W$, a trick he often used thereafter. Instead of finding the entropy S from a known probability W, he read off the unknown probability from the entropy of radiation obeying Wien's law. This leads to the probability W for the total energy E of the radiation, at any given moment, to be concentrated in a subvolume v of the total volume V:

$$W = (v/V)^{E/h\nu}\,.$$

Now he introduces the analogy between radiation and matter. Consider a gas in equilibrium consisting of n particles in a volume V, each moving according to any law of motion, so long as it does not single out a preferred point or direction in V. Then the probability w that any one particle will be in the subregion v is clearly: $w = (v/V)$. If the particles are *statistically independent* of each other (as classical particles in an ideal gas in equilibrium always are), then the probability that all n particles are in v at a given moment is clearly the product of the probabilities of n statistically independent events:

$$W = w^n = (v/V)^n.$$

If we take the analogy between radiation and matter seriously, then comparison of the two formulae yields Einstein's conclusion: Radiation in the Wien region behaves like a gas of independent particles of energy $E = h\nu$.

Given these considerations, one might wonder why Einstein did not immediately apply a similar argument to Planck's law. Why did he not use the well-known form of the entropy of black body radiation to calculate the probability of a certain distribution of light quanta in phase space—in other words, why did Einstein not derive Bose statistics? I believe he probably did make some such attempt, but failed. Why publish what he calls "a *heuristic* point of view," based on the Wien limit of Planck's law, without first trying—and failing—to account for the entire spectrum covered by Planck's law? Such a failed attempt would explain an otherwise curious remark near the beginning of Einstein's next (1906) paper on radiation, "On the Theory of Light Production and Absorption" (reproduced as Doc. 34, Einstein 1989, pp. 350–358). After briefly summarizing his light quantum hypothesis in the first paragraph, Einstein goes on: "At the time [of his 1905 paper] it appeared to me as if Planck's theory of radiation in certain respects was counterposed to my work." That is, he originally believed that Planck's law did *not* support the light quantum hypothesis; now he was able to give an argument showing "that Planck's theory makes implicit use of the above-mentioned light quantum hypothesis" (ibid., p. 350).

If there was such a failed attempt to understand Planck's law in 1905, it could only have reinforced several quite independent reasons why Einstein did not pursue a "quantum gas" model of black-body radiation around 1905:

1) Aside from energy, he had no warrant to attribute other particle-like properties to light quanta. Only between 1909 and 1916 did Einstein develop arguments that finally led him to attribute momentum and spatial localization to them.

2) He was already well aware that, whatever they might be, light quanta could *not* be statistically independent of each other, as "normal" particles should be. Inverting the argument of his 1905 paper: if light quanta *were* independent, a light quantum gas would obey *Wien's* law and not Planck's.

He made both these points explicitly in 1909:

> Indeed, I am not at all of the opinion that one should think of light as composed of quanta localized in relatively small spaces and independent of each other. This would indeed be the most convenient explanation of the Wien end of the radiation formula. But, by itself, the division of a light ray at the surface of a refracting medium completely forbids this conception. A light ray can divide, but a light quantum cannot divide without change of frequency (Einstein to Lorentz, 23 May 1909, Doc. 163, Einstein 1993, p. 193).

3) If light quanta are *not* statistically independent, even in an ideal radiation gas, Einstein had no idea how to calculate the probabilities for an ensemble of light quanta to be distributed over various states. Einstein had strong views about the nature of probability: The time-ensemble definition of the probability of the state of a physical system was primary for him. Any *combinatorial* argument, based on counting the number of complexions of the elements composing the system that correspond to a given physical state, must be supplemented by a *dynamical* argument proving that these complexions are equally probable before such a count could be used to define the W needed in Boltzmann's principle.[5]

Planck's original derivation of the black-body spectrum applied such combinatorial considerations to the energy of an ensemble of material oscillators in thermal equilibrium with the radiation field, just assuming all complexions to be equiprobable; and a decade later Debye applied similar considerations to the radiation field itself, considered as an ensemble of normal modes in the cavity. In both cases, the problem was to divide the fixed total energy into quanta of magnitude $h\nu$ and distribute them among the material oscillators (Planck) or the normal modes of the radiation field (Debye). Complexion here is just the name for one such way of distributing the total energy. The counting of complexions is based on the assumption that, while the material oscillators or the normal modes of the radiation field are *distinguishable*, the quanta of energy of a given frequency, being just aliquot portions of the total energy at that frequency, are *indistinguishable*. If there are seven quanta of energy associated with one oscillator and five with another, it makes no sense to ask *which* seven and *which* five quanta of energy are involved.

Planck and Debye just *assumed* that the number of such complexions is a measure of the probability of a state, that is, that each complexion is equally probable. And here Einstein's criticism applies. They could offer no dynamical argument why such complexions should be equiprobable in *his* (time ensemble) sense of probability. Rather, their procedure was justified *a posteriori* by its success in producing Planck's formula. An attempt to turn this argument, based on quanta of energy, into an argument based on energy quanta, i.e., on a light-quantum gas, faces two problems. How to justify the count of the number of possible states of the gas corresponding to a given energy (or frequency)—which is the problem of the first factor; and how to justify the counting of equal probabilities—the problem of the second factor. If the first problem is solved, the second only becomes more acute. For, in addition to Einstein's *general* objection to the counting of complexions, a *new* objection arises when such a count of complexions in the Planck–Debye sense is interpreted as a count of complexions for *particles*. While elements of the total energy may be indistinguishable, in the sense that it makes no sense to ask *which* quanta of that energy are in a given state, how can one avoid this question for energy quanta, i.e., particles? Classically, one *cannot*, and it must have seemed by the nature of the particle concept that such a question makes sense. No wonder that even Einstein paused before entering such a statistical morass.

But Bose did so, with a courage perhaps born in part from an incomplete awareness of the perils[6]—the very reason that young men are supposed to make the best soldiers. (More seriously, it is often the case that knowing too much inhibits imaginative thinking.) As we have seen above, he solved the first problem—explaining the first factor in Planck's formula—fairly straightforwardly, by dividing the phase space into cells of size h^3 and counting them. To solve the second problem, he turned the problem of partitioning *quanta of energy* among normal modes of the field into one of distributing *energy quanta*—that is photons, considered as particles—among cells in phase space; but continued to count with the *old* method to solve the *new* problem.[7] Now the cells, like the normal modes, are certainly independent of each other; but, if the old counting method is used for the new problem, then the photons *cannot* be statistically independent. For if they were, as Einstein had known since 1905, Boltzmann's standard counting method for independent particles, accepted by everyone, would lead to Wien's radiation formula rather than Planck's.

As we have seen, Einstein's reaction was to accept the force of Bose's method applied to radiation—after all, it did lead directly to Planck's law—and combine it with his belief in the analogy between radiation and matter: If one wants the quantum theory of an ideal (monatomic) gas, the same method must be applicable. The only differences between the cases of radiation and matter are: 1) the relation between energy and momentum for a massive particle must be used; 2) as well as the total energy, the number of particles must be kept constant in maximizing the entropy.

It must have been clear from the start to Einstein that the use of the new counting method for distributions of particles among cells in phase space amounts to the assumption of a certain kind of statistical dependence between them: If they were counted as if they were independent—that is, in the Boltzmann way—this would have lead to the classical ideal gas law.

I say it must have been clear to Einstein from the outset, although he did not make the point in his first paper. But when it was pointed out to him, he responded in his second paper in a way that suggests he regarded the point as obvious:

> Ehrenfest and other colleagues have reproved Bose's theory of radiation and my analogous one of the ideal gas because in this theory the quanta and the molecules are not treated as statistically independent entities without any special indication of this circumstance in our papers. This is completely correct. If one treats the quanta as statistically independent of each other in their localisation [in cells in phase space–JS], one arrives at Wien's radiation law; if one treats the gas molecules analogously, one arrives at the classical equation of state of the ideal gas, even if one proceeds otherwise exactly in the same way as Bose and I have done (Einstein 1925a, p. 5).

Einstein proceeds to show this in detail, but rather than quote the paper, I shall quote a contemporaneous letter to Schroedinger, which makes the same point more graphically and less technically.

> In the Bose statistics employed by me, the quanta or molecules are not treated as being *independent of one other*. ... A complexion is characterized through giving the number of molecules that are present in each individual cell. The number of complexions so defined should determine the entropy. According to this procedure, the molecules do not appear as being localized independently of one another, but rather they have a preference to sit together with another molecule in the same cell. One can easily picture this in the case of small numbers. [In particular] 2 quanta, 2 cells:

Bose-statistics	1st cell	2nd cell
1st case	o o	–
2nd case	o	o
3rd case	–	o o

independent molecules	1st cell	2nd cell
1st case	| ||	–
2nd case	|	||
3rd case	||	|
4th case	–	| ||

> According to Bose, the molecules stack together relatively more often [two out of three cases] than according to the hypothesis of the statistical independence of the molecules [two out of four cases] (Einstein Archives, Doc. 22 002, translation cited from Howard 1990, p. 67).

Indeed, this pecularity of Planck's counting was pointed out in Ehrenfest and Kammerlingh Onnes 1914 as an argument *against* light quanta (presumably known to Einstein, who was in close contact with Ehrenfest). Contrasting Planck's method of counting quanta of energy with Einstein's energy quantum hypothesis, they state:

> Planck does not deal with really mutually free quanta ε. The real *object which is counted* remains the number of all different distributions of N resonators [oscillators] over the energy grades [levels] 0, ε 2ε, ... with a given total energy $P\varepsilon$.

After contrasting the two counting methods for $P = 3$ and $N = 2$, setting up a table similar to Einstein's, they conclude:

> *Planck's formal device* (distribution of P elements of energy ε among N resonators) *cannot be interpreted in the sense of Einstein's light quanta* (Ehrenfest and Kammerlingh Onnes 1914, cited from Mehra and Rechenberg 1982, p. 560).

But apparently, until Einstein reacted to Bose's work, it never occurred to anyone to make a virtue of this "defect" by interpreting it as evidence for a new type of statistical *dependence* between light quanta, peculiar to quantum theory.

In short, what Bose called "essentially the same calculation" of the second factor in Planck's formula was actually a revolutionary act: in going from quanta of energy to energy quanta, he introduced a type of statistical dependence between the latter peculiar to quantum particles. This feature is often characterized as the *indistinguishability* or *identity* of particles. While these terms express certain features of the situation, as Howard 1990 points out there are other features that they do not. Using Einstein's example quoted above, while in the 1st and 3rd cases of Bose-statistics, the particles are indistinguishable, they are *not* identical (in at least one sense of the word); for if they were, there would be only *one* particle in the occupied cell. On the other hand, in the 2nd case they are certainly distinguishable (in at least one sense of the word): even if we cannot tell which is which (as we can in the classical 2nd and 3rd cases), one occupies a *different* cell of phase space from the other. What term, then, best describes this peculiar property of the Bose particles? Howard 1990 suggests *non-separability*, assimilating this property to that chacterizing any composite quantum system after interaction between its elements. Others, including myself, prefer the more general term *quantum entanglement* (but what's in a name?), a topic to which I shall return shortly.

From the beginning, Einstein saw the use of a new statistics in counting complexions as mysterious and disturbing. Two days after submitting his first paper on the quantum gas, he wrote to Ehrenfest that "the Indian Bose has given a beautiful derivation of Planck's law, including the constant ... but the essence remains obscure" (Einstein to Ehrenfest, 12 July 1924, Doc. 10 089). In his second paper, he notes that Bose's formula for the counting of complexions of equal probability

> thus expresses indirectly a certain hypothesis about how the molecules mutually influence each other that is temporarily of an enigmatical nature, which [hypothesis] determines just the equal statistical probability of the cases defined here as complexions (Einstein 1924b, p. 6).

In line with his earlier critique of the method of determining "probabilities in the sense of Planck" (as he almost always characterizes them), it was clear to Einstein that something more fundamental must lie behind Bose statistics.

This problem was still bothering Einstein when he met Bose and, as you may remember, "proposed two problems to Bose to work on: first, the question of whether the new statistics implied a novel type of interaction between the light-quanta; second, what the new statistics meant within the scheme of quantum mechanics" (Mehra and Rechenberg 1982, p. 571).

A word on the second problem, even though it is not strictly germane to my topic. If Bose or Einstein had been able to solve it, they would have discovered the need to totally symmetrize the wave function of N bosons discovered by Dirac in 1926. As Pais 1982 points out (pp. 424–425), the historical development of quantum statistical mechanics reversed the logical sequence. Bose, followed by Einstein, discovered how to do the course-grained counting for bosons before Dirac's fine-grained counting justified it. But there was considerable confusion over when to apply Bose–Einstein and Fermi–Dirac statistics, and for a time they were rivals until it became clear they peacefully divide the quantum kingdom on the basis of spin.

But, to return to the first problem, what Einstein had in mind is clear from a talk of September 1924 "On the Aether" (Einstein 1924), to which Howard 1990 drew my attention:

> Recently a paper of the Indian Bose on the derivation of Planck's formula has appeared, which is of particular importance for our theoretical outlook for the following reason: Previously all complete derivations of Planck's formula made use of the hypothesis of the undulatory structure of radiation. Thus, for example, the factor $8\pi h\nu^2/c^3$ in this formula was obtained in the well-known Ehrenfest–Debye derivation by counting the number of proper vibrations in the cavity in the frequency range $d\nu$. Bose replaced this counting based on the concepts of the wave theory by a gas-theoretical calculation, which he relates to the light quanta present in the cavity, the quanta being thought of as a kind of molecule. The question then arises, if the phenomena of diffraction and interference might not be connected with the quantum theory in the following way: the field-like concepts of the theory only represent expressions of the interactions between quanta, an independent physical reality no longer being ascribed to the field.[8]
>
> The important fact that, according to Bohr's theory, the frequency of the emitted radiation is not determined by electric charges that undergo periodic processes of the *same* frequency can only strengthen us in this doubt about the independent reality of the undulatory field" (Einstein 1924b, p. 93).

Einstein was looking for some way of assimilating *entanglement* of quantum systems, which, in retrospect, we can see first manifested itself in 1900 in Planck's radiation law—even though it was not understood as such until a quarter of a century later—to some more traditional type of *interaction* between particles. The circumstance that entanglement occurs even in the case of an ideal gas (of massive particles or photons), in which there are *no* interactions between particles except for collisions needed to randomize them, tended to militate against such a possibility from the start. The discovery of Fermi–Dirac statistics made the problem of quantum entanglement of otherwise-free particles of the same species more acute, by showing that it comes in two brands.

But there is more to entanglement than this. The work of Einstein, Podolsky and Rosen and subsequent tests of Bell-type inequalities show that, once two quantum systems of any type interact (in the ordinary sense of the word), they are quantum-entangled forever; they suggest that, even if quantum mechanics is modified or replaced, some such entanglement feature will almost surely characterize the new theory.

So far, I have discussed quantum entanglement of two or more particles. The Feynman approach suggests that, even in the one-particle case, entanglement is *the* feature distinguishing quantum theory from classical physics. This approach comes as close as can be to fulfilling Einstein's hope of eliminating the field and explaining interference effects in terms of "interactions" between particles. In the one-particle case, this can be done by introducing a new type of "interaction," or entanglement: one between all *possible* paths the particle could take between its initial preparation and final registration that are compatible with, and indistinguishable from each other within, the given physical arrangement. This peculiarly quantum entanglement between possible paths is implemented by assignment of a probability amplitude to each such path, and addition of all such amplitudes in order to get a total amplitude for the entire process.[9] This addition of amplitudes, rather than probabilities as for distinguishable paths, produces a non-local entanglement of the paths, entirely distinct from any classical interaction, but which accounts for all quantum interference effects. Perhaps it is not stretching things too much to assimilate this one-particle quantum entanglement to the many-particle entanglements discussed above, and say that entanglement is *the* distinctive feature of quantum mechanics.

As discussed in Stachel 1986b, even more than the probabilistic element it was entanglement, and the attendant non-locality that it introduced into the heart of physics, that most bothered Einstein about quantum mechanics. It is ironic that he played a crucial role in introducing both these troubling features of quantum mechanics.

Notes

[1] I am much indebted to Mrs. Enakshi Chatterjee for explaining to me the importance of situating Bose in this context.

[2] Here, I also differ with Bergia 1987, which follows Mehra and Rechenberg's account. Bergia's article contains a very interesting account of the prehistory of Bose–Einstein statistics, including the contributions of Natanson, Wolfke, Krutkow, and DeBroglie, which I shall not discuss.

[3] I am extremely grateful to Professor Parthe Ghose of the Bose Institute, Calcutta, for providing me with a photocopy of this letter.

[4] Interview with (May 21, 1994) and letter from (July 8, 1994) Mme. Jacqueline Eisenmann, Paris. I am grateful to Mme. Eisenmann (née Zadoc-Kahn) for her kindness in pro-

viding information on Bose's visit to Paris and copies of surviving letters from him. At the time of the visit, she was working in Aimé Cotton's laboratory.

[5] Einstein's 1909 criticism of Planck's derivation of his radiation formula shows how Einstein saw the problem:

If one adopts the standpoint that the irreversibility of natural processes is only *apparent*, and that irreversible processes consist of a transition to a more probable state, then one must first of all give a definition of the probability W of a state. The only such definition that can be considered, in my opinion would be the following:

Let $A_1, A_2, \ldots, A_l$ be all the states that an externally closed system with a definite energy is able to assume, or more precisely stated, that we are able to distinguish with certain instruments [i.e., coarse graining]. According to classical theory, after a definite time the system assumes one of these states (for example A_ν), and persists thereafter in this state (thermodynamical equilibrium). According to statistical theory, however, the system assumes all of the states $A_1 \ldots A_l$ in irregular sequence over and over again. If one observes the system for a very long time Θ, then there will be a certain part of this time τ such that during τ and only during τ the system is in the state A_ν. τ/Θ will have a definite limit which we call the probability W of the state in question A_ν.

Starting from this definition one can show that the following equation for the entropy S must hold

$$S = R/N \lg W + \text{const},$$

where the constant is the same for all states of equal energy.

Neither Boltzmann nor Planck have given a definition of W.

Proceeding quite formally, they set $W =$ the number of complexions of the state in question.

If one now requires that these complexions be equally probable, where the probability of a complexion is defined analogously to how we defined the probability of a state [above], then one arrives at exactly the definition of the probability of a state given [above]; except that the logically unnecessary element complexion has been used in the definition.

Although the relation given between S and W only holds when the probability of a complexion is defined in the indicated way or an equivalent one, neither Boltzmann nor Planck has defined the probability of a complexion. ... In the oscillator theory of radiation Planck was not free to choose his complexions. He could only postulate the pair of equations $S = R/N \lg W$ and $W =$ number of complexions if he added the condition that the complexions must be so chosen that, in his theoretical model, they are found to be equally probable on the basis of statistical considerations. In this way, he would have been led to the Jeans formula. (Cited from reproduction in Einstein 1989, Doc. 56, pp. 544–545).

[6] Bose is reported to have told Mehra: "I had no idea that what I had done was really novel ... I was not a statistician to the extent of really knowing that I was doing something which was really different from what Boltzmann would have done, from Boltzmann statistics" (Mehra and Rechenberg 1982). In addition, Bose's very isolation from the physics community in Europe shielded him from the skepticism there about the photon concept that prevailed before Compton's work.

[7] Since he was considering cells in phase space, this amounts to what is now called a coarse-grained counting problem.

[8] This formulation has a Machian ring to it, and indeed Einstein had originally applied what he called "Mach's principle" to the gravitational field. But in the final paragraph of this article, Einstein explicitly excepts general relativity, "because general relativity, to whose fundamental viewpoint physicists will probably always adhere, excludes an imme-

diate action-at-a-distance; but every local-action theory presupposes continuous field, and thus the existence of an 'aether' " (*ibid.*).

[9] The total amplitude must be squared in order to get the probability for the given process. For *distinguishable* alternative processes, the probability of each must be added in the usual (classical) way. For a discussion of the Feynman approach, with references to the original papers, see Stachel 1986a, pp. 314–315 and 331–333. [Note added for this edition: see Stachel 1997 for further elaboration of the relation between quantum entanglement and the Feynman approach.]

References

Bergia, Silvio (1987): "Who Discovered Bose–Einstein Statistics?," in Manuel G. Doncel et al., eds., *Symmetries in Physics (1600–1980)* (Barcelona; Siminari d'Historia de las Ciencies Universitat Autonoma de Barcelona), pp. 222–250.

Blanpied, William A. (1972): "Satyendranath Bose: Co-Founder of Quantum Statistics," *American Journal of Physics* 40: 1212–1220.

Bose, Satyendra Nath (1924a): "Plancks Gesetz und Lichtquantenhypothese," *Zeitschrift für Physik* 26: 178–181.

—— "Wärmegleichgewicht und Strahungsfeld bei Anwesenheit von Materie," *Zeitschrift für Physik* 27: 383–393.

Chatterjee, Enakshi (1994): "The Business of Freeing India," *Miscellany*, January 9, 1994, pp. 7, 15.

Dirac, Paul Adrian Maurice (1926): "On the Theory of Quantum Mechanics," *Proceedings of the Royal Society* A112: 661–667.

Ehrenfest, Paul and Kammerlingh Onnes, Heike (1914): "Vereinfachte Ableitung der kombinatorischen Formel, welche der Planckschen Strahlungstheorie zugrunde liegt," *Annalen der Physik* 46: 1021–1024.

Einstein, Albert (1924a): "Quantentheorie des einatomigen idealen Gases," *Sitzungsberichte der preussischen Akademie der Wissenschaften (physik.-math. Klasse) [n.v.]*: 261–267.

—— (1924b): "Quantentheorie des einatomigen idealen Gases. Zweite Abhandlung," *ibid.*, 3–14.

—— (1924c): "Ueber den Aether," *Schweizerische naturforschende Gesellschaft, Verhandlungen* 105: 85–93.

—— (1925): "Quantentheorie des idealen Gases," *Sitzungsberichte der preussischen Akademie der Wissenschaften (physik.-math. Klasse) [n.v.]*: 18–25.

—— (1987): *The Collected Papers of Albert Einstein*, vol. 1, *The Early Years, 1879–1902*, John Stachel et al., eds. (Princeton: Princeton University Press).

—— (1989): *ibid.*, vol. 2, *The Swiss Years: Writings, 1900–1909.*

—— (1993): *ibid.*, vol. 5, *The Swiss Years: Correspondence, 1902–1914*, Martin J. Klein et al., eds. (*ibid.*).

Einstein Albert and Ehrenfest, Paul (1923): "Zur Quantentheorie des Strahlungsgleichgewichts," *Zeitschrift für Physik* 19: 301–306.

Howard Don (1990): "'Nicht sein kann was nicht sein darf,' or the Prehistory of EPR, 1909–1935: Einstein's Early Worries About the Quantum Mechanics of Composite

Systems," *Sixty-Two Years of Uncertainty*, Arthur Miller, ed. (New York: Plenum), pp. 61–111.

Mehra, Jagdish (1975): "Satyendra Nath Bose," *Biographical Memoirs of Fellows of the Royal Society* 21: 117–154.

Mehra, Jagdish and Rechenberg, Helmut (1982): *The Historical Development of Quantum Theory*, vol. 1, part 2, *The Quantum Theory of Planck, Einstein, Bohr and Sommerfeld: Its Foundation and the Rise of its Difficulties 1900–1925* (New York/Heidelberg/Berlin: Springer-Verlag).

Pais, Abraham (1982): *'Subtle is the Lord . . .' The Science and the Life of Albert Einstein* (Oxford: Clarendon Press/ New York: Oxford University Press).

Pauli, Wolfgang (1923): "Über das thermische Gleichgewicht zwischen Strahlung und freien Elektronen," *Zeitschrift für Physik* 18: 272–286.

Stachel, John (1986a): "Do Quanta Need a New Logic?," in *From Quarks to Quasars/ Philosophical Problems of Modern Physics*, Robert Colodny, ed. (Pittsburgh: University of Pittsburgh Press), pp. 229–347.

—— (1986b): "Einstein and the Quantum: Fifty Years of Struggle," in *From Quarks to Quasars/ Philosophical Problems of Modern Physics*, Robert Colodny, ed. (Pittsburgh: University of Pittsburgh Press), pp. 349–385.

—— (1997): "Feynman Paths and Quantum Entanglement: Is There Any More to the Mystery?", in *Potentiality, Entanglement and Passion-at-a-Distance*, Robert S. Cohen, Michael Horne and John Stachel, ets., (Dordrecht/Boston/London:Kluwer Academic), pp. 245–256.

Venkataraman, G. (1992): *Bose and His Statistics* (Hyderabad: Universities Press).

(538)

A conversation with Satyendranath Bose about five decades ago – Some recollections

S. Ramaseshan

Seventy-five years have passed since the epoch-making discovery of Bose statistics. My mind goes back to the memorable meeting I had with Satyen Bose about five decades ago. At that time when I was comparatively young I found meeting 'distinguished scientists' (young or old) in India was by no means an easy or enjoyable task. They were usually very busy, always going from one meeting to another, always wanting to impress, more often than not pompous. I was however extremely keen on meeting Satyen Bose and the meeting proved to be quite easy and was a success. I went prepared to discuss many topics with him.

When I was in the University College, Calcutta, I was told he was in, and that he would be leaving for his home within the next half an hour. I tried my luck and peeped into his room. A visitor was just leaving. Bose welcomed me with a smile. I introduced myself. He asked me whether I wanted to discuss with him any scientific problem. I said 'Not really. Since I am C. V. Raman's student I would like to know whether you have had any interaction with him. I would like to know about the remarkable discovery you have made. I have a big list. For example, now that you are turning your attention to X-ray crystallography and you are a recognized authority in statistics, I would like your views about the Hauptmann and Karle intensity statistics by which the phases of X-ray reflection could be determined'. Bose said: 'The statistics I am familiar with is taught in high schools. You should probably meet P. C. Mahalanobis about this matter. (I said I would.) In any case, I will not like to discuss with you this X-ray statistics problem as I have not read the paper by Hauptmann and Karle. However, I am not surprised that the phases could be determined by intensity measurements only. In fact Kedareshwar Banerjee, I think, and Ott and Avarami also tried this approach but they did not persist with it nor did they go far enough.' I then told him these are really minor matters. Frankly, I had longed to meet him – asked whether I could waste his precious time just talking and listening to him. Again he flashed his charming smile and said if I wanted to talk to him why not we go to his home and relax. 'I can give you tea, I can play you music, you can see my cats. I hope you are not allergic to them!' So we went to his house and sat in his room which was in a perfectly chaotic state but which strangely had a character of its own. It seemed to be overflowing with cats, every type that T. S. Eliot describes and more.

Tea was brought in (extremely sweet) with some Bengali sweets, which I love, and a couple of arrowroot biscuits. In this brief piece I shall try to present three or four of the many topics we talked about. It is very difficult to convey all that Bose said and implied, the spirit with which he presented everything and the incredible friendliness of the meeting. The words that I put into his mouth are of course not his exact words. I have tried to convey the spirit of what he said.

Knowing that I was Raman's student, Bose commented on Raman's work on the diamond. He was certain, all along, that Raman was not right in his conjecture of the four forms of carbon and that most of the experimental observations made by Raman and his school could well have been explained using some other simpler and not so revolutionary hypothesis as the four forms of carbon. I agreed with him and he said: 'Do not think that committing scientific errors is the monopoly of ordinary scientists. Even the greatest of them, Einstein himself committed a very grave one. As late as 1926 he proposed an experiment to distinguish classical and quantum views of light. If it had been made by a lesser man, one would have called it a foolish proposal. Raman had a very acute mind and it was he who first pointed out the error Einstein had made.'

'Do you think all the work done on diamond in Bangalore was a waste?' I asked. 'Oh no, all errors made, particularly by men of intuition always become useful and even fruitful. Just count the number of techniques which were evolved and results that these experiments yielded. I believe that science was enriched by these studies. I also feel that the later discovery of impurities in diamond was to some extent, provoked by Raman's wrong thesis!!' I was truly surprised that Bose had studied most of Raman's papers so well that he could comment on them with discrimination.

Later on the conversation turned to the Raman Effect. I asked 'As one who had a lot to do with the foundations of quantum mechanics, would you subscribe to Born's statement that Raman's discovery is predicted in its entirety by quantum mechanics and so it could be thought of as a proof for the same'. 'It is in a sense true, but it is very easy for theoreticians to make such statements after the experimental discovery had been made. They do not understand the immense struggle an experimenter has to go through.'

'Raman and I used to meet off and on and discuss some science. Since 1921, I had been continuously thinking about the reality of the light quantum. Raman, of course, was a firm believer in the photon. As he once humourously remarked, "You, I and Einstein were the only three who believed in the light photon in 1921–1922." Raman had an uncanny intuition and he could ask some very penetrating questions. He had read the paper by Kramers and Heisenberg on the dispersion theory, within a week of the arrival of the journal *Z-fur Physik* in Calcutta. Later when I had just read the paper, he said that he feels that Kramers and Heisenberg imply that there is another type of quantum scattering quite different from the classical one. "Am I right?, he asked. At that time, I could not give him a definite and precise answer. Later after 4 years, when he had discovered the Raman Effect I realized that his question clearly showed that he had hit the nail right on the head. In fact, at the time the Kramers–Heisenberg paper appeared, few of us understood its full implications. I was quite vague about it and I also realized that it would have been difficult for an experimenter to envisage the type of experiment to be done to verify the implication of what they said.'

'I must say the discovery of the Raman Effect did come to me as a surprise. I am also convinced that it was a surprise to most including the great Einstein himself as is indicated by what he said. (I quote – says Einstein, "C. V. Raman was the first to *recognise* and to *demonstrate* that the energy of the photon can undergo partial transformation in matter. I still recall the deep impression that this discovery made on all of us.") Raman had stated that while from his light scattering experiments he did expect to discover a quantum effect he did not realize how useful his discovery would be to chemistry, physics and engineering.'

'I was staggered that such an important quantum effect could come out of such simple scattering experiments. Actually thinking deeply I should not have been surprised as most major advances in physics since the turn of the century, i.e. in nuclear physics (Rutherford–Geiger experiment) in quantum physics (the Compton effect) and recently in particle physics – have come out of scattering experiments. Yes, Raman had a tremendous horse sense. I can say that it shows the uncanny or inspired intuition Raman had. No wonder he persisted in light scattering experiments over 7 years till he discovered the quantum effect. In fact, it is obvious that he had a prescience of it as he mentioned it clearly in the monograph which he published as early as 1922.'

'In fact, I was so impressed with the discovery that when we both went on a walk along the Hooghly I told him near the Prinseps Ghat "Prof. Raman, you have made a great discovery. You will surely get the Nobel Prize for it".'

After discussing a few other topics, he suddenly asked me, 'Can you deliver a scientific lecture in your mother tongue?' I shook my head and gave him a sheepish grin. Again after flashing a charming smile he said very seriously 'You should be ashamed of yourself. Did you know that K. S. Krishnan could'. He then talked about the work he and many others were doing to teach science in Bengali. 'I feel it is unnatural and immoral to try to teach science to children in a foreign language. They will know facts but they will miss the spirit. People forget that science is culture, a child must learn from science many things; a true scientific attitude not only engenders objectivity but also spiritual values. We MUST teach science in the mother tongue. Otherwise science will become a highbrow activity. It will not be an activity in which all people can participate. It will not percolate into the heart or blood of our people.' He spoke with great intensity and verve. He then read out some of his writings in Bengali. His reading was so sonorous – and it sounded like poetry – and I said so. 'Yes, Bengali is a beautiful language; I think it is one of the most beautiful languages in the world'. I said 'French too has this reputation'. 'Yes I agree, it is also a language of love. We have been inhibited far too much in the last 100 years. That is probably why exquisite love poetry is not being written in our languages. Bengali has directly descended from Sanskrit. Can anyone say Kalidasa's language is not the language of love? But I am digressing too much'.

We went on and on and finally I asked him whether he felt elated or happy or transported when he discovered what is called the Bose Statistics. 'If I were truly honest I can say NO. My feelings were entirely different. Let me put it this way: Suppose you had a severe headache or stomachache – and the ache suddenly stopped. That was the feeling I had. For all the previous derivations gave me ceaseless pain. If you call cessation of pain as happiness – then I can say I was happy. I must also add that I never felt I had done something great.

Everyone insists on saying that I have done something remarkable. Because of this insistence I have come to accept it. But still I feel that all I did was just to clear up a messy situation. So I was really satisfied. But because of my attitude people think I did not understand what I was saying. The few sentences which Einstein crossed out about the angular momentum of light quanta is evidence of this (Bose was referring to the later discovery of the spin of the photon which his reference to angular momentum implied).'

'Since you are interested in my interaction with Raman you may be interested to know that once Raman accosted me and asked me "You once told me that you were convinced that photon had angular momentum, i.e. spin." I told him about how I was unhappy that Einstein removed a couple of sentences I had written and substituted a sentence or two. I then explained to him my point of view. He said that he and Bhagavantam were doing some experiments which indicate that the photon had a spin. May I put in your views in the paper? I said yes and I think Raman also wrote about my conversation with him in his paper with Bhagavantam.' I said that I had read it. I quote from *Indian Journal of Physics*, 1931, **6**, 353.

'In his well-known derivation of the Planck radiation formula from quantum statistics, Prof. S. N. Bose obtained an expression for the number of cells in phase space occupied by the radiation, and found himself obliged to multiply it by a numerical factor 2 in order to derive from it the correct number of possible arrangements of the quantum in unit volume. The paper as published did not contain a detailed discussion of the necessity for the introduction of this factor, but we understand from a personal communication by Prof. Bose that he envisaged the possibility of the quantum possessing besides energy h and linear momentum h/c also an intrinsic spin or angular momentum ± h/2 round an axis parallel to the direction of its motion.'

I told Bose that it was the only statement in a scientific journal that he had proposed *for the first time* that photon had a spin. I asked him whether something should be done so that he be given the credit for proposing the existence of the spin of the photon, for the first time. Bose suddenly showed some annoyance. 'You are descending from the sacred to the profane – as the saying goes. What is important is whether the photon has a spin and it is not important as to who proposed it and when. The high aims of science get very debased by such talks of priority etc'. I told him that I stand rebuked but my admiration for the man went up greatly.

'Reverting to your question, you asked about elation. Yes, it would surely have come later; but many things happened almost immediately afterwards which made me very very sad'. I (SR) did not have the courage to pursue this further.

'You enquire about happiness.
When a man does what he *has* to do, does it bring him happiness? When a man does what he *wants* to do, does it bring happiness? It is so difficult to say when a man can experience happiness. Happiness can sometimes be experienced when you are in an environment different from the ordinary. Happiness can possibly come only when you feel free and unfettered. You never know when it can hit you. I think I had a glimpse it when I was once out of India.'. I started asking him some more questions. 'Enough' he said. 'I have talked too much today. You have asked me so many questions and these have provoked me into saying many things which I may not have said otherwise. Now I have to punish you. You have to listen to my playing the esraj'. It was a delightful ending to an unforgettable meeting. After half an hour of esraj I rose and said goodbye. For me it had been a great experience. As I wended my way back through the Calcutta lanes this man haunted me – his gentleness, his warmth, his incredible modesty, one whom fame had not spoiled and veneration had not touched. Neither could I stop speculating as to what happened to him after his famous discovery which made him so unhappy and what made him see a glimpse of happiness in a foreign land.

S. Ramaseshan is in the Raman Research Institute, Bangalore 560 080, India.

Edited and published by P. Balaram and S. Ramaseshan, Current Science Association, Bangalore 560 080.
Typeset by R D Typesetters, Bangalore (Ph: 3405672). Printed by Lotus Printers, Bangalore (Ph: 3209909).

Satyen Bose in Berlin

Purnima Sinha

Between 1950 to 1956 about ten of us were engaged in experimental research in the X-ray laboratory of the Physics Department of Calcutta University College of Science, under the guidance of Professor S.N. Bose. Apart from ourselves, students and teachers of other departments, especially the applied mathematics department, used to come to him and have long discussions with him. But it had not occurred to any of us to record systematically what this man of rare brilliance said in those discussions. He used to work out detailed and distinctly written out steps of calculations on sheets of paper with meticulous care when he gave us courses on X-ray crystallography in our mother tongue Bengali. He would bring out off and on the beauty of the broader perspective of physics, as he talked on and on for hours to the students sitting around his table. We did not keep those papers, which would have been a record of his way of teaching in a branch of physics which was not even his main field of interest. There is no doubt that the people around him were not prepared enough to take full advantage of his off beat genius.

No one in our lab used ready made imported equipment for his work. Every one had to set up instruments from locally available materials and construct them with the help of technicians of our lab workshop. Thus a demountable X-ray unit, differential thermal analyzer, fully automatic scanning spectrophotometers for thermoluminescence studies and some other equipment were fabricated in our lab with Professor Bose's active involvement.

In the thirties, in the Department of Physics of Dhaka University under the guidance of Professor Bose, a mechanic in the lab workshop fabricated a Weissenberg Camera, the most modern equipment of an X-ray crystallographer at that time. Professor Bose had built up a full fledged X-ray lb, and the students of his department and other departments of Dhaka University and, even the students of his friends from Calcutta, had utilized that lab for their work.

We had an egalitarian informal atmosphere in our X-ray lab in Calcutta Science College. Professor Bose was a very friendly and approachable to all his students. After completing my Ph.D, I worked in different laboratories, but my association with S.N. Bose continued up to the end of his life. We had discussions on topics covering a wide range in and beyond Science, of our common interest in music, art, philosophic and social aspects of science. He would often get involved in stories of ordinary human life flowing with wit, humor and compassion.

In the summer of 1973 the idea of writing a biography of Professor Bose came to my mind. On July 22, 1973 I went with a tape recorder to Professor Bose's house in north Calcutta. It was apparent that he still had a photographic memory and he talked with a lot of interest. The only regret I had was that we had not begun to record the story of his life in his own words earlier.

It was only possible for me to record three long interviews with him on 22 July, 25 July and 15 August, 1973 before he died on 4 February, 1974.

When he talked about his friends in Paris and Berlin, he especially mentioned two of his very close friends. One was **Professor Herman Mark** in Berlin and the other was Jacqueline Zadoc Kahn in Paris. Fortunately, I was able to make contact with both of them. I will describe some extracts from interviews and correspondences with them. I hope these records will help terminate the perpetuation of some of the speculative statements about S.N. Bose.

I requested Dr. Jagdis Sharma, my former colleague in Professor Bose's laboratory in the University College of Science, Calcutta who had been working in different laboratories in the U.S.A., to interview Professor Mark on the basis of a few questions sent by me. Dr. Sharma promptly sent me a taped interview in 1974.

Interview

Question: *Professor Mark, you have met Professor Bose in Berlin and that was about the period about which we know the least. My question would be: When did you meet Professor Bose for the first time, under what circumstances and what do you remember about the initial meeting?*

Hermann Mark: In 1924-25 there was in Berlin a famous seminar conducted at the Institute of Physics of the University of Berlin. At that time, the Director of that Institute was Professor Rubens. He is very well known as one of the originators of the infrared spectroscopy. At this seminar, there were such eminent people as **(Max) Planck**, **(Max von) Laue**, **(Walther) Nernst**, **Otto Hahn**, **Lise Meitner**, **(Fritz) Haber** and a very large number of younger physicists and physical chemists. At that time, I was a member of the **Kaiser-Wilhelm-Institute** in Berlin Dahlem [*editor: now known as the* ***Max Planck Institute for Chemistry - Otto Hahn Institute***], the Director of which was Professor Herzog and we all capitalized on the opportunity to go downtown Berlin and attend the physics seminar. There was a subway ride of about 25 minutes, and at our Institute and at the neighboring Institute of Professor Haber there were a number of younger physicists like **(Karl Friedrich) Bonhoeffer**, **(Theodore von) Karman**, **(Leó) Szilard**, **(Eugene) Wigner**, **(Michael) Polyani**, **(Karl) Weissenberg** and a few others. And we always traveled together and attended the seminars and in one of those seminars Bose gave a report on his work. The seminars were conducted in such a manner that usually two or three talks were given on some very interesting new article or work that was done by one of the members. And Laue conducted the seminar, he distributed the articles mainly among the younger people, and then he would give the talk and the older experienced professors would ask questions. Everything was done in German - I should say almost everything was done in German. Bose gave one of those seminars. That was when I saw him for the first time. That was based on his ideas about nonclassical statistics of a system particularly of low temperature characteristics -- and Nernst was sitting there, Einstein was there (with a hearty laughter) and everyone interested in low temperature and statistics -- and Planck was also there.

He (Bose) gave this seminar. I do not remember any more whether he gave it in German or English. But I think he gave it in German. Usually, after the presentation, Laue would get up and say, alright fine, this was nice, and is there any question? Then there would be a lengthy discussion on what the issue was. And in this case Laue said immediately, "I am not sufficiently familiar with the topic and I would like to ask Professor Einstein to comment." Then Einstein said, "Well I think this is one of the most worthwhile works of the last few years." He talked about it and asked a few questions and Bose answered the questions and then Nernst asked a few questions. In other words it was a very sensational event on that day. Because the discussion on this specific seminar took a few hours.

This was the first time I saw him, and I liked him. Then of course Karman, Szilard, Wigner also asked questions. We were young and he was young -- so we quickly became friends and we invited him to come to the Kaiser-Wilhelm Institute in Dahlem and give us the same seminar to continue the discussion -- and so he came out a few weeks later and we had there a seminar...

There were a number of people who established an atmosphere of friendship and since at the university it was not quite so lively as it was in the Kaiser-Wilhelm-Institute then he came to the Institute almost everyday and he was very much interested in our work. I worked together with Weissenberg and Polyani at that time on X-ray structure, study of crystals of metallic materials -- the way they formed -- and then we started to work on polymeric materials like cellulose and rubber and he was very much interested and he always asked sharp questions and kept us thinking. You know, the most real impression we all had was when Bose was facing some new of scientific endeavour or activity, he just sat there, listened for a while and he had a fantastic grasp over the fundamentals of the specific work and then he would ask questions and he would ask extremely relevant -- not only intelligent -- questions, he would ask questions which were piercing and pointing into future developments. Now, of course, we had a lot of fun when he and Szilard and Wigner argued about statistics. Szilard was a man who worked with Laue on statistics and dynamics and Wigner was another mathematician ... so they argued hours and hours and my wife Mimi, always invited them to come up to us. We lived in the Institute -- we had a little apartment in the Institute. So that was very convenient. "Now why don't you come up for a drink of tea of coffee?" and they all came up ... very informally ... and we had a very wonderful time.

Question: *Before that I would like to ask you a question. There is some confusion about his interaction with Einstein. For how long they interacted? When did he come to Berlin?*

Mark: During the end of 1924 to 26. Two years. But then in 1926 before he went back to India we invited him to come to Vienna for a few days and stay with my parents and son and we went to visit the University. That time at the University in Vienna there was **(Erwin) Schroedinger** and there was **(Hans) Thirring** and there was **(Ludwig) Flamm**. Three theoretical physicists and there was Smekal also and they invited Bose to give a seminar at the University.

Question: *So he interacted with Einstein but at that time he did not publish anything jointly with him.*

Mark: I don't remember, I did not read the publication.

Question: *Before he went to Einstein he had an intention of working with him. But there was no publication.*

Mark: Well you see, because of course they were like Socrates and like Greek philosophers, they did not publish anything. You see Einstein was also the same way. He walked out there -- there was a nice park you know in the neighborhood of the Institute in Dahlem and there he and Einstein would walk and they would talk (laughing) and I don't think anything was published.

Question: *And this question might sound a little strange, but there is some background. An article has been written by a man called Blanpied in this country and he thinks that Bose was a withdrawn and shy person in Europe...*

Mark: No! No! No! (laughing). Full of humour and always with good jokes and then he would be singing German songs -- of course, when we were having a party singing German songs and he was conducting!

Question: *How did he impress his colleagues in Europe as a thinker in theoretical physics and as a person?*

Mark: Well, as I said, Bose had to do with **(Paul Peter) Ewald**, Wigner, Szilard, Karman -- they were very much impressed and all of them established with him life long friendship -- all of them!

Question: *Do you know anything about Bose's association with Madame Curie and de Broglie? He was associated with X-ray work done my Maurice de Broglie in Paris.*

Mark: I think before he came to Berlin he was in Paris a year or so or a few months -- and we were in Paris at the same time but I did not meet him then but of course we met **Maurice de Broglie**. We met Trillat and Thibaud ... I think he worked with them.

Question: *So this question is unnecessary. While Bose was there in Germany and Europe he did utilise his time and it helped him to develop his scientific research interest later on.*

Mark: Yes, yes.

Question: *But one thing, after coming back to India in 1926 or so for a very long period he did not produce any papers of any kind. He had been teaching but he did not produce any papers.*

Do you think there is some reason behind?

Mark: I would think -- this is my opinion -- that he was not ... too much of a practical man interested in detail. We were after details of structural themes -- whether the atom is here or the atom is there and what would be the consequences ... he was not interested in that. He was interested in the very broad line of development of the fundamental physics and specifically of quantum mechanics that existed at that time and of quantum theory -- the quantum hypothesis that existed. That was really what attracted him.

Question: *Would you make some comments about him as a man?*

Mark: Oh! He was very friendly. He was a real gentleman. And he would tell his jokes to everybody and had excellent sense of humour. Ah! Wonderful! He was very benevolent! He could only do good things! We always called him our Buddha (laughter). He was sitting there on the couch and ...

Question: *Is there any special incident which you would like to mention?*

Mark: Well, maybe, when he was in Vienna you know, that was in 1926 and he gave a seminar and I mentioned that there was Schrödinger , and many very good physicists in Vienna. And that was already when Schrödinger was thinking on wave mechanics you know, because his paper came out in '26, there was a very interesting conversation. **(Louis) De Broglie's** thesis was published ... then of course, everybody realized that it was Bose that was the forerunner ... it was the statistical expression of the wave nature of an electron and then the question was "Okay, what are we going to do now?"

Question: *Was Heisenberg also involved at that time?*

Mark: No, he was not in Vienna at that time.

Question: *The application came with electron diffraction later.*

Mark: Well, the theoretical expression came with the Schrödinger equation -- and the experimental proof came with the electron diffraction.

Question: *Blanpied, who wrote the article a couple of years back, said that Bose had failed to fully grasp the significance of his own discovery.*

Mark: If somebody makes such a statement, one must ask him, "alright, give us the reason for the statement," and then he may say, "Well, after these two papers for a long time Bose did not publish any thing." Then I will answer him, "you can just as well say, what an idiot was de Broglie that he did not derive the Schrödinger equation!" (Laughter). Of course, he wrote his thesis and lost interest in it. This does not prove that de Broglie completely missed the importance of his own discovery! After all a lot of people have to contribute to progress, one single mind cannot do everything."

This was the end of the interview with Professor Mark.

I met Professor Bose's friend Madame Jacqueline Eisenmann (a retired scientist of CNRX) in Paris in the summer of 1980. In the course of my interview with her she gave me the following letter which she had received from Bose from Berlin in 1926.

Unter der Eichen 88
Bei Ehlert
Berlin Dahlem

...I am in my new rooms since the first day I arrived here; it is very nice and comfortable and I am really in love with the balcony ... my friends live very near me, about 5 minutes walk from here, in the very buildings of the laboratory and I go there almost every day...

Everybody (every physicist) seems to be quite excited in Berlin, about the way things have been going on with physics, first on the 28th last, Heisenberg spoke in the colloquium about his theory, then in the last colloquium, there was a long lecture on the recent hypothesis of the spinning electron (perhaps you have heard about it). Everybody is quite bewildered and there is going to be very soon a discussion of Schroedinger's papers. Einstein seems quite excited about it. The other day coming from the colloquium, we suddenly found him jumping in the same compartment where we were, and forthwith he began to talk excitedly about the things we have just heard. He has to admit that it seems a tremendous thing considering the lot of things which these new theories correlate and explain, but he is very much troubled by the unreasonableness of it all. We are all silent but he talked almost all the time, unconscious of the interest and wonder that he is exciting in the minds of other passengers". Bose

Although the letter is undated, it must be just after April 1926, because Heisenberg presented his lecture in Berlin colloquium on that date.

Madame Eisenmann wrote in a letter to me:

*I was then a young girl who had just finished her 'licence de sciences physiques' and who had just begun to work in Professor Cotton's lab ... **Sylvain Levi**, the great Indianist and Sanskritist was a friend of my father (Dr. Leon Zadoc-Kahn who was in 1943 assassinated by the Germans with my mother). Learning from my father that I intended to work in Physics, Levi said he would make me know 'un jeune physician genial'.*

*I was very impatient to meet this genius. When he came to my lab, accompanied by another Indian named Tendulkar, he did not tell me so as to tease me, who was the physicist. Bose was so unassuming that I didn't find out immediately who was who! From that day I saw him very often. He always went to **[Paul] Langevin's** lectures. Langevin gave many lectures. Louis de Broglie came later, Langevin told **Madame Curie** about him. Bose worked in Madame Curie's lab and in Maurice de Broglie's lab for sometime. He went very much to the museum, loved nature, particularly the alps, went to see and live in the countryside.*

He talked much about Bengali ... writing science in Bengali -- to teach the students in Bengali. He impressed me very much by his great love for his country. He never went to England until India was free. In 1953 he went to England and lived with Dirac.

He went to Germany in October 1924. Perhaps Bose could communicate more with scientists in Germany.

On 25 July 1973, I asked Professor Bose why he had gone to Paris in the first place. He answered, "I was informed that my friend Abani Mukherjee (a terrorist nationalist leader who was absconding) was in trouble. I had taken some money for him from the country. After meeting Abani, I thought that I will stay in Paris for a while. I had many friends there. They asked me to stay on, I got the idea of doing some experiments so that I would teach these to students in our country. I worked for a while in Maurice's lab. He had already read my paper. He told me that his brother did the kind of work that I did."

When I asked him about his encounters with Einstein in Berlin, he said, "... You know that Einstein was included towards "red" -- so that chauvinistic German students used to create trouble for him. This made him abandon class lecture. We used to go to his house. He had no research student either. He used to tell us what he thought and some times gave lectures too."

Madame Eisenmann said: "It was a great joy to know Bose at all. He was so wonderful, so gifted, knew so much about Hebrew literature and religion. He had an extraordinary heart! He had nearly feminine reaction! He had no ambition for himself, too modest and humble a young man." Referring to a letter by Bose she wrote, "The letter written in 1951...was sent a few days after we met in Paris after being entirely without news since 1929. This encounter was a wonderful shock for me and the first happiness after the death of my gather and my mother in 1943."

"He (Bose) told me in Paris, after the war (in 1951), as I asked him whey he had not published more work, that his surroundings were not favourable. He added, he had spent a great deal of time in preparing experimental research work for his pupils in Dhaka."

"Moreover, he said another time that he threw away most of his works that he judged not good enough."

Among his total 24 [sic] published papers, Professor Bose had published 17 papers after coming back from Europe, between 1936 and 1955. The context and content of these papers have not yet been analysed. The topics range from mathematics, theoretical and experimental physics to biochemistry. Why some of them were not of fundamental important towards the progress of physics? They were the outcome of resolving obstacles in handling problems by his students and friends, and also on problems which are of practical use for our country.

Sometimes he used to spend days after days in chemical laboratories. Under his guidance students were able to prepare some useful medicines and also to make significant contributions in the synthesis of important chemical compounds.

During the fifties, Bose followed Einstein in his research on 'Unified Field Theory'. Several other very eminent scientists such as Herman Weyl, Kaluza, and Schrödinger were also involved in the field for many years. Bose was drawn into this movement of handling very difficult mathematical problems and contributed five papers during 1953-55 which were published in French scientific journals.

Self taught Meghnad and Satyendranath were teaching untrained students in newly established Departments of Science. one of them knew the art of getting an aim and building up towards it, and the other expressed ideas like flashes of lightening from clouds spread all over in random clusters. Before their partnership could gather sufficient momentum, one was transferred to Allahabad and the other to Dhaka. Ours is a large country and each region has its own specific demands. Only a handful of men had to cope with the growing demands. They were always under pressure concerning everything. In the West the scene was different.

Following 'is an extract from an interview with Professor **André Weil**, Professor of Mathematics, Institute of Advanced Study, Princeton, taken by my friend Professor R. Bannerjy (Chemistry Department, Sorbonne) in Paris on 13.6.1979:

André Weil: I have known Bose in Dhaka in 1932. I left for India in 1930, with an appointment as head of the Department of Mathematics of Aligarh Muslim University. **T. Vijayaraghavan**, a brilliant young student of G. H. Hardy, left in 1931 for Dhaka University. I agreed to stay with Vijayaraghavan as his guest for a period of some weeks. We had become very close friends ... Apart from Vijayaraghavan, there was no one in the mathematics department in Dhaka who would have qualified as a mathematician.

I soon met Satyen Bose, and saw him quite frequently during my stay there. It was clear that Bose possessed a very brilliant and penetrating intelligence, he as unwilling to take an interest in any but the deepest problems, working in almost complete isolation. **(K.S.) Krishnan** was a brilliant experimentalist, but he was not much of a theoretician. As to Vijayaraghavan, he was a typical product of the school of G.H. Hardy, very deep but narrow, and could be of little use to Bose.

At that time, Bose was chiefly interested in "general theory". I do not agree with those who suggest that his career and reputation was based on a piece of luck viz., the fact that Einstein took up his early contribution to theoretical physics. The quality of is intelligence was such that he deserved whatever position and honours came to him. One can only deplore that, for lack of a suitable environment, he was unable to realize his potentiality fully. There is no doubt in my mind that, given more favourable circumstances, he was well fitted to play a most important role in laying the foundations for scientific research in India.

Reprinted from **The Physicist Bulletin of the Bangladesh Physical Society**,
January-February 1995, p. 17-21

Satyendra Nath Bose — Historical Notes and Reflections

E.C.G. Sudarshan

Our knowledge of the subatomic world distinguishes two kinds of particles — electrons, protons, and neutrons, which obey the exclusion principle,[1] and photons, mesons and even alpha particles, which besides not obeying the exclusions principle, also prefer to have any number of them in the same state. With the advent of the new quantum mechanics, multi-particle states were associated with anti-symmetric wave functions in the first case and symmetric in the second case. With the development of quantum field theory, they became associated with anti-commuting and commuting fields, respectively. The first species provides structure, which are kept in cohesion by the particles of the second kind.[2] We now know that, at the level of the Action Principle, this relates to the anti-symmetry or symmetry of the scalar product of two fields.[3]

All these developments began with the work of Satyendra Nath Bose, a thirty-year old physicist from Dacca University; who had completed his Masters' Degree in Mixed Mathematics nine years earlier at the Calcutta University. Bose gave a derivation of the Planck formula without recourse to classical physics, implying that Planck's quanta of black body radiation (photons) of the same wave number and polarization are strictly identical. This "indistinguishability" led him to introduce a new method of counting states of identical particles and a new statistics, now called Bose statistics. It not only gave Planck's formula but also overcame Gibbs paradox[4] for a Boltzmann gas.

This paper was published in *Zeitschrift für Physik*, after being translated into German by Albert Einstein, with the added note asserting that it was an important advance. In his paper, Bose had used the dispersion formula $E=ck$ and treated the number of photons to be non-conserving. This possibility of photons being created or destroyed independently led to a zero chemical potential. The situation in the case of a generic Bose gas, consisting of non-relativistic particles of finite mass, needs modification since the number of particles is conserved, leading to a non-vanishing chemical potential. The necessary modifications were implemented by Einstein in a paper that he sent for publication in June 1924. In a subsequent paper, Einstein arrived at a rather startling result: *At a sufficiently low temperature and nonzero chemical potential, a finite fraction of all particles should be in the lowest energy state.* This would give rise to a new phase, what is a now called the "Bose–Einstein Condensate". In recent years this has been a topic of both experimental and theoretical research.

It is interesting to recall that for Planck,[5] light-quanta were relevant only in the exchange of energy between radiation and the walls of the cavity. The quantum parameter (Planck's constant of action) was determined by an empirical fit. But when Einstein used the light quantum hypothesis in the study of photoelectric effect,[6] the light quanta had an independent existence. Niels Bohr employed the notion of light quantum in his theory of the hydrogen atom during transitions between the stationary states of the atom.[7]

In modern quantum field theory, the "complexions" used by Bose may be identified as a single state of the electromagnetic field. To make this correspondence, the field at distinct space points must commute: Nowadays such fields are called Bose fields.

This was also about the same time that we got a clear understanding of many electron systems. From the work of E.C. Stoner, it became clear that a new quantum number was needed for atomic systems. Pauli reformulated the finding as the "exclusion principle" in 1925.[8] Enrico Fermi extended the exclusion principle to free electrons. It is remarkable that the existence of two types of particles and two types of statistics was established before the advent of new quantum mechanics. Dirac introduced the terms, bosons and fermions, and associated them with symmetric and anti-symmetric wave functions within the framework of new quantum mechanics.

The computation of the sum over states, $Z(\beta)$, for a classical gas encountered the Gibbs Paradox. For a perfect gas this was "fixed" by including a factor of $1/N!$ A similar situation occurs in the use of the Boltzmann collision equation. It is usual to have the concentration of the components A and B but not of C and D in the collision process $A + B \rightarrow C + D$. In the case of photons colliding with atoms, we need for the absorption of photon a probability proportional to n_A, but for the emission a factor $n_A + 1$, the 1 corresponding to spontaneous emission. So $n_A/(n_A + 1) = e^{-\beta hw}$ or $n_A = e^{-\beta hw}/(1 - e^{-\beta hw}) = 1/(e^{\beta hw}_{-1})$. Both the Boltzmann collision equation and radiative equilibrium are automatically corrected by using Bose's statistical principle.[9]

The early decades of the last century produced a number of outstanding scientists in India. These include Jagadish Chandra Bose, (who discovered the long distance propagation of electromagnetic waves two years before G. Marconi and demonstrated it in the Royal Institute in London), Prafulla Chandra Ray, Meghnad Saha, C.V. Raman, G.N. Ramachandran, S.N. Bose, and Srinvasa Ramanujan to name a few. There were also people like Iswara Chandra Vidyasagar, Sir Asutosh Mookerjee, and Tarak Nath Palit who supported scientific and intellectual efforts and built the foundation of a vigorous group.

This intellectual awakening has continued to this day. Of all the places in India, Kolkota is the place that stands out — where the very best students choose to study science and mathematics in preference to administrative services, information technology, computer science, or business administration. This is helped by the considerable esteem that the sciences elicit from society at large. In this context, we note that Satyendra Nath Bose was chiefly instrumental in founding *Bangiya Bijnan Parishad* and publishing a monthly periodical in Bengali, *Jnan O Bijnan* (knowledge and science) to popularize scientific discoveries and developments, and promote a scientific outlook in the society. Elsewhere in the country also, effort was made and being made to popularize science. Sir C.V. Raman, for instance, used to invite himself to visit colleges in South India and lecture on various scientific topics. In recent times, we have as an example, the eminent scientist Jayant Narlikar, who writes on science for the layman both in Marathi and in English. The importance of an informed society is essential for the longevity of scientific traditions.

The eminent mathematicians Neelakanta and Madhava in Kerala who had discovered the Binomial Theorem and the beginnings of Differential Calculus in the 14th and 15th century died without much of a trace.[10] There was no general appreciation of what these men had accomplished. For centuries, their work was forgotten and it is only in recent times, it has come to the notice of contemporary mathematicians. The society was largely unaware and unappreciative of their contributions.

It is also curious that S.N. Bose and his contemporaries in Calcutta looked to the Continent rather than Britain for scientific ideas. Bose and Meghnad Saha translated Einstein's papers on the theory of Relativity from the original German, which was published by the Calcutta University. Rather than attempting publications in the *Philosophic Magazine* or the *Proceedings of the Royal Society*, Bose chose to send his article to the *Zeitschrift für Physik* to Einstein. Perhaps Bose knew the fate of his mentor Jagdish Chandra Bose, who not only pioneered many experimental devices in electromagnetic waves out of jute and other "raw" materials, and who took his entire apparatus to the Royal Institute in London to demonstrate the long distance propagation from Britain to the Continent.[11] Yet the priority (and the Nobel Prize) was given to G. Marconi. The official "filtering" of scientific publications to make the most continental scientific journals and books did not prevent the Bengali scientists from getting some of this information out. Bose himself got proficient in German, French, and Italian. It should also be noted that Dacca University granted S.N. Bose a two-year study leave in Europe.

Returning to the scientific activities of S.N. Bose, he took up the problem of the dynamic derivation of radiative equilibrium distribution for arbitrary number of species and energy levels in full generality. It generalized Einstein's work on the A and B coefficients (including spontaneous emission) but with a two-level system and monochromatic radiation. Bose's paper in *Zeitschrift für Physik* **27**, 384 (1924) made this more general and incidentally corrected the deficiency of Boltzmann's collision equation. But, Einstein, who was very happy about the spontaneous emission process, did not approve of these arguments and that meant no one else would. But the simple fact is that his work with two-level atom and monochromatic radiation was put into full generality by Bose.

Incidentally, there are two curious items worth mentioning. Louis de Broglie thought of the possibility of atoms of light (light quanta) aggregating into molecule to explain both the quantum nature of radiation consistent with Planck's law and at the same time the wave nature to explain the phenomena like interference, diffusion and dispersion.[12] Louis de Broglie was fond of making composites of elementary objects. [After all he wanted to view the photon as a compound of a neutrino and an antineutrino.] Secondly, I came across an article by the distinguished physicist biologist, Max Delbruck,[13] claiming that Bose made a "mistake" and so the discovery of Bose statistics was a fluke! This reminded me a story of a husband and his wife returning after a Bridge game with friends. The husband told his wife that she played the last game quite incorrectly. She retorted that "But we won!"; He said, "Yes, but if you had done it right, we would not have won." Delbruck also points out the "absurd" conclusion of Bose condensation. This was published in the *Journal of Chemical Education*. However, in this case it has been proved that Bose was right and Delbruck was wrong, for Bose–Einstein condensates are an experimental reality.

After his retirement from Calcutta University in 1956, Bose became the Vice Chancellor of Viswa-Bharati University. He was made a National Professor in 1959. He was elected belatedly to a Fellowship of the Royal Society in 1958. While Bose is celebrated for his discovery of Bose statistics, he was also involved in many other areas of physics. Meghnad Saha and Bose discovered a modified equation of state for gases. He was quite interested in general relativity and the problem of unifying electromagnetism and general relativity. He was also involved in showing that the "horpolhode", has no point of inflection; and the stress Equations of Equilibrium. Bose, during the course of his work in Physical Chemistry, was instrumental in the construction of a special camera for the study of X-ray Fluorescence.[14] His interest in science, and science-based technology did not exclude the study of philosophy. He had a special interest in the study of Yoga Vasista, an interest that I shared with him. I had only one chance to spend an evening with him shortly before his passing away. He was indeed a Lion in Winter.

Bose was widely respected not only in Bengal but throughout the country. The Indian Institute of Science celebrated, "50 years of Bose Statistics", largely due to the efforts of J. Pasupathy and the enthusiastic support of the then Prime Minister Indira Gandhi and the then Minister of Science and Technology, C. Subramanian. A commemoration volume was published celebrating his 80th birthday.

References and Footnotes

1. I.M. Duck and E.C.G. Sudarshan, "Wolfgang Pauli and the Spin-Statistics Theorem," World Scientific, Singapore (1998).
2. S. Weinberg, "Quantum Theory of Fields," Cambridge University Press, Cambridge (1994).
3. L.J. Boya and C.G. Sudarshan, "The Spin Statistics Connection in Arbitrary Dimensions," Int. J. Th. Phys. **46**, 3245 (2007).
4. J.W. Gibbs found that the entropy of an ideal gas computed on the basis of standard counting of states had the property that when two identical volumes of gas at identical density and temperature were mixed, the entropy increased.
5. M. Planck, "Temperature and Entropy of Radiant Heat," Annalen der Physik **1**, 719 (1900).
6. A. Einstein, "A Heuristic Interpretation of the Radiative Transfer of Light," Annalen der Physik, **17**, 1332 (1905). A. Einstein, "Emission and Absorption of Radiation according to Quantum Theory," Deutsche Physical Geselle Verh. **11**, 318 (1916).
7. N. Bohr, "On the Constitution of Atoms and Molecules," Phil. Mag. **26**, 4 (1913).
8. W. Pauli, Zeitschri. f. Phys. **31**, 765, 1925. E. Fermi extended the exclusion principle to many electron atoms.
9. J. Mehra and E.C.G. Sudarshan, "Classical Statistical Mechanics of Identical Particles and Quantum Effects," Int. J. Theor Physics **3**, 245 (1970).
10. George Ghevargese Joseph, "The Crest of the Peacock: Non-European Roots of Mathematics." Since the origins of differential calculus could be traced to these Indian mathematicians more than a century before Newton or Leibnitz were born, the bitter controversy between Leibnitz and Clark (proxy for Newton) seems ridiculous.
11. A chronology of the works leading to J.C. Bose and of G. Marconi maybe of interest. I am indebted to Professor Rodger Walser for his efforts to make IEEE (Institute of Electrical and Electronics Engineers) declare the priority of J.C. Bose: 1865 James Clark Maxwell's equations; 1888 Hertz generate first electromagnetic waves; 1895 J.C. Bose uses 5mm radiation to remotely ring bell and explode gunpowder; 1897 Marconi 1st signaling on Salisbury plains; 1909 Marconi — Nobel Prize for inventing radio. However, the Nobel Prize for the discovery of long distance propagation of electromagnetic waves went to G. Marconi. This is not the only instance the Nobel Committee has erred in its awards.

12. L. de Broglie, "C.R. Hebd. Seances," Acad. Sci. (France), **175**, 811–813 (1922). Note that while the obtained result is the same as in Bose's paper, the physical pictures are quite different. Instead of postulating "giant atoms of many photons," Bose uses the state of his photon field.
13. Max Delbruk, "Was Bose–Einstein Statistics Arrived at by Serendipity?" Journal of Chemical Education **57**, 467 (1980). [Ed. Note. See Partha Ghose: In Addendum]
14. S.N. Bose and M.N. Saha, "On the Equation of State," Phil. Mg. Ser 6, **39**, 456 (1920); S.N. Bose, "Certain Consequences of the Existence of the Tensor g in the Affine Relativistic Field," Le Jour. de Phys. et le Radium **14**, 645–647 (1953). S.N. Bose, "On the Herpolhode," Bull, Cal, Math. Soc. **11**, 21–22 (1919); S.N. Bose, "The Affine Connection in Einstein's New Unitary Field Theory," Ann. Math USA **59**, 171–176 (1954); S.N. Bose, J. Sharma, and B.C. Dutta, "Report on the Study of Thermoluminescence," Transactions of the D.M. Bose Research Institute **X**, 177 (1955). [Ed. Note These papers are included in this volume]

Satyendra Nath Bose — Some Vignettes of a Many-sided Personality

Dwarka N. Bose

My father, Sunil Chandra Bose, was a contemporary of Satyendra Nath Bose at Presidency College, Kalkaska (1909–11), and later his doctor and friend. They were born within a few days of each other, Satyendra Nath on 1st January and Sunil Chandra on 12th January 1894. Their friendship was to last till my father passed away in November 1953. My two uncles, Sarat Chandra Bose and Subhas Chandra Bose were also students of Presidency College. Sarat, a few years senior to Satyendra, joined the Presidency College in 1905 and was a resident of the Hindu Hostel. Subhas was a few years junior. Both Sarat and Subhas would play dominant roles in India's struggle for Independence. Subhas was the left wing leader of the Congress party and its President for two terms. At the outbreak of the Second World War, he escaped from India to form the Indian National Army and fought against the British on the eastern front. Sarat became an eminent lawyer and the President of the Bengal Congress before forming his own socialist party. Sarat, Sunil and Subhas, the Bose brothers as they were called, were not related to Satyendranath Bose, but they were close friends. My siblings and I were thus brought up on tales of his genius.

The particular batch of 1911 at the Presidency College was notable for its extraordinary brilliance. It included the future eminent chemists, Jnan Chandra Ghosh and Jnan Mukherjee, the eminent economist-cum-musicologist Dhurjoti Prasad Mukherjee, and the noted mathematician Nikhil Ranjan Sen. While pursuing the B.Sc, they were joined by Meghnad Saha from Dhaka. Science students had the well-known inspirational personalities of Acharya Jagadish Chandra Bose in physics and Acharya Prafulla Chandra Ray in chemistry. J.C. Bose was a student at Christ's College, Cambridge having studied under Lord Rayleigh and obtained his D.Sc from London while Prof. Ray obtained his doctorate from Edinburgh University. Thus they were exposed to science at the highest level and were imbued with a spirit of nationalism. They were also eager that India be not left behind in the development of science and in its application to the improvement of the life of common people.

Satyendra, known for his brilliance, acted as a mentor to many of his juniors. Dilip Kumar Roy, who was the son of the great poet Dwijendra Lal Roy, was one among them. Roy was to go on to become an outstanding musician and vocalist, who studied western classical music in Germany. The two would maintain close friendship throughout their lives, love for music being the strong bond between them. Satyendra was omnivorous in his quest for knowledge. He had deep knowledge of English and Bengali literature and joined the literary group *Sabuj Patra* ("The Green Leaves"), which included leading poets and writers as members. His favorite poets were Tennyson and Tagore. An example of his prodigious memory — he could recite the former's "In Memoriam" and Kalidasa's "Meghdoot" by heart. His love for music was also well-known — just as the violin was dear to Einstein — and he played the esraj with élan. Although he was primarily a student of physics, Satyendranath had opted to take physiology as the fourth subject for his I.Sc examination and would often borrow and devour books on anatomy and physiology from my father.

Medicine was looked down upon as a profession for bright students in those days. My father, who was among the top ten in the university in his sophomore year was certainly bright, but chose medicine as a profession. He used to recall Satyendranath often chiding him for choosing to be a doctor. "Only *gadha* become doctors," my father used to respond "Geniuses like you will have to come to us *gadha* as the last resort." My father went on to receive his M.R.C.P in London and then was trained under Sir Thomas Lewis, King's Physician and pioneer cardiologist. He was the first Indian cardiologist, returning to India in 1928, with the first ECG machine in the country. Among his patients were Mahatma Gandhi, Rabindranath Tagore, M.N. Roy, his friend Satyendra and many other luminaries.

In later years, my father would often drop in on Professor Bose either at Science College or at his home on Ballygunge Circular Road. In my student days, I often accompanied him and was a witness to their banter. After a quick check-up with a stethoscope and the measurement of blood pressure, my father who was against the random use of drugs, would advise moderation in eating and a little exercise. This would invariably be followed by a debate as to who would outlive the other. On one occasion, Bose asked as to what I wanted to become when I grew up, I replied hesitantly that I wanted to be a physicist. "Become a doctor like your father," he said, "You'll make lots of money!"

The first general election after India's Independence was held in 1952. An indelible ink was used to mark the fingers of voters to identify the voter and to prevent the casting of vote more than once. After casting his vote, my father, with me in tow, decided to visit Bose in the Science College. After the usual enquiries about each other's health, my father mentioned that he had just cast his vote and his finger had been marked with indelible ink. Curious about such things, Bose called a laboratory assistant and asked him to prepare a solution. When this was brought, he proceeded to dip a matchstick in the solution and to rub the proffered finger. Imagine his childish delight when gradually the spot disappeared! "Baby [Sunil Bose's nick name], you can go and vote again!" he exclaimed.

As I recall, the many cupboards in his office were topped with models of crystal structures. He had developed a well-equipped crystallographic laboratory in Dhaka after his return from Europe in 1926 where he had worked on the subject in the laboratory of Maurice De Broglie. He was obviously interested in X-ray diffraction for the study of materials. When in 1947–48 transistors using germanium had been invented at the Bell Laboratories in the U.S.A., Bose thought it was worth while investigating the possible sources of germanium in India and find how India could have access to this rather precious raw material. Satyen had collected sphalerite crystals during this visit to Nepal. Along with R.K. Dutta, he subjected the material to a spectroscopic analysis and found significant amount of germanium.[1] Silicon displaced Germanium soon after as the preferred transistor material on account of its superior properties. No doubt this was in my subconscious mind when my group in the seventies undertook and developed a method of extracting Silicon from rice-husk in an attempt to reduce the cost of the raw material for photovoltaic energy conversion.

After his return to Calcutta from Dhaka in 1945, Bose took up the study of thermoluminescence in solids with his students Harsha Narayan Bose and Jagadish Sharma.

This was a simple and well-known technique in which a solid was excited by radiation and on subsequent heating the stored energy was released as light. What was unique at that time was that Bose and his group developed a method of spectral analysis of the emitted radiation using a rotating mirror.[2] Thermoluminescence in alkali halides was subsequently studied in depth at the newly formed Indian Institute of Technology, Kharagpur.

Bose wielded great influence in the selection of well qualified candidates. He was the Chairman of the Selection Committee of the Department of Physics at I.I.T. for a number of years. On one occasion, when the unanimous choice of the Committee was sought to be overturned by the Chairman of the Board, who was also formally the Chief Minister of West Bengal, the Director of the Institute S.R. Sengupta was able to sustain the Selection Committee's choice because of Bose's eminence and his integrity. The Chairman of the Board was thereafter replaced by a non-political person, and that has been the rule ever since.

As is well known, Bose never took the trouble to submit a doctoral thesis. Of interest to the physics community, however, is the role he played in mentoring generations of students and researchers. In the early forties, his student Shyamadas Chatterjee set up an experiment with completely indigenous apparatus at the Bose Institute (named after Jagdish Chandra Bose) in Kolkata to study the newly-reported uranium fission when it was bombarded with neutrons. While setting up the experiment, he found that distinct counts were recorded even without the presence of the neutron source. He was puzzled by this observation, since he had taken precaution to shield uranium from cosmic rays. He reported the phenomenon to Bose, who quickly arrived at the correct conclusion that this must be due to spontaneous fission of Uranium.

However, Chatterjee needed help to calculate the half-life of spontaneous fission of Uranium from his experimental observations. Nikhil Ranjan Sen, Head of the Dept of Applied Mathematics, Calcutta University, lived nearby. One day he came to Chatterjee with a valve radio in need of repair, as Chatterjee was known to be handy with such things. When the job was completed, Chatterjee instead of accepting token payment for the job he had done, seized the opportunity to broach Sen with his experimental results and requested assistance in the calculation of the half-life. After a few weeks of frenzied calculations, the half-life they calculated, which later proved to be correct, was significantly lower from that attributed to Edward Teller. As a result, Chatterjee had to withdraw his paper by the orders of the Director whose permission he had failed to obtain. Chatterjee recounted to me with suppressed emotion the Director's reaction to his "insubordination" — "There's fission in your brain!" he had said. The phenomenon was discovered almost simultaneously by Georgii Flerov and Konstantin Petrazhak in the Soviet Union. He published his work later and was recognized by the Russian authors, who presented an inscribed copy of their book on the subject to him. Thus it was Chatterjee's experimental diligence coupled with Bose's insight and Sen's timely help that provided this significant result.

Another anecdote relates to Bose's first meeting with P.A.M. Dirac in the early fifties in Kolkata at the Institute of Nuclear Physics. As Dirac spoke, Bose sitting in the front row, appeared to doze off. When writing an equation on the blackboard, Dirac appeared to hesitate and looked in the direction of the white-haired Bose for help. Bose lumbered to his feet, scribbled the rest of the equation and resumed his earlier somnolent posture.

Dirac was surprised because he had never met Bose, although Bose was in England during his European visits. He was also surprised that he was not a Fellow of the Royal Society of London. It was shortly after Dirac's visit that S.N. Bose and S.K. Mitra, the latter well-known for his pioneering researches on radio and wireless, were elected belatedly in 1958 as Fellows of the Royal Society. In 1958, Prof. Bose went to London to attend his first meeting as a Fellow. In the course of his visit to several laboratories in England, he went to the University of Reading, which had hosted the first International Conference on the Physics of Semiconductors. He met the Head of the Department, who arranged a junior faculty Philip Banbury to take Bose around the laboratory. Banbury, later my doctoral guide, did as requested and, as he later recounted to me, found little or no response from the visitor. He was finally reduced to explaining to him energy levels in solids, Fermi–Dirac and Bose–Einstein statistics — with not a glimmer from his venerable visitor. It was only later that he discovered, to his huge embarrassment, that his visitor was none other than Bose himself! My surname was thus a source of amusement to him when I joined the laboratory a few years later.

Another interesting story is connected with the discovery of Helium at the hot springs at Bakreshwar near Kolkata. Dr. B.C. Roy, the Chief Minister of West Bengal and a distinguished physician, had gone to Vienna for eye surgery. He then recuperated at Badgastein where there were sulfurous hot springs with radioactive emanations. On his return he called Triguna Sen, Vice-Chancellor of the newly formed Jadavpur University to enquire whether similar hot springs may be present in West Bengal. Sen passed on the query to Shyamdas Chatterjee, the Head of the Department of Physics. Chatterjee carried out measurements at Bakreshwar and confirmed the presence of trace elements of Radon as well as Helium, which was the first such source in India. The concentration of Helium was found to be much above average. Bose sponsored the scheme for making a systematic analysis of the spring water and evolved gases and Chatterjee supervised the construction of a field laboratory at Bakreshwar. It was later taken over by the Department of Atomic Energy. Bose also encouraged Chatterjee to set up a small Helium purification apparatus at the Indian Association for the Cultivation of Science in the seventies, where he was the National Professor.

Satyendra Nath Bose was a legend in his lifetime but a more unassuming "legend" would be difficult to find. His warm, gregarious nature and accessibility to all marked him out as a truly exceptional savant.

References

1. S.N. Bose, "Extraction of Germanium from Sphalerite Collected from Nepal" Part II, J. Sci. Indust. Res. **9B**, 27 (1950).
2. S.N. Bose, "A Report on the Study of Thermoluminescence," Dr. D.M. Bose Seventieth Birthday Commemoration Volume, Trans. Bose. Res. Inst. **20**, 177.

Contributors

Bandied, William A.
Senior Research Scholar in the Science and Trade Policy Program at George Mason University since his retirement in 2003 as the Senior International Analyst at the National Science Foundation. He is the author of several books, numerous articles on physics, history of science, international science and science policy including both the national and international aspects.

Bergia, Silvio
Silvia Bergia currently teaches Relativity and Philosophy of Physics in the physics department of the University of Bologna in Italy. During his professional career, he worked in Particle Physics, Quantum Mechanics and History of Physics. He is the author, *From the immutable cosmo to the evolving universe* (Bollati Boringhieri, 1995) and one of the editors of *Modern Cosmology in Retrospect* (Cambridge University Press, 1991).

Bose, Dwarka N.
Distinguished Emeritus Scientist at the University Science Instrumentation Center, Calcutta University. He completed his undergraduate and post-graduate studies at Calcutta University and as an I.C.I (India) scholar at the J.J. Thompson Physical Laboratory, University of Reading. He obtained his doctorate degree in semi-conductor physics. He has made seminal contributions in several areas of condensed matter physics, held academic and administrative positions in several institutions including Penn State University, Indian Institute of Science, Bangalore, and I.I.T. Kharagpur.

Ghose, Partha
A student of Satyendranath Bose, he taught at Viswa-Bharathi and S.N. Bose Center for Basic Sciences till his retirement. A theoretical high energy physicist, he has made fundamental contributions to the foundations of quantum mechanics (Testing Quantum Mechanics on New Ground). Involved in popularizing science, he is the winner of Indira Gandhi prize from Indian National Science Academy. Besides physics and history of physics, he is deeply interested in music and philosophy.

Pais, Abraham (1918–2000)
A renowned theoretical physicist of the twentieth century and an eminent historian of science, he is the celebrated author of *Subtle is the Lord: The Science and Life of Albert Einstein.*

Ramaseshan, Sivraj (1923–2003)
One of India's most accomplished scientists and a main contributor to the development of science in India. Editor of *Current Science*, the journal of the Current Science Association and founder of *Promina*, a physics journal. Editor of C.V. Raman's papers on Light Scattering, he has written perceptively on several noted scientists of India, S. Chandrasekhar, Harish Chandra, Bhabha, Sarabhai, Mahalanobis among them.

Sinha, Purnima
A student of Bose in the 1950s at Calcutta University, she did her doctoral research under his guidance. She became the first Bengali woman doctorate in physics. A pioneer in her field of research (structure and material properties of clay minerals), she has explored over the years various other fields including biophysics, music, sculpture and painting. A regular contributor to *Jnan Bijnan*, she has written extensively, both in English and in Bengali, popular articles on science and music. She is a noted biographer and has written several articles on Bose.

Stachel, John
Well-known Einstein scholar, historian and philosopher of science, he was the first editor of "Einstein Papers Project," at the Institute for Advanced Studies at Princeton.

Sudarshan, George
One of the foremost theoretical physicists, whose V-A theory of weak interactions laid the foundations of the Unified Theory of Weak and Electromagnetic Interactions. Likewise, his seminal contributions range over a wide variety of physics that include Spin-Statistics connection, Quantum Optical Coherence (Sudarshan Representation), Quantum Zeno Effect, Theory of Tachyons, Quantum Mechanics of Open Systems and Axiomatic Field theories.